Handbook of Research on Urban-Rural Synergy Development Through Housing, Landscape, and Tourism

Aleksandra Krstić-Furundžić
University of Belgrade, Serbia

Aleksandra Djukić
University of Belgrade, Serbia

A volume in the Advances in Civil and Industrial Engineering (ACIE) Book Series

Published in the United States of America by
IGI Global
Engineering Science Reference (an imprint of IGI Global)
701 E. Chocolate Avenue
Hershey PA, USA 17033
Tel: 717-533-8845
Fax: 717-533-8661
E-mail: cust@igi-global.com
Web site: http://www.igi-global.com

Library of Congress Cataloging-in-Publication Data

Names: Krstić-Furundžić, Aleksandra, editor. | Djukić, Aleksandra, editor.
Title: Handbook of research on urban-rural synergy development through housing, landscape, and tourism / Aleksandra Krstić-Furundžić and Aleksandra Djukić editors.
Description: Hershey, PA : Business Science Reference, an imprint of IGI Global, [2019]
Identifiers: LCCN 2019013251| ISBN 9781522599326 (hardcover) | ISBN 9781522599340 (ebook)
Subjects: LCSH: Rural-urban relations--Case studies. | Regional planning--Case studies. | City planning--Case studies.
Classification: LCC HT361 .U71792 2019 | DDC 307.1/2--dc23 LC record available at https://lccn.loc.gov/2019013251

This book is published in the IGI Global book series Advances in Civil and Industrial Engineering (ACIE) (ISSN: 2326-6139; eISSN: 2326-6155)

British Cataloguing in Publication Data
A Cataloguing in Publication record for this book is available from the British Library.

All work contributed to this book is new, previously-unpublished material. The views expressed in this book are those of the authors, but not necessarily of the publisher.

For electronic access to this publication, please contact: eresources@igi-global.com.

Advances in Civil and Industrial Engineering (ACIE) Book Series

Ioan Constantin Dima
University Valahia of Târgovişte, Romania

ISSN:2326-6139
EISSN:2326-6155

Mission

Private and public sector infrastructures begin to age, or require change in the face of developing technologies, the fields of civil and industrial engineering have become increasingly important as a method to mitigate and manage these changes. As governments and the public at large begin to grapple with climate change and growing populations, civil engineering has become more interdisciplinary and the need for publications that discuss the rapid changes and advancements in the field have become more in-demand. Additionally, private corporations and companies are facing similar changes and challenges, with the pressure for new and innovative methods being placed on those involved in industrial engineering.

The **Advances in Civil and Industrial Engineering (ACIE) Book Series** aims to present research and methodology that will provide solutions and discussions to meet such needs. The latest methodologies, applications, tools, and analysis will be published through the books included in **ACIE** in order to keep the available research in civil and industrial engineering as current and timely as possible.

Coverage

- Structural Engineering
- Engineering Economics
- Optimization Techniques
- Materials Management
- Earthquake engineering
- Production Planning and Control
- Urban Engineering
- Construction Engineering
- Ergonomics
- Hydraulic Engineering

IGI Global is currently accepting manuscripts for publication within this series. To submit a proposal for a volume in this series, please contact our Acquisition Editors at Acquisitions@igi-global.com or visit: http://www.igi-global.com/publish/.

The Advances in Civil and Industrial Engineering (ACIE) Book Series (ISSN 2326-6139) is published by IGI Global, 701 E. Chocolate Avenue, Hershey, PA 17033-1240, USA, www.igi-global.com. This series is composed of titles available for purchase individually; each title is edited to be contextually exclusive from any other title within the series. For pricing and ordering information please visit http://www.igi-global.com/book-series/advances-civil-industrial-engineering/73673. Postmaster: Send all address changes to above address.

Titles in this Series

For a list of additional titles in this series, please visit: www.igi-global.com/book-series

Re-Coding Homes Through Flexible Interiors Emerging Research and Opportunities
Nilüfer Saglar Onay (Istanbul Technical University, Turkey) S. Banu Garip (Istanbul Technical University, Turkey) and Ervin Garip (Istanbul Technical University Turkey)
Engineering Science Reference • ©2020 • 165pp • H/C (ISBN: 9781522589587) • US $175.00

Green Building Management and Smart Automation
Arun Solanki (Gautam Buddha University, India) and Anand Nayyar (Duy Tan University, Vietnam)
Engineering Science Reference • ©2020 • 312pp • H/C (ISBN: 9781522597544) • US $215.00

Handbook of Research on Implementation and Deployment of IoT Projects in Smart Cities
Krishnan Saravanan (Anna University Chennai – Regional Office Tirunelveli, India) Golden Julie (Anna University, India) and Harold Robinson (SCAD College of Engineering and Technology, India)
Engineering Science Reference • ©2019 • 415pp • H/C (ISBN: 9781522591993) • US $295.00

Handbook of Research on Digital Research Methods and Architectural Tools in Urban Planning and Design
Hisham Abusaada (Housing and Building National Research Center, Egypt) Carsten Vellguth (German Academic Exchange Service, Germany) and Abeer Elshater (Ain Shams University, Egypt)
Engineering Science Reference • ©2019 • 445pp • H/C (ISBN: 9781522592389) • US $265.00

Recycled Waste Materials in Concrete Construction Emerging Research and Opportunities
Jahangir Mirza (Research Institute of Hydro-Québec, Canada) Mohd Warid Hussin (Universiti Teknologi Malaysia, Malaysia) and Mohamed A. Ismail (Miami College of Henan University, China)
Engineering Science Reference • ©2019 • 169pp • H/C (ISBN: 9781522583257) • US $165.00

Ambient Urbanities as the Intersection Between the IoT and the IoP in Smart Cities
H. Patricia McKenna (AmbientEase, Canada)
Engineering Science Reference • ©2019 • 333pp • H/C (ISBN: 9781522578826) • US $195.00

Optimizing Current Strategies and Applications in Industrial Engineering
Prasanta Sahoo (Jadavpur University, India)
Engineering Science Reference • ©2019 • 382pp • H/C (ISBN: 9781522582236) • US $245.00

IGI Global
DISSEMINATOR OF KNOWLEDGE

701 East Chocolate Avenue, Hershey, PA 17033, USA
Tel: 717-533-8845 x100 • Fax: 717-533-8661
E-Mail: cust@igi-global.com • www.igi-global.com

List of Contributors

Table of Contents

Detailed Table of Contents

The spatial complexity of metropolitan territories is the result of the accumulation of various forms of urban and infrastructural development, some of which can be characterized as being based on a fragmentary mode of production. The chapter aims at building an empirical approach to a broader discussion regarding spatial fragmentation as part of metropolitan territorial development using an example on an important suburban territory in the Lisbon metropolitan area (Portugal). As such, it will provide a conceptual review of contemporary urban and morphological approaches to metropolization and the issue of urban-rural relationships, an overview of Lisbon's metropolitan development and the specific changes identified in the case study, the overview and results of morphological analysis regarding land subdivision processes, and a discussion on a design-oriented description of territorial components with potential to reinforce the spatial synergies of urban fabrics, infrastructures, and open space.

The field of housing has undergone significant changes during the challenging post-socialist transition. Serbia has been a somewhat different case in the way that its transition was postponed and more complex. Conversely, its housing also carries the legacy of the more developed former socialist Yugoslavia. This dichotomy can easily be observed in the case of the newly-built, "transitional" multi-storey housing. However, the data relating to its characteristics is rare and insufficiently accurate. The aim of this chapter is to improve the research of the urban dimension in newly-built multi-family housing in Serbia, regarding housing planning in the new master and detailed urban plans for Serbian cities. This has been researched by analyzing the professional knowledge and experience of local experts. Therefore, the method used in the research was survey, in which urban experts had participated. It had been carried out in May 2015.

The cities are characterized by the rapid development of technologies and the inevitable expansion, which requires looking at ways to address new challenges and overcome the problems that have arisen. Living conditions may vary in many ways, depending on where people live, but the advantages and disadvantages in terms of comfort of human life can be recognized in both urban and rural environments, as analysis of case studies shows. Factors that occur in both areas and affect human life, comfort, and health are presented. The positive effects of rural areas and the advantages that cities have in terms of preserving human health are discussed in this chapter. By comparing the good characteristics of the urban and rural environment, the goal is to indicate some postulates that can be applied in preventive medicine.

Decades of regional polarity in Serbia conditioned the adoption of a polycentric territorial development as one of the main strategic goals of the country's spatial development. Politics of polycentrism specifically highlights the role and importance of small urban centers in generating a more balanced regional development. In constellation with their rural surroundings, they can serve as strong agents of change towards sustainable spatial development, functional cohesion, and transformation of the rural hinterland, and thus, contribute to the balanced spatial development. The main aim of the chapter is to examine the potential of small urban centers to generate spatial and functional integration, that is, to explore the possibility for small towns to integrate and transform their rural surroundings. The chapter analyses the field of functional influence of small towns on their rural surroundings on the example of Blace Municipality through determining the intensity of daily migrations which are a well-known indicator of spatial and functional integration of rural areas and urban centers.

This chapter will examine the transition zones around the borderline of the legal perimeter defined by law and master planning of Skopje that have always played a role of interactive borderland in which the concept of interactivity and urban-rural-exurban synergy is applicable. Of particular interest is how edge position of certain abandoned structures, industrial areas in particular, in the period of post-socialist transition have affected use and movement patterns changing. These structures have originally influenced the surrounding area by their users, and now they are largely neglected or underused and need interventions of rehabilitation into beneficial use. The authors support the view that there is significance in conceiving of a "edge effect" of the re-habilitated social and industrial fringe structures. They find that they should be given more consideration in the context of new economies and innovative technologies which will enable rehabilitation and adaptive reuse of the existing premises.

The morphogenesis of the urban territory and its contact with the non-urbanized, natural environment of the wider metropolitan area distinguish issues of the synergy between landscape and spatial patterns in order to achieve their balance, optimization, and harmony. This chapter highlights the conceptual framework of landscape ecology as linking to place-based design approach for studying the synergy of landscape and housing spatial patterns in order to improve their integration in future perspective. The territory of the city of Belgrade is recognized as a specific spatial-morphological system that is a consequence of the urban-rural synergy between socialist housing settlements and environmental processes. The chapter points at the environmental and functional values of nature with a particular focus on housing typology in the process of urban planning and architectural design.

The most common action for the buildings' energy upgrade across Europe is the addition of thermal insulation on the external walls. Such interventions, although simple on their construction, cause significant changes on the building's behavior, not only on its energy needs, but also on the hygrothermal and visual performance. The effects are not always positive; for example, thicker insulation may result in lower thermal transmittance and better thermal energy performance, but on the other hand the thermal bridging effect is amplified, and the daylight levels are decreased. This research intends to quantify these impacts by analyzing the relevant parameters for different regions of Europe. The analysis aims at explaining the complicated interrelationships on the building physics' aspects encountered through interventions on the building envelope, but also at identifying appropriate measures that could counterbalance the negative impacts and enhance the overall building performance.

Yielding from an overall quantitative study of the residential sector of Bosnia and Herzegovina (B&H), this chapter concentrates on the ratio between single-family and collective housing, as well as on the urban-rural ratio of the single-family housing. Based on the data from the existing building stock (buildings built by 2014) and the statistical estimates, 23% of the buildings belong to the urban areas and 77% belong to the rural areas. The main goal was to study the correlation between the characteristics of the building envelope, the shape factor (A/V ratio) and the energy savings potential for the application of conventional measures of refurbishment of the building envelope of the single-family houses (type of buildings, which dominate in rural and urban areas). The chapter wraps up with recommendations for the adequate level of the energy performance indicator (energy need for heating) for the approved energy class for single-family houses located in the climate zone of the northern B&H.

The chapter provides an insight into the relationship between synergic (r)urban systems, information networks and climate change, discussing the emerging ideas, and concepts related to the increasing use of information networks in the process of climate adaptation and mitigation. Emphasizing the strategic role of both digital and material information flows, supported by ICT tools, the chapter focuses on two main domains of data exchange and knowledge transfer: the public communication of climate change and the connectivity and interaction within (r)urban hybrid systems. Underlining the issues of effectiveness, accessibility, and low-carbon outcomes of synergic (r)urban reactions to climate shift, recent environmental and technological trends are considered in accordance with the preferred spatio-functional flexibility of emerging (r)urban hybrid settlements. The chapter also identifies and analyses three areas of ICT applicability, targeting the role of information networks in the anticipated climate-friendly development: human behavior, ecological awareness, and general efficiency.

The solar radiation and energy potential in Serbia is 30% higher than in Central Europe, and the intensity of solar radiation is among the highest in Europe. Specifically suitable are the condition for usage of solar energy in southeast Serbia where the annual average of global radiation on a horizontal surface amounts to more than 4.2 kWh/m2 a day. This chapter discusses four photovoltaic solar power plants, two having been installed in the territory of the city of Leskovac and two in the Bosilegrad territory. The research is based on geographical location of solar power plants, orientation, and inclination of the panels in respect to the horizontal surface. All the processed solar power plants have the capacity 30-40 kW. The chapter is projection of two years of researching and measuring data on the initial investments in order reach profitability and investment return period. The solar power plants that have fitted angle of 33° (Fortuna) give higher production of electric power than in those panels that have angle inclination of 10° (Domit).

Section 4
Urban-Rural Cooperation in Territorial Development

The chapter presents and discusses the process and results of collaboration between an academic institution and local government in students' masters' theses and integrated urban development projects for urban-rural continuum of the territory of the City of Pančevo, Serbia. The chapter has four main parts: 1) a brief overview of the theoretical framework for multilevel governance; 2) the background, focusing on the national policy; 3) applied educational model; and 4) 11 students' projects dealing with the urban-rural continuum. Recommendations are made for continuing application of and research into potential teaching methods that promote meaningful interaction and practical and problem-oriented instruction that contribute to an understanding of contemporary problems of balanced urban-rural development. The conclusion discusses process and product outcomes, differentiating benefits for all those involved, as well as the obstacles and challenges encountered.

Throughout history, built environment developed intuitively on empirical experience led by trial and error. Such approach provided resilience and evolvement of patterns that guided spatial organization. Newer interventions in the rural environment resulted in disconnected spatial fragments. By comparing vernacular and contemporary planning and construction practice in the natural park Blidinje, the authors tried to identify the reason for which contemporary interventions resulted in new patterns in spatial planning, ones completely unfamiliar for this geographical area. They identified the reason in the fact that men started to treat the natural space as a commodity, forcing stakeholders to be led primarily by economic principles. Such principles are rigid and linear, instead of contextual in this matter. At the same time, the environmentally sound approach should respect complexity of whole endeavor aiming to achieve diversity and variability. The models developed based on complexity theory and self-organization should preserve continuity and integrity of the place and man.

Many regions worldwide are faced with the problem of shrinkage, manifested through demographic decline, economic loss, and perforation of the urban tissue. Starting from the assumption that the shrinking phenomenon is present in Bosnia and Herzegovina, the research deals with the diagnosis of the shrinking condition in urban and rural areas of the northwest region of the country. In the light of the shrinkage problems, the chapter defines a new alternative model—e-co—which should enable the recovery from the crisis and restore the vitality of the region. The chance for recovery is recognized in the processes of rural urbanization and urban ruralization, the integration of rural values and urban advantages, the modern globality, and traditional locality.

The zone of post-socialist suburbia was predefined even during socialism, when rural settlements surrounding large cities were joined with city territory. After the fall of socialism, new intensive transformations unfolded in suburban areas. This chapter analyzes the centers of suburban settlements by using a segment of the administrative area of the City of Niš, Serbia as a case study. Selected settlements have high values of demographic growth in the post-socialist period: four settlements of rural origin, one sprawling settlement, and one settlement of former weekend-residences. This chapter is set to investigate current development patterns in suburbia and to formulate guidelines for designing suburban centers. Urban matrix for mixed-use centers is created, which shapes the center spatially and functionally. Vital centers of suburban neighborhoods that are integrated into the city network represent the only sustainable form of development of suburban areas and city as a whole.

Preface

URBAN-RURAL COOPERATION IN HOUSING, LANDSCAPE, AND TOURISM TO FOSTER SYNERGY DEVELOPMENT

The chapters in the *Handbook of Research on Urban-Rural Synergy Development Through Housing, Landscape, and Tourism* contribute to the discussion about the places and technologies related to the future of urban-rural synergy development in line with new urban concepts related with housing, landscape and tourism, and discussions with respect to strategy for sustainable development and human welfare. Regions are continually developing, and climate change and ICT have strong effects on urban-rural interactions. Most articles are about urban development and about rural too, but there are also those focused on interactions and dependencies between towns and its surrounding rural area and their synergetic development.

The book covers different themes connected to the urban and rural synergy, from different scientific fields in humanities and social sciences and engineering and technical sciences, with the overall goal to provide a wide perspective on how to plan and develop towns and cities with its rural surroundings in a synergetic way. The indispensable theme of protecting human health and achieving human well-being has led the field of medical sciences to be included. The stated general objectives point to the necessity of a multidisciplinary approach to this matter, highlighting relationships between issues of technological development, rural areas, social change and tourism. However, the main focuses of the articles are related to what extent the technologies could provide sustainable development of a rural area and on the innovative forms and methods of sustainable development. Integrative methods of development are realized as successful urban-rural development strategies, supported with incorporation of new models and active involvement of citizens during the process.

A selection of articles has been done after "The Fourth International Conference Places and Technologies – Keeping up with Technologies in the Context of Urban and Rural Synergy" which was held in Sarajevo June 2017 and where more than 120 papers from 17 countries were presented. The organizers of the conference were: University of Sarajevo (Faculty of Architecture), University of Belgrade (Faculty of Architecture) and Professional Association Urban Laboratory (UrbanLab). The aim of the conference was raising the questions about the future development of cities and rural area, understanding from the critical aspect, the importance and role of technology in designing creative ideas to improve places, providing social progress and finding the models for achieving the regional competitiveness and successfulness. Urban and architectural practice shows that new urban and architectural concepts require new methods for urban and rural planning and design as well as applying the new technologies, for sustainable future development of the regions.

The book includes 18 chapters structured into 4 main parts/sections.

The first part, "Urban-Rural Cooperation in Housing and Landscape," covers the topics of a) developing knowledge on urban spatial fragmentation and metropolitan regeneration providing smart combinations between spatial development, network cohesion, landscape regeneration, and alternative land-use, b) urban dimension of housing and the improvement of relations between architectural design, urban planning and landscape ecology, with particular emphasis on the perspective for the future development of socialist housing settlements, c) activities to enable the urban settlement to integrate and transform its rural environment and to influence its further qualitative transformation, d) new perspectives for preventive medicine based on the knowledge of the advantages and disadvantages of urban and rural environments that affect the quality of life, human health and human well-being.

In the second part of the book, "Urban-Rural Cooperation in Tourism," it is indicated that rural tourism represents the drive for urban–rural synergy and it is one of the key factors of rural development. The idea of promoting local values and cultural heritage is recognized through the architecture of public buildings in rural areas, especially tourist centers, which aims to improve the quality of life in such areas by highlighting major advantages in space, such as the natural environment. Importance is given to an information system that can increase efficiency of rural tourism, as it gather a rich database that will allow users to identify their next destination and identify most valuable assets for each location in the region.

A significant influence on the development of new and improvement of existing technologies has been conducted by the society commitment to protect the environment, reducing pollution and using renewable energy sources. Several chapters in the third part, "Urban-Rural Cooperation Through Technologies," are devoted to this topic, stressing that improving the energy efficiency of the entire building stock, especially the housing sector, and upgrading the facades of buildings, as well as building solar power plants are crucial for the protection of the environment.

The forth part, "Urban-Rural Cooperation in Territorial Development," includes topics related to the theoretical framework for multilevel governance, integrated urban development, balanced urban-rural development, complexity theory in spatial planning and design, sustainable systems to manage hazardous household and farming waste, mapping environmental polluters and creating a GIS database, as well as introducing new concepts and instruments through education. A new governance model is noted, characterized by greater flexibility in decision making and greater involvement of different stakeholders in articulating and executing decisions, including an increased range of modalities for stakeholder participation. The integration of architecture and nature is imperative, the achievement of which is characterized by an open dynamic design that has evolutionary qualities, and interventions in space based on complexity, diversity and variability.

According to the contemporary relations between architecture and technology new urban and building concepts have been created mostly relying on ICT. In this regard, certain chapters discuss the development of ideas and concepts relating to the increasing use of information networks in the process of climate adaptation and mitigation.

We are proud to share selected articles by authors from all over Europe in one place, something made possible by IGI Global, a leading international academic publisher that recognized the importance of the topic and value of the articles, deciding to publish this book. As editors of this book, we want to express our appreciation to all the authors and our partners from IGI Global for their kind invitation, cooperation and assistance in the preparation of the book.

Editors,

Aleksandra Krstić-Furundžić
University of Belgrade, Serbia

Aleksandra Djukić
University of Belgrade, Serbia
12th September 2019

Section 1

Urban–Rural Cooperation in Housing and Landscape

Chapter 1
Decoding (Sub)Urban–Rural Fragmentation Processes:
A Morphogenetic Approach in Lisbon Metropolis

João Rafael Santos
CIAUD, Faculdade de Arquitetura, Universidade de Lisboa, Portugal

ABSTRACT

The spatial complexity of metropolitan territories is the result of the accumulation of various forms of urban and infrastructural development, some of which can be characterized as being based on a fragmentary mode of production. The chapter aims at building an empirical approach to a broader discussion regarding spatial fragmentation as part of metropolitan territorial development using an example on an important suburban territory in the Lisbon metropolitan area (Portugal). As such, it will provide a conceptual review of contemporary urban and morphological approaches to metropolization and the issue of urban-rural relationships, an overview of Lisbon's metropolitan development and the specific changes identified in the case study, the overview and results of morphological analysis regarding land subdivision processes, and a discussion on a design-oriented description of territorial components with potential to reinforce the spatial synergies of urban fabrics, infrastructures, and open space.

INTRODUCTION

As a process, spatial fragmentation reflects an accelerated dynamic of transformation, a common feature of the 20th century urbanization around the world. It mirrors changing patterns of land use, land value and economic rationale, as well as new functional and spatial configurations brought by strongly disruptive physical elements such as heavy traffic and transport infrastructure. To a large extent, the spatial complexity of metropolitan territories is the result of the accumulation of various forms of urban and infrastructural development, some of which can be characterized as being based on a fragmentary mode of production. The understanding of this phenomenon is a difficult and still widely unexplored task,

DOI: 10.4018/978-1-5225-9932-6.ch001

especially when considering the spatial, morphological and procedural perspectives, on one hand, and the variety of empirical evidence. The call to address the everyday landscapes of residential sprawl and dispersion is not new. Seminal studies regarding the spatial configurations of today's metropolises, both at an overall scale (Font, 1999; Oswald & Baccini, 2003; Panerai, 2008) and at the scale of specific elements and structural spaces (Boeri, Lanzani, & Marini, 1993; Font, 2012; Muñoz, 2004; Vecslir, 2007; Viganò, 1999) are important contributions. However, limited work has been made exploring the variety of land transformation processes involving the rural parcel-to-urban plot change. In Portugal, urbanization processes have significantly relied on piece-meal rural land subdivision, making it a fundamental topic to understand them.

This chapter explores findings from the research project entitled *Adaptpolis - Beyond urban fragmentation - infrastructure, landscape and territorial design for Lisbon Metropolitan Area* (adaptpolis.fa.ulisboa.pt) based at Lisbon School of Architecture of the University of Lisbon, aimed at developing knowledge on the topic of urban spatial fragmentation and metropolitan regeneration to tackle ongoing socio-economic adjustment. In order to develop further urban and territorial regeneration strategies, the project's goal is to consider and decode the processes behind the multiple phenomena of fragmentation: in the process of urban growth, in the laying of heavy infrastructures, in the disruption of landscape and environmental continuities, in the clashes of everyday's mobility. With a forward-looking perspective, the project aims at providing smart combinations between spatial development, network cohesion, landscape regeneration, its financing, and alternative land-uses.

A specific case-study in Lisbon Metropolitan Area (LMA) – *Cacém-Massamá* (Figure 1), in the Municipality of Sintra – is being used as a test-bed for a number of analytical tools, critical assessment procedures, strategy and policy inception, and prospective design – all with actor involvement through partnering with local authorities. This site is located in a suburban corridor (Figure 2), 20 km west of Lisbon, in which the research critical issues can be found: a peripheral location originally developed around a suburban railway line, later followed by a metropolitan radial motorway – IC19; low quality residential development, poorly developed network of public space and urban facilities; spatial and scalar fragmentation between urban fabric, open spaces and natural features; poorly integrated mobility networks, with weak links between urban settlement and public transport infrastructures.

The chapter aims at sequentially build an empirical approach to a broader discussion regarding spatial fragmentation as part of metropolitan territorial development. As such, it will provide 1) a conceptual review of contemporary urban and morphological approaches to metropolization and the issue of urban-rural relationships, 2) an overview of Lisbon's metropolitan development and the specific changes identified in the case study, 3) the overview and results of morphological analysis regarding land subdivision processes, and 4) a discussion on a design-oriented description of territorial components with potential to reinforce the spatial synergies of urban fabrics, infrastructures, and open space.

FRAGMENTATION AS A PROCESS AND TERRITORIAL PRODUCT

The concept of fragmentation applied to urban research has had a strong relationship with the social and economic impact of globalization and neoliberal economic rationale, namely in the production of segregation and social inequities. The concept has been used to characterize the tendency to develop cities and metropolis focusing on bounded, controlled and detached packages of space (Mangin, 2004; Marcuse, 1994), to be planned, programmed, built and operated as a market commodity, such as gated-

Figure 1. Aerial photography and overall perspective of study area
Source: Bing Maps. Two aspects of the study area – São Marcos viewed from Massamá (below left) and Cacém with the railroad and highway (below right). Source: author

communities, high-tech campuses, logistic platforms or mega-shopping malls. This move has been in tandem with the concept of *splintering urbanism* (Graham & Marvin, 2001), which decodes the transition from the *modern infrastructural ideal* of a unitary, centralized and integrated urban organism, to a fragmented, polycentric and highly differentiated metropolitan configuration. These two approaches share the acknowledgment of a key role played by contemporary technological possibilities of ubiquitous digital connection and advanced production, management, and distribution systems, in the structural restructuring and recombination of global, national, regional and local scales. These approaches have been particularly grounded on empirical evidence from North-American, Latin American and Asian cities, which have undergone a large-scale restructuring process in the wake of post-industrial globalization transitions after the 1980s.

The argument for this article follows a different line of conceptualization, more specifically engaged with the spatial, morphological and landscape dimensions of urbanization processes. On the other hand, its geographical contextualization in the Southern European metropolises also differs from the above-mentioned cases and requires specific insight (Font, 2004, 1997; Llop & Bosc, 2012; Solà-Morales,

Figure 2. Evolution of urban fabric and infrastructure in western Lisbon metropolitan corridors, with the identification of Massamá-Cacém study area (1944, 1995, 2017)
Source: author

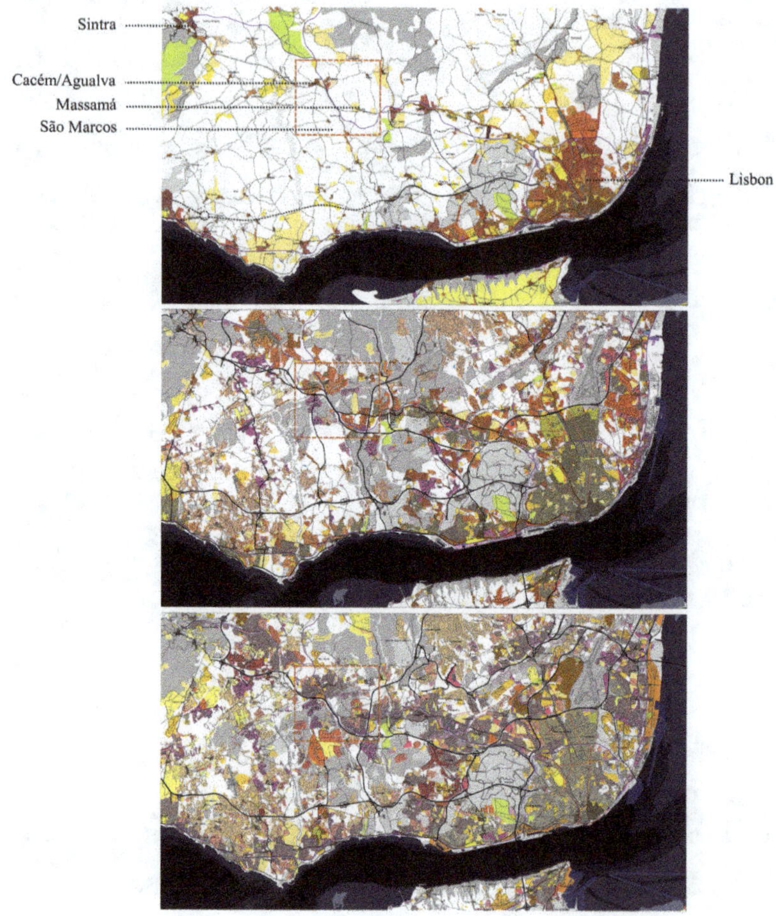

2004). In this context, and especially in Portugal, a number of specific factors need to be considered when discussing the forms of urban fragmentation. Firstly, the introduction of comprehensive urban and spatial planning systems to guide fast metropolization processes occurred on comparatively late stage, when compared to other European countries leaving a wide margin of urban development to be managed by private, market-driven and informal sectors, with limited public sector involvement; secondly the new technological infrastructure and industrial development that shaped European cities' spatial, economic and social modernization throughout the 20th century was far less intense and disseminated in Portugal, with the heavy weight of a poor economy and strong rural patterns remaining embedded in the socio-cultural and economic geography of all major cities (Domingues, 2006; George, 2001; Portas, Domingues, & Cabral, 2003); and thirdly, while most European economies were growing hand-in-hand with a Welfare State apparatus in the post-war periods until the oil crisis of the 1970s, Portugal was captured under a conservative, authoritarian and elite-driven political regime, suspicious of the cosmopolitan and proletarian ideas that flourished in the modern City, with low concern and investment on equitable and generalized redistribution of housing, urban infrastructure and social facilities. Even when formal government-driven urban planning was in place, its implementation capacity was very limited, especially at the municipal level. Aside from some exceptions, such as some expansion sectors for the city of Lisbon, Portuguese cities largely grew in response to basic needs of internal migration and not as planned and social-oriented process (Barata Salgueiro, 1992, 2001; Silva, 2008).

A significant part of these territories faces problems arising from the multiplication of infrastructural barriers and frequent failures and shortcomings in the provision of urban services. Spatial integrity is also compromised in the face of various levels of morphological fragmentation, as well as the proliferation of vacant spaces without a clear use and purpose. These forms of fragmentation are also reproduced at the socio-spatial level, translated into situations of exclusion and inequity, as well as of environmental risks.

Urban, Rural, and the Contemporary Metropolitan Patchwork

As widely acknowledged in literature, contemporary urbanization has changed the classic duality between the urban and the rural realms. The dichotomy between city and countryside has been fundamentally challenged by new societal and economic processes through which *limits* between them have been largely eroded; seminal works by Lefebvre (1970) and Brenner (2014) take that argument further in the sense that they see all parts of the globe are subject to spatial, economic, cultural and social process of *planetary urbanization*.

The transition towards a dominantly urban society meant the overlapping of its spatial, social, economic and infrastructural rationales over the hitherto rural spaces. From two relatively defined realms – city and countryside –, often defined by clear limits such as walls, a far more complex intermixing emerge with the industrial metropolises of the 19th century and the horizontal growth of cities. On the reverse direction, many pockets and patches of open space located *inside* or *within* the city have gained a new rationale and importance, from the parks and natural continuums, to the requalified waterfronts and riverfronts or to the reuse of large deactivated infrastructural, industrial or military facilities. As well illustrated by Mangin (2004), Ingersoll (2006) and Muñoz (2008), contemporary urbanization processes are also the result of multiple individual demands, trying to meet and to recombine functional, social and imagetic desires in new territorial products: the golf and residential enclaves, the regional theme parks and outlets, the countryside villas, ecolodges, and the artificial lake condominiums are some of these new expressions of the *countryside in the city*.

Although they make the larger portion of current metropolises worldwide, the areas produced by these processes have remained aside from the morphological attention for a long time, hidden by the interest on historical districts and monumental city fabrics (Moudon, 1998). The need to address them through the lens of canonical morphological studies or through the inception of new specific lens remains a challenge only met by few scholars (Stanilov & Scheer, 2004; Tatom, 2004). To some extent, this is the result of a bias against suburban development, often associated with a negative and less interesting feature when compared to the City (Pavia, 1996).

As argued by Marot's (1995) historical perspective on the urban-rural thresholds, *suburban* areas are far from a contemporary idiosyncrasy. On the contrary, since Roman times, very heterogeneous and sophisticated territorial arrangements have existed around the denser urban settlements, closely related through infrastructure and functional connections, and integrated into highly valued systems of agricultural production, defense, leisure or residence. This complex and organized interplay between different pieces of a heterogeneous territorial patchwork is the basis for a call to maintain a specific look on the specific process and structures that shape and explain today's urban territories.

Two directions could be synthesized on this point: one advocating for a gradual and incremental transition from the urban space towards the open space, with a *transect* of progressively lighter and thinner footprint of artificial activities and built space; another, arguing for a networked, intermingled and *patchworked* conceptualization, in which the key is not the limit or its declination into a sequential filter, but a multi-layered and networked field of ecosystemic, urban and infrastructural systems. The former is usually associated with New Urbanism thinking (Duany & Talen, 2002). The latter resonates strongly in the Landscape Urbanism perspective (Corner, 2006; Hough, 2007) and admits a variegated, less hierarchical and polycentric metropolitan configuration. The introduction of the concept of *landscape* as a matrix for contemporary planning and spatial design is seen by its sponsors as bringing in operative tools to address timely challenges, such as the capacity to address large scale, heterogeneous and complex spatial realities, the understanding of ecological processes and their dynamism and potential in shaping space with the energy of nature, and finally, the conceptualization of a dialectic construction between natural and cultural realms (Corner, 2006).

This perspective is also in line with European scholars, who bring into the debate the thick territorial palimpsest (Corboz, 1983; Secchi, 2007), from contemporary urbanization patterns emerged. As a result of the century-old layering of spatial signs and cultural constructions, contemporary territorial forms are the expression of accumulated fragments of wider socio-spatial constructions; they are rarely the result of comprehensive and unitary design, but often the product of partial, auto-organized, diversified and spatially and timely uncoordinated initiatives. As such, they are hardly understandable in simple terms, with clear outlines or coherent geometric layouts - particularly so with the accelerated 20th century territorial transformation.

Having both the idea of a palimpsest and patchwork as a background, authors such as Manuel de Solà-Morales (1997) inquire as to the need of a more structural approach to the multiple *fragments* that result in today's urban landscape. Acknowledging a rather elusive and fragmented system of *interesting distances* as a location rationale for the multiple activities and land uses across today's territory, he argues for a more intentional and designed exploration of infrastructural and ecological spaces – as mediators between such local activities and the larger scales. Bringing a cultural meaning to these *interesting distances* can only be attained if one finds the opportunities and positive values in those spaces often seen as negative, disjointed, splintered and fractured. To go beyond this idea of an *evil* periphery (Pavia, 1996; Solà-Morales, 1997), one needs to find the rhythms and the distances, along with the structures of continuity (paths, infrastructures, corridors) which allow for a more coherent landscape fabric, while allowing for the autonomous, flexible, adaptive and efficient decisions of individual players. This idea builds on Secchi's (1986) call for a project of the ground (*progetto di suolo*), through which a system of public space and active open space becomes a structural framework in extensive, dispersed and discontinuous urban landscapes.

Reading the Processes of Fragmentation With the Lens of Metropolitan Morphology

Despite an acknowledged difficulty in reading, understanding and providing coherent pictures of these forms of metropolitan landscapes (Choay, 1994; Mangin & Panerai, 1999; Sieverts, 2003; Tatom, 2004), it is possible to identify representative patterns, based on territorial integration and on typological differentiation. Many of the authors engaged in the explanation of contemporary urbanization processes and forms acknowledge the need to have a productive linkage between reading, description, and design.

The idea of descriptive geography (Dematteis, 1995) as a synthesis between the dimensions of analysis and of intervention calls for the need to develop common tools to both. This synthesis puts significant weight on the interpretative framework of values and its evolution. The described reality is thus an ever-changing entity, according to the specific lens being used. They provide different meanings and propose various roles, relations, and opportunities to the various social, physical or ecological elements being described. In this sense, the idea of description carries within the seed of an *implicit project* (Dematteis, 1995): the way how territory is described, interpreted and recalled hides (or reveals) a project in itself, thus bridging past, present and future into not only *scientific* but also *political* practice.

When addressing the form of contemporary metropolization, the acknowledgment of interest, value and opportunity in fragmented spaces, usually labelled as suburban and peripheral, entails an ethical position of commitment to address them not only in their shortages and frailties, but also with the imagination to provide them with the conditions to actively participate in the spatial, socio-economic, political and ecological landscape of the metropolis. On the other hand, acknowledging the form of the territory – following the seminal challenge by (Gregotti, 1981) – as a key component to address their integration, requalification, and integration, one needs to bring in the disciplinary fields of urbanism, urban morphology, architecture and landscape architecture. In common, these disciplines share a key concern with the formal structure of built landscape over time and of *drawing* and *mapping* as fundamental methodological tools.

Within the field of urban morphology, Moudon (1994, p. 289) emphasizes the complex nature of typo-morphological studies, by integrating site-specific, time-framed and open space-aware issues:

A typo-morphological approach to defining type differs from other approaches in three ways. First, type in typomorphology combines the volumetric characteristics of built structures with their related open spaces to define a built landscape type. This approach is in opposition the monumental, siteless typology of Durand, for instance. The element that links built spaces to open spaces is the lot or parcel, the basic cell of the urban fabric. Second, the inclusion of land and its subdivisions as a constituent element type makes land the link between the building scale and the city scale. Third, the built environment type is a morphogenetic, not morphological, unit because it is defined by time – the time of its conception, production, use or mutation.

Five morphological approaches can be highlighted in this regard, all of them resorting to a strong drawn and mapped methods:

- the Venetian school, in which Bernardo Secchi and Paola Viganò stand prominently, researching into the dispersive, porous and multi-scalar fabrics of various Italian regions and major Europeans metropolises (i.e. Paris, Antwerp), also with master-planning and design interventions (Secchi & Vigano, 2009; Secchi & Viganò, 2011; Viganò, 1999);
- the Parisian school, around the works of David Mangin and Philippe Panerai, offering a decoding the evolution and paradigmatic spaces of metropolitan change in Paris (Mangin, 2013; Panerai, 2008), of the franchising, sectorization and generic layouts of globalization used in contemporary urbanization processes (Mangin, 2004) and of the discrete characters and opportunities for a more balanced organization of low-density urban-rural territories (Brès, 2015);
- the Barcelona perspective, mainly developed at the *Laboratório de Urbanismo de Barcelona*, with regard to the mapping and spatial interpretation of the main processes of urban growth and ter-

ritorialization in the Catalan metropolis (Font, 1999, 2004, 2012; Llop, 1997; Solà-Morales, n.d.; Vecslir, 2007);

- the contributions from the USA regarding suburban morphological analysis by Moudon (1998), Tatom (2004) or Scheer (2004);
- the Lisbon and Oporto schools in Portugal, explored through the work of URBinLAB, FormaUrbisLab, and MDT research groups, with specific approaches to Lisbon metropolitan development patterns (Leite, 2016; Santos, 2012, 2013, 2017a, 2017b; Travasso & Domingues, 2015).

Four main methods can be highlighted from these main contributions:

- the study of elements and their decomposition in coherent systems, either in simple groupings (streets and squares, plots, buildings) or in more individualized or even *reversed* types (Viganò, 1999), such as the elements of everyday materiality and in-between open spaces;
- the importance of evolution and transformation as a layered process of both interdependency or rupture between older and newer territorial strata;
- the understanding that the key processes of (1) land subdivision, (2) infrastructure delivery and (3) building are combined in different ways, according to the specific economic and institutional context; these processes lead to multiple urban development patterns which can be categorized and systematized in morphological types;
- the need to address the relational role of open space, large scale infrastructure and new and complex spatial assemblages (i.e. nodes, polarities, corridors) as fundamental systems to understand the heterogeneous and multi-scalar features of the metropolis; such analysis can be done either by isolating them and looking into their inner components or to frame them as networked territorial systems.

From these contributions, this study engages with the notion of fragmentation as a morphological process of loss of integrity and continuity of territorial systems and matrices. In order to provide a conceptual synthesis, a framework regarding *fragmentation* is defined. At the first level, two fragmentation realms are identified:

- the fragmentation of open spaces;
- the fragmentation of collective space.

Open space fragmentation is mainly concerned with the pre-metropolitan matrix, on which fast urban growth was grounded. It can be still subdivided into three types: 1) the *fragmentation of the ecological structure*, namely of continuities and habitats (Forman, 1995); 2) the *fragmentation of the rustic land plot pattern*, as a result of the disruption of the rural societal and economic model; and 3) the *fragmentation of landscape units*, namely by through the loss of visual continuity and coherence of open space.

Collective space fragmentation is mainly concerned with the elements present in urban space, although it may include network spaces that go beyond the boundaries of the most densely built urban environments. It can be divided into 1) the *fragmentation of public space*, mainly impacting the continuous elements that support soft mobility flows; 2) the *morphological fragmentation*, seen as the loss of cohesion and intelligibility of the spatial and volumetric armature (Shane, 2005) that defines, bounds and provides guidance, balance, and proportion to the urban space; and 3) the *fragmentation of infrastructural networks*, mainly seen in the multiple gaps, missing links and poor connectivity of infrastructure or the spatial clash between metropolitan pieces of infrastructure (i.e. highways, railroad lines) and local spaces.

Project and Design for Building Synergies Between Urban Fabrics, Open Space, and Metropolitan Landscape

Secchi (2005) and Tatom & Kahn (2009) points to the need to draw new methodologies and conceptualizations in face of a more open, diffuse and centrifugal territorial system built from fragments and individual elements (Viganò, 1999), shaped by interstitial uncertain spaces (Font, 1999) and crossed by arterial elements (Llop, 1997). From this perspective, open space is a fundamental system for design, as put by Sieverts (2003, p. 121) in his discussion of the *urbanized landscape* or *Zwischenstadt: The open space of the landscape will become the actual creative field, which must preserve and restore the identity, the unique character of the Zwischenstadt.* Some of the guidelines for such path is further exploited by Llop (2016) in a number of actions: 1) consolidate open spaces into a networked ecological infrastructure, 2) recover territorial continuity, 3) refunctionalize natural and environmental *drosscapes*, 4) design the edges and promote interaction with ecotones, 5) promote urban and territorial regeneration, 6) reconfigure metropolitan streets as axes of territorial vertebration, 7) foster mobility network intermodality, 8) integrate infrastructures.

It is therefore hypothesized that these landscapes of metropolitan fragmentation contain within themselves the elements for a potential multi-scale and multidimensional regeneration, namely at the level of 1) containment of urban sprawl/dispersion, 2) redefinition of limits and forms of transition between the urban and the rural, 3) public space qualification as a tool to promote active mobility, 4) innovation in the field of agricultural and forestry uses in a metropolitan context, and 5) strengthening of infrastructure and ecosystem services.

THE PROCESS OF SUBURBANIZATION IN THE MAKING OF LISBON'S METROPOLIS

The Metropolitan Area of Lisbon

Lisbon Metropolitan Area is the most populous region in Portugal, with a population of 2.8 million inhabitants. It is the country's most important political, economic and cultural urban area, with good infrastructural national and international connections, namely through its two main port cities of Lisbon (in the Tagus river estuary) and Setúbal (in the Sado river estuary). Its administrative area includes 18 municipalities, among which Lisbon, the capital city with 550.000 inhabitants. The metropolitan territory is highly diverse in its natural features, land uses and demographic dynamics. With important environmental areas, such as Tagus and Sado rivers natural reserves and Arrábida and Sintra mountain ranges, along with significant agricultural, forestry and fisheries activities, this region displays complex interplays between urban and rural territorial patterns. Up until the 20th century, these activities had a strong relationship with the multiple region's urban settlements, of various size and configuration. Industrialization and profound changes in the economic realm during the last century had a strong impact on this relationship, with a sharp decrease on agriculture-related population; nevertheless, the fundamental pattern of rural activities distribution in the metropolis remains unaltered.

Another aspect of the urban-rural interplay is found in the urban development pattern that shaped the metropolitan region since the mid-19th century. At that time, Lisbon played a significant role as an undisputed urban centre in the region (see Table 1). With the development of railroad transport, modern shipping and industrial production, both Lisbon and smaller settlements started to grow as new migrants from the countryside were looking for better life conditions. In 1960, Lisbon reaches its peak with 800.000 inhabitants, but still accounting for more than half of the total region around it. However, major change occurs from 1960 to 2011, with Lisbon losing around 350.000 inhabitants while the rest of the metropolitan area growing 1.3 million: from a mostly rural territory, it becomes a complex metropolitan organism, in which urban space is intermixed with an extraordinarily heterogeneous mosaic of open space. In this intermix, the area of agriculture and forestry decreased dramatically, under the pressure of urban growth; however, and on the other hand, new economic rationales also allowed for specialization and selective improvement of some agricultural sectors, combining them in innovative added-value urban, industrial and touristic chains.

The transformation of the metropolitan territory of Lisbon between the mid-1960s and mid-1990s was shaped by various levels of disruption in the relationship between the development of infrastructural support and the production of urban space: disruption between the programming and the delivery of mobility, utility and facility networks; disruption of the territorial matrix based on the rural parcellation; disruption of socio-economic and political layers at the metropolitan and national levels. From an urban structure based on defined on 19th century radial lines, more diffuse urban growth patterns emerge, with new configurations and variable intensities. These patterns are supported by new road and electrical networks which, together, enable residential and industrial deconcentration away from conventional railroad-oriented lines of development. Tensions between functional requirements and the territorial support result in a fragile and coalescing development of urban fabric, supported by minimum levels of infrastructural and environmental support. It is only after the 1990s that some of the metropolitan networks designed in the 1960s to meet booming urban growth begin to take shape.

Also, significant legislative changes had a strong impact on a splintered urban development pattern (Santos, 2013). On one hand, with the approval of Decree-Law No. 46673 of 29/11/1965 - known as the Law of *Loteamentos* (urban subdivision) - the government allowed the entry of private actors in urban land production, giving away the exclusive role it held in this area since the 1940s (Nunes da Silva & Pereira, 1986). On the other hand, the legislation extends the private appropriation of urban capital gains to the process of urbanization; until then it was basically limited to the building process (capital gains associated with urbanization and subdivision were retained by the public sector). Henceforth, virtually

Table 1. Evolution of resident population in Lisbon Municipality, Sintra Municipality and total Lisbon Metropolitan Area in 1864, 1900, 1960 and 2011

	1864	1900	1960	2011
Lisbon	190 311,00	351 210,00	802 230,00	545 245,00
Sintra	20 766,00	26 074,00	79 964,00	377 249,00
Lisbon Metropolitan Area	316 483,00	549 945,00	1 505 982,00	2 815 851,00

Source: National Census – National Institute of Statistics

all urbanization cycle is inscribed within a market rationale, mostly speculative in nature and, above all, informed by investment profitability criteria (Nunes, 2011). Urban development operations are therefore understood as real estate products. Finally, the establishment of horizontal property[1] in the Civil Code 1966 (although it was legally defined since 1955) introduces fundamental changes in investment rationale and in the social and spatial distribution of property and its development, especially in the field of housing.

A Paradigmatic Territory in Lisbon Metropolis: Cacém, Agualva, Massamá, São Marcos, and the Spaces In-Between

Although highly diverse and heterogeneous, the development process of Lisbon's Metropolitan Area in the second half of the 20th century is commonly regarded as highly fragmentary, discontinuous and disruptive in terms of infrastructural provision, morphological coherence and open space/ecological continuity. To have a deeper understanding of these processes, a specific study area had to be chosen. In order to have a relevant and representative study case, four main criteria were taken into consideration:

- a territory showing considerable tracts of open space, associated with natural/ecological networks, agricultural/forest land uses, or vacant land;
- a territory shaped by fragmentary and discontinuous processes of urban growth with multiple development rationales, resulting in heterogeneous forms of urban fabric and complex land-use and spatial configurations;
- a territory in which large scale metropolitan infrastructures play a relevant role in its spatial and functional organization, both by its connective capacity and by its spatial splintering effects;
- a territory struggling with large areas of derelict, abandoned or fallow land, with unclear prospects regarding its future, to bring into the discussion major societal challenges and the agenda of the multiple transitions (energy, mobility, economic, territorial).

Additionally, a prospective criterion was also taken into consideration: the acknowledgment of high potential regarding territorial regeneration, through open experimentation and scenario development. This acknowledgment should manifest itself in the form of planning tools, urban and landscape studies or implemented projects in which interesting ideas for the site's future – based on the project's topics and thematic interest – could be devised.

The chosen study area is defined by the outer boundaries of four urban settlements – Cacém, Agualva, Massamá and São Marcos – and the open spaces in-between, in which a deep valley and its waterline (known as *Ribeira das Jardas*) feature prominently. This area, from this point on designated as Cacém-Massamá (see Figure 1), is located along Lisbon-Sintra multi-infrastructural axis, one of the most important suburban corridors in Lisbon Metropolitan Area in terms of population, commuting intensity, functional diversity and historical interest. Shaped by the combination of a suburban railroad line and one of the main metropolitan highways, it became the test-bed for most of the urban change processes developed around Lisbon in the post-WWII period. Unlike the city of Lisbon and the Lisbon-Cascais corridor (known as *Costa do Sol*), in which urban planning had already made consistent steps in the first half of the 20th century and in the 1940s, Lisbon-Sintra corridor was mostly developed without a binding and coherent infrastructural and urban development scheme. As such, development occurred mostly as an accumulation of piece-by-piece processes led by private initiative. Other than two urban development plans in Amadora and Sintra (two settlements along the corridor) – largely based on formal principles of urbanism – most of the territory in-between was seen as rural, and thus not needing any specific planning tool. However, urbanization was allowed on rather casuistic criteria, if some basic infrastructural conditions were met by the developer.

As a result of such piece-meal process, large scale infrastructure, social facilities and a balanced distribution of residential and employment areas were not properly provided, resulting in poor built environment quality and inadequate urban network services. This trend shaped a territory that became not only known for its frailties and low-standard amenities, but also for housing lower-income residents. The process of suburbanization, in Lisbon's case – and unlike the middle and affluent class flight of United States cities' suburbs – was also a socio-spatial process of peripheralization. Commuting to Lisbon remained at very high levels, despite the creation of some industrial employment along the corridor. Since the early days of suburban growth and still today, congestion and weak transport services are seen as critical weaknesses of an unbalanced territory. The booming increase on car-ownership has reshaped the urban and functional landscape, with new motorways and looser urban development patterns, now polarized by large shopping malls and more dispersed geography of metropolitan centralities, but the class between regional infrastructure and local fabrics remains critical.

From a morphogenetical point of view, Cacém-Massamá shows some of the most common features of the metropolitan fragmentation processes: 1) persistence of rustic land patterns as a matrix for urban subdivision operations; 2) induction of urban growth by improved accessibility (railroad line and IC 19 highway); 3) use of a legal framework that favors fragmentary urban development, namely through piece-meal and unarticulated subdivisions; 4) presence of multiple barriers and discontinuities in public space and pedestrian connections, discouraging the use of soft mobility and public transportation solutions; 5) use of standardized, low quality and poorly designed building types, as a result of laissez-faire, liberal market-driven approach to urban development, resulting in the impoverishment of spatial and landscape qualities.

Lastly, the study area includes various opportunities for territorial regeneration based on open space. Featuring considerable areas with fallow land, steep slopes and an important stream of water, it is a clear example of a pattern found in many other locations in Lisbon metropolis. These open spaces, now embedded in a sharply urbanized territory, are remnants of a now lost rural territorial organization. They showcase, nevertheless, both the problems and the potential fields for the morphological, ecological and connective suture. They stand as promising fields for some of today's major national and international urban agenda concerns: the blue and green infrastructures, the urban-rural transitions and the innovative land uses for vacant land. In a framework of potentiating urban-rural synergies, the inception of new economic rationales for open space is of the utmost importance as part of urban sprawl control, improvement of environmental resilience and optimization of soft mobility networks strategies.

The Evolution of a Suburban Territory

Figure 2 shows the mapped evolution of the studied territory in relation to Lisbon and its western metropolitan corridors. In the 1940s this was still a rural landscape, on which Cacém, Agualva, and São Marcos stood still as small scale and compact settlements, governing a clearly agriculture-oriented territory. Located on the mid-level of low-rise elevations, they were served by a thin structure of roads to which main buildings were faced. Their edges and backyards, on the other hand, were turned to the water-irrigated gentle slopes on which vegetables for local consumption were grown. That time's road structure was largely the same used in the 19th century, on which the Road of Sintra stood prominently.

The railroad line navigates through the valley's lower levels, avoiding unnecessary bridges and land-forming. The stations were located on hitherto unbuilt areas; they were connected to the existing settlements through small roads, later to become axes of growth.

The 1960s were characterized by the first meaningful patterns of urban growth, still much grounded on the rustic land matrices. The electrification of Sintra railroad line in 1957 boosts commuting to Lisbon. Improved accessibility together with relatively low housing costs favored the development of the first urban subdivision operations. These were still rather frail and small-scale processes, mostly aimed at delivering detached housing and small low-rise rent housing buildings. The roads converging on the railroad station played an important role, providing the spatial armature for the first generation of urban development. The rustic parcels and the land pattern along them remained, however, mostly unchanged. Of greater importance, the building of a throughway bypass (later widened to become today's IC 19 highway) to the old Road of Sintra reorganized the road system and provided fast access to Lisbon by car and bus.

The 1990s were the years of fast-paced suburbanization. In a wider context, the entire length of the Lisbon-Sintra suburban corridor underwent major changes as a result of multiple private-led urban development initiatives. The area around Cacém stood prominently in this process, becoming one of the densest urban spaces in the region. Just after not more than twenty years, the first generation of urban growth was overlapped by a new generation shaped mostly by higher multi-level apartment development. With the legal establishment of horizontal property in 1966, higher-rise apartment building (up to 12 storeys high) replaced the former detached-houses, as demand for housing in the metropolitan area rose exponentially and a new real-estate development model started to emerge. Industrialized and medium-to-large scale developers, supported by a more dynamic banking and credit system, started to shape extensive residential areas in greenfield areas, as is the case in Massamá and São Marcos. Public space, social facilities, and local amenities are consistently inadequate and insufficient for the sudden demand increase. Besides housing, new industrial estates also emerged, taking advantage of better accessibility and electrical supply infrastructures. Jobs and production start to relocate from the late 19th century industrial locations towards a more dispersed pattern on the metropolis.

The first two decades of the 21st century can be seen as a stage of stabilization and infilling. Some of the previous trends continued its path during the first decade, namely new with large scale residential developments in greenfield and urban edges. However, a trend to limit urban growth starts to emerge as a result of major changes in demographics (aging and limited population growth with the trend to decline) and economy (a significant impact of the 2002 and 2008 financial crises on the construction and real-estate industries). Urban infill and replacement of older buildings are also further continued in the older suburban fabrics. Retail and industrial spaces also undergo significant shift with a rise on the large pavilion-shaped shopping centers, surrounded by parking space and good connection to the road system, often away from consolidated urban areas. To the south of the study area, the large urban development schemes of *Taguspark Science & Technology campus* and the *Universidade Católica campus* are examples of a new knowledge-economy metropolitan organization, based on the combination of advanced services, R&D, university facilities and high-end residential estates.

TRACING PLOT SUBDIVISION PATTERNS: SHAPING A FRAGMENTED URBAN LANDSCAPE

The analysis of the urban development process through the lens of subdivision patterns was carried out resorting to rustic plot cartography available at the Directorate-General for Territory (DGT - Direção-Geral do Território). These maps were surveyed in the decades of 1940 and 1950 and overlapped land plots over the topographical survey. Its aim was mostly concerned with the land value and tax system. In the legal framework of that time, *rustic* land differed from *urban* land for not having the infrastructure or land development permit. Therefore, many spaces within the urban settlements and its perimeters were designated as rustic land. Such cartography is useful to trace the first stages of urban development in most Portuguese cities, as it was surveyed just before the intense expansion process of the second half of the 20th century. On the other hand, it allows to trace back the land plot patterns on which subdivision processes were grounded. As will be shown in the following lines, the vast majority of urban subdivisions in Portuguese cities and in the study area were carried out piece-by-piece, using the existing land plot layout as a basis. Therefore, the rustic patterns are a sort of genetic code that pre-exists the forms of urban development.

The Persistence of the Rustic Land Plot Pattern

The overlapping of the 1940's and 1950's rustic plot cartography with other imagery such as historical aerial from various dates, allows the analysis of the status of such plot genetic code as an evolving morphological stratum. The analysis of the rustic land plot maps also brings forward different organizational patterns in regard to the topographical condition and agricultural exploitation. In the better-irrigated areas and in the vicinity of the older villages (Cacém, Agualva, and to a lesser extend São Marcos), the plots were smaller and usually elongated in shape. This layout is the result of the optimization of plot sides served by water streams and road infrastructure. On the dryer hill tops and mild slopes, land plots are usually bigger and irregular in shape. When they meet steep slopes, their shape is clearly defined by the topography edges.

The designation of places and agricultural settlements is also relevant as it translates specific forms of territorial organization. In the study area, two main designations are found: 1) the large and complex agricultural estate (*quinta*), and the simpler but more common farm estate (*casal*). The *quinta* is usually are more sophisticated structure, bounded by high walls and associated with a prominent main landlord building, surrounded by complementary constructions and labor housing. The complex is often served by hydraulic systems of sourcing, transport, storage, and irrigation, allowing for multiple forms of agricultural production: leisure gardens, fruit trees plots, vegetable growing plots and small woods for lumber. Livestock can also be part of the agricultural system of the *quinta* estates. Six *quintas* stand out as important elements in the study area: *Quinta da Bela Vista* (Cacém), *Quinta do Castelo* (next to *Ribeira das Jardas*), *Quinta da Fidalga* (Agualva), *Quinta da Barroca* (Agualva), *Quinta do Porto* (Massamá) and *Quinta da Tascoa* (in the study area's eastern edge).

The *casal* structures are usually associated with the dryer soils. They are the offspring of the centuries-old *villa* roman settlements, with their longstanding parcel configurations. They form an organized exploitation system of various parcels centered around a building nucleus, occupied by resident farmers to which the lands were leased. As such, they lack complexity, sophistication, and erudition of the *quinta* structures, but play a fundamental role as one of the most common elements in the pre-metropolitan rural landscape. Four of them are identified in the study area: *Casal do Olival* and *Casal do Bareta* (Massamá), *Casal de Colaride* and *Casal do Cotão* (São Marcos). Whereas the main buildings of *quintas* often had a direct relationship with the road system – they played a representational role as focal points of social, economic and political power –, the main building of *casais* were usually set away from the roads, but with good control conditions of the surrounding parcels.

Much of this rural territorial structure is still detectable in the parcel lines that resisted the passage of time and the profound metropolitan transformations (Figure 3). Despite the many changes, there are two forms of geometric land plot pattern persistence:

- the ones still observable by built elements such as walls, fences, and well-defined property lines;
- the ones that are only partially recognizable by the edge configurations of the urban fabric, usually corresponding to the outer polygons of subdivision operations.

From another point of view, the overlapping of the 1940s and today's land plot patterns also allows the identification of erasures, segmentations, and cuttings through. The most expressive cases are those associated with the railroad and highway infrastructures which split apart large tracts of former agricultural parcels.

Figure 3. Identification of plot pattern persistence in Cacém-Massamá through the overlapping of land plot cartography and aerial photography
Source: author.

A Typological Systematization of Subdivision Processes

The morphological analysis was developed through the review of fifty-six urban subdivision processes identified in the study area, approved between 1967 and 2003, available on Sintra Municipal Archives. This sample represents approximately 60% of the subdivision processes approved in the study area. Since 1965, urban subdivision (*loteamento*) is the most common urban development legal tool, not only in the study area but in most of Lisbon metropolitan area and most of the Portuguese cities. In the study area, there were no new subdivision processes approved after 2003, attesting to a changing territorial rationale, from urban growth to consolidation.

Various cartographical representations were developed in order to systematize the analysis of subdivision operations:

- location regarding the older rural settlements
- location regarding the main roads and railroad infrastructure
- size of subdivision operation
- date of approval
- morpho-typological organization scheme

In order to decode the morphological processes underlying the building boom, it is possible to outline fourteen different cases of plot transformation. These cases are systematized in terms of the morphological process and reflect not only forms of spatial transformation but also changing realms of development rationale (Figure 4).

1. Small scale plot transformation
   - 1.1 Regularization and road/street widening + transfer to the public domain
   - 1.2 Linear and regular subdivision along road/street + row buildings
   - 1.3 Single plot + Contiguous and linear aggregation along road/street + stand alone building
   - 1.4 Piece-meal building in single plots, without urban alignment
2. Small-to-medium scale plot readjustment
   - 2.1 Piece-meal subdivision + laying of new internal road/street
   - 2.2 Plot consolidation + regular subdivision + laying of new internal road/street
   - 2.3 Articulation between individual plots/developers + Basic urban layout rules + laying of new common road/street
   - 2.4 Unplanned and informal subdivision + detached building
3. Medium-to-large scale subdivisions
   - 3.1 Partial subdivision of old farming estates + new urban layout [keeping old buildings and part of land]
   - 3.2 Complete subdivision of farming plots + new urban layout
     - 3.2a Complete subdivision of farming plots + new urban layout (industry)
   - 3.3 Aggregation and subdivision of rustic plots + New urban layout
   - 3.4 Phased aggregation and subdivision of rustic plots + phased new urban layout

Figure 4. Types of plot subdivision found in the study area (numbering according to the list above)
Source: author

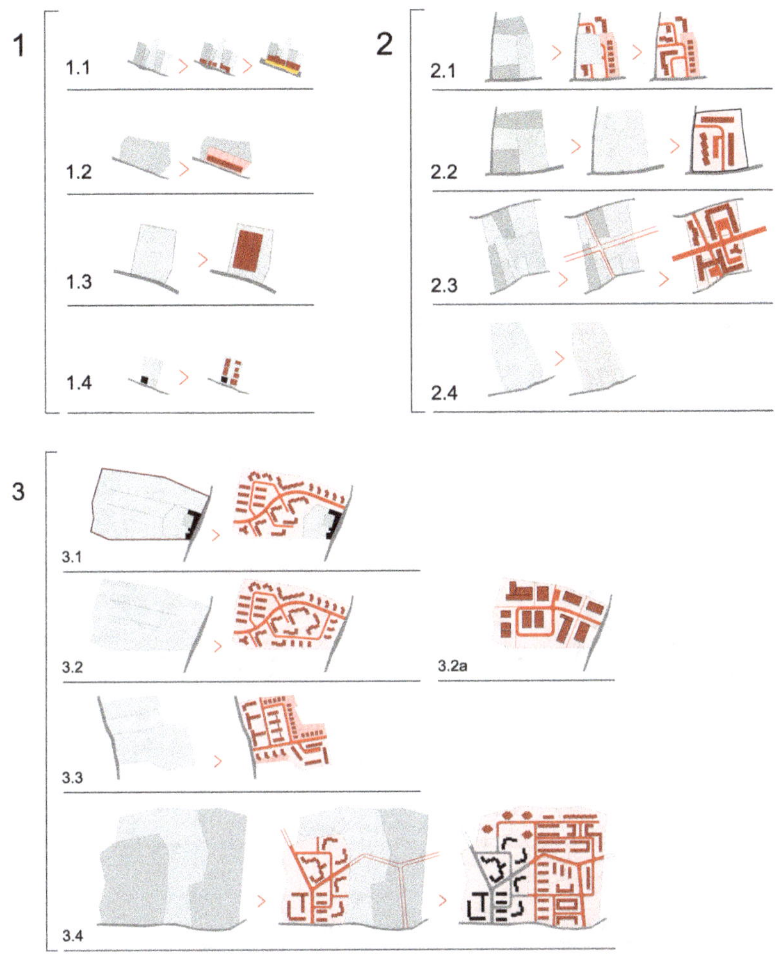

Having these types and their location (Figure 5), it is possible to build an interpretative framework at a larger scale, according to 1) location, 2) relationship with structural elements (roads, waterlines, train stations) and 3) type of morphological configuration (building type, urban layout, organization of public space).

This analysis provides a morphogenetic basis to understand the various processes of urban development and their rationale. If crossed with a diachronic perspective, one can see a change from smaller scaled transformation processes to larger scaled ones, especially after the late 1990s. This change is noticeable geographically:

- The predominance of types 1 and 2 in the first generation of suburbanization (1950's-1980's), namely in Cacém and Agualva;
- The predominance of type 3 in the second generation of suburbanization (1980's-2000's), namely in Massamá and São Marcos;

Figure 5. Mapping of subdivision types in study area
Source: author

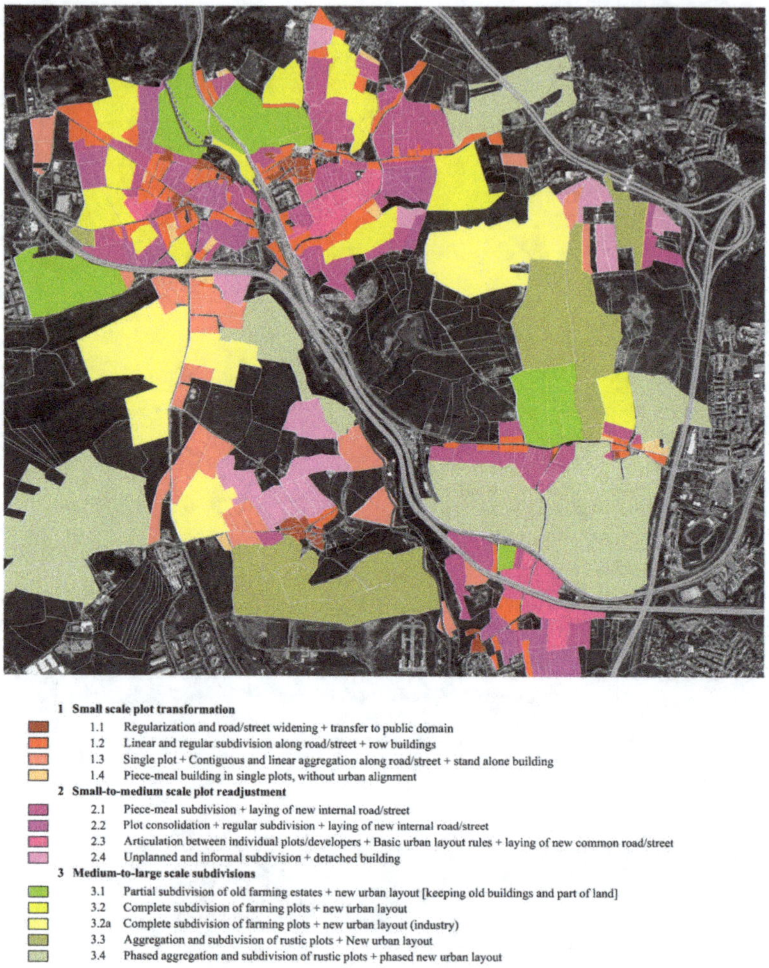

- Presence of overlaid transformation processes in the older settlements of Cacém and Agualva (types 1, being replaced by types 2), building a more complex and stratified urban matrix;
- The predominance of types 1 in the parcels adjacent to the oldest roads;
- The introduction of medium-to-high rise housing typologies in the types 2 and 3;
- The prevalence of large-scale urban layouts in types 3, although often keeping some of the former parcel boundaries in phased development schemes.

The Emergence of Type-Volumetric Spacing: The Clashing Roles of Infrastructure, Public Space, and Building

A new and diffused morphological configuration emerges from the combination between 1) a fragmented urban development initiative (in space, in time, in infrastructural support), 2) a focus on real estate optimization of residential products and 3) a development scale-up, associated with the construction industrialization. In this type of fabric-weaving, the key element is the collective housing building, understood as a module unit and optimized in its internal structure. Seen as a repeatable and addable module, it is often laid over a pre-existing rustic plot pattern, maintain its frail geometric layout. Reparcellation operations or land consolidation for urban transformation, if and when occur, remain apart from any effective effort of integration into wider settlement structure (Arrabaça, 2003; Cruz, 2008; Santos, 2012). A morphological gap between the new urban condition and the original land-shaping rationales is thus opened.

In addition, aggregation forms tend to mirror the building process rationale. The free-standing model is repeatedly used as an effective and operative tool. It focuses on the building itself and allows for an easy and phased construction process. As a result, however, it also erodes clear traces of order, hierarchy, regularity, and quality of open space – so dear to the modern urban discourse –, resulting in an otherwise vague and undifferentiated *type-volumetric spacing* (Santos, 2012). This model's capacity to adapt to very different situations can also be seen as its main setback, as it relinquishes on more sophisticated forms of adjustment to local spatial context. These *type-volumetric spacings* are the result of a volume-oriented design, detaining itself at the object scale, with limited concern with the spatial result of modular repetition patterns over a neutral open space surface.

With very little involvement of architectural skilled design, the development of reinforced concrete construction technologies, such as pillar-beam and prefabrication, result in a rather unsophisticated appropriation of typological innovations such as the free-standing building and the *plan libre*. The outcome is a new kind of urban and architectural spatial order: on one hand, structural metrics based on pillar-and-beam allow for greater freedom in design; on the other hand, standardization and prefabrication often introduce repeated, monotonous, undifferentiated patterns, with little concern for urban integration and architectural composition. Public-private boundaries, such as small walls with back and front yards, give way to a permeable space, often lacking legibility and hierarchy (Mangin & Panerai, 1999). The dual relationship between façade and backdoor is blurred. Kitchen and laundry areas, hitherto turned away from streets and public space, become visible. The prominent role of façade alignment, front doors and ground floors as thresholds between the building and public space is eroded in an often-uncontrolled addition of shops storefronts, housing windows, garage doors and faint references to modern *pilotis*. A rather atopic verticality emerges, replacing a predominantly horizontal grounding of buildings as a spatial ordering device (Santos, 2012).

The resulting urban spaces face, therefore, a dual fragmentation process. On one hand, the collective space is splintered, discontinuous and lacking intelligibility; on the other hand, the architectural spacings is focused on the built units individually, with little concern for the overall coherence and composition qualities. In the midst of such a fragmentary landscape, multiple patches of open space persist, some of them still related to the pre-metropolitan territorial organization. If they are seen from a positive design-oriented point-of-view, one can find hidden features for potential recomposition. As traces of a former stage of landscape pattern, they can be seen as a sort of *territorial heritage* – keys to acknowledge, read and understand the spatial processes that shaped today's metropolis through time.

A TERRITORIAL MOSAIC: POTENTIAL SPACES FOR A SYNERGIC TERRITORY

The territorial interpretation of Cacém-Massamá grounded on its major geographical and morphological features resulted in a set of specific components:

As part of the open space system:

1) the valley and the waterlines, 2) the slopes and steep topography, 4) the vegetable gardens and agricultural patches, 5) the in-between open spaces, 6) the special geological features and stone quarries, 7) the fragments of the rustic land pattern, 8) the formal urban green spaces.

As part of the urban and infrastructural system:

1) the old roads and paths, the railroad and stations, 2) the archaeological sites, 3) the historical urban squares and listed heritage buildings, 4) the old agricultural estates (*quintas*), 5) the first urban subdivisions, 6) the laying of new streets and partial land redevelopment, 7) the large urban subdivisions, 8) the industrial subdivisions, 9) the commercial street fronts, 10) the pavilion retail and shopping, 11) the social and educational facilities and precincts, 12) the campuses and enclaves, 13) the heavy highways and intersections, 14) the urban requalification projects.

Out of these components, six are shown in figure 6 and can be seen as both analytical and design-oriented mappings. As such, they not only illustrate identifiable morphological entities useful in the understanding of today's territorial configuration, but also by providing lines of potential action for intervention. These entities or components hold an *implicit project* – in Dematteis (1995) sense. They are the threads upon which minimum action can provide maximum impact in terms of urban and landscape regeneration. They are the traces that bridge the urban, the rural and the geo-ecological strata that continuously (re)shape this territory.

Waterlines and Vegetable Production Patches

The valley of *Ribeiras das Jardas*, the study area's main *continuum naturale*, is a fundamental physical element that simultaneously shapes the site, its hydrological cycle, air flow patterns and supports micro-agriculture activities. On the other hand, the valley is a territory prone to multiple natural risks, such as flooding, soil erosion and, to a lesser degree, landslides.

The water lines that cross through this territory are, in large tracts, channeled under the ground, namely in the densest areas of urban development. However, when they are in the open, they not only play a vital ecosystemic role but are also a support for human uses and appropriations such as vegetable growing or green urban space. Continuity and porosity are features to which water lines relate prominently, enabling the combination of structural lines (continuity) with the richer intermix between different kinds of open space and urban fabric (porosity).

Four types of vegetable growing patches are identified in the study area, attesting both to the existing territorial diversity and to its potential in terms of the urban landscape:

- linear spaces associated with *Ribeira das Jardas*, taking advantage of good soil fertility, but highly constrained by urban and infrastructural pressure, namely those related to untreated wastewater;
- linear spaces associated with secondary waterlines, usually of larger size and heterogeneity, and less constrained by urban pressure – most are located to the north of Cacém and Agualva;
- backyards of small parcels located within informal detached-house urban fabrics, such as those in São Marcos;
- dispersed patches within medium sized parcels, without a clear relationship with the urban fabric.

Figure 6. Selection of key territorial components in Massamá-Cacém (from left to right, top to bottom: 1 - the valley, the waterlines, and the old road network, 2 - the first urban subdivisions, 3 - the laying of new streets and partial land redevelopment, 4 - the large urban subdivisions, 5 - the urban requalification projects, 6 – the in-between open spaces, steep slopes, special geological features, stone quarries and fragments of rustic land pattern)
Source: author.

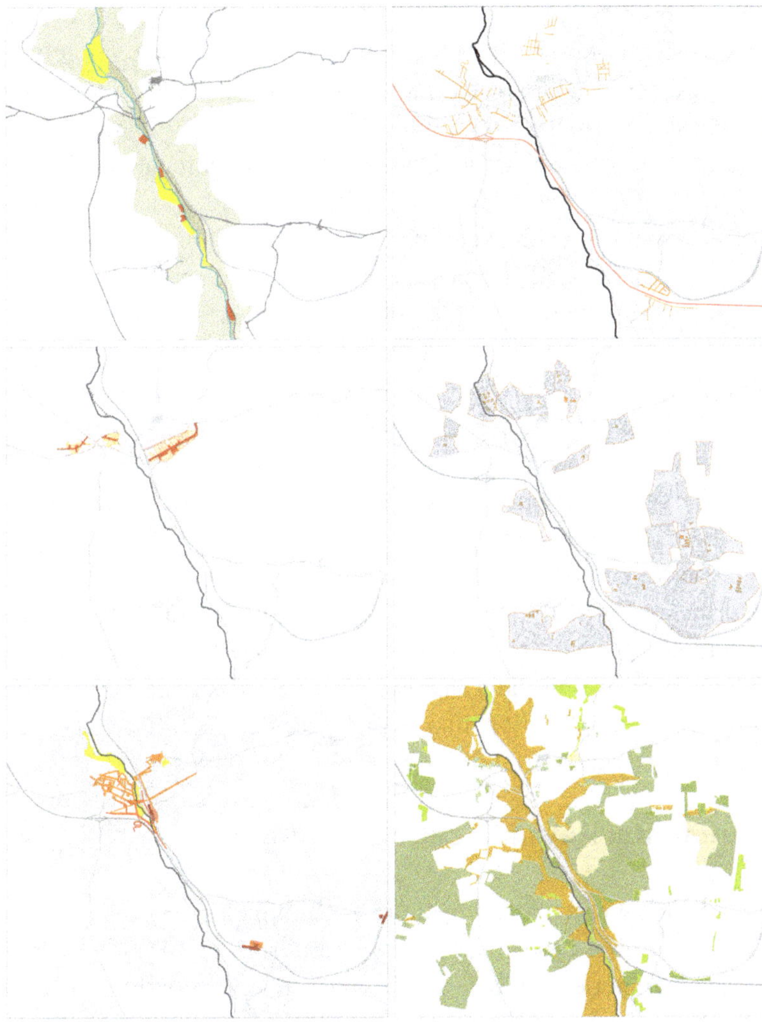

Slopes and Large Open Spaces: Topographical Condition and Factor of Discontinuity

A significant part of unused open spaces in the study area is explained by the steep slopes on which they're located. Many are at risk of soil erosion and, in some cases, of landslide, as is the case in some slopes around *Ribeira das Jardas*. Urban planning documents regard these spaces as green urban spaces, a classification that brings protection regarding urban development but needs further attention in order to guarantee good articulation with the surrounding spaces. On a closer level, steep slopes are important spatial discontinuities and obstacles to safe and comfortable pedestrian flow, requiring specific and creative solutions to overcoming such gaps. However, they are important scenic features and ecosystemic patches, as they can provide safe and stable places for biodiversity and dense vegetation. When absorbed into recent urban development areas, steep slopes are often configured as small open space pockets, but lacking continuity and a structural role in the wider urban spatial configuration strategy.

In the spaces between the compact settlements of Cacém, Agualva, Massamá, and São Marcos, considerable patches of unbuilt open space persist. Most of these parcels have a history of cereal production, with an average size above 3 hectares. Despite some forage crops, they are largely underused and with little agricultural-based economic value. In fact, up until the early 2000s, they were labeled for future urban development, awaiting good market conditions to be urbanized. When one considers the evolution of these open spaces through time, it is clear the downsizing, retrenchment and fragmentation processes, as former parcels are subdivided, either one by one or by groupings.

In this process, retrenchment is neither continuous nor contiguous, and thus fragmentation has a strong impact in the landscape, the agricultural/forestry production and the ecological systems. Adding to the urban subdivision process, there were also planned infrastructural corridors in the 1999 Sintra Municipal Master Plan, which would cross through these open spaces, and were never built. The ongoing revision of this Master Plan is no longer considering these infrastructures and re-labeled many open fields as rustic land, defined as natural and forest spaces.

On the other hand, some of the open spaces are also labeled as strategic permeability areas, aimed at the protection and recharge of aquifers, opening another door to the acknowledgment of ecological service values. One of the main issues in terms of territorial design and regeneration for these areas is the nature of their edges with urban fabrics and infrastructural spaces. As fragmentary splinters of urban development, the open spaces have a poor relationship on their edges. They are often seen as urban voids, defined by barriers, limits, and frontiers. The same indifference occurs in the built spaces, with little or no-acknowledgment regarding the transition between urban and open space. Thick, complex, multi-functional edges are, therefore, critical in providing access, visibility and opportunity to discover, use and acknowledge a far richer territorial mosaic.

Urban and Environmental Requalification Projects

The transition to the 21st century was marked by an incremental change in Portuguese urban policy and public funding regarding territorial cohesion. From a priority on solving basic and much needed infrastructural and social facility amenities, left behind after years of fast growth and limited public resources, the new generation of public investment became mainly concern with the aspects of environmental and urban qualification. With basic needs largely met, the focus shifted towards a new benchmark of public space, transport interfaces, restoration of water lines, urban rivers and waterfronts and improvement of pedestrian spaces. In the last twenty years, several projects were carried out in Cacém-Massamá under this new agenda.

The first and largest was *Programa Polis*, a national government-led project to improve urban and environmental conditions in several Portuguese cities and critical urban areas. Cacém was chosen for a 20 million Euro investment, in recognition of its substantial problems as one of the densest and most critical urban areas in Lisbon Metropolitan Area. The project's strategic plan placed *Ribeira das Jardas* waterline as the central focus of the intervention, along which a linear park with four hectares was built. Wastewater decontamination and stormwater design were key components, as part of environmental resilience objectives. This park establishes a complex relationship with its edges, bridging hitherto fragmented spaces, opening new paths on the site and providing new leisure and sports facilities. The connection to the surrounding areas was complemented by the spatial requalification of various streets, thus connecting the lower areas of the park and a railroad station with the old urban center in the higher slopes. One can say that *Ribeira das Jardas* and its park became the new city center, confirming open space as a fundamental shaper of today's metropolitan regeneration strategies.

Heavier infrastructural modernization works were also carried out a few years later, with the refurbishment of Cacém and Massamá railroad stations. The small depots that endured since the early 20th century, were restructured into larger multi-functional transport hubs, with a significant upgrade on the adjacent public space quality. The stations were acknowledged also as a sort of urban plaza, on which daily life converges and became a representational feature of a rather anonymous urban landscape. Since the middle of 2010s, new initiatives have also targeted specific public space and street amenities, in order to provide better pedestrian and bicycle accessibility, connectivity and comfort, along with alternative roads connecting different urban areas. This process is improving the network potential of existing infrastructures, which were still very much reliant on old roads. An intermediate level of connection was – and still is – acknowledged as a need and a challenge for the future integration and balance between the various urban settlements and the open space patches between them.

An interesting idea coming out of the Sintra Municipal Master Plan revision is the definition of a large-scale park with 391 hectares, expanding the linear park in Cacém to the north and to the south, along *Ribeira das Jardas*. This park is seen as a new typology of urban and green space planning. It includes agricultural patches, natural slopes, and crosses through the multiple open space patches that persist around Cacém and Massamá. Its components are less dependent on conventional arrangements of green space, resorting to nature-based low-cost, low-irrigation and low-maintenance solutions, open to agricultural appropriation and to the passage of seasons. Its scale also differs from conventional urban parks and gardens. It can be better understood as a metropolitan park, in which ecological flows play a far more important role than the human-shaped spaces and facilities usually provided in densely built environments. Additionally, this park is materialized as a part of wider vital corridors of the metropolitan ecological structure. This not only defines a multi-level, trans-scalar conceptual approach to metropolitan design and planning but also enables EU program funding aimed at improving territorial resilience through nature-based solutions.

CONCLUSION

Spatial fragmentation is a common phenomenon in the development of many metropolises. It can be observed in various dimensions, such as open space and ecological networks, public space, and morphological armatures or infrastructural networks. Fragmentation results from the combined effect of multiple factors, namely the lack of coherence between planning and implementation processes, piece-meal urban development, conflicting interests in the use of land and territorial resources or unbalanced or inefficient provision of infrastructural amenities. One of the clearest expressions of spatial fragmentation can be seen in the complex interrelationship between rural and urban fabrics, as they continuously evolve in a metropolitan framework. Lisbon, as a complex metropolis shaped by heterogeneous processes of urban development, is a paradigmatic example of this fragmentary territorial pattern, observed at both the regional and the local levels.

The article offers a morphological approach to identify, map and decode the process of spatial fragmentation as part of a wider contribution to understand and explain Lisbon's fast and intense process of suburban development in the second half of the twentieth century. A specific focus on the urban subdivision processes (*loteamento*) provides a typological systematization which can be useful to understand the basis of Lisbon's metropolitan landscape. On the other hand, the article describes and highlights the territorial components that play a potential structural role, thus offering clues to inform future design-oriented strategies.

It is argued that fragmented spaces can no longer be seen under the conventional urban-rural and center-periphery dualities, as they ignore the powerful forces shaping contemporary urbanization patterns. Instead, greater attention must be paid to the in-between spaces that permeate through the heterogeneous and splintered urban and rural fabrics. In this sense, urban-rural synergies can be fostered not only by activating and linking the fragmented patches of open space, public space, and infrastructure, but also by engaging with the specific features and processes that shaped each one of those patches. Future research would gain from a design-oriented approach, that is, through the analysis of interesting examples of urban-rural transition projects and their land readjustment processes. The potential for such approach is clearly found in architectural and urbanism schools, in the development of speculative and innovative design solutions to face these territorial challenges.

REFERENCES

Arrabaça, P. (2003). *Formas Urbanas na Cidade Construída: Agualva-Cacém, 1953-2001*. Universidade de Lisboa.

Barata Salgueiro, T. (1992). *A cidade em Portugal: uma geografia urbana*. Porto: Afrontamento; doi:10.18055/Finis1874

Barata Salgueiro, T. (2001). *Lisboa, Periferia e Centralidades*. Oeiras: Celta. Retrieved from http://www.bulhosa.pt/livro/lisboa-periferia-e-centralidades-teresa-barata-salgueiro/%5Cnhttp://www.bulhosa.pt/livro/do-comercio-a-distribuicao-roteiro-de-uma-mudanca-teresa-barata-salgueiro/livro/lisboa-periferia-e-centralidades-teresa-barata-salgueiro/

Boeri, S., Lanzani, A., & Marini, E. (1993). *Il territorio che cambia : ambienti, paesaggi e immagini della regione milanese*. Milano: Abitare Segesta. Retrieved from http://www.worldcat.org/title/territorio-che-cambia-ambienti-paesaggi-e-immagini-della-regione-milanese/oclc/717038600&referer=brief_results

Brenner, N. (2014). *Implosions/Explosions. Towards a study of planetary urbanization* (N. Brenner, Ed.). Berlin: Jovis.

Brès, A. (2015). *Figures Discretes de l'Urbain – À la reencontre des réseaux et des territoires*. Genéve: MètisPresses.

Choay, F. (1994). La mort de la ville et le règne de l'urbain. In *La Ville: art et architecture en Europe, 1873-1993* (pp. 26–35). Paris: Éditions du Centre Pompidou; doi:10.1093/ndt/12.4.839

Corboz, A. (1983). Le Territoire comme palimpseste. *Diogene, 121*, 14-35.

Corner, J. (2006). Terra Fluxus. In C. Waldheim (Ed.), *The Landscape Urbanism Reader* (pp. 21–33). New York: Princeton Architectural Press.

Cruz, C. M. (2008). A importância do cadastro no desenvolvimento urbano português. In XI Colóquio Ibérico de Geografia. Alcalá de Henares.

de la Solà-Morales, M. (1997). Territoris sense model. *Papers: Regió Metropolitana de Barcelona*, (26), 21–27. Retrieved from http://www.raco.cat/index.php/PapersIERMB/article/viewArticle/101847/0

de la Solà-Morales, M. (2004). Contra el modelo de metrópolis universal. In Á. M. Ramos (Ed.), *Lo urbano en 20 autores contemporáneos* (pp. 99–104). Barcelona: Edicions UPC.

de la Solà-Morales, M. (n.d.). Las formas de crecimiento urbano. Las formas de crecimiento urbano. *Universitat Politècnica de Catalunya.*

Dematteis, G. (1995). *Progetto implicito : il contributo della geografia umana alle scienza del territorio.* Milano: Franco Angeli.

Domingues, Á. (2006). *Cidade e Democracia. 30 Anos de Transformação Urbana em Portugal.* Lisboa: Argumentum/Ordem dos Arquitectos.

Duany, A., & Talen, E. (2002). Transect Planning. *Journal of the American Planning Association*, (68): 345–266.

Font, A. (1997). Anatomia de una metrópoli discontinua: la Barcelona Metropolitana. *Papers*, (26), 9–19.

Font, A. (1999). *La construcció del territori metropolità. Morfogènesi de la Regió urbana de Barcelona.* Barcelona: Mancomunitat de Municipis de l'Àrea Metropolitana de Barcelona.

Font, A. (Ed.). (2004). L'explosió de la ciutat. Morfologies, mirades i mocions sobre les transformacion territorials recents en les regions urbanes de l'Europa Meridional. Barcelona: Universidad Politècnica de Catalunya, Institut d'Estudis Territorials, Generalitat de Catalunya.

Font, A. (Ed.). (2012). Patrons urbanístics de les activitats econòmiques. Regió Metropolitana de Barcelona / Urban patterns of economic activities. Barcelona Metropolitan Region. Barcelona: Universidad Politècnica de Catalunya, Institut d'Estudis Territorials, Generalitat de Catalunya.

Forman, R. (1995). *Land Mosaics: The Ecology of Landscapes and Regions.* Cambridge, UK: Cambridge University Press. doi:10.1017/9781107050327

George, P. (2001). Paisagem – do urbano ao rural notas a partir de Lisboa…. *Finisterra*, *36*(72), 115–121. doi:10.18055/Finis1626

Graham, S., & Marvin, S. (2001). *Splintering Urbanism. Networked infrastructures, technological mobilities and the urban condition. Technology.* London: Routledge; doi:10.4324/9780203452202

Gregotti, V. (1981). La forme du territoire. *AA; l'Architecture d'Aujourd'Hui*, 218.

Herce Vallejo, M., Magrinyà, F., & Miró, J. (2004). *Construcció de ciutat i xarxas d'infraestructures.* Barcelona: Edicions UPC.

Hough, M. (2007). Nature as Infrastructure: Strategies for Sustainable Regional Landscapes. *Places*, (19).

Ingersoll, R. (2006). *Sprawltown – Looking for the City on its Edges*. New York: Princeton Architectural Press.

Lefebvre, H. (1970). *La Révolution Urbaine*. Paris: Éditions Gallimard.

Leite, J. S. (2016). *Ruas emergentes: Interpretação morfológica do contexto urbano português*. Universidade de Lisboa. Retrieved from https://www.repository.utl.pt/handle/10400.5/12443

Llop, C. (1997). Lògiques i espais projectuals d'una metròpoli, Barcelona 1976-1992. *Papers*, (26), 37–52.

Llop, C. (2016). *Ciudades, territorios metropolitanos y regiones urbanas eficientes: estrategias y propuestas de proyecto para la regeneración de la Ciudad_Mosaico_Territorial después de la Explosión de la Ciudad: la Región Metropolitana de Barcelona como laboratório*. Barcelona: Universitat Politècnica de Catalunya. Retrieved from https://upcommons.upc.edu/handle/2117/97155

Llop, C., & Bosc, S. (2012). *Working with the territory: strategies for the new territorialities / Travailler avec le territoire : stratégies pour les nouvelles territorialités*. Barcelona: CREPUD-MED / ACTAR.

Mangin, D. (2004). *La ville franchisé : Formes et structures de la ville contemporaine*. Paris: Éditions de la Villette.

Mangin, D. (2013). *Paris/Babele – Une mégapole européenne*. Paris: Éditions de la Villette.

Mangin, D., & Panerai, P. (1999). *Projet urbain*. Marseille: Parenthèses.

Marcuse, P. (1994). Not Chaos, But Walls: postmodernism and the partitioned city. In S. Watson & K. Gibson (Eds.), *Postmodern Cities and Spaces* (pp. 243–254). Cambridge, MA: Blackwell. doi:10.1111/j.1365-2184.1980.tb00458.x

Marot, S. (1995). L'alternative du paysage. *Visiteur: Ville, Territoire, Paysage, Architecture. Revue Trimestrielle de Critique Des Situations Construites*, *1*, 54–80.

Moudon, A. V. (1994). Getting to Know the Built Landscape: Typomorphology. In L. S. Karen Frank (Ed.), *Ordering Spaces: Types in Architectural Design* (pp. 289–311). New York: Van Nostrand Reinhold.

Moudon, A. V. (1998). The Changing Morphology of Suburban Neighborhoods. In A. Petruccioli (Ed.), *Typological Process and Design Theory* (pp. 141–157). Cambridge, MA: Aga Khan Program for Islamic Architecture.

Muñoz, F. (2004). *urBANALización. La producció residencial de baixa densitat a la província de Barcelona, 1985-2001*. Universitat Autònoma de Barcelona.

Muñoz, F. (2008). Paisajes de la movilidad: de los espacios multiplex a los aeropuertos low cost. *Papers*, (47), 68–73.

Nunes, J. S. (2011). *Florestas de Cimento Armado. Os Grandes Conjuntos Residenciais e a Constituição da Metrópole - Lisboa, 1955-1981*. Lisboa: Fundação Calouste Gulbenkian.

Nunes da Silva, F., & Pereira, M. (1986). Ilusões e desilusões das periferias na área Metropolitana de Lisboa. *Sociedade e Território*, (5), 14–24.

Oswald, F., & Baccini, P. (2003). *Netzstad: Designing the urban*. Basel: Birkäuser.

Panerai, P. (2008). *Paris Métropole. Formes et échelles du Grand-Paris*. Paris: Éditions de la Villette.

Pavia, R. (1996). *Le paure dell'urbanistica. Disagio e incertezzaa nel progetto della città contemporanea*. Gènova: Costa & Nolan.

Portas, N., Domingues, Á., & Cabral, J. (2003). *Políticas Urbanas - Tendências, estratégias e oportunidades*. Lisboa: Fundação Calouste Gulbenkian.

Raxworthy, J., & Blood, J. (2004). Introduction. In *The MESH book: Landscape/Infrastructure* (pp. 10–17). Melbourne: RMIT University Press.

Santos, J. R. (2012). *Espaços de mediação infraestrutural: Interpretação e projecto na produção do urbano no território metropolitano de Lisboa*. Universidade Técnica de Lisboa. Retrieved from https://www.repository.utl.pt/handle/10400.5/5429

Santos, J. R. (2013). Planning a Metropolis from Infrastructure: Lisbon from 1940 to 1966. *Journal of Planning History*, *12*(4), 312–333. doi:10.1177/1538513213487503

Santos, J. R. (2017a). AdaptPolis - Descodificar a fragmentação urbana: infraestruturas, paisagem e projeto territorial para a Área Metropolitana de Lisboa – Relatório Intermédio. Lisboa.

Santos, J. R. (2017b). Discrete Landscapes in Metropolitan Lisbon: Open Space as a Planning Resource in Times of Latency. *Planning Practice and Research*, *32*(1), 4–28. doi:10.1080/02697459.2015.1028253

Scheer, B. C. (2004). The radial street as a timeline. A study of the transformation of elastic tissues. In B. C. S. Kiril Stanilov (Ed.), *Suburban Form: An international perspective* (pp. 102–122). London: Routledge.

Secchi, B. (1986). Progetto di suolo. *Casabella*, (520), 19–23.

Secchi, B. (2000). *Prima lezione di urbanistica. Prime lezioni*. Roma: Laterza.

Secchi, B. (2005). *La città nel ventesimo secolo*. Roma: Laterza.

Secchi, B. (2007). Rethinking and Redesigning the Urban Landscape. *Places*, (19), 6–11.

Secchi, B., & Viganò, P. (2009). *Antwerp, Territory of a New Modernity*. Rotterdam: Sun.

Secchi, B., & Viganò, P. (2011). *La ville poreuse, un projet pour le Grand Paris et la métropole de l'après-Kyoto*. Paris: MètisPresses.

Shane, D. G. (2005). *Recombinant Urbanism: Conceptual Modeling in Architecture, Urban Design and City Theory*. West Sussex, UK: Wiley Academy. doi:10.1016/j.numecd.2017.10.006

Sieverts, T. (2003). *Cities Without Cities: Between Place and World, Space and Time, Town and Country*. London: Spon Press/Routledge. Retrieved from http://www.amazon.com/dp/0415272602

Silva, C. N. (2008). Urban Planning in Portugal in the Twentieth Century. *The International Journal of Regional and Local Studies*, *4*(2), 23–39. doi:10.1179/jrl.2008.4.2.23

Stanilov, K., & Scheer, B. C. (2004). *Suburban form: an international perspective*. London: Routledge. Retrieved from http://books.google.com/books?id=fso45Od_jUYC

Tatom, J. (2004). Making the metropolitan landscape: Lyons 1812-1994. In *Suburban Form: An international perspective* (pp. 77–101). London: Routledge.

Tatom, J., & Kahn, A. (2009). Programs for a metropolitan urbanism. In *Making the Metropolitan Landscape: Standing Firm on Middle Ground* (pp. 195–202). New York: Routledge; doi:10.4324/9780203872048

Travasso, N., & Domingues, Á. (2015). *Território: Casa Comum*. Porto: FAUP.

Vecslir, L. (2007). Paisajes de la nueva centralidad. *Urban*, (12), 34–55. Retrieved from http://dialnet.unirioja.es/servlet/extart?codigo=2857098%5Cnhttp://dialnet.unirioja.es/descarga/articulo/2857098.pdf

Viganò, P. (1999). *La città elementare. Biblioteca di Architettura, 7*. Milano: Skira.

KEY TERMS AND DEFINITIONS

Land Plot Persistence: The expression of stable spatial elements associated with older land parcel configuration which resisted territorial transformations overtime.

Plot Subdivision Patterns: The morphological expression of different subdivision process, providing the basis to understand the relational rationales between land plots, street, and infrastructural configuration and building typologies.

Rural: A socio-economic system primarily based on the exploitation of natural resources with pre-industrial technology, hierarchical geographic configurations, and local fields of the relationship between markets, government, and community.

Spatial Fragmentation: The result of multiple processes converging in the break of continuity, contiguity and morphological coherence of urban, rural and natural spaces and networks.

Urban: Sophisticated socio-economic systems primarily based on the industrial, financial and knowledge-based transformation of resources, information, and capital, organized at territorial levels through highly connected technological systems.

Urban Splintering: A socio-spatial process characterized by the fragmentation, deconcentration, and differentiation of access to urban services, infrastructural networks, and spatial amenities, associated with the economic rationales of liberalization, privatization and globalization.

Urban Subdivision: A legal process by which one or multiple rustic land parcels are subdivided, readjusted in geometric layout and served by infrastructural amenities to support urban land uses.

ENDNOTE

1 *Horizontal property* is a legal figure by which the ownership of multi-apartment buildings is subdivided in fractions. Until this figure was approved, buildings could only be entitled to a single owner, which could rent apartments to tenants, usually as a source of revenue and investment. With the introduction of *horizontal property*, the housing market undergoes a deep change, with a major increase of home-ownership. This allows for a new urban development rationale based on credit-mortgage, with the banking system playing an important role in the real-estate industry. Larger multi-apartment buildings become popular products: on one side they are affordable for majority of the medium-class; on the other hand, they have a short investment life-cycle, offering short-medium term returns to the developer and contractor.

Chapter 2
Trends in New Multi-Family Residential Development in Serbia:
Overview by Local Urban Professionals

Aleksandra M. Djukić
Faculty of Architecture, University of Belgrade, Serbia

Branislav M. Antonić
https://orcid.org/0000-0002-2675-9264
Faculty of Architecture, University of Belgrade, Serbia

ABSTRACT

The field of housing has undergone significant changes during the challenging post-socialist transition. Serbia has been a somewhat different case in the way that its transition was postponed and more complex. Conversely, its housing also carries the legacy of the more developed former socialist Yugoslavia. This dichotomy can easily be observed in the case of the newly-built, "transitional" multi-storey housing. However, the data relating to its characteristics is rare and insufficiently accurate. The aim of this chapter is to improve the research of the urban dimension in newly-built multi-family housing in Serbia, regarding housing planning in the new master and detailed urban plans for Serbian cities. This has been researched by analyzing the professional knowledge and experience of local experts. Therefore, the method used in the research was survey, in which urban experts had participated. It had been carried out in May 2015.

INTRODUCTION

The urban dimension of housing as a research topic can be studied through several relevant aspects. First, housing is often considered as a spatially-prevalent land use, occupying the major area in an urban settlement. Second, housing is also known as the most sensitive general urban function, because it is

DOI: 10.4018/978-1-5225-9932-6.ch002

strongly related to the private sphere of human life. Finally, every flat or house is also a commodity, i.e. housing is also a market-dependent urban function. Thus, the complexity of housing can be easily observed (Petrović, 2004).

This complexity especially comes to the fore in the periods of sudden and tempestuous changes, when it is difficult to balance between all the mentioned aspects. A good example of this phenomenon is a post-socialist transition from a centrally-planned economy to a capitalist market economy in the countries of Central and Eastern Europe. This transition has been usually described with less or more negative attributes: sudden, dramatic, hard to understand, etc (Hamilton, Dimitrovska Andrews & Pichler-Milanović, 2005; Pickvance, 2008; Hirt & Stanilov, 2009). The post-socialist transformation has brought the myriad of new socio-economic changes (and challenges) for states in Central and Eastern Europe. All such great changes mirror in urban space (Stanilov, 2007). Considering these attributes, the qualitative transformation of urban housing in Central and Eastern Europe has been one of main challenges during the transitional period (IUDK, 2013).

In the case of Serbia, these post-socialist changes in housing have been more severe due to postponed and more complex transition. On the other side, the uncommon model of qualitative socialist housing in the former socialist Yugoslavia has left a positive legacy to present-day situation. Hence, Serbian post-socialist housing has some unique features comparing to other post-socialist countries (Antonić, 2016). This uniqueness is reflected in the ongoing development of Serbian housing, too. A good illustration is a newly-built, "transitional" multi-storey multi-family housing as a dominant type in local typology after the fall of socialism (Jovanović Petrović et al, 2013).

Despite the dominance of this type in situ, there is the general lack of adequate and accurate statistical data. For example, there is no clear division between homes and flats in statistical data – the division is made on the number of dwelling units per building[1] (SORS, 2013), regardless if it function as a single- or multi-family housing. The general shortage in the housing statistics is consequently followed by a shortage of adequate scientific and research analyses and projects. Therefore, the main problem is how to observe the macro-urban characteristics of housing, especially those which are crucial to general urban development.

In addition to statistical data, it is necessary to include the other resources based on the irreplaceable knowledge and experience of local urban experts. Many indicators point that their role in has become (even) more challenging since the start of the post-socialist transition (Hirt & Stanilov, 2014). This can also generally apply to Serbian urban experts, but the fact is they are generally pretty aware about ongoing urban processes and their actors, including those related to housing planning and its numerous problems (Vujović & Petrović, 2007). This is the reason why the research interlocutors are Serbian experts in the field of urbanism. To interpret objectively their professional views and opinions about the main issues of urban dimension in newly-built multi-family housing in Serbia, a survey and questionnaire have been chosen for methodological approach.

The findings of this research should present which characteristics of newly-built multi-family housing are crucial for general urban development. In accordance with the expected findings, the contribution of this research is directed to the formation of guidelines for the general improvements of the urban planning and practices in housing sector in Serbia. Final contributions are how they can apply to international level, to add a value for housing planning in its essence.

METHODOLOGY

The methodology presented in this paper is based on a survey. This method was chosen because it is the proper methodological approach for the fields where professional estimation is important. The feature of qualitative estimation is certainly significant for urban planning – it is in the essence of every planning process (Saroff & Levitan, 1969). Therefore, the foci of this survey are on the macro-urban issues, which are more related to the urban-planning level. Before them, historical background about housing policy in 20th-century Serbia is explained, to properly introduce the topic for prospective readers.

Since a survey is chosen as a methodological instrument for such comprehensive research, more than a dozen questions are arranged. The prepared questionnaire has four groups of questions. The first group of the questions was introductory and differed from the other groups, i.e. it referred to the respondents' professional background. The others refer to the main topic of the research. All 20 questions in the questionnaire are organized as closed-ended ones. Most of them were multiple-choice questions, where only one answer was required. A few are "yes/no" questions. Finally, one question belongs to scaled ones, where respondents need to grade all choices on a continuum.

The aforementioned survey belongs to qualitative research methods. It was conducted by relevant questionnaire in the Summer School of Urbanism 2015, which took place in Kragujevac, Serbia, in May, 14-16, 2015. This is the main annual conference and professional meeting of urbanism-related experts in the country. For this research, it was also important that this conference is always based on topics current for 'Serbian urbanism-related professional milieu'. Serbian experts in this field are the main participants of the conference. These experts, who numbered 114, took part in the survey. Most of them completed the entire questionnaire, but some of them omitted, or wrongly answered some questions.

The overall majority of the respondents (92%) were professionals in the field of urbanism. The others worked in related sectors, such as the cadastre or building construction in situ. A visible majority of the respondents (82%) were primarily professionals who participated in the creation of the urban and spatial plans and urban projects, as well as other strategic documents. A small percentage (5%) included those who were primarily engaged in the sector of building provisions, or in education and research in urbanism. Those employed in the public sector accounted for 76% of the respondents. Apart from one respondent (an NGO employee), all the others came from private companies.

Observing the spatial determination of their jobs, the respondents were mainly professionally engaged in important urban centers in Serbia (82%). This fact was in correlation with the aim of the research; major concentration of multi-family housing is generally in urban areas. Nevertheless, the issue of primary spatial jurisdiction of their professional work was diverse; the most prominent cases were local-level jurisdiction (47%), followed by district-level (29%), and, finally, national and regional levels were selected in minor cases (13% and 7%).

HISTORIC BACKGROUND – MODERN HOUSING IN 20TH CENTURY SERBIA

The history of modern housing in the territory of present-day Serbia began with the formation of the Kingdom of Serbs, Croats and Slovenes in 1918, after the collapse of Hapsburg Empire at the end of the World War I. This kingdom was better known as the Kingdom of Yugoslavia, after the change of its name in 1929 (Bataković, 2005). Before the war, multi-family housing, an indicator of modernity in housing, was almost non-existent in this territory. By the latest research, less than 0.4% of existing multi-family housing in Serbia belongs to the housing built before 1918 (Jovanović Popović et al, 2012). This oldest stock of multi-family housing was limited to relatively scarce buildings in Belgrade, the former capital of the Kingdom of Serbia (Milić, 2006), and major cities in the northern Serbian Province of Vojvodina, then in southern Hapsburg Empire (Radovanović, 2015). All these buildings were specially built; for workers of industrial of mining companies (so-called 'worker colonies') or for state employees in the main public companies (railways, for example).

With the establishment of the new kingdom, much larger territory with bigger population was unified than in the territory of the previous Serbia, which consequently brought a lot of political power and general economic and administrative concentration to Belgrade, a 'new-old' capital. This impulse had an enormous impact on its demographic situation and eventually on housing development. Belgrade population tripled in a relatively short period of 22 years between two world wars (Marković, 1992). This caused a boom in construction, especially in housing sector. Many new multi-family 'row houses' in the form of a condominium or for rent were built, especially in the city center. Then, several new affordable housing estates were established and built by state institutions for their employees (Milić, 2006). However, interwar Yugoslavia was a young capitalist and underdeveloped state (Borsay, Brown & Palairet, 1993), which could not afford inexpensive and decent solutions in multi-family housing for all new city dwellers. This was particularly visible in booming Belgrade, where many dwellers were trapped in illegal residential construction at the city edge (Vuksanović-Macura & Macura, 2018).

Similarly to Belgrade, the general concentration of power and activities influenced the rise of regional centers throughout Yugoslavia. This situation was only emphasized with the formation of the administrative system of large provinces (Serb. *banovina*) from 1929. For illustration, the territory of present-day Vojvodina inherited the relatively balanced organization of middle-size district cities with a significant local autonomy from Hapsburg period (Pušić, 1987). Nevertheless, this was sharply disrupted by the establishment of the Danube Province (Serb. *Dunavska Banovina*) with its seat in centrally located Novi Sad. Local economy flourished by this decision and Novi Sad demographically doubled during interwar period thereof. This was mirrored in the appearance of the first modern multi-family housing in the city, especially in locations connected to the new status of Novi Sad. A very good portrait for this is a completely new urban quarter with multi-family residential development around new 'Banovina' (provincial-seat) building. In contrast to this, the other, previously similar-size cities in Vojvodina were mainly in socio-economic stagnation and without noticeable development in multi-family housing (Djukić, Stupar & Antonić, 2018). The last observation indicated a polarization between bigger and smaller urban settlements in interwar Yugoslavia and its spatial reflection in multi-family housing.

This polarization in general urban development and multi-family housing was radically changed with a new political and economic order in the country after the World War II. The second Yugoslavia (1945-1991) became a socialist state, but gradually switched to a different model than typical communist states in the Eastern Block. This process started already in 1948, when the Yugoslavian leadership broke relations with the USSR and J. V. Stalin and opened doors for cooperation with western countries in the 1950s (Stupar, 2015). During the 1960 and the 1970s, the whole country was fully decentralized, which was a sharp contrast to the other, centralized socialist countries (Petrović, 2004). Such process allowed a better balance between different regions and cities. Then, the mixture of planned and market economy enabled the better living standard in Yugoslavia (Hirt & Stanilov, 2009). This 'in-between' position of the country had further a deep impact in the fields of architecture and urban development (Stupar, 2015).

All these specificities of the socialist Yugoslavia made ultimately housing policy in Yugoslavian cities very unique (Antonić, 2016). First, it better to say that Yugoslavia had many housing policies, because decentralization process profoundly gave a lot of freedom to regional and local levels to form their own housing policies with separate standards and norms (Petovar, 2003; Hirt & Stanilov, 2009). For example, on-site research in middle-size cities in northern Serbia confirms that analyzed cities have the very different patterns of multi-family housing neighborhoods relating their number, position, size, and concentration (Antonić & Vaništa Lazarević, 2019). Similarly, the implemented typology varies; in some cities multi-family housing was encouraged in new residential development; in the other ones raw-houses or detached family houses were preferred. Nevertheless, mass-housing estates with prefabricate residential buildings, typical for the communist cities of Eastern bloc, were built pretty rarely outside the biggest cities. Even there, these estates have the higher standard regarding the general quality of construction and infrastructure (Djukić, 2015). For example, the majority of urban blocks in New Belgrade, the largest mass-housing settlement in the former Yugoslavia, were uniquely designed, from block to block, to adequately present the prestige of unique Yugoslavian socialism (Blagojević, 2007).

Despite all these differences, Yugoslavian political and economic system preserved socialist fundamentals. The production of new multi-family housing was fully in the hands of socialist state and new apartments were predominantly a public property and mainly assigned to public servants – 'apartments with tenancy right' (Milić, 2006). These apartments (actually, the whole apartment buildings) were also exclusively built by state construction companies (Petrović, 2004). Nevertheless, there were noticeable differences than in the rest of socialist world. For example, Yugoslavia did not have a simple state-directed provision of such apartments like a typical communist state. Instead of it, a specific 'quasi-market' of new multi-family housing existed, where different state companies were actors. State construction companies were free to form the price of flats and the other state companies were obligated to transfer at least 4% of their annual net revenue for housing investments for their employees (Petrović, 2004). This model was known as "solidarity housing funds" (Antonić, 2017). In this situation, for predominant and very numerous state servants, the successfulness of their company was an at least equally important criterion to obtain a new apartment as their education degree or the position in the company.

At the end, it had to be underline that multi-family housing, despite its status as an ideal dwelling type for proletariat, was not the prevalent type of housing in socialist Yugoslavia. This was just a case in the main cities (Petrović, 2004). In smaller cities and towns, single-family housing was dominant. Moreover, the significant part of single-family houses was even built illegally at urban periphery, which had proved to be a huge challenge during the post-socialist transition (Antonić & Djukić, 2018).

NEW MULTI-FAMILY HOUSING IN SERBIA – CURRENT ESTIMATIONS

The explained housing legacy has had an influence on the present situation in housing sector, which has been further transformed during the post-socialist transition, shaping unique "local" characteristics.

A post-socialist transition as a recent phenomenon in the eastern half of Europe similarly differs between countries and regions. This can also be applied to housing in urban areas. These differences have been shaped by both historical heritage and current conditions (Stanilov, 2007). Housing in Serbia was included in "south-eastern transition sub-region" (Hamilton, Dimitrovska Andrews & Pichler-Milanović, 2005). The specificity of Serbia and this region is its postponed transition. It was the last country to start with the "real" post-socialist transition in 2000, i.e. after more than one decade of the so-called "blocked socio-economic transformation of society" during the Yugoslav wars (Petrović, 2004, p. 149).

This postponed transition has had apparent and usually negative reflections in the development of urban housing. During the Yugoslavian crisis in the 1990s, the most active segment in housing sector was illegal construction (Petovar, 2003). Nevertheless, the new elements of illegality appeared, such as illegal multi-family projects, absent during socialist period (Antonić, 2016).

After 2000, legal multi-storey multi-family housing has taken dominance in housing sector. The latest figures from the early 2010s reveal that post-socialist projects represent more than 20% of total multi-family housing in Serbia by area (Jovanović Popović et al, 2013). However, this new legal multi-family construction, developed by private investors, has also appeared as problematic, because the maximization of plot capacities has become a key factor for new construction, regardless the quality of dwellings or the respect of urban surrounding (Figure 1). The reason is simple – "the construction of new apartments is one the most profitable activities [in Serbia]" (Mojović & Žerjav, 2011, p. 10). This is certainly one of the consequences of weak post-socialist governance across post-socialist space, dependent of private interests (Stanilov & Hirt, 2009). Unfortunately, the practice of qualitative multi-family housing from the post-socialist countries with more experience has not been used to overcome the weaknesses of the existing situation.

Figure 1. New controversial interventions in multi-family housing in Serbia: a completely new building with the maximization of building plot capacities in Belgrade (Author: B. Antonić)

The other controversial albeit widespread practice has been the 'refurbishment' of multi-family residential buildings from socialist period. The problem with these buildings is mainly related to their dilapidated flat roofs, initially built in functionalist manner. The refurbishment was implemented in very market-driven way, with the parallel addition of one of more floors above the existing building (Vranic, Vasilevska & Haas, 2015). In many situations, this housing interventions have been poorly examined and controlled by public authorities, producing unattractive buildings (Figure 2).

The explained inconsistencies in multi-family housing development reflect the common problem of inadequate and underdeveloped housing policy and legislation (Antonić, 2015). This is followed with relevant statistical data, which is generally rare and insufficiently accurate. As it was mentioned, official statistics differentiates multi- and single-family buildings differently that it is accepted internationally[2]. Available data are mainly numbers and based on official statistics; this approach is questionable considering the latest estimations about the huge number of illegal buildings in Serbian cities, which are further mostly residential (Antonić & Djukić, 2018).

ANALYSIS OF PROFESSIONAL OPINION BY SURVEY

In this situation, one of solutions to acquire data is to redirect focus from quantitative data to qualitative analysis. This research has been rare, but it can be done efficiently if it targets the right focus groups, such as professionals related to urbanism and architecture and close disciplines: urban planners, urban designers, architects, transport engineers or landscape architects.

Newly-Built Housing vs. Renovated Housing

The second group of questions was the first ever dedicated to the essential analysis of newly-built multi-family housing in Serbia. It tries to explore the importance of this type of housing in urban areas and its relations to other forms of recent housing development, such as housing renovation or new affordable/

Figure 2. The refurbishment of an old building with an extra new floor at its top in Pančevo (Author: B. Antonić)

social housing. The first question was the key for further questions, since it referred to the basic presence of multi-family housing in the area of professional jurisdiction. All the respondents answered positively, i.e. they confirmed that new multi-family housing is resent in areas where they work.

The next question is related to the ratio between older and newer multi-family housing. Older housing, thereby, refers to those buildings built during the period of former Yugoslavia, before 1992. New housing was built after this year, during the post-socialist transition. It was expected that the majority of participants would select the choices connected with older housing. It was mentioned previously that this generation of housing was firmly supported by the socialist state, through the well-known model of mass construction of large housing estates – open housing blocks (Hegedüs & Tosics, 1996; Petovar, 2003; Hirt & Stanilov, 2014)

Nevertheless, the result (Figure 3) demonstrated that the majority of the respondents recognized both older and newer housing as equal in their areas (68%). The choices supporting the (pre)dominance of older housing were less represented (20% and 9%). Only 3% of the answers were linked to the dominance of new housing and they are probably "specific by origin". The indication for this statement can be explained by the "self-added view" of one respondent, who, assuming that his/her choice will be unusual, answered that he/she was from a specific town, which has been an important tourist destination in the recent years.

The issue in the third question is the ratio between the types of interventions on multi-family housing during the transitional period (Figure 4). Two major types of interventions are specified: the building

Figure 3. The ratio between older and newer multi-family housing (Author: B. Antonić)

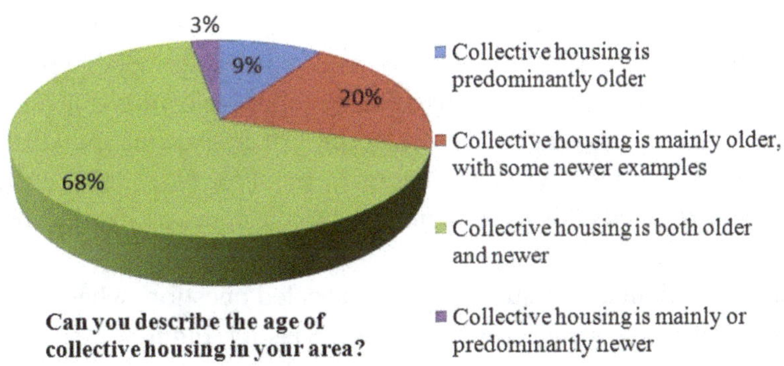

Figure 4. The ratio between two types of interventions in housing (Author: B. Antonić)

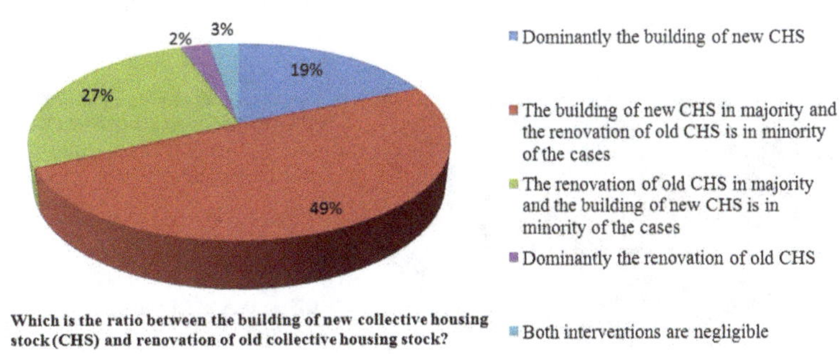

of new multi-family housing stock and the renovation of the older one. The second type (renovation) has been noticed to be especially problematic, due to poor maintenance of many multi-family housing estates during the transition, which has consequently led to new areas of urban poverty (Petrović, 2014).

The results confirm this stance; almost half of the answers give prevalence to the newly-built housing stock rather than the renovated one. Then, 19% of the answers are fully oriented to the newly-built housing stock as the predominant type. Only 29% of the answers are connected to the prevailing renovation of the old multi-family housing stock.

The last question in the first group (Figure 5) is concerned with the theme of public housing in Serbia, officially named as a social housing by national legislation (NARS, 2012). This theme, nowadays, is very popular in Serbia in both academic and professional circles. For example, there are many related strategic, guidance and similar documents at national and local level. Contrary, there are few examples *in situ* (Ristić, 2010), which are often new multi-storey multi-family buildings with social-housing units. The result of the question (Figure 3) is similar to this observation. Most of the respondents (48%) answered that new social housing is rare in their jurisdiction areas; many of them also gave a negative answer, i.e. they claimed that this kind of housing does not exist in their areas (32%); finally, only 1/5 of all the respondents claimed that social housing is adequately or even significantly represented.

Localization of Newly-Built Housing in Urban Areas

The third group of questions is associated with spatial distribution and location of newly-built housing in urban areas in Serbia.

The opening question refers to the possible concentration of newly-built multi-family housing in urban areas (Figure 6). This feature is relevant for a wider topic of urban gentrification and ghettoization of post-socialist cities, which are both considered as considerable social problems (Kovács, 1998). However, unsolved property rights (privatization and restitution) are common in post-socialist countries and they, thereby, can inhibit or even postpone these kinds of transformation (Petrović, 2014). In the case of the survey, the concentration is visible in the majority of the cases (45%), but it is not prevalent. Significant percentages of selected choices are connected to the cases with rare concentration (15%), or even with any concentration (24%).

The second question is unique by its structure; it is a scaled question, where respondents needed to grade urban factors relevant for the location of newly-built housing (Figure 7). These factors are based on the accessibility of important personal needs: workplace, public services, transport, leisure places, and specific attractions and landmarks. Therefore, cumulative gradation is presented in the graph as a better

Figure 5. The importance of newly-built social housing (Author: B. Antonić)

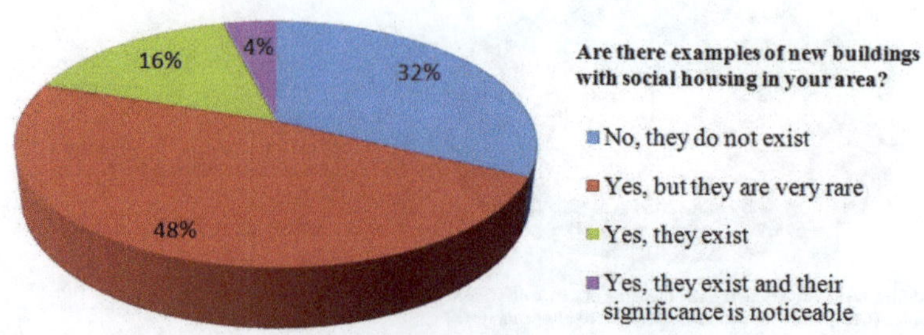

Figure 6. The existence of the concentration of newly-built housing (Author: B. Antonić)

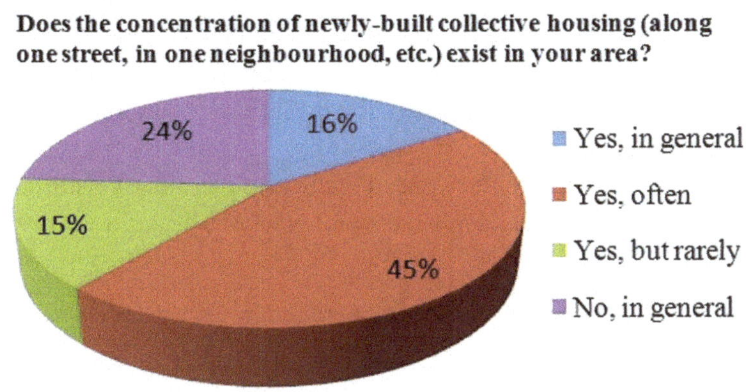

Figure 7. The relevance of different urban factors relevant for the location of newly-built multi-family housing (Author: B. Antonić)

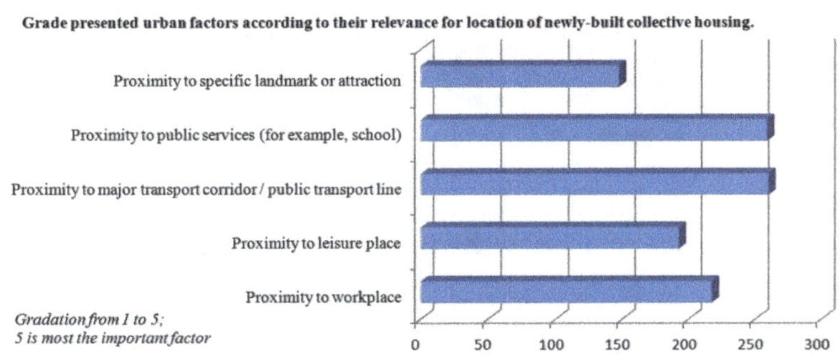

explanation of the selection process. A relatively equal division of grades among five choices means that there is no dominant pattern, which would be suitable for application in urban areas in Serbia. In other words, this can be interpreted as a real image of various urban situations in Serbian urban areas, which are caused by the different size and density of urban areas, street and block matrix, the size and shape of building plots, etc.

The next two questions are linked to the issue of spatial (de)centralization of newly-built multi-family housing. Firstly, one is directly connected to the "classical" monocentric model of the city. Such a model was the unofficial "standard" for socialist countries. This situation has been rapidly transformed by retail initiative, more flexible to the new capitalist economy (Nagy, 2001). The retail sector requires easily accessible locations without the problems of restitution and strict planning and conservation rules. In some cases, this initiative has led to the decline or even emptiness of city centers and to the emergence of new centers with new shopping malls and other retail facilities in well-accessed peripheral urban or suburban areas with more free land (Petrovic, 2005). It is clear that post-socialist transformation has incited the formation of the polycentric urban model.

Thus, the first question (Figure 8) was related to the monocentric model of the city, more relevant to socialist inheritance; in contrast to the first question,, the second tries to touch upon the issue of decentralization and polycentricism, offering the options of main and local centers (Figure 9). In both cases, the bulk of the selected options makes three evenly formed graph sectors, which lead to the conclusion that there is no specific pattern for newly-built housing in Serbia in relation to the issue of spatial (de) centralization of its urban areas.

The fifth question supervenes to the previous one, which examines the streets/urban transport corridors as linear centers (Figure10). The link between streets and newly-built multi-family housing is its focus.

Figure 8. Newly-built multi-family housing in relation to the socialist monocentric urban model (Author: B. Antonić)

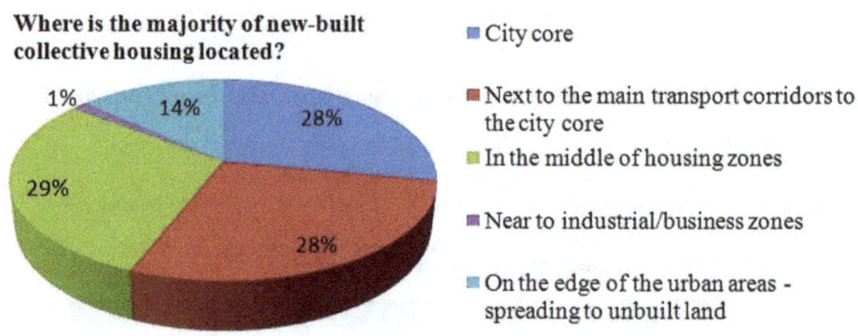

Figure 9. Newly-built multi-family housing in relation to the post-socialist polycentric urban model (Author: B. Antonić)

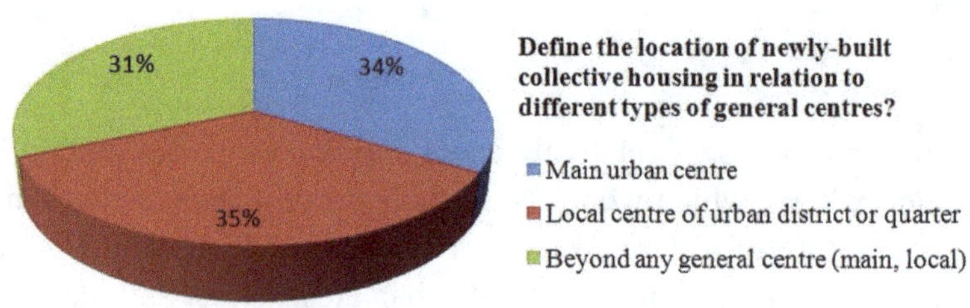

Figure 10. Newly-built multi-family housing in relation to street types (Author: B. Antonić)

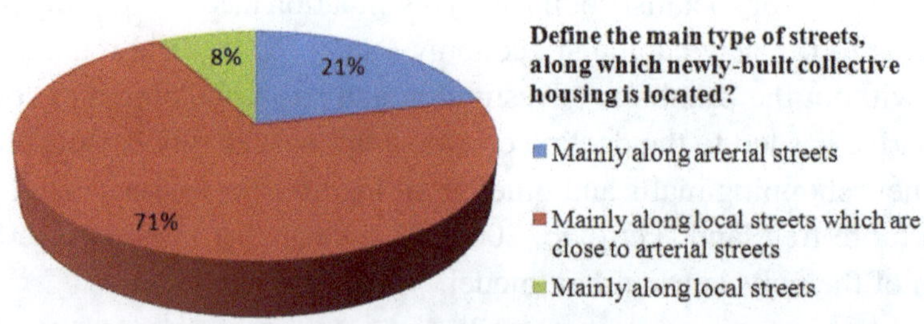

Here, it is easy to notice the prevalence of the second option (71%) – newly-built multi-family housing, which is mainly located along local streets close to arterial transport corridors. The other two choices are appreciably less represented. It seems that newly-built multi-family housing in Serbia "strives" to balance the two mentioned aspects of urban housing dimension; the economic advantage is the proximity of arterial streets as linear centers; contrary, local streets are more appropriate for private life, as a sensitive "element" of housing.

The last question in this group is in the form of a "yes/no" question. It asks if some great spatial attractor or landmark exists in the vicinity of the concentration of newly-built multi-family housing. This attractor can be a shopping mall, a huge sport complex, an attractive riverfront or park, etc, but it is not specifically explained. The question is included in the questionnaire because it is well-known that many post-socialist cities have bidden to initiate new development with flagship projects (Temelová, 2007). This question was answered positively by 63% of the participants. Nevertheless, there is also the noticeable minority (37%) that supports the view that flagship projects are still not crucial for the new urban development of multi-family housing in Serbia.

Functional Aspect of Newly-Built Housing in Urban Areas

The functional aspect of newly-built multi-family housing in Serbia, herein, is understood in two ways: how general urban zoning in Serbian urban areas has influenced housing and how it has influenced some accompanied urban-development projects, such as the improvement and renewal of urban infrastructure.

The main area for the construction of newly-built multi-family housing during the socialist period was generally in the peripheral part of urban areas. However, this pattern has gradually changed since the post-socialist transition, mostly leaving peripheral and suburban areas for the construction of new single-family housing for the emerging middle class (Stanilov & Sykora, 2014). The urban core, in contrast, has begun to be primarily the place for the construction of new luxurious non-residential facilities (usually in the retail sector), displacing the average, middle-class housing (Petrović, 2014). This process can be very intensive in some examples – it can be described as a "centre discharge" (Keivani, Parsa & McGreal, 2002). M. Petrović (2009) generally identified three parallel processes in the post-socialist transformation of urban housing:

1. The process of suburbanization into new single-family residential areas;
2. The gentrification of central urban areas into luxurious, closed housing projects; and
3. The marginalization of socialist housing estates with open-block structure and mass construction.

The survey presents a slightly different situation in Serbia's case. For example, newly-built multi-family housing is usually developed without clear spatial relation to older housing (Figure 11a), which was built during the socialist period (59%), but the percentage of opposite cases is not negligible (41%). Then, the tendency to build new multi-family housing near the recently built ones is very visible (Figure 11b). This is generally supported in 79% of the cases and in 23% of the cases it is strongly supported.

To conclude, as the usual model for post-socialist countries, the gap between the newly-built and existing multi-family housing is not very noticeable in Serbia's case as.

Another issue is the position of new multi-family housing to present single-family housing, which can be discussed taking into account both functional and physical-spatial aspects (Figure 12). This issue remains clear in many post-socialist cities, owing to the legacy of the mono-functional zoning

Figure 11. The tendencies of newly-built multi-family housing in relation to the older one (Author: B. Antonić)

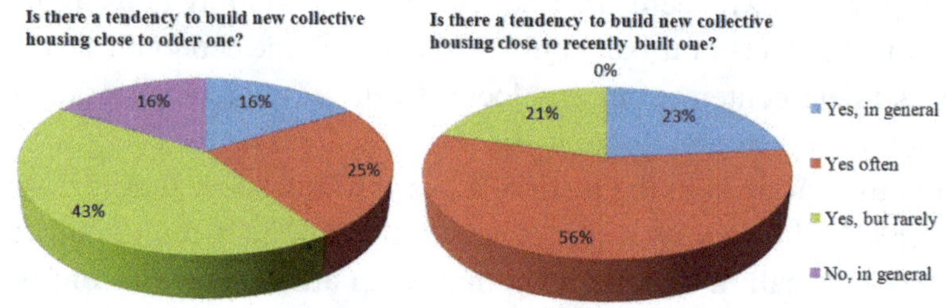

Figure 12. The tendency to build new multi-family housing near the zones of single-family housing (Author: B. Antonić)

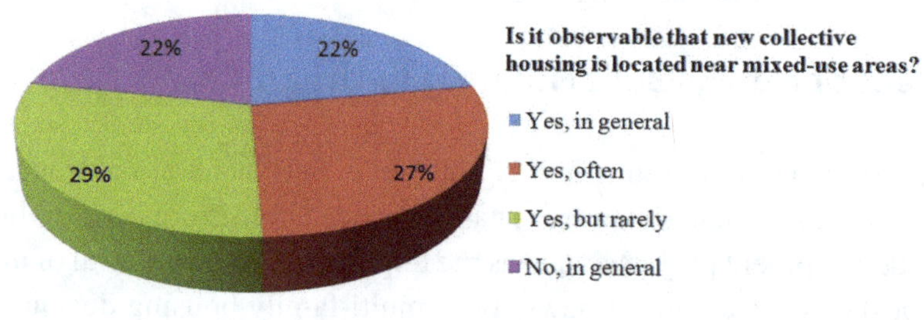

system (Petrović, 2014). Nevertheless, Serbia as well as other countries in south-eastern Europe has had traditional urban development with less control, which produced a less organized zoning system in the urban areas at the end of the socialist era. It seems that this "pattern" has continued throughout the transition. The survey results support this; the spatial "mixture" of newly-built multi-family and single-family housing is very visible. More than 2/3 of the answers in the survey confirm the statement that newly-built multi-family housing is located in the zones of single-family housing. Moreover, 23% of the answers strongly support this statement.

Following the previous observation, it is also interesting to understand the relation of newly-built multi-family housing and other urban facilities, especially those which are connected to quality of life. The development of the retail sector has been especially important in post-socialist cities (Nagy, 2001). The focus is on the importance of mixed-use areas (Figure 13). The survey clearly shows that expert opinion is greatly divided regarding this question.

Another issue is the importance of functional attractors for the locations of newly-built multi-family housing. They can also be treated as flagship projects (Temelová, 2007). The issue is covered by a "yes/no" question, where respondents needed to confirm or reject the statement that some new multi-family housing in their area was built exclusively because of the vicinity of some specific facility as an attractor. The majority of the respondents (63%) confirmed this.

The last two questions are connected to the possible influence of newly-built multi-family housing on accompanying urban-development projects in the field of urban infrastructure and services, as basic amenities for decent and qualitative housing. Their importance was even accented for the entire European Union housing policy in the documents of the main EU bodies [20].

Figure 13. The tendency to build new multi-family housing near mixed-use areas (Author: B. Antonić)

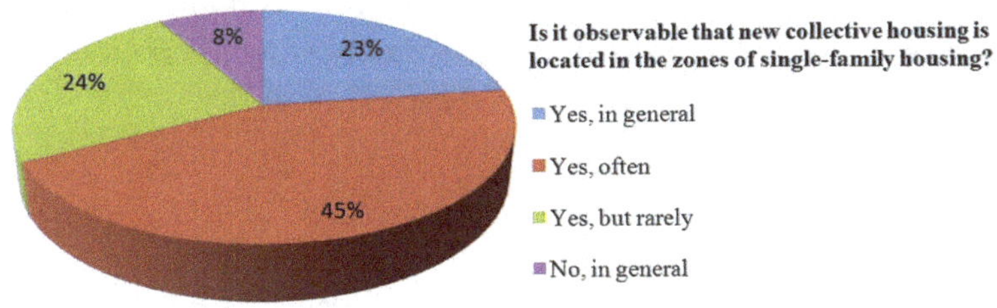

One of the questions was related to the possible influence of newly-built multi-family housing in Serbia on the interventions (renewal, construction or improvement) in the nearest public spaces, as an inevitable element of social infrastructure (Figure 14). Most of the answers are negative. Respondents of which 27% claimed that there was no intervention in the public spaces and 52% considered it to be rare. Only 21% of them were generally positive.

The other question (Figure 15) refers to the general problem of weak technical infrastructure (water supply, sewers, electrical grids, telecommunications, etc.) in south-eastern Europe (Tsenkova, 2005; IUDK, 2013). The results of the question confirmed this view. The respondents (78%) said that newly-built multi-family housing in Serbia has not caused either any or very little interventions in technical infrastructure.

Figure 14. The influence of newly-built multi-family housing on interventions in the nearest public space (Author: B. Antonić)

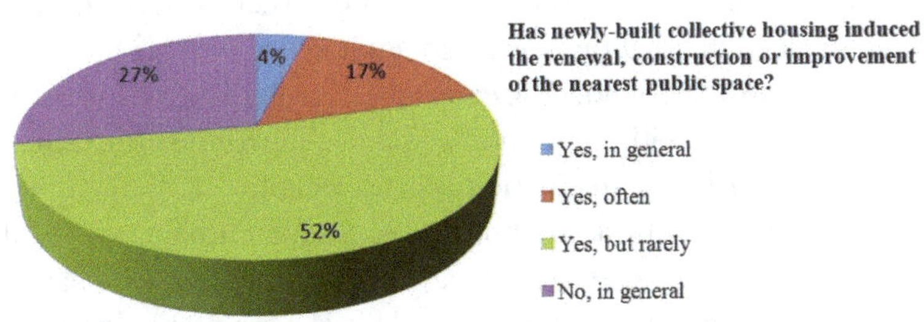

Figure 15. The influence of newly-built multi-family housing on the interventions in accompanying communal infrastructure (Author: B. Antonić)

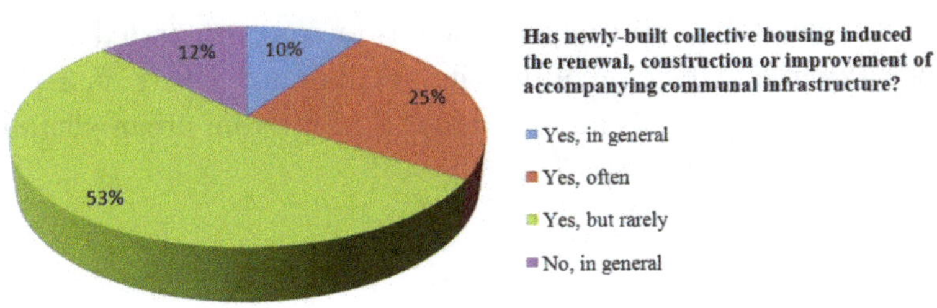

Both questions demonstrate that the construction of new multi-family housing has not been a driving-force for the qualitative interventions in urban infrastructure. This indirectly clarifies that newly-built multi-family housing is still a weak factor to affect remarkably general urban development.

DISCUSSION

Analyzing almost 30 years of the transition in Serbia, new multi-family housing, built during this period, has become the visible element of urban housing stock in the country. This stance is firmly supported by the survey, where the majority of respondents (68%) equated the older, mostly socialist housing with the newer or post-socialist housing in their presence. These opinions question the widespread opinion about the total predominance of housing from the socialist period of the former Yugoslavia, with mass construction of huge housing blocks. Different circumstances have certainly led to this equation. Professional daily routine and permanent contact with new buildings, perhaps, makes the presence of new housing more visible. The second reason can be linked to the result of one of following questions - the location of new multi-family housing buildings is usually in central areas and near to transport corridors, where they are in everyday spotlight; to the contrary, socialist housing blocks were usually built on inexpensive locations, so it is situated in the urban periphery today.

The mentioned visibility of new multi-family housing can be a crucial factor for the next question, which refers to newer interventions in the housing sector. New housing is more present (68%) than the renovation or the renewal of older multi-family-housing stock. The motive is probably associated with the current situation in Serbia, where the overall level of development as well as national planning and organization is still below those in many post-socialist countries in Central and Central-Eastern Europe. Therefore, it is easier to build new building than to organize and manage the renovation of old ones.

This view can also be expanded to the barely present social/affordable housing, which generally requires better organization and planning of public and non-profit sectors in addition to financial support. Almost 50% of the respondents think that there is no social housing in their Serbian cities. It is presumably underestimated, because this kind of housing exits albeit rarely. The reason could be the unaccented physical appearance of the buildings with social-housing units, or maybe the examples where social and commercial housing are mixed make the distinction less noticeable.

The localization of newly-built multi-family housing in Serbian urban areas is also unique by some elements. To clarify, there is no visible pattern of localization in relation to urban (de)centralization or specific urban factors (workplace, public services, etc.). This is presumably linked to a strong influence of different regional and local conditions in Serbia, as well as 'decentralization legacy' from socialist period, rather than to some nationally developed pattern or approach.

In contrast, the importance of good transport accessibility is very noticeable – it was confirmed in more than 90% of the answers in the survey. The interpretation can be connected with the transitional circumstances in Serbian cities, where new development is more concentrated along the main transport corridors. This is especially true in the case of the vital retail sector, where the issue of accessibility is significant. Finally, this clearly represents how the process of relevant urban planning and localization of new multi-family housing has been simplified.

The importance of the functional element of newly-built multi-family housing in Serbia comes right after the issue of localization. There is no visible approach of adequate "pattern" in the zoning or the functional concentration of new multi-family housing. Thus, according to the survey, this new multi-family housing can be commonly found in the zones of both older multi-family housing (79%) and single-family housing (68%). Instead of these tendencies, the expected tendency to build new multi-family housing near the places with a mix of urban facilities is generally not identified. Previous findings in the survey prove that there is still no process of expected polycentric development in Serbian cities. Other explanations can be linked to the undeveloped urban-planning process and its weak implementation *in situ*. Therefore, retail and public services usually follow the construction of new housing instead of acting as the "initiators" of new-housing development.

Given these points in the discussion, the general deficiency of a strategic and planned approach and fragile regulation of newly-built multi-family housing in Serbia should also be understood as a significant problem for the general development of urban areas. Solid proof can be found in the last two questions in the survey, which are dedicated to the possible influences of new multi-family housing on the improvement or renewal of accompanied urban and technical infrastructure as a basic element of qualitative urban life. In both cases, it is obvious that new housing is not a sufficient factor for the improvement of the nearest public space (79%), or technical infrastructure (65%). The financial side certainly is the main contributor to this situation, but it probably is not the only one. The causes can also be inadequate objectives, solutions, and measures in planning and strategic approaches at local level. For instance, almost all local governments in Serbian cities have invested immense resources in the renewal of relatively small pedestrian zones in their inner centers. If there had been better redistribution of these resources, more resources would have been realized for the improvement of infrastructure in other zones with new housing.

CONCLUSION

It is obvious that a general lack of good organization at national and local level and the visible instability of the emerging market economy in Serbia have had an inevitable influence on new multi-family housing. This can be described by two concise conclusions: the oversimplified implementation of all the important urban-planning elements (localization, functional zoning, connection to public infrastructure, etc) and the prevalence of local and regional adaptations over national guidelines or patterns. In line with this, newly-built multi-family housing in Serbia has developed specific characteristics, which have consequently shaped the contemporary urban reality in Serbian cities and towns, but with the question whether it can be considered as a "creative factor" for general urban development. It may better be described as an element of simple urban growth in many cases.

The results of the survey indirectly reveal that urban professionals are quite aware about the state of new multi-family housing in Serbia. They are at the most negative in the questions where public interest and intervention are the most desirable, such as infrastructure and the relation with urban surrounding. The results of the questions from the first thematic group, about new and old housing, also concur with the general data about post-socialist multi-family housing, presented before the survey. This confirms that their knowledge and experience are a good resource for the strengthening of the national and local institutions and bodies pertained to housing policy and planning.

These scrutinized observations lead to the final proposals for possible solutions and approaches in the planning of multi-family housing sector:

1. The better accommodation of urban zoning to desired housing patterns in an imperative in urban planning, to prevent the controversial confrontation of new multi-family housing with surrounding urban fabric;
2. The more prominent focus on the planning of mixed urban zones (housing/non-housing, single-/multi-family housing) along the main transport corridors as currently the key generators of new multi-family housing in Serbia;
3. Similarly to the previous proposal, the proper research and strategy for outer centers with functional and physical mixture is also important for multi-family housing, because these zones are those with the expected expansion of the future multi-family housing; and
4. The adoption of the polycentric development of Serbian cities in urban planning is also necessary, because it is indisputable that new multi-family housing concentrates by complex, multi-factor conditions that it was in the previous periods with the clear dominance of monocentrism in urban structure.

These solutions and approaches can work only if they are linked with the other sources of new housing development, such as financial, organizational, and fiscal mechanisms. If all these elements are integrated, this synergy certainly will create a positive and supportive environment for the innovative and qualitative development of multi-family housing.

The underpinned synergy can be reached by the formation of the agencies or inter-sector bodies directly dedicated for the improvement of housing policy and housing standards from different level: national, regional, and/or local. It is crucial to enable their developmental character in the future of Serbian housing. This is not a new approach in Serbia; for example, such agencies exist for rural development at local and national level. Moreover, the formation of such agencies or bodies can have an international contribution. At the end, this research can conclude by using its very first statements, about the complexity of housing and an urban function – the dealing with housing upgrading and standards cannot be done from one perspective due to this complexity. Instead of this, any body and agency that has housing planning in its agenda had to be multi-sector and integrative by its essence.

ACKNOWLEDGMENT

This paper was done within National research projects No 36034 and No 36035, financed by the Ministry of education and science of the Republic of Serbia.

REFERENCES

Antonić, B. (2015). Место становања у савременој урбаној политици: како искористити европска искуства у Србији [The Place of Housing in Contemporary Urban Policy: How to Use European Experince in Serbia]. In V. Trifunović, D. Minić i Z. Popović (Eds.), *Zbornik radova: I Planski osnovi / II Urbane politike / III Izmene i dopune Zakona o planiranju i izgradnji i Zakona o državnom premeru i katastru* (pp. 125-134). Beograd: Udruženje urbanista Srbije.

Antonić, B. (2016). How to understand the history of housing planning in modern Serbia to achieve new quality in housing? In C. Hein (Ed.), *Proceedings of 17th IPHS Conference* – Volume 02 *The Urban Fabric* (pp. 165-174). Delft, the Netherlands: TU Delft - Faculty of Architecture. DOI: 10.7480/iphs.2016.2.1234

Antonić, B. (2017). The Role of housing Agencies in the Future Development of Serbia. *plaNext – Next Generation Planning, 5*, 30-46. Doi:10.24306/plnxt.2017.05.003

Antonić, B., & Đukić, A. (2018). Phenomenon of the Shrinking Illegal Suburbs in Serbia: Can the Concept of Shrinking Cities be Useful for their Upgrading. *Habitat International, 75*, 161–170. doi:10.1016/j.habitatint.2018.03.010

Antonić, B., & Vaništa Lazarević, E. (2019). Decentralised Mass Housing Policy in Socialist Yugoslavia. In M. Benkő & K. Kissfazekas (Eds.), Understanding Post-Socialist European Cities – case studies in urban planning and design (pp 172-187). Budapest: L'Harmattan.

Bataković, D. (Ed.). (2005). Histoire du peuple serbe [History of the Serbian People]. Lausanne: L'Age d'Homme.

Blagojević, Lj. (2007). *Нови Београд: оспорени модернизам* [New Belgrade: Contested Modernism]. Belgrade: Zavod za udžbenike.

Borsay, P., Brown, V. C., & Palairet, M. R. (1993). Pedja J. Marković, Beograd i Evropa 1918–1941 [Belgrade and Europe 1918–1941]. *Urban History, 20*(2), 261–262. doi:10.1017/S0963926800010580

Djukic, A., Stupar, A., & Antonic, B. (2018). The Orthogonal Urban Matrix of the Towns in Vojvodina, Northern Serbia: Genesis and Transformation. In G. Carlone, N. Martinelli, & F. Rotondo (Eds.), *Designing Grid Cities for Optimized Urban Development and Planning* (pp. 128–156). Hershey, PA: IGI Global; doi:10.4018/978-1-5225-3613-0.ch008

Djukic, A., Stupar, A., & Antonic, B. (2018b). The consequences of urban policies in socialist Yugoslavia on the transformation of historic centres: The case study of cities in Northern Serbia. *Journal of Housing and the Built Environment, 33*(3), 555–573. doi:10.100710901-018-9612-7

Đukić, A. (2015). New Belgrade: visions, plans and realizations 1950-2014. In G. Doytchinov, A. Đukić & C. Ioniță (Eds.), Planning Capital Cities: Belgrade, Bucharest, Sofia (pp, 160-172). Graz: TU Graz Verlag.

Hamilton, I., Dimitrovska Andrews, K., & Pichler-Milanović, N. (Eds.). (2005). *Transformation of cities in central and Eastern Europe: Towards globalization*. Tokyo: UN University Press.

Hegedüs, J., & Tosics, I. (1996). *Transition of the housing sector in the East Central European countries*. Washington, DC: The Urban Institute. doi:10.1111/j.1467-940X.1996.tb00113.x

Hirt, S., & Stanilov, K. (2009). *Revisiting Urban Planning in the Transitional Countries*. Nairobi: UN Habitat.

Hirt, S., & Stanilov, K. (2014). *Twenty Years of Transition: The Evolution of Urban Planning in Eastern Europe and the Former Soviet Union, 1989-2009*. Nairobi: UN Habitat.

Institute of Urban Development Krakow (IUDK). (2013). *The State of European Cities in Transition 2013: Taking Stock after 20 Years of Reform*. Nairobi: UN Habitat.

Jovanović Popović, M., Ignjatović, D., Radivojević, A., Rajčić, A., Đukanović, Lj., Ćuković Ignjatović, N., & Nedić, M. (2012). *Atlas of Family Housing in Serbia*. Belgrade: Faculty of Architecture & GIZ.

Jovanović Popović, M., Ignjatović, D., Radivojević, A., Rajčić, A., Đukanović, Lj., Ćuković Ignjatović, N., & Nedić, M. (2013). *National Typology of Residential Buildings in Serbia*. Belgrade: Faculty of Architecture & GIZ.

Keivani, R., Parsa, A., & McGreal, S. (2002). Institutions and Urban Change in a Globalising World the Case of Warsaw. *Cities (London, England)*, *19*(3), 183–193. doi:10.1016/S0264-2751(02)00015-X

Kovács, Z. (1998). Ghettoization or gentrification? Post-socialist scenarios for Budapest. *Netherlands Journal of Housing and the Built Environment, 13*(1), 63-81. Doi:10.1007/BF02496934

Marković, P. (1992). *Београд и Европа: 1918-1941* [Belgrade and Europe 1918-1941]. Belgrade: Savremena Administracija.

Milić, V. (2006). *Urbanistički aspekti socijalnog stanovanja* [Urban Aspects of Social Housing]. Belgrade: Arhitektonski fakultet.

Mojović, Đ., & Žerjav, B. (2011). *Stanovanje pod zakup* [Rented Housing]. Belgrade: Urban Development Program.

Nagy, E. (2001). Winners and Losers in the Transformation of City Centre Retailing in East Central Europe. *European Urban and Regional Studies*, *8*(4), 340–348. doi:10.1177/096977640100800406

National Assembly of the Republic of Serbia (NARS). (2012). *Law on Social Housing*. Official Gazette of RS No. 72/09.

Petovar, K. (2003). *Urbana sociologija: Наши градови између државе и грађанина* [Urban Sociology: Our Cities between State and Citizen]. Belgrade: Geografski fakultet – Arhitektonski fakultet – IAUS.

Petrović, M. (2004). *Sociologija stanovanja: Stambena politika: Izazovi i mogućnosti* [Housing Policy: Challenges and Possibilities]. Belgrade: Institut za sociološka istraživanja.

Petrovic, M. (2005). *Cities after Socialism as a Research Issue*. London: LSE E-prints. Retrieved from http://eprints.lse.ac.uk/23378/

Petrović, M. (2014). *Transformacija gradova: Ka depolitizaciji urbanog pitanja* [Transformation of Cities: To Depolitization of Urban Issue]. Belgrade: Institut za sociološka istraživanja.

Pickvance, C. (2008). Chapter 9: State State Socialism, Post-Socialism and their Urban Patterns: Theorizing the Central and Eastern European Experience. In J. Eade & C. Mele (Eds.), Understanding the City (pp. 183-203). Oxford, UK: Blackwell. DOI: 582.ch9 doi:10.1002/9780470693

Radovanović, D. (2015). Индустријско наслеђе – стамбене колоније [Industrial Heritage – Worker Colonies]. In Индустријско и техничко наслеђе Војводине [Industrial and Technical Heritage of Vojvodina]. Zrenjanin: Zavod za zaštitu spomenika kulture.

Ristić, S. (2010, June 21). Архитектонски вишак на скандинавски начин [Architectural Surplus on Scandinavian Way]. *Politika Online*. Assessed 28.07.2015 from http://www.politika.rs/rubrike/ Kulturni-dodatak/Arhitektonski-vishak-na-skandinavski-nachin.sr.html/

Saroff, J. & Levitan, A. (1969). *Survey Manual for Comprehensive Urban Planning*. University of Alaska: Institute of Social, Economic and Government Research.

Stanilov, K. (Ed.). (2007). *The Post-Socialist City Urban Form and Space Transformations in Central and Eastern Europe after Socialism*. Dordrecht: Springer. doi:10.1007/978-1-4020-6053-3

Stanilov, L., & Sykora, L. (2014). *Confronting Suburbanization. Urban Decentralization in Postsocialist Central and Eastern Europe*. Chichester, UK: Wiley Blackwell.

Statistical Office of the Republic of Serbia (SORS). (2013). *Dwellings by the Type of Building: Data by municipalities/cities*. Belgrade: SORS. Retrieved from http://pod2.stat.gov.rs/ObjavljenePublikacije/ Popis2011/ Knjiga27.pdf

Stupar, A. (2015). Cold War vs. architectural exchange: Belgrade beyond the confines? *Urban History*, *42*(4), 622–645. doi:10.1017/S0963926815000528

Temelová, J. (2007). Flagship Developments and the Physical Upgrading of Postsocialist Inner City: The Golden Angel Project in Prague. *Geografiska Annaler: Series B, Human Geography, 89*(2), 169-181. DOI: .00246.x doi:10.1111/j.1468-0467.2007

Tsenkova, S. (2005). *Trends and Progress in Housing Reforms in South Eastern Europe*. Paris: Council of Europe Development Bank.

Vranic, P., Vasilevska, Lj., & Haas, T. (2015). Hybrid spatialities: Multi-storey extensions of socialist blocks of flats under post-socialist transition in Serbia, the case of Nis. *Urban Studies (Edinburgh, Scotland)*, *53*(6), 1261–1277. doi:10.1177/0042098015571887

Vujović, S., & Petrović, M. (2007). Belgrade's post-socialist urban evolution: Reflections by the actors in the development process. In K. Stanilov (Ed.), *The Post-Socialist City Urban Form and Space Transformations in Central and Eastern Europe after Socialism* (pp. 361–384). Dordrecht: Springer. doi:10.1007/978-1-4020-6053-3_18

Vuksanović-Macura, Z., & Macura, V. (2018). The right to housing: Squatter settlement in interwar Belgrade – defense and demolition of Jatagan-mala. *Journal of Urban History*, *44*(4), 755–774. doi:10.1177/0096144216632747

KEY TERMS AND DEFINITIONS

Banovina: Basically a historical term – a territory ruled by a ban, a noble title used mainly in several in Central and Southeastern Europe between the 7th century and the 20th century. In the context of this research, the term refers to nine provinces in the former Yugoslavia that lasted from 1929 to 1941. These provinces bear their names by geographical features.

Communist or Workers' State: A state that is administered and governed by a single (communist) party, guided by Marxist–Leninist philosophy. Such state is usually termed by Marxists as dictatorships of the proletariat/working class. Communist states usually have a planned economy and the strong limitation or even the absence of money and the market mechanism. The most prominent socialist state was the former Soviet Union.

Housing Policy: A deliberate system of principles adopted by competent governance to guide decisions and achieve rational outcomes in relevant housing sector (national, regional, local, etc.). The essence of housing policy is the right allocation of key resources in housing, aiming to properly control and direct housing market by governance interventions.

Multi-Family Housing or Collective Housing: A type of housing where multiple separate housing units for residential inhabitants are contained within one building or several buildings within one complex. This type of housing usually has parts that are jointly used and/or managed (entrance, stairways, corridors, roof, etc.).

Planned Economy: A type of economic system where investment and the allocation of capital goods take place according to economy-wide economic and production plans. This economy is usually commanded by high-level state government and administration. Typical planned economies were found in the former Soviet Union and Eastern Bloc.

Post-Socialist Transition: The process of multiple the transition of the entire system of the former socialist/communist states in Eastern Europe and the Commonwealth of independent states (CIS). Politically, it is a transition from a single-party system to a multi-party democracy. In economy, it is a transition from a planned economy to a market economy. In general organization of a state, it is a transition from a centralized political system to decentralized one.

Serbia: Officially the Republic of Serbia (Serb. *Република Србија*), is a country situated at the crossroads of Central and Southeast Europe in the southern Pannonian Plain and the central Balkans. It was the largest republic of the former Yugoslavia.

Socialist State: A state declared themselves Socialist or in the process of building socialism. In essence, such states can be both communist and of social democratic states, depending on the loose definition of the word socialism. The most prominent socialist state was the former Soviet Union.

Vojvodina: Officially, the Autonomous Province of Vojvodina (Serb. *Аутономна Покрајина Војводина*), is an autonomous province of Serbia, located in the northern part of the country, in the Pannonian Plain, physically located in Middle Europe. It was the only part of Serbia that was the integral part of the former Habsburg Empire.

ENDNOTES

1 The threshold for this division is four (SORS, 2013).

2 See: Key Terms and Definitions > Multi-family housing.

Chapter 3
Urban–Rural Synergy Toward Human Well-Being

Nikola Zoran Furundžić
Furundzic ordinacija, Serbia

Dijana Petar Furundžić
Furundzic ordinacija, Serbia

Aleksandra Krstić-Furundžić
https://orcid.org/0000-0001-9126-9336
Faculty of Architecture, University of Belgrade, Serbia

ABSTRACT

The cities are characterized by the rapid development of technologies and the inevitable expansion, which requires looking at ways to address new challenges and overcome the problems that have arisen. Living conditions may vary in many ways, depending on where people live, but the advantages and disadvantages in terms of comfort of human life can be recognized in both urban and rural environments, as analysis of case studies shows. Factors that occur in both areas and affect human life, comfort, and health are presented. The positive effects of rural areas and the advantages that cities have in terms of preserving human health are discussed in this chapter. By comparing the good characteristics of the urban and rural environment, the goal is to indicate some postulates that can be applied in preventive medicine.

INTRODUCTION

Health is of the same importance for people wherever they live and whatever they do, regardless of origin and nationality, as well as the age group they belong to. From time immemorial human beings are living in communities which are nowadays generally defined as specific two types: urban and rural. Regarding the population of these areas, relevant facts have been provided by the UN (2013), indicating that more than half of the world's population now lives in urban areas, compared with 29% in 1950

DOI: 10.4018/978-1-5225-9932-6.ch003

and 15% in 1900, as well as predicting that in 2050, seven out of every ten people will live in towns or cities - more than 6.25 billion people, while in the period from 2000 to 2050, developing regions could add 3.2 billion new urban residents, which is the figure larger than the entire world's population in 1950.

When it comes to quality of life, in the life of the city and the village there are many advantages and disadvantages. Regardless of the differences, in rural and urban areas in the human population appear similar health disorders. It would be interesting to determine which factors that occurs in the village and in the city disrupt human life, comfort and health. Given that there are many facilities for people in the city, city dwellers have more opportunities to progress in their lives. Cities provide great facilities for education, medical treatments, job opportunities, etc. However, many disadvantages are present (Figure 1), mainly the air of poor quality polluted from traffic and factories, often poor water quality, the environment is polluted with dust, smoke, garbage and greenhouse gases, the streets are mostly dusty and unclean, intense traffic and therefore noise. It appears that it is difficult to lead a healthy life in cities. The lifestyle in villages has other specifics in comparison to lifestyle in the city. People know each other very well and are more closely connected to each other than people in the city. As village has not lot of vehicles the pollution is less so the air is clean, less noise and rush is present. The environment of the village is pleasant and silent and it has scenic beauty (Figure 2). But, the village also has bad points. Often lifestyle is not as advanced as in the cities. Many people have difficulties to keep up with new developments in their field or profession. Since the functioning of the village is closely linked with the city, traveling problems village people often have to face. But certainly, the rural environment can be considered healthy.

Figure 1. Air pollution, intense traffic, noise and building density are disadvantages of urban areas (Author: A. Krstić-Furundžić)

Figure 2. Scenic beauty of rural areas (Author: A. Krstić-Furundžić)

Mutual understanding, connections and collaboration between communities, disregarding how large they are, is crucial for the development in all fields and overcoming the problems and challenges that arise. In this respect advantages and disadvantages of both areas for human health are discussed in the paper. Cooperation in sharing of information and experience can overcome differences and lead to better life for all. The synergy of urban and rural areas can provide new knowledge and skills that combine the advantages that these areas possess. The paper points to the positive effects that provide the rural environment and the presence of the advantages that city have in terms of preserving people's health, thus drawing the essence of these advantages. The aim of the paper is perception of new knowledge and perspectives that arise for preventive medicine from the synergy of urban and rural areas. It should be noted that some knowledge and perspectives have universal character, while some related to specific regions.

BACKGROUND

'Urban' and 'rural' are terms that mark two types of populated areas, which are different in many aspects. Defining the concept of urban and rural areas is a complex task, because it is hard to find a unique criterion by which the division performs. These terms are often used by policymakers, researchers, national administrations and international organizations. To account for differences among rural and urban regions, the international Organization for Economic Cooperation and Development established a regional typology (OECD, 2010; Brezzi, Dijkstra, & Ruiz, 2011), based essentially on the percentage of regional population living in urban or rural communities, classifying regions in three classes: 1. predominantly urban (PU-if the share of population living in rural local units is below 15%), 2. intermediate (IN-if the share of population living in rural local units is between 15% and 50%) and 3. predominantly rural (PR-if the share of population living in rural local units is higher than 50%). 'Urban' and 'rural' terms are readily understood by the general public, but a clear definition at the international level has remained elusive (Dijkstra & Poelman, 2014). In this respect Dijkstra and Poelman (2014) comment that the UN publishes data on cities, urban areas and rural areas, but relies almost entirely on national definitions of these areas, and that the UN principles and recommendations state that due to different characteristics of urban and rural areas across the globe, a global definition is not possible.

The population density is the dominant difference between rural and urban areas (Figure 3), but for the population are also important the parameters that vary for different communities, such as unemployment, crime, availability of different kinds of jobs, incomes, health services and educational potentials, infrastructure, typology of housing, energy efficiency and quality of connections with the surrounding areas. These parameters affect the quality of human life.

It is worth noting the following comment that relates to the common confusion regarding what constitutes urban:

The term conjures up images of skyscrapers, crowded sidewalks, subways, and concentrated poverty; although these conditions exist, they are not representative of the overall urban experience. Urban includes many community types ranging from city centers to suburban villages. Most urban residents live in moderate-density neighborhoods that contain a mixture of single-family and low-rise apartments, and

Figure 3. The population density is the dominant difference between rural and urban areas (Author: A. Krstić-Furundžić)

rely on a combination of walking, cycling, public transit, and automobile travel, and most cities have a mixture of low, middle- and high-income households. People who imply that most urban residents live in high-rise apartments, forego automobile travel, and are poor, are using atypical examples (Litman, 2017, p. 3).

Howard et al. (2002) point out the concepts of the village:

A village may be a small group of people living in a settlement who practice subsistence agriculture, with no specialization or division of labour, and who are isolated from national development agencies. A village may also be a large and differentiated conurbation where some people work in agriculture, some work in small-scale industries and others provide education, health care, administration and a variety of services. It is also recognized that many villages do not operate independently from cities, in that cities require sustained interaction with rural communities for their food and natural resources (including land for waste disposal) (p. VIII).

Urban areas have their own advantages and disadvantages, but urban areas are definitely more attractive for people to live. This trend will continue in the future, as indicated by statistics, (UN, 2013). Urban area can refer to towns, cities, and suburbs. An urban area includes the city itself, as well as the surrounding areas. For centuries, cities have been the centres of political decision-making and economic growth as well as the breeding ground for cutting-edge research, cultural and technological development, (Broman, 2017). These are some of the strong reasons why cities attract people to leave rural and immigrate to urban areas. Very important connection is between urbanization and economic growth. In order to explain this connection, the terms of industrialization, urbanization and globalization are commonly used. According to Chen, Zhang, Liu, and Zhang (2014), industrialization leads to economic growth that gives impetus to the process of urbanization in developed and newly industrialized countries, mainly via a specialization of labor and development of non-agricultural sectors. In the same publication, the authors note that it is generally accepted that economic growth promotes the expansion of modern industries and an increase in the urban population, and in turn, urbanization also promotes economic growth to some extent. Various programs of accelerated urbanization and rapid economic growth have, therefore, been embarked upon in many developing countries, (Chen, Zhang, Liu, & Zhang, 2014). Certainly, the goal of economic development is also to contribute to the improvement of the quality of life in cities. This is recognized as important for developing countries, and for them are introduced various programs of intensification of urbanization and economic growth. Policies pursuing positive urbanization, with the goal of boosting economic growth, are widely found in the developing world, (Chen et al., 2014; Friedmann, 2006; Hope, 1998; Pugh, 1995). Chen et al. (2014) point out that world urbanization is changing quickly and the rate of change has been rising faster in the last three decades than previously, in this age of globalization.

The theme of economic growth and urbanization today is focused on emphasizing the importance of urban regions. It is convenient to mention some interesting quotes presented in the World Economic Forum concerning how much economic growth comes from cities:

As Christopher Chase-Dunn has pointed out, it is not population or territorial size that drives world-city status, but economic weight, proximity to zones of growth, political stability, and attractiveness for foreign capital. In other words, connectivity matters more than size. Cities thus deserve more nuanced treatment on our maps than simply as homogeneous black dots.

Great and connected cities, Saskia Sassen argues, belong as much to global networks as to the country of their political geography. Today the world's top 20 richest cities have forged a super-circuit driven by capital, talent, and services: they are home to more than 75% of the largest companies, which in turn invest in expanding across those cities and adding more to expand the intercity network. Indeed, global cities have forged a league of their own, in many ways as denationalized as Formula One racing teams, drawing talent from around the world and amassing capital to spend on themselves while they compete on the same circuit (Khanna, 2016, para. 2, 10).

Since urbanization is closely related to economic growth, this issue has become a subject of interest and research in many disciplines. According to Asian Development Bank (n. d.), "Governments around the world have a growing interest in how to make cities more competitive – by attracting investment and by creating jobs that help reduce poverty and unemployment" (p. 3). Competitiveness can be considered by means of the assessment of the potential of an urban region or city to implement the production and sale of products and services that have a good but not the lowest value in relation to comparable products of other urban areas, (Webster & Muller. 2000; Kwon, Kim, & Oh, 2012). Competitiveness is a significant feature of urban regions that gives them the advantage over rural areas. The competitiveness of cities is directly related to the quality of life and their efficiency.

Webster and Muller (2000) point out that national political stability very much influences the competitiveness of cities and they believe that urban regions control only some of the factors that determine their competitiveness, "National policy frameworks and socio-economic conditions are also very important, e.g., national taxation, human resource development, tariff, macroeconomic, industrial incentives, policies, etc. However, in many countries, national factors are becoming relatively less important because of global forces, e.g., trade liberalization which makes tariff policies less important, or because of internal changes, e.g., decentralization which may result in devolution of responsibility for critical competitiveness factors, e.g., education of technical personnel, to the local level" (p. 2).

The comparison of regions can start from the point of location, physical and infrastructural characteristics of the region. However, for competitiveness assessment, it is important to take into consideration two things, to look at the activities that take place in the region (e.g., finance, tourism, computer manufacturing, the role of the informal sector) that is how cities compete in real world and for a place to look at non-tradeables, such as human resources, territorial endowments, and institutional milieus, as they are important for determining where the activities will be located, expanded or contracted, (Webster & Muller, 2000).

The ability of the city to create more and better wealth faster than other cities by engaging less resource, while at the same time ensuring the well-being of its residents, is recognized as a parameter for assessing the competitiveness of cities. Ni and Qinghu (n.d.) stated, "According to the definition, Long-term overall economic growth rate, comprehensive market share, comprehensive GDP per square of land, and general per capita income may synthetically indicate city's overall competitiveness" (p. 1).

According to the aforementioned, it is logical to conclude that people choose to live in urban areas, hoping to have a better quality of life. People want to be in the center of events, to have more opportunities for socialization, education, promotion, and to get a job and progress in business. These are the main pull factors.

As part of the mechanism of constituting city tissue, architecture often becomes a means to showing the power of a country or corporation, indicating their economic and political status and power. High-rise buildings have always been a challenge not only for the construction industry, but also for urban planners and city authorities. The development of advanced construction technologies made it possible for high-rise buildings to reach impressive heights, to become recognizable benchmarks in cities, symbolizing power and wealth. An interesting comment on tall buildings is given by Moldovan, Moldovan, and Ilies (2014), "As they slowly became generators of identity, architects and engineers aimed for the sky, setting their bars higher and higher. Skyscrapers are perpetual means of advertising, products of luxury and exclusivism in architecture and therefore their presence in the city's skyline is desired" (p. 1461). Kinds of poetic attitude on this type of buildings give Ali and Al-Kodmany (2012), "The tall building is the most dominating symbol of the cities and a human-made marvel that defies gravity by reaching to the clouds" (p. 304). Tall buildings can sometimes create the impression of a rich and pleasant life in cities that inspire people to come and live in cities, believing that there is a great potential for employment.

With the development of various technologies, the complexity and multiplicity of the expressive and narrative potentials of architectural achievements is becoming more pronounced, especially in urban areas whose further development can be (positively) affected. Formaneck (2013) gave an opinion on the symbolic potentials of architectural creations, "Skyscrapers were originally created in the United States of America as a symbol of economic strength and increasing global business operations. Most of the tallest buildings in the world were located in the US" (p. 350).

Kellert, Heerwagen and Mador (2008) note that the design of a modern urban built environment leads to massive transformation and degradation of natural systems and increases the separation of people from the natural world, and point out that "This design paradigm has resulted in unsustainable energy and resource consumption, major biodiversity loss, widespread chemical pollution and contamination, extensive atmospheric degradation and climate change, and human alienation from nature" (p. 5).

According to Kellert (2008), whose field of research is a biophilic design, in human nature is a connection with natural systems and processes, especially life and life-like features of the nonhuman environment. He points out that technological development has separated man from nature, and that it has an adverse effect on human well-being:

The emergence during the past roughly 5,000 years of large-scale agriculture, fabrication, technology, industrial production, engineering, and the modern city constitutes a small fraction of human history, a period that has not substituted for the benefits of adaptively responding to a largely natural environment. Most of our emotional, problem-solving, critical-thinking, and constructive abilities continue to reflect skills and aptitudes learned in close association with natural systems and processes that remain critical in human health, maturation, and productivity. The assumption that human progress and civilization is measured by our separation from if not transcendence of nature is an erroneous and dangerous illusion. People's physical and mental well-being remains highly contingent on contact with the natural environment, which is a necessity rather than a luxury for achieving lives of fitness and satisfaction even in our modern urban society (Kellert, 2008, p. 3-4).

FACTORS AFFECTING HUMAN HEALTH

The complexity of the impact of urbanization on human health is a complex and current topic that requires a multidisciplinary approach and is the subject of the consideration of experts from various professions. Economic, social and environmental conditions reflect on the living conditions and the health of the inhabitants. Cities have areas in which poor people live in bed living conditions, which results in many health problems. The risks to health are greatest for a billion people living in unhealthy, life-threatening conditions in informal settlements or 'slums' (Kjellstrom et al., 2008).

The effect of urbanization on health can be double-edged however, as a result of overcrowding, pollution, social deprivation, crime, and stress. Urbanization can also result in hypertension, heart disease, obesity, diabetes and asthma (Godfrey & Julien, 2005). Current rise of chronic non-communicable diseases (NCD) is characterized as an urban health crisis. Despite the many advantages of living in urban areas, for many millions of people cities are places of ill health (Pinoncely & Hartkoorn, 2014).

Public health interventions designed to reduce the risk of ill-health and promote feelings of well-being in a community must consider many social and environmental factors (Howard et al., 2002). Due to differences in communities, these factors will vary in importance.

Taking about non-communicable diseases, several authors pay attention to determinants that can provoke ill-health and can be recognized as "causes of the causes" of ill-health, such as transport, limited access to green space, pollution, noise, housing quality, access to food, unsatisfactory community participation, social isolation, violence and crime. Factors that influencing human health, particularly urban mental health, Litman (2017) classified in the self-selection, economic and social and environmental factors (Table 1).

Table 1. Factors affecting urban mental health (Source: Litman, 2017)

Self-Selection	Economic and Social	Environmental Factors
Differences in the types of people who locate in different community types. Does not reflect causation.	Differences in how people live and interact. May reflect causation, but often changes over time.	Factors innate to urban locations. These do reflect causation, but can change over time.
• Poverty and income • Age and life stage • Mobility (duration of residency) • Family & community connections	• Higher incomes and disparities • Higher costs of living • More subcultures • Higher crime rates	• More interactions with strangers • More racial and cultural diversity • Noise and air pollution • Less interaction with nature

Oral health contributes to general health, self-esteem and quality of life and although oral health may have a low priority in the context of mental illness, the impact of mental illness and its treatment on oral health must be addressed (Griffiths et al., 2000). Depression and stress increase the level of the hormone cortisol, which can contribute to periodontal disease (Rosania, Low, McCormick, & Rosania, 2009). Furthermore, medications, lack of motivation for maintaining good dental hygiene, dental phobia, dental care costs, geography and access to care have all been shown to contribute to oral diseases (Bardow, Nyvad, & Nauntofte, 2001; Selwitz, Ismail, & Pitts, 2007; Ramon, Grinshpoon, Zusman, & Weizman, 2003; Vargas, Dye, & Hayes, 2002).

Health Problems Associated With Urbanization

Today, urbanization has become a necessity. The attractiveness of cities is remarkable and it is hard to resist. Based on the insights into literature, the research of numerous scientists, architects, economic and sociological experts, urbanization is associated with an increase in the density of population in small areas, which reduces the quality of life. In this article, we want to point out the amount of health problems, cardiovascular, immunological and, of course, dentofacial, which have increased their level due to the actualization of urbanization and the reduction of conditions for the development of healthy life routines that can be used to overcome these health problems and diseases. Effects of rapid urbanization on lifestyles and on the development of non-communicable diseases was recognized by the World Health Organization, but also it should be noted that the results of such research are rarely documented globally. "The relationship between cities and health was an early consideration of public health, but the impact of local government policies on health gained prominence with the WHO-sponsored Healthy Cities Programme, starting in the mid-1980s" (WHO, 2012, p.5). The World Health Organization has developed the Healthy Cities Project, "The WHO Healthy Cities project is a long-term development project that seeks to put health on the agenda of decision-makers in the cities of Europe and to build a strong lobby for public health at the local level. Ultimately, the project seek to enhance the physical, mental, social and environmental wellbeing of people who live and work in Europe`s cities" (Tsouros, n.d., p.11).

It has been identified that the risk of developing cardiovascular disease (CVD) begins even before birth, during fetal development (IOM, 2010). In Urbanization and cardiovascular disease: Raising heart-healthy children in today's cities, published by the World Heart Federation, the importance of perceiving the disadvantages of exposure to risk factors and overcoming these problems is highlighted:

For many of today's children, this exposure is increased due to the negative impacts of urbanization. Children living in towns and cities, particularly in low- and middle-income countries (LMICs) face very real and growing health risks: crowded living conditions, air and water pollution, inadequate sanitation, limited green space and an overwhelming display of tobacco, alcohol and fast-food marketing, all disproportionately affecting certain populations and potentially imposing limitations on how they live, work, eat and sustain themselves.

Because the risk of CVD can be mitigated by changes to behaviour, there is a widespread perception that CVD – including heart disease and stroke – is a "lifestyle" disease. However, many of the challenges posed by urban environments cannot be addressed simply through individual lifestyle choices. For most of the world's people, and especially its children, where a person lives intrinsically affects their health and life options.

Action to address urban health risks is therefore essential, to protect the health of our children and the prospects of future generations. (Smith et al., 2012)

Consideration of the presence of cardiovascular diseases in urban areas, and pointing to an increase in the presence of diabetes mellitus in the future are very current in contemporary society. Worrying prediction give Allender et al. (2010) as well as Boutayeb A. and Boutayeb S. (2005), commenting that by 2020, it is expected that in developing countries, non-communicable diseases (NCDs) will account for 69% of all deaths, with cardiovascular diseases in the lead. However, urban dwellers are rapidly experiencing a transition in their coronary risk profile, and there are apprehensions that they may experience a rise in CHD rates in the future (Reddy & Yusuf, 1998, p. 598). Allender et al. (2010) also point out that the prevalence of diabetes mellitus will almost double in the next 25 years and at least 75% of those affected will be in developing countries. The lower socio-economic groups have greater prevalence of risk factors, higher incidence of disease and higher mortality (Boutayeb A. & Boutayeb S., 2005). Such perceptions are alarming for cities, as they contain areas inhabited by poor people. It was recognized that industrialization and urbanization caused lifestyle changes. Also, immigration to cities alters the lifestyle and habits of people who are immigrating. Certain authors note that there is a substantially higher level of CVD risk factors in urban population groups compared with rural population groups (Reddy, 1993; Yao, Wu Z. & Wu J., 1993).

Dental Problems Associated With Urbanization

In the field of dental medicine, there are also attitudes that some diseases can be characterized as "diseases of civilization". It is possible to notice authors who deal with the effect of urbanization on the development of malocclusion, the frequency of caries and periodontal diseases. Proffit, Fields, Ackerman, Bailey, & Tulloch, (2000) indicate the impact of urbanization on the frequency of malocclusion, stating that there is some evidence that malocclusion increases within well-defined populations after a transition from rural villages to the city. In the same publication it was stated that Corruccini (1991) reported a higher prevalence of crowding, posterior crossbite, and buccal segment discrepancy in urbanized youths compared with rural Punjabi youths of northern India. Proffit et al. (2000) stated, "One can argue that malocclusion is another condition made worse by the changing conditions of modern life, perhaps resulting in part from less use of the masticatory apparatus with softer foods now. The increase in malocclusion in modern times certainly parallels the development of modern civilization, but a reduction in jaw size related to disuse atrophy is hard to document, and the parallel with stress-related diseases can be carried only so far. Although it is difficult to know the precise cause of any specific malocclusion, we do know in general what the etiologic possibilities are" (p. 14-15). The authors of this chapter in some of their publications, based on their research by surveying employees in buildings with modern facades, indicate the presence of stress generators in work spaces and their impact on the development of depression and anxiety, as well as the relationship with the development of caries and periodontal diseases. Since most offices are located within urban areas, this indicates some other possible effects of urban environment on the occurrence of dental problems.

Figure 4. Light pollution in cities (Author: A. Krstić-Furundžić)

Environmental Factors

Environmental factors as noise, light pollution and interaction with nature are discussed.

Urban areas tend to have more ambient noise and light pollution, which can induce stress and interrupt sleep (WHO, 2011). Common urban noise sources include vehicle traffic, sirens and alarms, construction, loud music and voices, etc. (Jaffe, 2015).

Modernization of populated areas provided artificial light at night. Night light makes urban area secure for living but also to misalignment among physiology, behavior and the environment. Electrical light in living and working spaces and on streets is indispensable for activities of population that lives in urban areas, but it can also be aggressive acting as light pollution (Figure 4). But, over 99% of individuals living in the United States and Europe experience nightly light pollution (Cinzano, Falchi, & Elvidge, 2001). New researches suggest that any unnatural timing of light exposure, or lack of appropriate light/dark cues in the environment, can cause misalignment between internal biological processes and the external environment, putatively leading to impaired mood (Bedrosian & Nelson, 2013). As modern life has allowed humans to manipulate lighting easily and has led to unnatural exposure to night light, the prevalence of major depressive disorder (MDD) has increased in parallel (Kessler et al., 1993).

Frequent street lighting and decorative lighting of building facades are the features of urban areas. Streetlights and external decorative lighting illuminate the bedroom spaces, TV and computer screens, causing glow in the home at night. Nocturnal light exposure may have serious consequences for circadian timing and contributes to depressed mood. Depression has a complicated mechanism which effects on oral health, as already mentioned above.

Common urban noise sources include vehicle traffic, sirens and alarms, construction, loud music and voices, etc. The noise and crowds can make the environment unfriendly (Figure 5).

Noise and light pollution exposure tends to increase with density, and so can be considered as inherent to cities, but can often be managed and reduced with improved design (Litman, 2017).

In recent years, because of the rapid urbanization of the world can be noticed the increasing number of scientific articles which suggest professional community about the positive effects that nature has on human beings. For achievement of healthy lifestyle, scientific papers indicated the necessity of interaction between human and nature. Some people argue that people require regular interactions with nature, and that urban living leads to nature deficit syndrome (Berto, 2014; Hartig & Kahn, 2016).

Figure 5. Crows can cause the feeling of an unpleasant environment. (Bikers on the streets of Ferrara) (Author: A. Krstić-Furundžić)

Natural environment has a restoring effect on people's mental health that is conclusion of 20 years of research of Rachel and Stephen Kaplan. Based on this research they have conceived theory "Attention Restoration Theory" (Kaplan, R. & Kaplan, S., 1989) that describes the restorative effect of natural environments on human mental fatigue. Some studies have indicated a positive effect of nature on the healing process. One of the research approaches relating to health effects of landscapes is known as "Healing Gardens". If the view from a hospital window is important in terms of improving recovery (Ulrich, 1984), then the design of hospital gardens becomes a new and significant topic. There are different and sometimes conflicting ways to provide urban residents access to nature, as summarized in Table 2.

To be biophilic, cities should devote sufficient area (generally more than 15% of their total area) to public greenspace, provide public parks and recreational facilities within a five-minute walk of most houses, incorporate landscaping such as street trees and planters, and offer community gardens, green infrastructure (such as plants incorporated into buildings), and nature visiting programs in order to ensure that residents have frequent and significant exposure to natural environments and associated benefits (Green, 2016).

The aforementioned facts point to the positive effects that provide the plants in the areas of life and work, as well as the positive effect of window view at the landscape. Greenspaces in the areas of life and work allow for window view at the landscape (Figure 6). Urban and architectural design should take it into consideration as an imperative. Appropriate dimensions, orientation and views through windows in high density urban areas are crucial (Figure 7). In such areas streets are usually narrow, buildings are

Table 2. Three ways to increase interaction with nature (Source: Litman, 2017)

Approach	Advantages	Disadvantages
Lower density development, such as housing with large private gardens, or near farms and forests.	Private gardens tend to provide more privacy, plus physical and emotional involvement.	Increases per capita land consumption and costs associated with dispersed development.
Higher density development with public greenspace, such as apartments near urban parks.	Reduces land consumption per capita and provides benefits of compact urban development.	Increases costs associated with compact development. Requires more planning.
Green infrastructure, such as green roofs, and street trees.	Provides greenspace within developed areas.	Can increase infrastructure costs.
Natural area visiting, such as parks programs and holidays.	Allows more people to experience natural environments.	Tend to be infrequent. Increases transport costs.

Figure 6. Greenspaces in the areas of life and work allow for window view at the landscape. (left-Housing settlement Konjarnik, Belgrade; right- Main office of SOLON AG für Solartechnik, Berlin-Treptow) (Author: A. Krstić-Furundžić)

Figure 7. The view from the window connects with the external environment (Author: A. Krstić-Furundžić)

Urban environment-uninspiring view	Rural environment-interaction with nature

close to each other, facade distance between neighboring buildings improper which disrupt intimacy and window view is often poor, directed into the walls and skylights, which can be recognized as urban mental health impacts, and might result in stress and depression. Such situations are not present in rural areas, villages.

An increasing body of international research details the benefits that arise when people have a direct or indirect relationship with the natural world. Evidence is clear that positive physical benefits occur such as reduced blood pressure, reduced stress, and increased immunity (Gullone, 2000). Evidence of psychological benefits includes positively modified behaviour (particularly in terms of social interaction and reduced violence), decreased rates of depression, and increased ability to concentrate (Browning, Ryan, & Clancy, 2014; Zari, 2017).

UNDERSTANDING THE CHARACTERISTICS OF URBAN AND RURAL AREAS AS AN AID TO PREVENTIVE MEDICINE

The scientific and empirical knowledge that is related to advantages and disadvantages of urban and rural life can be used for illness prevention and also for treatment of patients.

Prevention is usually about prevention of disease, sometimes it is promotion of health and sometimes it is about medication or other agent used for prophylaxis. From medical point of view it is important to understand how urban areas can affect human health (Figure 8), and how rural advantages are healthful. Considering the effects of environment on human health, it is important to remind that in order to make a medical diagnosis in routine conversation with patients there is need for asking questions related to the quality of living space.

In everyday dental practice conversation with patient is crucial, to understand his problem and need and visualize cause of his dental problem. Using a larger number of information related to lifestyle of patient and handling this information in right way it can be useful for finding the right therapy for him. In order to achieve the results from therapy patient should be motivated to stick to therapy procedures. From the previous text it can be concluded that absence of natural environment and interaction with

Figure 8. Absence of natural environment and exposure to night light affect human health (Author: N. Furundžić and D. Furundžić)

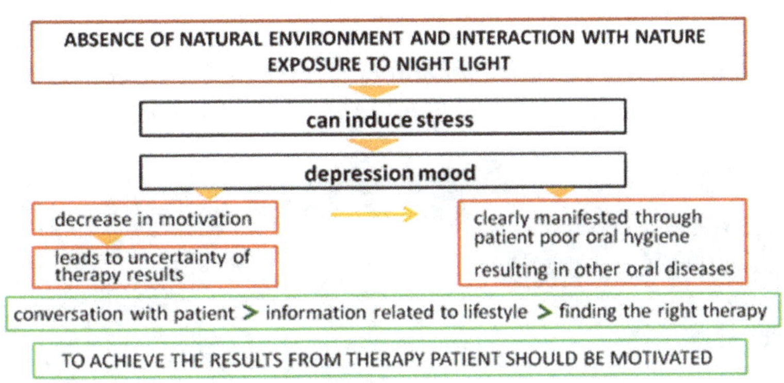

nature as well as exposure to night light can induce stress and depression mood. Patients with depression mood have decrease in motivation, which leads to uncertainty of therapy results. Depression mood is often clearly manifested through patient poor oral hygiene which results in other oral diseases.

SOLUTIONS AND RECOMMENDATIONS

Embracing experiences is often a way to prevent or overcome problems and create a solution. If we look back to the age of antiquity, it is possible to notice that the Asclepius cult in Epidaurus is the view that a part of the quality of life is related to the environment, which must have the content for performing the physical and spiritual nature of the activity, while everything is surrounded by beautiful, colorful landscapes (Figure 9). It can be noticed that when we want to check whether the scattering goes according to an imagined plan, it would be useful that sometimes we take into account the concepts of creators from the old Ancient times. They gave great importance to factors that affect urban mental health and achieving interaction with nature.

The reaction to factors that emanate from the environment and have a major impact on the overall human health can also be indicated through oral health. In order to create the concept of preventive measures, it is welcome to look at previous experiences, such as ancient postulates. In order for a human society to enjoy health, it is necessary to make its own sacrifice in the context of the consolidation of everyday activities, as did the cult of the first doctor-demigod in Greek mythology in Epidaurus, which the main therapeutic procedures were based on the patient's expressed faith in therapeutic treatment because the therapeutic procedure can be successful if the patient makes a sacrifice by keeping the day-to-day activities prescribed by the therapist. This also applies to the creation of urban areas so that they have proper contents that will enable people to have areas for everyday physical and spiritual activities regardless of age and gender. In order to achieve this it is necessary that the process of urban planning and renewal seeks to reduce the build-up density in the corresponding area in order to provide a public space for everyday people's activities, which can have a positive impact on urban mental health.

Urban agriculture or urban gardening is one of the solutions that benefits at different levels and can reflect economic and social development. Urban gardening can be planned in suburban areas, urban courtyards, courtyards of buildings, but also on terraces and balconies of flats, and within certain public buildings such as schools, nurseries, and the like (Figure 10). It is particularly useful to form them in the context of schools, kindergartens, elderly homes, hospitals, etc. Urban gardening helps elders and sick

Figure 9. Performing the physical and spiritual nature of the activity in beautiful natural environment. (left: Epidaurus Theater, right: Epidaurus Stadium) (Author: N. Furundžić)

Figure 10. Urban gardening for the well-being of the inhabitants. (left: urban gardening in the glazed courtyard of the building, Institut National de L'Energie Solaire, Le Bourget-du-Lac; middle and right: urban gardening in Freiburg) (Authors: A. Krstić-Furundžić and N. Furundžić, left figure).

people to relax, communicate and work with each other, which create a sense of usefulness. Children learn about agricultural products while they are at the same time having fun and playing. All this has a positive impact on urban mental health.

Given that predictions indicate the majority of the population in the future will live in cities, governments and citizens should develop an understanding of how to make life in cities more sustainable (Furundžić N., Furundžić D., & Krstić-Furundžić, 2019).

CONCLUSION

Cities are the main place of social development. Given the increased need for people to live in urban areas and reducing the desire for living in the villages, the analysis of the good and bad sides, that is, the advantages and disadvantages of both types of areas gives inputs to creating adequate plans for something called the ideal environment for living. The great challenge for today's architects is the sustainability of very urban areas. Many contemporary experts have faced the dilemma of creating the concept of a human modern city with all the modern technologies, of which it is dependent, while fulfilling the primordial needs of human beings. It is necessary to take into account factors that affect human health, particularly urban mental health, and environmental factors are highlighted in this paper.

In order to create urban and rural areas that provide the welfare of the population, based on the considerations conducted in the paper, the following conclusions can be given:

- The future resilience of urban and rural areas depends on the ability to predict problems that might be encountered.
- In this respect should be considered prevention of human health. Factors that can promote or harm health and wellbeing in urban and rural areas are discussed in the paper. The paper points to the positive effects that provide rural areas and the presence of the advantages that cities have in terms of preserving people's health. It is noticed that the synergy of urban and rural areas can provide new knowledge which opens up new perspectives for preventive medicine.
- Exhaustive interviews with patients are important preventive and prophylactic measures in medicine and dentistry and the information thus obtained may be directed to experts in other fields. It can be concluded that the progress of society largely depends on mutual cooperation between different professions and multidisciplinary work is the key to success in all domains.

- Urban areas should have proper content that will enable people to have space for everyday physical and spiritual activities, regardless of gender and age, which can have a positive impact on urban mental health.
- Problems can be managed and reduced with improved design.

REFERENCES

Ali, M. M., & Al-Kodmany, K. (2012). Tall Buildings and Urban Habitat of the 21st Century: A Global Perspective. *Buildings, 2*, 384-423. Retrieved February 16, 2019, from https://www.researchgate.net/publication/307762382_Tall_Buildings_and_Urban_Habitat_of_the_21st_Century_A_Global_Perspective

Allender, S., Lacey, B., Webster, P., Rayner, M., Deepa, M., Scarborough, P., ... Mohan, V. (2010). Level of urbanization and noncommunicable disease risk factors in Tamil Nadu, India. *Bulletin of the World Health Organization*, *88*(4), 297–304. https://www.who.int/bulletin/volumes/88/4/09-065847/en/ RetrievedFebruary162019. doi:10.2471/BLT.09.065847 PMID:20431794

Asian Development Bank. (n.d.). *Competitive Cities: City Cluster Economic Development* (Brochure). Retrieved February 16, 2019, from https://www.adb.org/sites/default/files/competitive-cities-brochure.pdf

Bardow, A., Nyvad, B., & Nauntofte, B. (2001). Relationships between Medication Intake, Complaints of Dry Mouth, Salivary Flow Rate and Composition, and the Rate of Tooth Demineralization in situ. *Archives of Oral Biology*, *46*(5), 413–423. doi:10.1016/S0003-9969(01)00003-6 PMID:11286806

Bedrosian, T. A., & Nelson, R. J. (2013). Influence of the modern light environment on mood. *Molecular Psychiatry*, *18*(7), 751–757. doi:10.1038/mp.2013.70 PMID:23711982

Berto, R. (2014). The Role of Nature in Coping with Psycho-Physiological Stress: A Literature Review on Restorativeness. *Behavioral Science*, *4*(4), 394–409. doi:10.3390/bs4040394 PMID:25431444

Boutayeb, A., & Boutayeb, S. (2005). The burden of non communicable diseases in developing countries. *International Journal for Equity in Health*, *4*(1), 2. doi:10.1186/1475-9276-4-2 PMID:15651987

Brezzi, M., Dijkstra, L., & Ruiz, V. (2011). *OECD Extended Regional Typology: The Economic Performance of Remote Rural Regions. OECD Regional Development Working Papers, 2011/06.* Paris, France: OECD Publishing. Retrieved February 24, 2017, from doi:10.1787/5kg6z83tw7f4-en

Broman, M. (2017). Introduction. In K. Gunnartz (Ed.), Drivers and consequences of urban development, IVA project Good cities of the future. Stockholm, Sweden: The Royal Swedish Academy of Engineering Sciences (IVA).

Browning, W. D., Ryan, C. O., & Clancy, J. O. (2014). *14 Patterns of Biophilic Design. Improving Health & Well-Being in the Built Environment*. New York: Terrapin Bright Green, LLC.

Chen, M., Zhang, H., Liu, W., & Zhang, W. (2014). The Global Pattern of Urbanization and Economic Growth: Evidence from the Last Three Decades. *PLoS One*, *9*(8), e103799. doi:10.1371/journal.pone.0103799 PMID:25099392

Cinzano, P., Falchi, F., & Elvidge, C. D. (2001). The first World Atlas of the artificial night sky brightness. *Monthly Notices of the Royal Astronomical Society*, *328*(3), 689–707. doi:10.1046/j.1365-8711.2001.04882.x

Corrucini, R. S. (1991). Anthropological aspects of orofacial and occlusal variations and anomalies. In M. A. Kelly & C. S. Larsen (Eds.), *Advances in dental anthropology*. New York: Wiley-Liss.

Dijkstra, L., & Poelman, H. (2014). *A harmonised definition of cities and rural areas: the new degree of urbanization. Regional and Urban Policy,* Working Papers 01/2014. Brussels: European Commission. Retrieved February 25, 2017, from-https://ec.europa.eu/regional_policy/sources/docgener/work/2014_01_new_urban.pdf

Formaneck, S. (2013). Skyscrapers and Economic Strategy and Development: The Case of Bahrain"s Proposed Murjan Tower. *Journal of Economics. Business and Management*, *1*(4), 350–354.

Friedmann, J. (2006). Four theses in the study of China's urbanization. *International Journal of Urban and Regional Research*, *30*(2), 440–451. doi:10.1111/j.1468-2427.2006.00671.x

Furundžić, Z. N., Furundžić, P. D., & Krstić-Furundžić, A. (2019). Chapter Three – New Technology for the Promotion of Health in the Cognitive City. In E. Vaništa Lazarević, M. Vukmirović, A. Krstić-Furundžić, & A. Đukić, Keeping (Eds.), Keeping Up with Technologies to Create the Cognitive City (pp. 31-43). Newcastle upon Tyne, UK: Cambridge Scholars Publishing.

Godfrey, R., & Julien, M. (2005). Urbanisation and health. *Clinical Medicine*, *5*(2), 137–141. doi:10.7861/clinmedicine.5-2-137 PMID:15847005

Green, J. (2016). *Biophilic Cities Lead the Way to Urban Sustainability*. The Dirt, American Society of Landscape Architects. Retrieved January 20, 2017, from https://dirt.asla.org/2016/06/06/biophilic-cities-forge-new-paths

Griffiths, J., Jones, V., Leeman, I., Lewis, D., Patel, K. K., Wilson, K., & Blankenstein, R. (2000). *Oral Health Care for People with Mental Health Problems Guidelines and Recommendatio*ns. British Society for Disability and Oral Health. Retrieved February 20, 2017, from http://www.bsdh.org/documents/mental.pdf

Gullone, E. (2000). The Biophilia Hypothesis and life in the 21st Century: Increasing Mental health or Increasing Pathology? *Journal of Happiness Studies*, *1*(3), 293–321. doi:10.1023/A:1010043827986

Hartig, T., & Kahn, P. H. (2016). Living in cities, naturally. *Science*, *352*(6288), 938–940. doi:10.1126cience.aaf3759 PMID:27199417

Hope, K. R. (1998). Urbanization and urban growth in Africa. *Journal of Asian and African Studies*, *33*(4), 345–358. doi:10.1177/002190969803300403

Howard, G., Bogh, C., Goldstein, G., Morgan, J., Prüss, A., Shaw, R., & Teuton, J. (2002). *Healthy Villages: A guide for communities and community health workers*. Geneva: World Health Organization.

IOM - Institute of Medicine. (2010). *Promoting cardiovascular health in the developing world: A critical challenge to achieve global health.* Washington, DC: The National Academies Press. Retrieved January 18, 2012, from http://www.iom.edu/Reports/2010/Promoting-Cardiovascular-Health-in-the-Developing-World-A-Critical-Challenge-to-Achieve-Global-Health.aspx

Jaffe, E. (2015). *Why City Noise Is a Serious Health Hazard: An audiologist explains why it's so much more than a mere annoyance.* City Lab. Retrieved December 20, 2016, from http://www.citylab.com/housing/2015/04/why-city-noise-is-a-serious-health-hazard/391194

Kaplan, R., & Kaplan, S. (1989). *The Experience of Nature: a Psychological Perspective.* Cambridge, UK: Cambridge University Press.

Kellert, R. S. (2008). Dimensions, Elements and Attributes of Biophilic Design. In S. Kellert, J. Heerwage, & M. Mador (Eds.), *Biophilic Design: The Theory, Science and Practice of Bringing Buildings to Life* (pp. 3–4). Hoboken, NJ: John Wiley & Sons. Inc.

Kellert, R. S., Heerwagen, H. J., & Mador, L. M. (2008). *Biophilic Design: The Theory, Science and Practice of Bringing Buildings to Life.* Hoboken, NJ: John Wiley & Sons. Inc.

Kessler, R. C., McGonagle, K. A., Swartz, M., Blazer, D. G., & Nelson, C. B. (1993). Sex and depression in the National Comorbidity Survey. I: Lifetime prevalence, chronicity and recurrence. *Journal of Affective Disorders, 29*(2-3), 85–96. doi:10.1016/0165-0327(93)90026-G PMID:8300981

Khanna, P. (2016). How much economic growth comes from our cities? *World Economic Forum.* Retrieved February 20, 2019, from https://www.weforum.org/agenda/2016/04/how-much-economic-growth-comes-from-our-cities/

Kjellstrom, T., Friel, S., Havemann, K., McGranahan, G., Mercado, S., & Sattherthwaite, D. (2008). *Our cities, our health, our future: Acting on social determinants for health equity in urban settings. Report to the WHO Commission on Social Determinants of Health from the Knowledge Network on Urban Settings Acting on social determinants for health equity in urban settings.* World Health Organization, Centre for Health Development.

Kwon, S., Kim, J., & Oh, D. S. (2012). Measurement of Urban Competitiveness Based on Innovation Indicators in Six Metropolitan Cities in Korea. *WTR- World Technopolis Review, 1,* 177-185.

Litman, T. A. (2017). *Urban Sanity: Understanding Urban Mental Health Impacts and How to Create Saner, Happier Cities.* Victoria Transport Policy Institute.

Moldovan, I., Moldovan, V. S., & Ilies, M. N. (2014). Iconic Architecture: Skyscrapers. In *Proceedings of 5th International Conference "Civil engineering – science and practice"* (pp. 1461-1468). Montenegro: Žabljak. Retrieved January 17, 2019, from https://www.researchgate.net/publication/308099136_ICONIC_ARCHITECTURE_SKYSCRAPERS

Ni, P., & Qinghu, H. (n.d.). *Comparative Research on the Global Urban Competitiveness.* Retrieved February 14, 2019, from http://www.stats.gov.cn/english/specialtopics/iaos/papers/concurrent_session_30b.pdf

OECD - Organization for Economic Cooperation and Development. (2011). *OECD Regional Typology.* Directorate for Public Governance and Territorial Development.

Pinoncely, V., & Hartkoorn, I. (2014). *Planning Horizons no. 3 - Promoting Health Cities. The Royal Town Planning Institute*. RTPI.

Proffit, R. W., Fields, W. H., Ackerman, L. J., Bailey, J. L'T., & Tulloch, J. E. C. (2000). Contemporary Orthodontics (3rd ed.). Mosby Inc.

Pugh, C. (1995). Urbanization in developing countries: An overview of the economic and policy issues in the 1990s. *Cities (London, England)*, *12*(6), 381–398. doi:10.1016/0264-2751(95)00083-X PMID:12319994

Ramon, T., Grinshpoon, A., Zusman, S. P., & Weizman, A. (2003). Oral Health and Treatment Needs of Institutionalized Chronic Psychiatric Patients in Israel. *European Psychiatry*, *18*(3), 101–105. doi:10.1016/S0924-9338(03)00023-3 PMID:12763294

Reddy, K. S. (1993). Cardiovascular disease in India. *World Health Statistics Quarterly*, *46*, 101–107. PMID:8303903

Reddy, K. S., & Yusuf, S. (1998). Emerging Epidemic of Cardiovascular Disease in Developing Countries. *Circulation*, *97*(6), 596–601. doi:10.1161/01.CIR.97.6.596 PMID:9494031

Rosania, A. E., Low, K. G., McCormick, C. M., & Rosania, D. A. (2009). Stress, Depression, Cortisol, and Periodontal Disease. *Journal of Periodontology*, *80*(2), 260–266. doi:10.1902/jop.2009.080334 PMID:19186966

Selwitz, R. H., Ismail, A. I., & Pitts, N. B. (2007). Dental Caries. *Lancet*, *369*(9555), 51–59. doi:10.1016/S0140-6736(07)60031-2 PMID:17208642

Smith, S. (2012). *Urbanization and cardiovascular disease: Raising heart-healthy children in today's cities*. Geneva: The World Heart Federation. Retrieved February 20, 2019 from https://www.world-heart-federation.org/wp-content/uploads/2017/05/FinalWHFUrbanizationLoResWeb.pdf

Tsouros, D. A. (Ed.). (n.d.). *World Health Organization Healthy Cities Project: A Project becomes a movement, review of progress 1987 to 1990*. Sogess. Retrieved January 23, 2019, from http://www.euro.who.int/__data/assets/pdf_file/0016/101446/WA_380.pdf

Ulrich, R. S. (1984). View through a window may influence recovery from surgery. *Science*, *224*(4647), 420–421. doi:10.1126cience.6143402 PMID:6143402

UN. (2013). *World Economic and Social Survey 2013*. New York: UN.

Vargas, C. M., Dye, B. A., & Hayes, K. L. (2002). Oral Health Status of Rural Adults in the United States. *The Journal of the American Dental Association*, *133*(12), 1672–1681. doi:10.14219/jada.archive.2002.0120 PMID:12512669

Webster, D., & Muller, L. (2000). *Urban Competitiveness Assessment in Developing Country Urban Regions: The Road Forward*. Paper prepared for Urban Group, Washington, DC. Retrieved January 23, 2019, from https://www.adb.org/sites/default/files/competitive-cities-brochure.pdf

WHO. (2011). *Burden of Disease from Environmental Noise: Quantification of Healthy Life Years Lost in Europe*. Copenhagen, Denmark: WHO Regional Office for Europe. Retrieved March 14, 2017, from http://bit.ly/2btWBXy

WHO. (2012). *Effects of urbanization on incidence of noncommunicable diseases.* WHO, Regional Office for the Eastern Mediterranean. Retrieved January 23, 2019, from http://applications.emro.who.int/dsaf/EMPUB_2012_864.pdf

Yao, C., Wu, Z., & Wu, J. (1993). The changing pattern of cardiovascular diseases in China. *World Health Statistics Quarterly*, *46*, 113–118. PMID:8303905

Zari, M. P. (2017). What makes a city 'biophilic'? Observations and experiences from the Wellington Nature Map project. In M. A. Schnabel (Ed.), *Back to the Future: The Next 50 Years, Proceedings of the 51st International Conference of the Architectural Science Association (ANZAScA)*, (pp. 105–114). Architectural Science Association (ANZAScA).

KEY TERMS AND DEFINITIONS

Biophilic: The human tendency to affiliate with nature, to interact or be closely associated with other forms of life in nature. Biophilic design incorporates natural materials, natural light, vegetation, natural views and involving experience of the natural world in the creation of the modern built environment.

Built Environment: An environment created by a human being, which ensures the realization of human activities, consisting of buildings, parks, and infrastructure.

Health Promotion: Activities that enable people to increase control over their own health, to become aware of activities that damage their health.

Human Well-Being: Well-being is understood as a state of health, happiness and/or prosperity, and can be evaluated by objective and subjective factors that constitute the health and quality of life.

Natural Environment: The natural environment includes all living and non-living beings that are naturally occurring, which is not man-made.

Prevention: Prevention is a set of measures and actions that are used in order to stop or reduce the incidence of a disease, disorder or behavior that causes negative effects on the health of a person or group of people.

Preventive Medicine: The aim of preventive medicine is to improve health and prevent disease. It consists of a set of medical techniques and treatments and their combinations intended to prevent disease before it happens, rather than curing it.

Urban Gardening: Also known as urban horticulture or urban agriculture is the process of growing plants of all types and varieties in an urban environment.

Urban-Rural Synergy: A number of different connections and interactions between rural and urban areas, a common thread that connects the enormous diversity of connections and interactions between the given places.

Urbanization: The process by which towns and cities are formed, and which, by natural or mechanical increase in the population, become larger because more and more people live and work in the central areas.

Chapter 4
Small Urban Centers as Drivers of Daily Migrations and Agents of Transformation of Rural Hinterland:
Example of Blace Municipality

Eva Vaništa Lazarević
Faculty of Architecture, University of Belgrade, Serbia

Milena Ivanovic
National Spatial Data Infrastructure, Republic Geodetic Authority, Serbia

Dušan Ristić
https://orcid.org/0000-0002-0061-5190
Faculty of Natural Sciences and Mathematics, University of Priština, Serbia

Marjan Marjanović
https://orcid.org/0000-0001-6741-8477
College of Urban Planning and Public Affairs, University of Illinois, USA

Mario Miličević
GDi Solutions, Serbia

ABSTRACT

Decades of regional polarity in Serbia conditioned the adoption of a polycentric territorial development as one of the main strategic goals of the country's spatial development. Politics of polycentrism specifically highlights the role and importance of small urban centers in generating a more balanced regional development. In constellation with their rural surroundings, they can serve as strong agents of change towards sustainable spatial development, functional cohesion, and transformation of the rural hinterland, and thus, contribute to the balanced spatial development. The main aim of the chapter is to examine the potential of small urban centers to generate spatial and functional integration, that is, to explore the possibility for small towns to integrate and transform their rural surroundings. The chapter analyses the field of functional influence of small towns on their rural surroundings on the example of Blace Municipality through determining the intensity of daily migrations which are a well-known indicator of spatial and functional integration of rural areas and urban centers.

DOI: 10.4018/978-1-5225-9932-6.ch004

INTRODUCTION

Polarized development and significant regional disparities in Serbia directly affect the adaptation of planning documents to the concept of polycentric spatial development. The European Spatial Development Perspective – ESDP, an informal document endorsed by European Commission (EC), promoted this concept back at the turn of the century with the aim to ensure regionally balanced development of European territory, especially in relation to the prospect of future enlargements (EC, 1999). Davoudi (2003) suggests that polycentricity was increasingly shaping the spatial policy discourses in European countries at the time. Therefore, although a well-known idea in Serbian planning, following the adoption of the ESDP, polycentric development was introduced as a leading concept of the national planning discourse which it remains to be today.

The spatial and functional organization of the Serbian territory and the development of urban centres are based on European models and instruments which are dominated by the concept of nodal region with a strongly emphasized role of urban (primarily small- and middle-sized) centres and functional urban areas[1]. Within the network of settlements, smaller urban centres are expected to take on the role of nodes, i.e. of promoters of regional development and drivers of transformation of undeveloped and weakly integrated rural hinterland, while together making functionally coherent system. These nodes, together with the surroundings they generate, would serve as a counterweight to big urban centres or 'big city narrative' (Cox & Longlands, 2016).

Commitment to the policy of polycentrism[2] and a necessity of decentralization with an aim of establishing balanced regional development impose the need for exploring development potentials of small urban settlements in Serbia and evaluation of demographic flows. The main hypothesis of this paper is that municipal centres and small urban settlements within a rural area can integrate their rural hinterland and impact its functional and socio-economic transformation. It is based on the importance of 'strengthening small and medium-sized towns in rural areas as focal points for regional development' promoted in various EU policy documents and initiatives (EC, 1999, p. 25; Atkinson, 2001), but also in academia (Davoudi & Stead, 2002; Davoudi, 2003; Zonneveld & Stead, 2007).

A small urban settlement such as Blace in this case, possesses a capital (residential, functional, economic, etc.) and a capacity to integrate neighbouring space, generate labour, and become an important element and driver of sustainable spatial development and transformation of rural hinterland. It is expected that this will be achieved through strengthening spatial and functional interdependencies in the relation of rural and urban settlements. The further development of daily migration systems and the reconstruction of the existing housing stock constitute the backbone of the modernization of the society and the continuation of urbanization of the settlement. Likewise, existing transport connections, the proximity and accessibility of the urban settlement to the rural environment, and the existing and planned housing are also important factors in the creation of urban-rural synergy. The paper will thoroughly examine these elements while aiming at estimating Blace's potential and the role as a driver of regional integration and sustainable spatial development.

METHODOLOGY

The first part of the paper gives an overview of inconsistencies in the administrative division of districts and municipalities within the formal territorial organization of the Republic of Serbia, all in order to better understand the position of the Blace municipality in relation to the Republic of Serbia and the Toplica District to which it belongs. The conclusions in this section are based on statistical and economic analyses.

The present paper analyses development potentials and the current level of development of Blace municipal center by identifying its economic capacities and availability of public and social services. The study applies Schmook's method of determining the degree of centrality of municipal centers (Đorđević, Tošić, 2013; Đerčan et al., 2017) as drivers of the transformation of their rural surrounding. The index of urban primacy, first defined by American geographer Jefferson in 1939, is also used as a quantitative method (numeric indicator) for estimating dominancy and centrality of the central urban area (Đorđević, Tošić, 2013; Đerčan et al., 2017).

The integration and coherence of the main urban centre in the municipality and its rural hinterland, as well as dynamics of urban/rural structure are defined on the basis of transport connections and, more specifically, on the basis of the coefficient of transport network. The accessibility of the municipal centre from the position of rural settlements is evaluated in Geographic Information System (GIS) by determining relevant isochrones.

The scope and intensity of the municipal centre's gravitational reach is explored by analysing indicators on the mobility of labour force, pupils and students (daily migrants) from rural areas to the municipal core. This segment of the work relies on the method of determining the gravitational field of urban centres through the evaluation of intensity, routes and directions of daily migrations (Tošić, Krunić & Petrić, 2009).

The effects of urban settlement on its rural hinterland are investigated by analysing functional and socioeconomic types of settlements and by defining the degree of functional and socio-economic transformation. The study used published and specially processed data of Statistical Office of the Republic of Serbia, collected by the administration of Blace Municipality.

Overall, the data used in the analysis was collected from the following sources:

- publicly-available data of the Republic Statistical Office, published or specially processed, for the period between 1948 and 2018,
- relevant publications of the Development Agency of Serbia,
- relevant regulations and legal acts of the Government of the Republic of Serbia and the Blace Local Government Unit
- data collected by the officials of the Blace Municipal Administration.

CONTEXT OF THE SYSTEM OF SMALL URBAN SETTLEMENTS: PROBLEMS OF SIZE, TERRITORIAL ORGANIZATION, AND DEVELOPMENT IN THE REPUBLIC OF SERBIA ON THE EXAMPLE OF THE MUNICIPALITY OF BLACE

The organic connection of urban settlements with the environment created an inseparable system. Urban settlements are a magnet that constantly attracts the population from rural areas. Accordingly, urban settlements and cities cannot be seen as individuals without their environment. The environment is usually defined as administrative (spatial) or functional.

In Serbia, there always existed an administrative territorial division that corresponded to the functionality of a certain area. However, following the more active involvement of Republic of Serbia to join the European Union, it began with more frequent procedures for adjusting the regulations of the European Union. This resulted in the adoption of new norms that also applied to territorial division. The former division into the north and south of the country now has a legal basis. Despite the great appeals and efforts to come to decentralization of Serbia, this did not happen.

The administrative-territorial division of the territory of the Republic of Serbia is defined by division into Nomenclature of territorial units for statistics - NUTS[3] 1 (*Serbia - North and Serbia - South*) regions, NUTS 2 (*Belgrade Region, Region of Vojvodina, Region of Sumadija and Western Serbia, Region of South and East Serbia, Region of Kosovo and Metohija*), NUTS 3 which defines spatial areas in the republic that are equivalent to administrative districts. In Serbia, there are 30[4] areas (NUTS 3) and 29 administrative districts. Toplica Administrative District is located in the Region of South and East Serbia (NUTS 2), which is at the same time one of the most underdeveloped regions in the Republic. The administrative district in the Republic of Serbia regulation is a regional center of the state administration which includes district regional units of all state administration bodies that are educated for its area (Official Gazette of Republic of Serbia No.15/2006).

According to the data of the Development Agency of Serbia, more precisely according to the Regulation on the establishment of a single list of regional development and local self-government units for 2014 (Official Gazette of Republic of Serbia No.104/2014), there were formed four groups of municipal development and a special subgroup that includes devastated areas. The scope of the Regulation covers 164 municipalities, of which 63 belong to the 4th and 5th group that account for 60%, or 50% of the development below the republic average. Calculation of the degree of development of municipalities in Serbia is based on the value of gross domestic product per capita in the region in relation to the republic average for the reference period.

It is important to point out that all municipalities from the most undeveloped groups 4 and 5 belong to the region Serbia - South, which includes the Toplica District in which the Municipality of Blace is located, with the exception of Opovo Municipality, which belongs to the region of Serbia – North. The Toplica District located in the south-eastern part of Serbia, bordering with the Jablanica, Nišava and Rasina Districts, and partly with Kosovo and Metohija. It has an area of 2,231 km 2 . The total population according to the census of 2011 amounted to 91,754 with an average density of 41.13 per km ^{2}. Toplica District includes 4 municipalities: Prokuplje (city - the seat of the region), Blace, Kuršumlija, and Žitorađe. The Administrative District itself (Toplica District) is the most undeveloped district within the region. Until 2018, this was one of only two districts in the Republic of Serbia that did not have a city as a territorial unit, but only municipalities. With the latest decree on territorial organization of the Republic of Serbia from 2018, Prokuplje became a city, while Blace, Kuršumlija, and Žitorađe remained the city settlements and centers of their municipalities.

AREA OF RESEARCH

In Serbia, all underdeveloped areas boast some common characteristics. The region of South and East Serbia (division by NUTS 2) has always faced the problem of small, undeveloped, and spatially devastated settlements (Spatial Plan of the Republic of Serbia, 2010). The least developed region in the country has the largest number of municipalities, which is a major problem in maintaining and preserving their population levels. It is considered that the system of settlements in this region is inadequately articulated for three reasons (Videnović, 2017):

- Lack of lower-ranked central settlements,
- Inadequate distribution of urban functions among smaller towns in the area,
- Lack or weak links between settlements in the region.

In the process of industrialization of the former Yugoslavia, the settlements in the Region of South and East Serbia, to which Toplica District belongs, were formed as small urban centers. They had the tendency to develop a strong production base that was accompanied by the influx of workers from rural areas, and hence the need for the expansion of the housing stock (Petovar, 2003). In that way, urban settlements developed while rural households rapidly emptied. This was also the case with the Municipality of Blace.

Blace Municipality is located in the south of the Republic of Serbia and in the northern part of the Toplica District. The municipality of Blace is a small municipality, 306 km^2 of surface, with 11.754 inhabitants distributed in 40 settlements, of which one is urban and 39 are rural. The Blace Municipal Center is the largest settlement in the municipality with 5353 inhabitants. It boasts a well-developed tertiary-quaternary activity sector and represents the administrative, economic, and cultural center of the municipality.

Although the Municipality of Blace has an area smaller than average for the region of South and East Serbia (which is 504 km^2), it belongs to the municipalities of Serbia with small population (11754 inhabitants according to the 2011 census) and a low population density (38.43 inhab/km^2) compared to the municipalities of the same region (61.43 inhabitants/km^2). In relation to its immediate environment, which has very similar geographical and socio-economic predispositions, the population density of the Municipality of Blace (38.41 inhab/km^2) is lower than in most other municipalities of Toplica District (Žitorađa 76.48 inhab/km^2 and Prokuplje 58.52 inhab/km^2) (Statistical Office of the Republic of Serbia -SORS, 2013).

The *Table 1* shows the comparative number of inhabitants of Blace municipality according to the censuses of the Republic of Serbia in the period between 1948 and 2011. The municipality of Blace reached the peak of its population in 1953. In that period, the rural environment also followed the trend of growth, and recorded a maximum of 22427 inhabitants. However, ever since the first census year, the Blace municipal center has recorded population growth and reached the peak of its population in 2002, after which the number has been declining gradually. In the same period, a drop in the population of rural areas became rather evident, mainly due to the rising urbanisation trends, especially in the period between 1961 and 1981, i.e. in the time of rapid industrialization of the area.

From 1959 to 1981, several public companies and institutions were created in Blace, namely: Blačanka knitwear (1959), "Napredak" ready-made clothes (1961), rubber industry "Braća Vuksanović" (1691), grammar-school (1961), metal industry "5. December 1965, TIB (1965), School for the Vocal Qualifi-

Table 1. Population of the municipality of Blace 1948 – 2011(Data source: Statistical Office of the Republic of Serbia, 1948-2011)

Year	1948	1953	1961	1971	1981	1991	2002	2011
Total population of Blace Municipality	23676	24608	22099	19331	17338	15709	13759	11754
Population of the main urban center	1824	2181	2564	3373	4409	5228	5465	5253
Population of rural settlements within the municipality	21852	22427	19535	15958	12929	10481	8294	6501

cation of Workers (1968), Department Store (1973), Jezero Hotel (1978), Cultural Center (1981), MIB (1981) Children's kindergarten "Our joy" (1982), sports hall (1988), and "Fruto" plant (1993). Although the majority of population left the municipality of Blace in the period of construction of these economic units (from 1961 to 1971), the population of the Blace settlement (i.e. main urban center) in the period between 1961 and 1991 almost doubled.

For the purposes of this paper, the mathematical indicators for the settlement system (*Table 2*) are calculated. They indicate mutual relations and relationships, general dispersion, as well as conditionalities induced by spatial-functional connections and expressed through unequal values of parameters of area, number, and demographic size of settlements (Stamenković & Bačević, 1992).

The density of the settlement network represents the ratio of the area of the territory to the number of settlements and defines the dispersion of settlements in space. The average settlement network density is determined by the formula Gn = P / n (Gn - settlement network density; P - territory area; n – number of settlements) and is expressed in km^2.

The average density of a settlement network is a ratio between the number of settlements and the area of a given territory (usually 100km2). The value is calculated by the formula $T = (n \cdot 100) / S$ (T - mean density of settlements; n - number of settlements; S - area of the territory) and is expressed by the number of settlements per 100 km2.

The medium distance (spatial distance) between settlements determines the spatial relationships and the connections between them. The square root of the surface area and number of settlements represent the values of the mean distance between settlements, and is determined by the formula $d = 2\sqrt{P / n\pi}$ (d - mean distance; P - area of the territory; n - number of settlements).

Table 2. Indicators of the structure of territorial organization and concentration of settlements

Indicator	Blace Municipality	Toplica District
Medium sized settlements (number of inhabitants)	294	344
The density of the settlement network (km^2)	7.65	8.35
Medium density of the settlement (100 km^2)	13.07	11.97
Average density of population (inhabitants/km^2)	38.41	41.13
Medium distance (km)	3.12	3.21
Coefficient of variation in settlement size	0.45	0.45
Coefficient of gathering	21.57	28.19

The coefficient of variation in settlement size (V) is used to determine the heterogeneity of the population per settlement and indicates deviations from the average population. It is calculated by the formula:

$$V=\frac{\sqrt{\frac{(Ni-N)^2}{n}}}{N}$$

Where is: V - coefficient of variation of settlement size; Ni - population of an individual settlement; N- average population per municipality; n - total number of settlements in the municipality. Higher values of the coefficient of variation have municipalities that have larger differences in the population of their settlements.

The coefficient of gathering is used to determine the distribution of the population of municipalities and indicates the degree of orientation or gathering of the population around the municipal center. The higher the coefficient, the greater the level of assembly. It is determined by the formula:

$$K=\frac{E\cdot N}{T}$$

Where is: K - the coefficient of gathering; E - the number of inhabitants of the municipality less the number of inhabitants of the municipal center; N - number of municipal settlements less the municipal center; T - population of the municipality. The lowest gathering rate in the Toplica District is in Blace Municipality (21.57).

ECONOMIC DEVELOPMENT AND PROVISION OF PUBLIC AND SOCIAL SERVICES AS FACTORS DRIVING DAILY MIGRATIONS

In order to determine the hierarchy of settlements in the municipality of Blace and estimate the role of the Blace urban center within the municipal network of settlements, the following indicators were used: the index of urban primacy and the degree of centrality. Hierarchical characteristics within a settlement system are quantitatively analyzed using the urban primacy index (Vresk, 2002). The focus of the analysis is on the size of urban centers, assuming that the size reflects the functions of urban centers in the settlement system (Vresk, 1990). Qualitative analysis of the hierarchical characteristics of a settlement system, in practice, is most often done by applying the Schmook's method for determining the degree of centrality of a settlement.

The occurrence that a capital city usually boasts the highest population in a given country, was first spotted by American geographer Jefferson (1939) and called the phenomenon the Law of the Primary City. Based on this law, the Urban Primary Index (I) is defined, which is used in the analysis of the hierarchy of urban systems by quantifying the phenomenon of primacy. The index shows the relation between the largest and the second largest city in terms of population, and is determined by the formula (Vresk, 1990a):

$$I = \frac{G1}{G2} = \frac{5253}{499} = 10,53$$

I - index of urban primacy; **G1** - population of the largest settlement (Blace); **G2** - population of the next largest settlement in the Municipality (Trbunje) - according to 2011 Census of Population of Republic of Serbia.

According to this law, the ratio from the equation should be 2 (Vresk 1990; Vresk 1990a), and deviations can be both positive and negative (Vresk, 1990). In other words, urban primacy is high if the index value is greater than 2 (Vresk, 1990a).

In practice, the three largest cities in size are the most commonly used (Vresk 1990; Vresk 1990a):

$$I = \frac{G1}{G2 + G3 + G4} = \frac{5253}{499 + 367 + 358} = 4,29$$

Wherein:

- G3 = number of inhabitants of the third city by size,
- G4 = number of inhabitants of the fourth city by size.

Schmook's quantitative method (Schmook, 1968), based on Kristaler's method of centrality, uses the employed population in the tertiary-quaternary sector of activity (Đorđević, Tošić, 2013) to determine the centrality of settlements in the region. The degree of the centrality of a settlement in the region (S) is determined by the ratio of the active population in service activities to the total active population in the settlement, all in relation to the same indicators in the region, so mathematically can be written in this way (Đorđević, Tošić, 2013):

$$C = T_n - A_n \left(\frac{T_r}{A_r} \right)$$

- C - the center
- Tn - number of active residents in tertiary-quaternary activities in the settlement
- An - number of active residents in the settlement
- Tr - the number of active residents in tertiary-quaternary activities in the region
- Ar - the number of active residents in the region

The urban primacy index (Jefferson, 1939), which refers to highly unbalanced values and significant differences of settlements sizes (in terms of population) in which one settlement predominates, points to a dominant role of the Blace municipal centre with the index value of 10.53 in relation to the second most populated settlement in the municipality . Moreover, when compared to the total population of the next three biggest settlements (in demographic terms), Blace still expresses strong urban primacy with the index value of 4.29.

Schmook's method of determining the degree of centrality (Schmook, 1968) refers to the level of development and to the position and importance of the main urban centre in the settlements network and gravitational environment. The degree of regional centrality (C) expresses the relation of active population in the service sector and total active population in the settlement, compared to the same indicators in the municipality. In the case of Blace municipality, this degree varies from 0.26 in Više Selo to 1.21 in the municipal centre (Blace). In accordance with the number of active citizens in tertiary and quaternary sectors, Blace is the settlement with the highest degree of centrality in the whole municipality.

In addition to the quantitative indicators of the development of the urban settlement - the municipal center, qualitative approaches have also been applied. They have been used to evaluate the importance of the urban settlement - the municipal center according to the concentration production activities and available public services.

After the World War II, municipal centre recorded a strong surge in economic development with the establishment of numerous economic institutions and enterprises. After 1980, metal industry is in stagnation, and the industrial sector is based on the processing of agricultural products. Industry was a dominant economic activity until 1990s. According to the data from 2011, secondary sector employs 27% of working population, while only 4% work in the primary sector activities (Stamenković at al., 2001, 2002). Economic development was followed by the improvements in infrastructure and public and social functions. Blace as the municipal centre has a relatively considerable level of public functions provision (*Table 3*).

Rural settlements are less equipped with public services than the municipal center. Only the settlements Barbatovac and Gornja Draguša boast a relatively higher number of available public services. Due to the relatively small distance of the municipal center from all rural settlements, local municipal offices in rural settlements have been abolished since 2010 (Official Gazette of Blace Municipality No. 4/11). Post offices are located in two rural settlements: Gornja Draguša and Barbatovac. The facilities of the preschool establishments function in elementary schools in the settlements of Barbatovac, Prebreza, and Međuhana. In the village of Trbunje, there is a home department for accommodating adults, who functions as a unit of the Center for Social Work. Health care in rural settlements takes place in outpatient clinics located in 5 rural settlements: Barbatovac, Gornje Grgure, Gornja Draguša, Međuhana, and Gornja Jošanica. Primary education functions within two primary and secondary schools in Barbatovac and Gornja Dragusa and 16 combined classes for 1st to 4th graders (Trbunje, Cungul, Gornje Grgure,

Table 3. Public and social function facilities in the Blace municipal centre (Data source: Municipality of Blace)

Public and social facilities in the municipal centre	
Health care facilities	Health Center (1), Pharmacy stores(4)
Education facilities	Preschool education (central facility); Main elementary school; High School; College.
Local government and other administrative facilities	Local government building; Police station; The post office; Public utility companies; National Employment Service; Tax Administration; Real estate cadastre service; Local court unit; Pension Fund and Health Insurance Fund.
Facilities for Social Welfare	Center for Social Work (Department for elderly and adults); Red Cross.
Cultural facilities	Cultural Center (multifunctional cultural facility), the National Library.

Rasica, Međuhana, Gornja Josanica, Kasevar, Sibnica, Prebreza, Gornja Draguša, Gornje Svarče, Popova, Pridvorica, Suvi Do, Višeselo, Đurevac) (Official Gazette of Blace Municipality No. 4/11).

Concentration of activities, functions and population in the municipal centre has predisposed Blace, as a settlement of highest hierarchical level, to become the destination point of daily migrations from all settlements of lower hierarchy. There are different causes of daily migrations, but the most common one is usually related to the functions of work and education, i.e. to basic socio-economic needs.

TRANSPORT NETWORK AND TRAVEL TIME AS DETERMINANTS OF URBAN-RURAL INTEGRATION

Main causes of increased mobility of daily migrants from rural settlements towards municipal centre can be found in its high functional capacity, favourable morphological position and good transport connection. Modern transport network, communication flows and transport vehicles are important factors of developing daily migrations and enabling their greater time efficiency (Lukić, 2011). Quality of transport connections between rural and urban areas, spatial distance and travel time directly affect the daily mobility and prevent rural depopulation as people remain in rural areas (have a place of residence) and temporary (daily) migrate to the main urban centre instead of permanently relocating. Most authors agree that spatial distance, time and travel expenses play a key role in determining the type of migration: daily migrations or permanent emigration to urban settlements (Wheeler, 1971; Reitsma & Vergoossen, 1987). Reaching certain travel distance, i.e. the critical isochron, travel time and expenses exceed acceptable values which leads to decreased interaction of urban and rural settlements and to a decline in daily migrations as people emigrate from rural areas outside this isochron. Likewise, availability of jobs and employment opportunities within and close to the area of the central settlement, considerably increase the possibility of daily migrations as a modality of the population mobility (Lukić, 2013).

Based on the census data (*Table 4*), we can conclude that the road network in the territory of the municipality of Blace has been constantly developing, except in the last few years, when it has been increased by only one kilometer. The largest leap is recorded in the process of industrialization of the

Table 4. The construction and the quality of the road network in the Municipality of Blace 1973 – 1948 (Data source: Statistical Office of the Republic of Serbia, 2011,2018)

Census year	Total roads (km)	Modern roads (km)	State roads I category		State roads II category		Municipal roads	
			Total roads (km)	Modern roads (km)	Total roads (km)	Modern roads (km)	Total roads (km)	Modern roads (km)
1973	123	30	/	/	18	18	105	12
1981	182	69	/	/	30	27	152	42
1991	188	104	/	/	30	29	158	75
2002	187	111	/	/	29	29	158	82
2011	199	156	/	/	29	29	170	127
2018	202	157.39	18.18	18.18	12.8	10.9	171	128.3

urban settlement (period 1973-1981), where the total network of roads increased by 59 km, which today accounts for 29.8% of all roads in the territory of the municipality of Blace.

Blace municipality has a well-developed road transport, which takes place on the 202 km of categorized roads. The 77.91% of roads is under a modern surfacing. The municipal centre, Blace, is a traffic hub where two categories of national roads intersect (National Roads No IB and IIA) . Around 15% of road network is consisted of high quality roads of national importance. Municipal roads make 85% of road network of which 93.8% has modern surfacing (Municipalities and regions in Republic of Serbia, 2018).Following the new regulation on the categorization of roads from 2013, part of the state roads of the second category passed to the state roads of the 1st category. This phenomenon is clearly evident in (*Table 4*).

In accordance with the adopted Decree on the categorization of state roads and the reference system of the state roads network of the Republic of Serbia, the most important traffic routes in the territory of the municipality are the state roads IB of the line number 38 (Kruševac-Blace-Belolin) and the state road of the IIA line 214 (Blace- Kuršumlija- Kursumlijska Banja-Podujevo).

A significant part of the road network on the territory of the municipality is a system of municipal roads that connects the settlements with each other and with the municipal center of Blace. There are 34 roads in the category of municipal roads. This network is relatively modern, but it characterized by a small average width of the carriageway (3.8m) (Official Gazette of Blace Municipality No. 4/11).

The coefficients (*Table 5*) show that the road network density in Blace municipality (K=0.66) even without highways is higher than in the total territory of Serbia (K=0.5) and of Toplica District (K=0.54). The density of the categorized road network in the territory of Blace municipality is 66 km/ 100 km^2 and is above the average in relation to the same indicators for Central Serbia and Toplica District. Both qualitative and quantitative characteristics as well as coefficients of density point to the quality transport connection of the municipal centre and its rural hinterland in the Blace municipality, which is a precondition for the population mobility and establishing of a daily urban system. All settlements are mostly located next to the main roads.

Using the method of spatial analysis in which the input parameters were the position of the municipal center and the traffic network, and by applying the GIS tool, the time distance between the main urban centre and rural areas is calculated, where the travel time by car expressed in minutes serves as the principal indicator. The adopted travel speed used in calculations is 60 km per hour for state roads and 50 km per hour for municipal roads.

Based on the accessibility of the municipal centre, 5 travel zones are determined, with the limiting values based on the following isochrones: 1) 0-5 minutes; 2) 5-10 minutes; 3) 10-15 minutes; 4) 15-20 minutes and the zone without access to categorized roads.

Table 5. Coefficient of the road network density

	State road I category	State road II category	Municipal roads	Total
Blace municipality	0.06	0.04	0.56	0.66
Toplica District	0.04	0.12	0.38	0.54

Figure 1. Accessibility of the Blace municipal centre

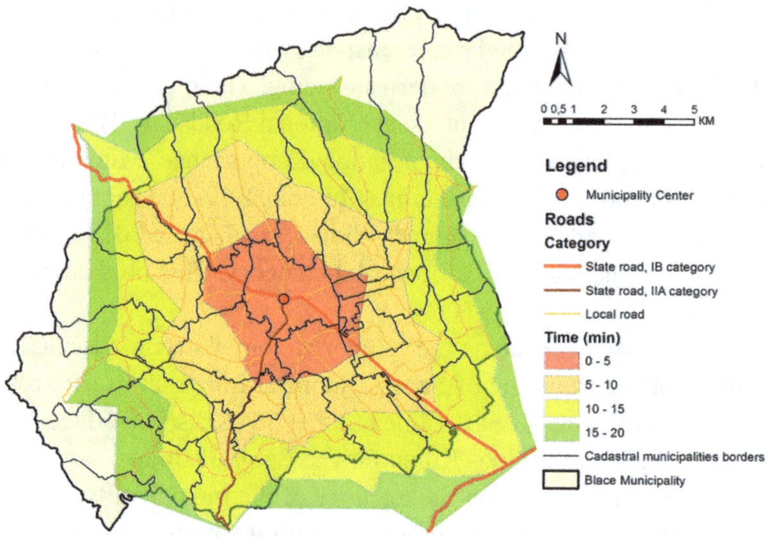

According to the obtained zones we can conclude that the centers of rural settlements in the Blace municipality are only 20 minutes away by motorized transport from the center of the main urban settlement - Blace. The largest number of rural settlements is actually within the isochron of 5-10 minutes which is an exceptional convenience and confirms the favorable spatial position of the main urban center Blace in relation to the municipality of the same name. According to that, it can be concluded that the municipality of Blace has great potential for the development and survival of daily migrations on its territory.

DAILY URBAN SYSTEM OF THE BLACE MUNICIPAL CENTRE

Daily migrants are consisted of workers (active population in work without those that work in agriculture), pupils and students, who make everyday travels without tendencies of permanent relocation to a destination area. According to the official definition of the Republic Institute for Statistics of Serbia, daily migrants are "persons who work or study outside their place of permanent residence, but return to it on a daily basis or several times a week" (SORS, 2014).

Daily migrations are an important form of economic and social relations between rural and urban areas, as well as indicators of functional relevance of urban centres. Population of rural areas that participate in daily migration activities makes the functional area of the municipal centre and determine its gravitational area. In addition to the degree of urbanization of the area, Gross Domestic Product (GDP), population, and workforce, daily migrations are one of the key criteria used for the classification of functional urban areas in the Spatial Plan of the Republic of Serbia (i.e. international, national and regional ones) (Official Gazette of Blace Municipality No. 4/11).

The share of commuters at the level of the settlement, is determined on the basis of the total number of economically active population together with students and pupils, reduced by the agricultural population . In the Blace municipality, 49,65% (1.561) of workers, pupils and students commute. However, based

Table 6. Migration flows of commuters from Blace municipality by destination (adapted from the data of Statistical Office of the Republic of Serbia)

Total number of commuters	Towards municipal centre		Towards other settlements in the municipality		Towards other centres in Toplica District		Towards other centres outside the county	
	Number	%	Number	%	Number	%	Number	%
1561	1036	66,37	134	8,58	239	15,31	137	8,77

on the number of commuters that stay on the territory of their own municipality (where their residence is), the importance and strength of a municipal centre can be determined. In that sense, Blace urban settlement represents a strong gravitational centre within its own municipality towards which commutes the majority of municipal daily migrants (66,37%) as well as commuters from all settlements in the municipality. Due to functional dependencies, 74.95% of commuters (1.170) remains within the limits of the municipality, of which 66.37% gravitates towards the municipal center, while 8.58% commutes to other rural areas of the municipality. A small number of commuters from the Blace municipality, i.e. 15.31%, gravitates toward other centers of the Toplica District (Prokuplje and Kuršumlija), while 8.77% of them commutes to centers that are not part of the county (Niš, Kruševac, Brus). The direction of movement of commuters and quantitative indicators clearly show the importance of Blace as a strong center of gravity for all the rural areas of the municipality.

Determining the influence fields of the main urban centre was done on the basis of the intensity of commuting (Tošić, Krunić & Petrić, 2009):

1. **Zone of intensive influence** - over 70% of employees commute to the centre;
2. **Zone of strong influence** - 50-70% of employees migrate to the centre daily;
3. **Zone of medium influence** - 30-50% of employees commute to the centre;
4. **Areas of weak influence** - less than 30% of the employed migrates to the centre of work (there are three groups of settlements: 20-30% of people commuting, 10-20% of people commuting and 5-10% of the employed commuting to the centre), and
5. **Periphery of urban living system** - less than 5% of employees commute to the centre of the work.

In the zone of intensive influence of the municipal centre Blace, with the intensity of commuters from 70 to 100%, there are 10 settlements. In the zone of strong influence with the intensity of people commuting from 50 to 70%, there are 10 settlements as well, while 11 settlements records medium intensity (30-50%) of commuters. Areas of lesser influence (5-30% of commuters) were observed in 8 settlements. Areas that are close to the centre or those that have good accessibility provide most commuters to the main urban settlement, as evidenced by the spatial distribution of settlements in the zone of intensive influence. Zones of influence and spatial distribution of settlements in the gravitational field of Blace are displayed in the *Figure 2*.

Daily migrations are closely related to the hierarchy of settlements, and the fact that the influential spheres of functions of different settlements do not have the same significance and scope. The distance from the main urban centers and above all the importance of the centers in the realization of the function of work are some of the most important factors in the analysis of daily migrations.

Figure 2. Daily urban system of the Blace municipal centre

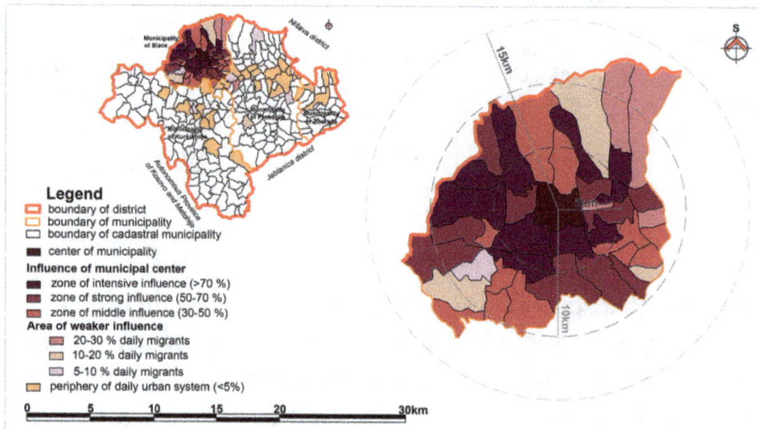

However, looking at the functional reach of the municipal centre from a regional perspective, and the intensity of its influence in the area of Toplica District, it can be seen that Blace has no force to develop and integrate a wider area. The field of influence with a stronger intensity is felt within the municipality, but with a much smaller one when it comes to settlements outside the municipal area. The development power and influence of Blace in lesser intensity is manifested in 22 settlements (20.56%) of Prokuplje municipality, from whose territory it attracts 4.27% (190) of commuters. Blace also attracts 10.15% (122) of commuters from 10 settlements (11.11%) of the municipality of Kuršumlija, while from the area of Žitorađa municipality, it attracts 0.22% (7) of commuters from 5 settlements in total (16.67). Majority of settlements in municipalities of Prokuplje, Kuršumlija and Žitorađa actually represent the periphery of urban living system of Blace.

In Blace Municipality there is a high concentration of both population and economic activities, as well as industrial capacities and services in the municipal center, so the flows of daily migrations have a dominant direction from other rural settlements to the urban center.

Daily migrations are a relevant quantitative and qualitative indicator of the spatial - functional integration and connection of rural areas to the main urban centre which serves as a potential driver and promoter of functional cohesion and transformation of the rural hinterland. Daily mobility of population reflects the demographic and socio-economic development, causing changes in the population structure and transformations in the labour market. The development of daily migrations enables the diversification of the activities of rural population, resulting in changing socio-economic structure which directly influences the transformation of rural areas and prevents the depopulation of the countryside.

FUNCTIONAL AND SOCIOECONOMIC TRANSFORMATION OF RURAL SETTLEMENTS AS CONSEQUENCES OF DAILY MIGRATIONS

On the basis of the parameters on the relationship of different activities and their respective sectors in the contingent of working population, a number of settlement classifications has been developed, which is dominated by the functional ones (Tošić, 2012). Model of activity sectors which was most often applied in Serbian spatial planning (*E.g., in cases of:* Stamenković, 1990; Tošić, 1999; Grčić, M. & Grčić, Lj.,

2002) is based on the method of Horst-Ferre's triangle (Ferre, 1961). The method was adapted to domestic conditions and the available data and includes abstracting settlement types based on the analysis of the population activity structure, i.e. share of activities of certain sectors in the working population. It was applied to determine the spatial and functional relationships that affect the functional transformation of settlements. The threshold for determining the dominant affiliation with a specific sector of activity is 60% (*Table 7*).

In the period from 1981 to 2011, there was a significant functional transformation of settlements in the Blace Municipality. Secondary and tertiary-quaternary activities have made the municipal centre a generator of change, from which these activities spread to rural areas. As a consequence, daily migrations caused deagrarization and rise of employment in the secondary and tertiary-quaternary activities. Thus, the process of diversification of the economy started and non-agricultural activities began to spread outside the municipal centre. Originally, it was the industry that was responsible for the development of this area, but at the same time, there was an increase in the share of service sector activities, which has become the driver of diversification after 2002. State roads also affected functional transformation of the settlements as, along these junctions, tertiary activities and small businesses have been developed, which classifies these settlements as a mixed type of agrarian settlements with the share of services and industrial activities.

In 1981, this was predominantly agrarian area, with 39 of the 40 settlements being distinctly agrarian function type. More than half, i.e. 66.48% of the population performed their activities in the primary sector. The fact that in 36 villages over 90% of the population was employed in the primary sector also tells about the dominance of agriculture in this area. However, these were mostly individuals who worked in the place of residence or on their own farm households. The influence of Blace, as the only industrial and service center of the municipality with a developed network of public services and partially developed infrastructure, started to be felt in the settlements of rural hinterland. In the first period of transformation, until 2002, suburban settlements found themselves under the influence of the municipal center, and this impact later spread to other rural areas. On the map of functional types of settlements from 2002 (*Figure 3*), it is noticeable that there was a small number of settlements that experienced functional transformation and these were mainly distributed around the municipal center or along state roads. However, in

Table 7. Model for defining specific types of settlements and their number in Blace Municipality

Functional type of settlements	Criterion	1981	2002	2011	Change
Agrarian	I >or = 60%	39	15	6	-33
Agro-industrial	I >II >III	0	13	8	+8
Agro-servicing	I >III> II	0	4	8	+8
Industrial	II >or = 60%	1	0	2	+1
Industrial-agrarian	II >I >III	0	2	0	0
Industrial-servicing	II >III> I	0	4	3	+3
Servicing	III>or = 60%	0	0	1	+1
Servicing-agrarian	III >I >II	0	0	4	+4
Servicing-industrial	III >II >I	0	2	8	+8
I – primary sector, II – secondary sector, and III – tertiary-quaternary sector of activities;					

Figure 3. Transformation of functional types of settlements in Blace Municipality

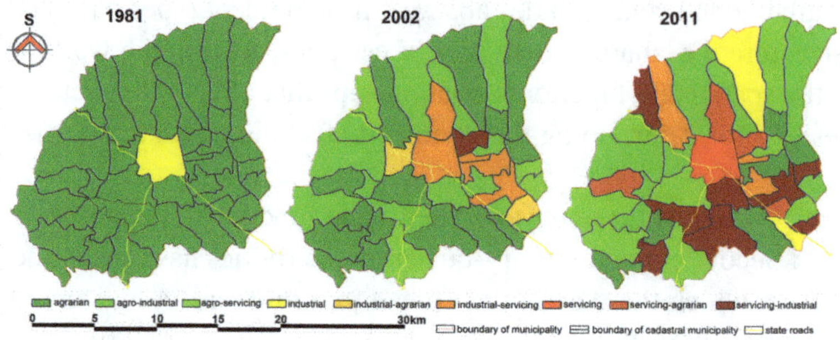

2002 most of the settlements still belonged to the agrarian type, i.e. 37.5% (15 settlements), but this is a significant decrease compared to 97.5% in 1981. The total of 8 settlements was transformed into one of the types of industrial or service settlements.

In the second phase of transformation, from 2002 to 2011, other distant settlements were affected by the municipal centre. Functional transformation was not recorded only in 6 villages that remained of a predominantly agrarian type, while in 34 settlements there was a change of functional type. In addition, another 16 settlements (40%) had a mixed function that was predominantly agrarian: 8 of them were agrarian-industrial, while the other 8 were of agrarian-service type. In 18 settlements, secondary or tertiary-quaternary sector was dominant in determining the functional type. The number of service-dominated settlements and mixed settlements with predominantly service function (service-industrial and service-agrarian) has significantly increased. There were not any settlements of this type back in 1981, while in 2002 there were only two of them, but in 2011 this number has risen to 13. The current structure of activities of population in Blace Municipality is shown in *Figure 4*.

Figure 4. Structure of activities in Blace Municipality (2011)

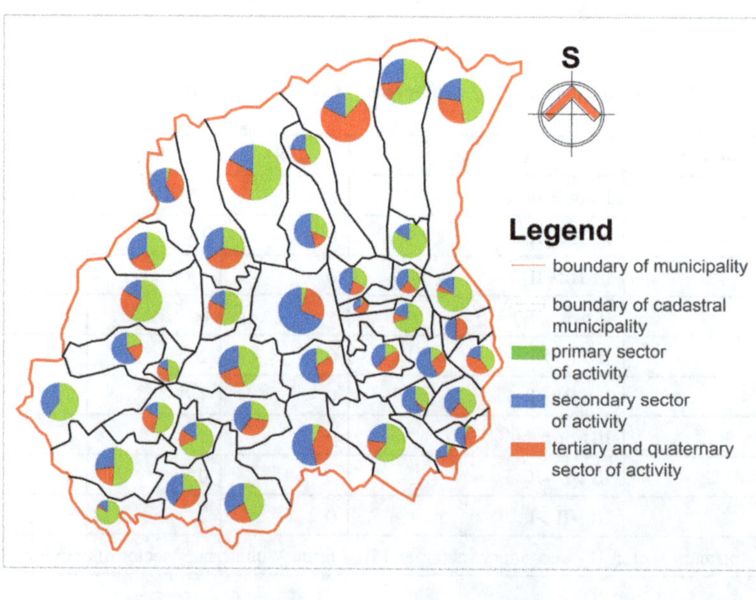

Table 8. Model for defining specific types of settlements and their number

Degree of urbanization	Participation of agricultural population in the total population (%)	Participation of households without agricultural holding in the total number of households (%)	Participation of employed persons in the active population (%)	1981	2002
Urban	≤10	≥ 70	≥ 70	0	0
More urbanized	≤15	≥ 20	≥ 50	0	2
Less urbanized	≤30	≥ 10	≥ 50	0	8
On the threshold of urbanization	satisfies two of the three conditions			1	8
Rural	doesn't satisfy two or all three conditions			39	22

The level of urbanity is established according to a set of indicators: share of agricultural population in the total population; share of non-farm households in the total number of households in the settlement; and the participation of the employed population in the total population (*Table 8*). Five groups of settlements are defined: urban, more urbanized, less urbanized, on the threshold of urbanity and rural settlements – villages (Tošić, 2004). Based on this model, it was found that the gradual spreading of urban influence from the main urban area to the surrounding villages has affected their transformation at the same time.

When applying this model to the municipality of Blace, it can be seen that in 1981 only municipal centre underwent a transformation, and that to a lesser extent, which classifies it as the village on the doorstep of urbanity. All other settlements were non-urban, rural settlements, which were 39 in total, or 97.5%. In the period from 1981 to 2002 the largest number of villages experienced a transformation from a rural to less urbanized (8) and to settlements on the border of urbanity (8). Two settlements that belonged to the category of rural moved into the category of more urbanized settlements. Number of villages (rural settlements) decreased by 17. Transfer from agricultural to non-agrarian activities had a crucial role in the socio-economic transformation of these settlements. Drivers of the transformation in the first stage were the processes of industrialization and urbanization. Due to urbanization induced by industrialization, functions and population started to be concentrated in the urban settlement, which required the development of a larger number of functions, while the increased population numbers facilitated the development of a variety of tertiary and quaternary activities, in order to meet their needs (Tošić, 2000). That makes tertiary and quaternary activities drivers of the transformation in the second phase.

The biggest changes took place in the settlements around the municipal center, while the degree of urbanity of individual settlements decreased when moving away from it. Settlements along state roads have also undergone transformation, while other settlements in the hinterland remained to be rural. They mostly boast households with a mixed source of income, while the number of agricultural population is being reduced, which indicates the increase in the level of urbanity of these settlements.

The degree of urbanity and functional development of the settlement can be also addressed through the network of settlements planned according to the current spatial plan of the municipality of Blace (on *Figure 5)* (Official Gazette of Blace Municipality No. 4/11). The spatial development of the settlement network is largely followed by the development of functions in the settlements of the Blace municipality (*Figure 2 and Figure 3*). The spatial plan of the municipality of Blace, the basis of the concepts of the

Figure 5. Types of settlements in Blace Municipality (2011) and the network of settlements in the Blace Municipality according to the Spatial Plan of Blace Municipality

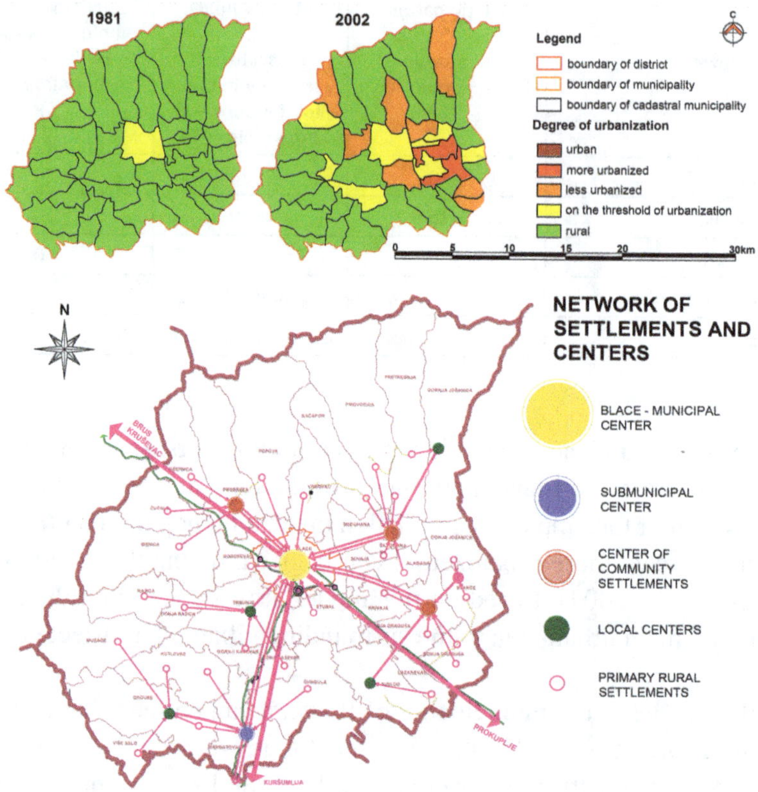

network of settlements, envisions communities of settlements[5], which allows the creation of complementary interests for the development of the whole area. The multistep hierarchy of the centers in the network of settlements is envisaged: one municipal / local center (the city settlement Blace), which is the center for all other settlements and towards which a sub-municipal center (Barbatovac) and the three centers of the settlement community (Međuhana-Siljomana, Prebreza and Gornja Dragusa) gravitate. These centers are further spatially and functionally connected with five local centers (Gornja Jošanica, Gornje Svarče, Suvi Do, Gornje Grgure, Trbunje) and 29 primary rural settlements without functions (Alabana, Brežani, Mest Selo, Vrbovac, Donja Draguša, Donja Jošanica, Donja Rašica, Donje Grgure, Donje Svarče, Drešnica, Đurevac, Kačapor, Krivaja, Kutlovac, Lazarevac, Mala Draguša, Međuhana, Muzaće, Popova, Pretežana, Prešnja, Pridvorica, Rašica, Sibnica, Stubal, Suvaja, Čungula, Čučale i Džepnica) . The network of settlements planned in this way has followed the trends in the development of the functions of the settlements and enabled the quality spatial and functional connection of the main urban settlement with the rest of the municipality.

CONCLUSION

Quality transport connections between the Blace Municipal Center and its rural hinterland, as well as low spatial and time distances, indicate that the municipal center is highly accessible to rural settlements, which is a prerequisite for the dominance of daily migrations instead of permanent emigration of rural population to the urban center. In accordance with the above, apart from daily (work) and administrative needs, the rural population has no need to leave their homes, and compensate for life in urban environment through daily migrations. The relative immutability of the place of residence affects the employment opportunities that are available at a daily distance from it.

The main aim of the paper is to examine the potential of small urban centers to generate spatial and functional integration, that is, to explore the possibility for small towns such as Blace Municipality can integrate and transform their rural environment to integrate and transform their rural surroundings. An analysis of commuting flows, housing patterns, and monitoring of functional and socio-economic transformation of settlements in the Blace Municipality, we came to the following conclusions:

- Blace is an underdeveloped settlement, which, due to the concentration of commercial and public facilities in the municipal center takes on the characteristics of the development of a city and attracts the majority of commuters from the territory of its municipality (66.37%) as well as a smaller number of them from the municipalities of the region;
- Daily migrations occurred as a result of the regression of agricultural activities and the development of secondary and tertiary-quaternary activities in the municipal centre;
- Daily mobility of rural population resulted in the transformation of the rural environment, both functional and socio-economic, which is reflected in: the diversification of occupations; prevention of depopulation in the countryside; decrease of employment in agriculture and of the percentage of households with farms; increase in households with a mixed source of income; development of services in rural areas; change in lifestyles; increase in the level of urbanity, which contributed to the overall development of the area.
- The most intensive functional and socio-economic transformation took place in the villages that were located along state roads and around the municipal center. Settlements in the hinterland remained of the agrarian and rural type.
- Functional transformation is reflected in reducing the number of agrarian settlements by 82.5%, the increase of the settlements of a service type and mixed settlements with predominantly service function (service-industrial and service-agricultural) by 32.5% and an increase in settlements of industrial and mixed type with predominantly industrial function by 10% .
- Socio-economic transformation is reflected in the reduction of rural areas by 42.5% and in the increase of settlements: at the threshold of urbanity by 17.5%, less urbanized by 20% and more urbanized by 5%.

Despite relatively limited economic and functional capacities, the sphere of influence of the Blace Municipal Center in the territory of its municipality prevails in relation to developed centres in the region (Prokuplje, Niš, Kruševac). Good transport links that reduce the length of travel time have caused an increase in daily migrations that have the municipal centre as a destination. Quantitative indicators presented in this paper show that even smaller urban settlement can adequately integrate their rural environment and be the drivers of spatial development, but also to instigate the transformation of rural hinterland.

Commuting is a clear proof of the existence of spatial and functional connectivity and integration of rural and urban settlements. Therefore, it is necessary to focus activities towards these settlements, strive for their functional and transport connection with the rural environment and settlements of a higher rank in a wider area, in order to encourage a polycentric territorial development and greater regional balance.

By accelerating and strengthening the spatial and functional synergy of the Blace Municipal Center and the rural settlements in its surroundings through the further development of daily migrations, it can influence the further qualitative transformation of the rural environment. This will undoubtedly lead to the rest of the population and make rural settlements vital. As this research has shown that traffic connectivity, temporal and spatial distances of settlements from the center, as well as the relative functional equipment of the center, have influenced the development of daily migration, these are factors that must be prioritized in the further planning of the municipal transport network and settlement network. Quantitative improvement of the municipal road network, building new roads to provide traffic access to even the outermost hamlets, and especially qualitative improvements to reduce travel time and improve the accessibility of central facilities, would lead to the preservation and potential increase in the intensity of daily migrations. Also, the development of local public transport and the organization of lines connecting the entire territory of the municipality to the municipal center would strengthen the connection between the urban center and the rural hinterland. This situation indicates that in the municipality when planning and organizing a settlement network, it should be taken into account that the former centers of the community of rural settlements have lost their position in the settlement hierarchy for several reasons. That is, spatial, traffic and other conditions enable rational spatial and functional-administrative functioning of the municipality in two hierarchical levels (urban settlement-rural settlements).

REFERENCES

Atkinson, R. (2001). The emerging 'urban agenda' and the European spatial development perspective: Towards an EU urban policy? *European Planning Studies*, *9*(3), 385–406. doi:10.1080/09654310120037630

Čavrak, V. (2016). *Razvoj metropolskih policentričnih regija i koncept pametnog grada*. Retrieved from: https://www.researchgate.net/publication/303413058_RAZVOJ_METROPOLSKIH_POLICENTRICNIH_REGIJA_I_KONCEPT_PAMETNOG_GRADA

Christaller, W. (1933). Die Zentralen Orte in Suddeutschland. Jena, Darmstadt: Wissenschaftliche Buchgesellschaft.

Cox, E., & Longlands, S. (2016). *City systems: The role of small and medium-sized towns and cities in growing the northern powerhouse*. IPPR.

Davoudi, S. (2003). European briefing: polycentricity in European spatial planning: from an analytical tool to a normative agenda. *European Planning Studies*, *11*(8), 979–999. doi:10.1080/0965431032000146169

Davoudi, S., & Stead, D. (2002). Urban-rural relationships: An introduction and brief history. *Built Environment*, *28*(4), 269–277.

Đerčan, B., Radaković, M., Ostojić, M., Mirković, M., Obrenov, S. & Vozar, J. (2017). Urban hierarchy in Serbia. *Researches Reviews of the Department of Geography, Tourism and Hotel Management*, *46*(1), 12-24.

Đorđević, J., & Tošić, B. (2013). *Regionalno planiranje i razvoj* [Regional Planning and Development]. Novi Sad: University of Novi Sad.

EC. (1999). European Spatial Development Perspective (ESDP). Towards Balanced and Sustainable Development of the Territory of the European Union. EC.

Government of the Republic of Serbia – GRS. (1995). Закон о планирању и уређењу простора и насеља [Law on Planning and arranging Settlements and Spaces of Republic of Serbia]. Belgrade: Official Gazette of Republic of Serbia No. 44/1995.

Government of the Republic of Serbia – GRS. (2003). Закон о планирању и изградњи [Law on Planning and Construction of Republic of Serbia]. Belgrade: Official Gazette of Republic of Serbia No. 47/2003.

Government of the Republic of Serbia – GRS. (2006). Уредба о управним округзима [Regulation on Administrative Districts of Republic of Serbia]. Belgrade: Official Gazette of Republic of Serbia No. 15/2006.

Government of the Republic of Serbia – GRS. (2010). Просторни план Републике Србије [Spatial Plan of the Republic of Serbia]. Belgrade: Official Gazette of Republic of Serbia No. 88/2010.

Government of the Republic of Serbia – GRS. (2014). Закон о планирању и изградњи [Law on Planning and Construction of Republic of Serbia]. Belgrade: Official Gazette of Republic of Serbia No. 72/2009, 81/2009, 64/2010, 24/2011, 121/2012, 42/2013, 50/2013, 98/2013, 132/2014 i 145/2014.

Government of the Republic of Serbia – GRS. (2014). Уредба о утврђивању јединствене листе развијености региона и јединица локалне самоуправе за 2014. годину [Regulation on establishing a single list of development of regions and local self-government units for 2014 of Republic of Serbia]. Belgrade: Official Gazette of Republic of Serbia No. 104/2014.

Grčić, M. (1999). Funkcionalna klasifikacija naselja Mačve, Šabačke posavine i Pocerine [Functional classification of the settlements of Mačva, Šabačka Posavina and Pocerina]. *Glasnik: Serbian Geographical Society*, *79*(1), 3–20.

Grčić, M., & Grčić, Lj. (2002). *Mačva, Šabavčka Posavina i Pocerina*. Belgrade: University of Belgrade - Faculty of Geography.

Jefferson, M. (1939). The Law of the Primate City. *Geographical Review*, *29*(2), 226–232. doi:10.2307/209944

Lukić, V. (2011). Commuting in the Settlement System of Serbia. *Stanovnistvo*, *49*(2), 25–50. doi:10.2298/STNV1102025L

Lukić, V. (2013). Veze između migracije i dnevne migracije [Links between migration and daily migration]. *Stanovnistvo*, *LI*(1), 69–90. doi:10.2298/STNV1301069L

Municipality of Blace. (2011). *Spatial plan of Blace Municipality* (Official Gazette of Blace Municipality No. 4/11). Blace, Serbia: Municipality of Blace.

Reitsma, R., & Vergossen, F. (1987). A Causal Typology of Migration. The Role of Commuting. *Regional Studies*, *22*(4), 331–340. doi:10.1080/00343408812331345010 PMID:12315224

Schmook, G. (1968). *Wiskundig Afgebakene Ommelanden en Hinterlanden van de Belgische Steden op de Basis van Geselektioneerde Diensten uit de Tertiare Sektor, Geogr.* Tijdschrift.

Stamenković, Đ. S. (2001). Geografska enciklopedija naselja Srbije, knj. I, II, III i IV [Geographical Encyclopedia of the Settlement of Serbia, Vol. I, II, III and IV]. University of Belgrade, Faculty of Geography, Belgrade:''Agena'' & ''Stručna knjiga'', Belgrade.

Statistical Office of the Republic of Serbia - SORS. (1982). *Municipalities in the Socialist Republic of Serbia 1981*. Belgrade: Statistical Office of the Socialist Republic of Serbia, Socialist Federal Republic of Yugoslavia.

Statistical Office of the Republic of Serbia - SORS. (2002). *Municipalities and Regions in the Republic of Serbia 2002*. Belgrade: Statistical Office of the Republic of Serbia, Republic of Serbia.

Statistical Office of the Republic of Serbia - SORS. (2011). *Municipalities and Regions in the Republic of Serbia 2011*. Belgrade: Statistical Office of the Republic of Serbia, Republic of Serbia.

Statistical Office of the Republic of Serbia - SORS. (2013). *2011 Census of Population, Households and Dwellings in the Republic of Serbia –2011 Census of population, households and dwellings in the Republic of Serbia; 23*. Belgrade: Statistical Office of the Republic of Serbia, Republic of Serbia.

Statistical Office of the Republic of Serbia - SORS. (2014). *2011 Census of Population, Households and Dwellings in the Republic of Serbia – Comparative Overview of the number of population in 1948, 1953, 1961, 1971, 1981, 1991, 2002 and 2011*. Belgrade: Statistical Office of the Republic of Serbia, Republic of Serbia.

Statistical Office of the Republic of Serbia - SORS. (2018). *Municipalities and Regions in the Republic of Serbia 2018*. Belgrade: Statistical Office of the Republic of Serbia, Republic of Serbia.

Tošić, D. (1999). Grad u regiji [City in the region]. *Glasnik, Glasnik. Geographical Society of Republic of Srpska, 4*, 47–85.

Tošić, D. (2000). Prostorno-funkcijske veze i odnosi u urbanim regijama [Spatial-functional connections and relations in urban regions]. *Arhitektura i urbanizam, 7*, 50-58.

Tošić, D. (2012). *Principi regionalizacije* [Principles of regionalization]. Belgrade: University of Belgrade - Faculty of Geography.

Tošić, D., Krunić, N. & Petrić, J. (2009). Dnevni urbani sistemi u funkciji prostorne organizacije Srbije [Daily urban systems in function of spatial organization of Serbia]. *Arhiterktura i urbanizam, 27*, 35-45.

Veljković, A., Jovanović, R., & Tošić, B. (1995). Gradovi Srbije centri razvoja u mreži naselja [Serbian cities development centers in the settlement network] (Special Edition of Book 44). Belgrade: Geographical Institute "Jovan Cvijić", Serbian Academy of Sciences and Arts.

Videnović, A. (2017). Obnova centara u ruralnim područjima [Renewal of centers in rural areas]. University of Belgrade - Faculty of Architecture, Belgrade: PRESSIA.

Vresk, M. (1990). *Grad u regionalnom i urbanom planiranju* [City in regional and urban planning]. Zagreb: Školska knjiga.

Vresk, M. (1990a). *Osnove urbane geografije* [An introduction to urban geography]. Zagreb: Školska knjiga.

Vresk, M. (2002). *Grad i urbanizacija* [City and urbanization]. Zagreb: Školska knjiga.

Wheeler, J. O. (1971). Commuting and the rural nonfarm population. *The Professional Geographer, 23*(2), 118–122. doi:10.1111/j.0033-0124.1971.00118.x

Wheeler, J. O., & Stutz, F. P. (1971). Spatial dimensions of urban social travel. *Annals of the Association of American Geographers, 61*(2), 371–386. doi:10.1111/j.1467-8306.1971.tb00789.x

Zonneveld, W., & Stead, D. (2007). European territorial cooperation and the concept of urban–rural relationships. *Planning Practice and Research, 22*(3), 439–453. doi:10.1080/02697450701666787

ENDNOTES

1 According to the ESPON classification, a functional urban area (FUA) represents a changing space that is constituted by a morphological urban area (MUA) of a city/settlement and its wider surroundings that generate a labour force of the city within the 45-minute isochron from the place of residence.

2 The term Polycentric Development refers to spatial development based on a hierarchically structured network of centers in a given area. The concept of polycentric development emphasizes the importance of the hierarchical structure of the network of centers in each country or province. It is based on the theory of central settlements, introduced by Walter Christaller (1933) in the 1930s, trying to explain primarily the role of service activities in society and the distribution of settlements and cities in space.

3 The Nomenclature of Territorial Units for Statistics or the Nomenclature of Units for Territorial Statistics (NUTS) is the European Geosciences Standard (Geospatial Entity Object Code) used for statistical purposes and by which the division of Europe into administrative regions has been made. In the Republic of Serbia, the division of NUTS was made according to the Decree on the Nomenclature of Statistical Territorial Units, however, although the European Union standard prescribes that the criteria for division by NUTS shall be based on the number of inhabitants, geopolitical position, natural resources, existing territorial organization, and cultural and historical heritage, in Serbia, the said decision covers only the existing territorial organization and a partial population.

4 Capital - The City of Belgrade is considered as an area within the NUTS3 but does not count in the administrative districts because it has a special status of the capital city.

5 Groups of settlements in a gravitational zone of a larger and more significant rural or urban settlement.

Chapter 5
Changing Territories and the "Edge Effect" of the Re-Habited Industrial Structures

Jasmina Siljanoska
https://orcid.org/0000-0001-7061-9771
Faculty of Architecture, University Ss. Cyril and Methodius, North Macedonia

Goran Mickovski
Faculty of Architecture, University Ss. Cyril and Methodius, North Macedonia

Ivana Angelova
Faculty of Architecture, University Ss. Cyril and Methodius, North Macedonia

ABSTRACT

This chapter will examine the transition zones around the borderline of the legal perimeter defined by law and master planning of Skopje that have always played a role of interactive borderland in which the concept of interactivity and urban-rural-exurban synergy is applicable. Of particular interest is how edge position of certain abandoned structures, industrial areas in particular, in the period of post-socialist transition have affected use and movement patterns changing. These structures have originally influenced the surrounding area by their users, and now they are largely neglected or underused and need interventions of rehabilitation into beneficial use. The authors support the view that there is significance in conceiving of a "edge effect" of the re-habilitated social and industrial fringe structures. They find that they should be given more consideration in the context of new economies and innovative technologies which will enable rehabilitation and adaptive reuse of the existing premises.

DOI: 10.4018/978-1-5225-9932-6.ch005

INTRODUCTION

Modern societies are constantly exposed not only to their own internal dynamics, but also to the impact of global changes, embracing an ideology of multiple urban transformations. This rapid transformation, economically and socially, as well as spatially, is shaping the new character and gradually is becoming recognizable in almost all the cities in the region of Western Balkans.

Within the context of the contemporary diffuse metropolis the concept of boundary changes and is materialized in new urban morphologies. The boundary between the end of the city and the beginning of nature or urban/rural fringe is subtle and cannot always be discerned. It defines discontinuous route which gradually perforated casing urban structure of the city as a direct influential on the development of the city and the dialectical relationship between inward and outward (the different typologies and morphologies). On the other hand, it is interesting from typological and morphological point of view to consider the constitution of space outside of the urban network, through the phenomenon of suburbanization, in order to clarify the vague spatial morphology and landscape between urban and suburban, as well as rural areas.

The method of forming the edges of the city, its quality and the subsequent consequences, is basically a result of the social system and spatial organization in today's cities. The concept of edge itself can be defined as both borderline of division and meeting place that allows interaction. Border lands besides the spatial aspect concerns include the social aspects, as well, such as segregation or integration between city residents and their activities and the residents of the exurban-suburban territories on the fringes.

The border space defined as borderland has been always an interest of urban planning in providing ways of how porosity and resistance can combine and defines both legal entities and limits, but also allows interactivity, porosity, flexibility and confluence of the urban and extra-urban territories and realities.

Skopje is a city with changing planning paradigms and conceptual misalignment of different periods of development. As a result of a rapid urbanization of Skopje during the XX century and constant increase of the territory former lines that define the border of the city are assimilated into its urban tissue and suburbs. The combination of its structural parts is no longer defined as homogenous mutually harmonious system, but rather its borders, barriers, edges are subject to a constant process of modification and reconfiguration of the dynamic image. Multiple layering and heterogeneity are present in the context of the changes within the social structure, to the interpretation on the architectural identity of the whole.

Comprehensive changes, after the establishment of new political and economic system in the early 1990s, inevitably caused changes in the urban structure inherited from the socialist period and again become the subject of transformation. The planning structures of public amenities were first to bear the negative consequences. Especially affected with these processes were small cultural centers and amenities with suburban position as part of the decentralized model of city planning. Even more affected were the border industrial zones allocated in the period of the industrialization process in the City of Skopje after the WWII, when industry was dominant in the economy and when Skopje has a status of the first industrial center with about 1/3 of the total industrial capacity of the Republic. The transformation confirms multifunctional fragmentation of the former industrial zones in the city. Both functional and spatial structures described above have lost their position in the urban fabric and meaning as important generator of activities and demographic exchange and interconnectivity of the urban-suburban-exurban territories.

Rehabilitation of these sites might enable redevelopment of the urban tissue of the post-socialist city, although it has been proven that in many cases it is not always possible and easy to implement. Unfortunately, the logic of capitalist economic restructuring of the industrial spaces is in the most of the cases inclined to destruction of the existing industrial structures and to building new capacities in newly designated industrial areas. The industries on the border which might have been treated as architectural heritage and rehabilitated by applying adaptive re-use method are rare in the dominantly negative practice. Creation of new policies and strategies for particular border land zones restructuring into places for beneficial use and development, such as creative industries, is necessitated.

We are interested in the transformation of industrial urban space in Skopje in the post-socialist period, and in particular in the industrial spaces' transformation of the abandoned structures on the fringe, some of them undergo intensive spatial restructuring of production activities, due to the effect of the processes of re-industrialization after the massive de-industrialization. The emergence of new special forms of production activities in transition period in Skopje, in the context of new economies and technologically beneficent development, and the new spatial distribution of industry appeared as a way the city should change its future development.

This is becoming the main research question. Our intention is to point on the values and possibilities of the socialist industry legacy, which should be reconsidered by the urban planners and developers, as a potential with values and advantages for future redevelopment and re-use of the abandoned industrial facilities. The multifunctional transformation of industry, as shown in the empirical research, verify the notion that production activity is no longer dominant factor for growth and development of the city. The new emerging industrial sites in Skopje, no matter of their position within or outside the urban area, demonstrate the tendency of decentralization of industry, concentrating economic activities this way into new focal points and poles of urban growth and development. The case study research emphases the context of new economic activities introduced into the socialist industrial structures and locations, by emphasizing their flexibility of production and organization, and the new cooperative relations, transactions and interactions between the organization, actors and the specific urban place.

Pursuing the main question in the research within the methodology involved three main subchapters are developed: Theoretical framework; Empirical research: Dynamics of borderland transformation and Project research: The case of factory "Kuprom".

THEORETICAL FRAMEWORK

From Mass to Flexible Production

The economic activity in capitalist and socialist countries in the XX century especially after the WWII took place within the regime of mass production that was supported by mass consumption, full employment and wide-range of social measures and politics of the state of welfare.

In the early 1970s capitalist countries after decades of continuous growth experienced period of uncertainty and encountered number of problems. Many authors tried to elaborate the reasons for the downturn. For Harvey (1990) the main cause was the rigidity of the system as its internal weakness, while Scott (1988) perceived the causes not only in the internal, but also in the external factors, such as possible alternatives for mass production in the newly industrialized countries.

The crisis of the system of mass production introduced the new model of production. Scott and Storper (1990) argued that the economic restructuring caused by the manufacturers search for "new models of industrial organization, labor relations and sites" led to "decentralization of branch plants to the periphery at a low cost" (p. 21). For Harvey and Scott (as cited in Soja, 2000) the new production model, so called "flexible specialization", describe the contemporary era of urban-industrial capitalism as period of transition from regime of mass production to regime of flexible accumulation. Flexibility generated changes in three interconnected areas of the production process: organizational, technological and spatial. Changes in the organization of production relate to the flexibility of the production process; changes in the technology of production relate to the introduction of new computer technologies, automated production lines, as well as many strategies for reducing production costs; while changes in the spatial aspect of production relate to disconnection of the production site and the production process, the decentralization and dispersion of industry and creation of new forms of organization of space.

The search for new industrial sites created new forms of the city or, as Soja (2000) explains, "... new clusters of industrial production, many located in sites remote from the old urban-industrial core, becoming the most symbolic demonstration of what could be redeveloped in the geographical anatomy of modern industrial urbanism" (p. 163).

In the socialist countries the means of production were in state or societal ownership while the market mechanism was replaced by state centralist or self-governing planning. Nevertheless, the production activity was fundamentally based on the same paradigms of the capitalist industrial society: economies of scale, industrial gigantism, centralization of production and decision-making and vertical integration of production process. Henry Lefebvre (1991) criticizing the socialist countries for not creating a proprietary model of production, equates the both models of industrial production. According to him, "State socialism aims to perfect capitalist growth strategies, supporting only the proven advantages of large-scale enterprises and big cities, which are at the same time large production centers and large centers of political power" (Lefebvre, 1991, p. 155). The symptoms of crisis and stagnation of industrial production appeared in late 1960s, but the rational restructuring of socialist production would lead to complete dismantle of socialist economic system. Hence, the industry continued to operate in highly irrational environment and conditions where the socialist ideology completely suppressed market rationality as driving force of change. Kumar (2005) notes that "Socialism was and remained until the very end obsessed with size, centralized planning, dampening of diversity and suppressing the momentum of centralism and authoritarianism" (p. 77).

Changes introduced with the decay of political and economic system in 1989 triggered the transformation of the socialist industry. While transformation of industry in capitalist countries took place in some sectors characterized by intensive workforce, high energy consumption and environment degradation, in socialist countries transformation was comprehensive, for all industries and for each individual enterprise. The transformation on the principles of free market economy has caused the largest number of state-owned industrial enterprises to either drastically reduce their volume (production and labor) or to close completely. Socialist industrial enterprises with outdated technology and uncompetitive, often very expensive and low-quality products could not withstand the imposed harsh conditions of competition driven by trade liberalization and opening of the countries to the world, while losing the domestic and traditional Eastern European markets. Accordingly, deindustrialization was the main outcome of socialist industry transformation. Yet, deindustrialization should not be regarded as destructive tendency of permanent lost of industry. It is necessity to enable the processes of tertiarization (diversification of services), re-industrialization (emergence of a new industry) and renewal of industry (changes and modernization in the existing industry) (Kiss, 2002).

Just as socialist societies and cities were not built overnight, it could not be expected that the post-socialist transformation would end expressly. As Hamilton et al. (2005b) argues "Transition does not take place in vacuum, and the role of the complex historical heritage that still influences the expectations and behaviors of different social groups cannot be ruled out" (p. 11). In this sense the development of post-socialist city Hamilton et al. (2005b) links with the "path dependency" or the three-tier historical legacy from pre-socialism, socialism and post-socialism that affects the overall success of reforms, the spatial model of development, and the urban trends of transformation in the post-socialist period.

The question is why some of socialist countries have been more successful in this reform process and achieved the transformation in relatively short period, while others are still behind, trying to complete the two-and-a-half-decade transition? Hamilton and Carter (2005a) explain that the outcome depends on the plurality and interplay of the following internal and external forces: (1) the government commitment to implement real institutional and market reforms, (2) the continuity of socialist culture, economy, organization and society, (3) the embedded legacies of pre-socialist socio-cultural and behavioral environments, and (4) the globalization forces, i.e. the inflows of capital, ideas, information, innovation, knowledge, technology and trade. In addition, Hamilton and Carter (2005a) emphasizes that the successfulness of the post-socialist city transition is determined by the geographical space itself that is subject to transformation, i.e. the location of cities in particular geographical regions of Europe. The rapid and inclusive transformation of socialist cities is characteristic for those positioned in areas "... adjacent to, or more accessible from, the European Union" (Hamilton & Carter, 2005a. p. 117). As the "... distance from the European Union eastward and south-eastwards into the Balkans and East Europe" increases, the transformation of socialist cities is slower and takes place in individual segments (p. 117). The North Macedonia and the city of Skopje, unfortunately are not a 'the best case scenario' of socialist transformation.

For Tsenkova (2006) the three aspects of transition "... to democracy (systemic political change), to market economies (systemic economic change), and to decentralized systems of local governance" (p. 24) at city level induced economic change, social change, change in urban governance and spatial change (transformation of spaces of production and consumption). Nedović-Budić et al. (2006) elaborates that spatial outcome of post-socialist transformation is "... rich mosaic of urban landscapes and experiences, bustling with innovative practices and creative solutions aiming to achieve higher standards of urban development, urban sustainability, and improve the quality of life for their inhabitants" (p. 15).

The spatial outcome of 'dysfunctional socialist economic policy' as Lux (2008) argues formed monofunctional urban centers where industry was primary and dominant economic activity, and transformation of socialist industry produced large abandoned industrial areas inside the urban territory. The reuse of these abandoned industrial sites would redevelop the post-socialist city inner space, although many empirical studies proved that not always this is feasible and easily achievable. In Stanilov's (2007) opinion, the so called 'push' factors limit the reuse of abandoned industrial sites: (1) complex ownership structures, with large number of small owners or unfinished privatization of the location, (2) the premises and infrastructure on the locations are unusable for modern production technologies, (3) the costs for cleaning-up the environment are potentially very high due to evident pollution of the locations, and (4) in the first years of transition, lack of real estate market and land price for the locations in the center and periphery of the city.

In post-socialist period following the capitalist economic restructuring logic, the new industrial locations are simultaneously created through destruction of the existing industrial structures and by building new capacities in new industrial areas. In the territory of the city, the new industries can be located either in the existing industrial sites or in the new locations beyond urban borders. Drivers for decentralization of industry in the post-socialist cities, besides the push factors elaborated previously, considered by Stanilov (2007) are the availability of large territories of rural land at low cost at the fringe of the urban territory and the new pattern of accessibility for rural and urban contexts, i.e. the proximity of locations, travel time and distances. For Barta (2006) spatial changes in post-socialist cities were result of processes of economic restructuring and economic suburbanization. The economic restructuring or replacement of production function with services and other non-industrial activities caused the segmentation of urban and industrial space. Formerly unique industrial areas in the city became 'mosaic in character'. At the same time, they are 'transience in character', since the scale and variety of functional use of these locations is temporary. Economic suburbanization is related with the "... occupation of space by the new economy, with multiplication of economic actors and the merger of enterprises into agglomerations and their networking and clustering …" (Barta, 2006, p. 58), thus increasing the demand for rural land, especially on the urban fringe.

What has to be particularly emphasized in the post-socialist transformation of industry is that the new industrial spaces in the cities cannot be equated with the new industrial spaces in capitalist economies. As Lux (2008) points out "... the dichotomy between old and new industries is a questionable one: even the new greenfield investments are closely linked to the previous production milieus, while old industries have often shown ability to be revitalized" (p. 5). The continuity of locating industry in or near the urban areas of post-socialist city has to be re-assessed for each city. Generally, as Lux (2008) notes "... the spatial structure of industry of Central Europe today is a patchwork of continuity and change, where convergence towards the core is just as possible as the recreation of traditional peripheral relationships" (p. 5). Although the new model of production is introduced, where manufacturing activity is based on knowledge, high technologies and innovations, in post-socialist countries, generally, the traditional model of production is dominant, since the competitive advantage derives from exploitation of raw materials, low or medium level of technology and cheap skilled labor force.

New Economy and the SenseAble City

As already elaborated, after the establishment of new political and economic system in the early 1990s, comprehensive changes in the urban structure happened. In context of current post-socialist transformation of cities in Western Balkan, the notion of Ann Markusen (1996) describes, at best, the lack of strategies for their redevelopment and reintegration into regional and global flows.

The post-socialist development is reflection of the changes in the economic system and the accession to global flows and networks. In today's connected world it is important to look at the global market changes and urban tendencies in order to predict the implications on the local context.

According to Soja (1989), the restructuring of the urban space in recent decades is the last of a series of changes that occurred in capitalist cities. Analyzing the traces of changes across the history of cities in the last 200 years, it can be argued that "… evolution of urban form has followed the same periodical rhythm of crisis-induced formation and reformation that has shaped the macro-geographical landscape of capital since the beginnings of large-scale industrialization" (Soja, 1989, p. 173). He recognizes five restructuring periods in which the evolution of urban form and the cities were changed: Mercantile City from 1770 to 1820, Competitive Industrial Capitalist City from 1820 to 1870, Corporate-Monopoly Capitalist City from 1870 till 1920, Fordist City from 1920 till 1970, and Post Fordist from 1970 until now. If we succeed in being informed by past periods of restructuring, as Soja (1989) says, "… we can outline some provisional expectations for the upcoming period" (p. 183).

Three decades after Soya's statement was made, we need to agree that global trends in energy consumption, climate change, aging population and migrations impact the concept of nowadays capitalism and make it unsustainable and ineffective in the long run as Mason (2016), Rifkin (2014), and Drucker (1993) discuss. The immensely complex system of the free market, as it is known today, is yet about to start changing and transforming itself because of many external and internal pressures. The advancements in automation will most certainly contribute to reduced need of labor work and change employment relationships, thus new social partnership between capital and labor is expected. In this process the new information technology plays a crucial role in the new economy. The question is: where is the next economic cycle taking us and how we can plan our cities now to be more resilient to keep up with the change?

Products, services and establishments do not play to the market commandments and the managerial hierarchy as they formerly did and in the present a spontaneous rise in the collaborative production is evident (Mason, 2016). An entire business subculture wherein new types of loaning, new lawful contracts, and many new ways of which people take ownership over goods have risen in the course of recent years. The media has named those appearances as 'sharing economy'. There are many positive repercussions deriving from this economy taking the aspect of social, economic and environmental sustainability: whereas "the improving employment, safety and effectiveness ... cost-saving and innovation for yielding economic benefits...helps in allocating resources and decreases the emission of greenhouse gases" (Wu and Zhi, 2016, p. 194).

The sharing economy combined with the artificial intelligence, virtual reality and automation in the production industries influence the built environment. It has become important for the city and its structure to keep up with the innovation in technology. We witness changes through the reuse and re-programming of old factories from production of goods and services to production of knowledge, culture and leisure, reshaping into new hybrid objects which meet the needs of citizens and the new economies, as well.

What moves capitalism in today's world is the abundance of information and it's free sharing, and as Rifkin arguments moving towards post capitalism (Rifkin, 2014). According to Mason (2016) the main driver of change, 'agent of change' is the 'educated and connected human being', implying the highest rate of tertiary educated humans in history. He argues that this 'connected, educated human' will find new effective ways to bring change. Florida's (2005) concern with why some regions prosper in relation to others has led him to conclude that it is all up to what he calls 'the creative class' as cities are vessels of creativity. Is the creative class that is a part of the creative sector: science and engineering, nascent technologies, the innovation-based enterprises in art, music and culture or in the knowledge based economy of medicinal services, finance and law that all together comprises of nearly one third of world economy. Florida's response to why some places are able to secure a greater quantity or quality of the creative flows is the openness, diversity, and a tolerance of a place. He is hypothesizing that where synergy between nascent technology and talent-nurturing places open to immigrants, artists, marginal groups and racial integration exists there is economic prosperity for all. The above mentioned 'creative class' (Florida, 2005) or 'the agents of change' (Mason, 2016) is a group leading the changes toward a more sustainable and democratic system and the sharing-economy activities that they are engaged in is a principal tool for urban regeneration and seems so be clustering regionally (Stevens, 2015).

However, the concept has attracted some criticism among scientists this class being used only as a magnet for market growth that is more often than not sustainable and leads to even more higher-class division and gentrification (Pratt, 2008). Taking into consideration the multilayered context of Skopje with big diversity in cultural, religious and social status, it is vital for all strata of people, and not just the 'creative class', to be engaged in the new economy of sharing where flexible processes of production and organization replace old industries of the past. This flexibility allows the possibility for all, being young or old, unemployed or highly educated, to actively engage in their community in a different most appropriate way for them.

Nevertheless, the trend of cities striving to become more 'creative' continues with new supporting policies and strategies. Such arrangements have been formed to extraordinary stretch out to the significance of a place for societal advancement, and greatly for the creation of innovative businesses. Stevens (2015), in his research on how the urbanity develop the creative clusters, comes to the conclusion that creative industries appear to bundle between boundaries of neighborhoods or on the edge of the city. The "zone in transition" was identified long ago by Burgess as "an interstitial area in the throes of change from residence to business and industry" which is home to a "mobile and mixed population of youth and old age, aspiring and defeated individuals, pleasure-seeking Bohemians and hard-working students ... and radical freethinkers" (as cited in Stevens, 2015, p. 1). The in-between urban-rural zone is full of idiosyncratic attributes like "typified by mixed land use, aging structures, general instability, and change, and by a wide range in type and quality of functions" (Preston as cited in Stevens, 2015,p. 1). Many regions have already invested in different combinations of technology exchange and commercialization, higher education and green industry partnerships, entrepreneurial incubation, and funding so as to keep and absorb the talented citizens (Florida, 2005).

The unique landscape of urban and rural convergence in Skopje can be identified as the above mentioned 'zone in transition'. It is the zone where citizens of different ethnical and social backgrounds meet. Once being re-evaluated and re-considered by the planners, investors and local authorities, the neglected places of abandoned socialist industry facilities on the outskirts of the city might become potential synergy points of human interaction and collaboration.

The other concept introduced is the concept of SenseAble city. When considering urbanites there is agreement on general standard for city quality of smartness. The former popular concept of a smart city integrates the ICT (information and communication technologies) infrastructure together with the human, social and relational capital in order to improve the environmental quality. Recent analysis show there are some ill points in the smart cities such as the formation of agglomerations, a collection of urban voids prone to ghettoizing with poor communication between citizens and institutions (Greco & Bencardino, 2012). This situation could be overcome if the idea of 'equity' is added, which leads to the SenseAble city of equality and inclusion of all parts and all citizens. With the social inclusion an equal power is given to every participant, and is avoided prioritization of one group over the other. The sensitive city promotes informed and fair discussions on urban development, a new approach to urban planning and the effective usage of networks (Greco & Bencardino, 2012). This is a shifting paradigm that should be used to affect the way cities are planned and governed. It is a paradigm that use data innovation, artificial intelligence along with human consciousness in order to advance the decision-making processes and strategies to use the available resources in the urban services delivery. This city is able to sense, understand and respond to changes in its environments.

If we add the "SenseAbility" to the concept of the "creative class", as a prerogative for urban regeneration, we can overcome the negative characteristics of regional competitiveness and class exclusion. It is important for the city of Skopje to emerge into a sensitive city of the future, but is also a challenge for the urban planning to point on development based on autonomous systems of energy and self-sufficiency. As part of the strategy to go "SenseAble" the urban-rural synergy development on the outskirts of the city should be taken into consideration, as well. The abandoned production facilities should bear a program based on the new economy as a result of transitioning from manufacturing-based economy to a service-based economy, such as business incubators and clusters, or establishment of innovation centers and technology parks, etc. This way, the concept of sensible city and new economies are seen as plausible way the city should change its future development. The new economy and technological performances are perceived as a new opportunity for rehabilitation of the social and industrial structures in revealing the possibilities for redefining the use and functionality of the abandoned and unused facilities on the city border. The answer to how those abandoned structures could rehabilitate and re-utilize, using different innovative scenarios, is provided in the case-studies project research subchapter.

EMPIRICAL RESEARCH: DYNAMICS OF BORDERLAND TRANSFORMATION IN THE CITY OF SKOPJE

Rapid expansion of socialist industry in Skopje lead to increase of industrial production output, industrial employment, diversification of production and general industrialization and urbanization of the city. However, the qualitative aspect of socialist industry assessed through industry's structure, production range and technical and technological features of production process confirm the notion that socialist industry in Skopje could not generate significant growth, rather continuity of the reached level of development (Mickovski, 2016).

Political and economic changes in the 1990s induced transformation of socialist industry. The industrial enterprises in Skopje transformed in several ways. There were companies that ceased to exist, closed the plants and suspended production activities, leaving behind abandoned buildings and locations (deindustrialization). Other continued with the production activity organized either within single enterprise, or within units of former industrial giants (industry renewal and reindustrialization). It is not insignificant the number of companies that changed its activity from production to other economic activitiy, mostly trade (functional transformation). Overall the common feature of socialist industry transformation in Skopje was drastic and dramatic decrease of industrial output and employment. The volume of industrial production in the country dropped from 100 index points in 1989 (base year) to 47.9 index points in 2010 (State Statistical Office, 2011), once mostly important factor of economic and urban development of Skopje, in the new business environment transforms at slower pace, thus becoming the most important factor for slow transformation of the entire society. The success of restructuring of socialist manufacturing activities in Skopje as Hamilton et al. (2005b) denote was "… influenced by the break-up of Yugoslavia, embargoes on trade with Serbia, and refugee and ethnic problems, not to mention political isolation from Greece. This environment of instability or potential instability combined with geographic isolation fostered specific conditions of city transformation …" (p. 13).

The transformation of socialist industry triggered transformation of industrial locations and urban space. Locations formerly used for manufacturing activity experienced deindustrialization, functional transformation, industry renewal or reindustrialization. Table 1 below summarizes the socialist industry transformation, present in Skopje as of year-end 2013, partially presented in Sinergi Project, Conference: Inclusive/Exclusive cities, held in Skopje (2016).

According to data in Table 1 it can be concluded that dominant outcome of of socialist industry transformation is deindustrialization (443.9ha of abandoned industry) representing 51.48% of the total industrial area in Skopje (862.4ha). Renewal of industry, i.e. continuity of industry on socialist locations is present in 366,18ha (42,46%), while functional transformation, i.e. introduction of non-production

Table 1. Overview of socialist industry transformation in Skopje, as at 2013 (Мицковски, 2015)

Industrial zone Transformation	West zone	Southeast zone	Rudnici and Zelezarnica	Northeast zone	Scattered industry	Total area by transformation (ha)
Deindustrialization (ha)	9,35	117,08	166,10	107,71	43,67	443,93
(% of Total zone area)	(12,92%)	(51,27%)	(60,51%)	(52,47%)	(53,32%)	(51,48%)
Functional transformation (ha)	1,79	3,21	0	12,07	19,03	36,11
(% of Total zone area)	(2,47%)	(1,41%)	(0%)	(5,88%)	(23,24%)	(4,19%)
Industry renewal (ha)	60,24	107,46	108,38	73,26	16,84	366,18
(% of Total zone area)	(83,26%)	(47,06%)	(39,49%)	(35,69%)	(20,56%)	(42,46%)
Reindustrialization (ha)	0,97	0,6	0	12,23	2,36	16,16
(% of Total zone area)	(1,34%)	(0,26%)	(0%)	(5,96%)	(2,88%)	(1,87%)
Total zone area (ha)	72,36	228,35	274,48	205,29	81,88	862,39

functions on former manufacturing locations and reindustrialization, i.e. new industrial sectors/production activities on locations formerly supporting the manufacturing activity, attribute to 36,11ha (4,19%) and 16,16ha (1,87%) respectively. The transformation of socialist industry in Skopje, in post-socialist period is illustrated in Figure 1.

Figure 1 reveals that the spatial transformation of industry in Skopje is fragmented and uneven and that deindustrialization is its dominant outcome. The abandoned industrial sites are located throughout the whole urban territory, but scattered industrial sites on the urban fringe are dominant. The renewal of industry and reindustrialization occur in areas with inherited socialist industrial activities and outside the city, in areas without any previous economic activity. The new industry is organized predominantly

Figure 1. Transformation of socialist industry in Skopje, 1991- 2013 (Мицковски, 2015)

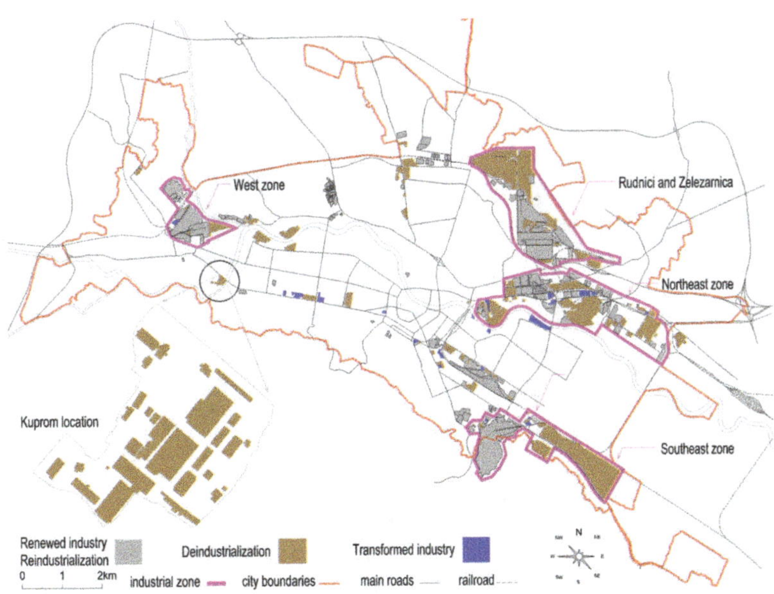

in commercial-economic zones and industrial-technological zones, being the new spatial forms of commercial and production activities (Figure 2).

The transformation of industrial locations, though very modest in scope, happened on smaller sized locations positioned in the central city areas surrounded by residential, commercial and recreational urban functions. Opposite, bigger industrial locations positioned in the industrial areas surrounded by manufacturing activities and services supporting industry (warehouses/distribution centers etc.) are transforming at slower rate (Mickovski, 2016). According to Kiss (2009) there are two ways to reuse former industrial locations. The first, so called 'adaptive reuse' refers to implementing new functions through renovation and restoration by creating necessary conditions for performing new activities. The second refers to 'urban renewal', demolition and complete removal of all production structures and premises and building a new facilities to meet new function requirements. Through adaptive reuse the existing industrial buildings remain unchanged and modification takes place within the inner space of premises, with no alterations of urban landscape. Urban renewal is more radical approach and function change is achieved through building of completely new facilities, thus determining entire transformation of location and space. The functional transformation of socialist industry locations in Skopje is shown in Table 2.

The data in Table 2 show that transformation of locations predominantly is made through adaptive re-use 21,83ha of transformed 36,11ha (no. of locations 18 out of 25), whereas transformed locations were positioned mainly in industrial zones 17,08ha (no. of locations 14 out of 18).

Identification of abandoned industrial locations within urban territories and borderline areas is crucial for the possibility for their re-use, since these locations are 'the golden reserves' for the future city growth (Kiss, 1989). Abandoned buildings and industrial sites are significant source of urban land that can be used for urban functions most needed in terms of urban living, whereas their transformation

Figure 2. Planned industrial zones in General urban plan for the city of Skopje 2012 -2022 and new industrial locations (Мицковски, 2015)

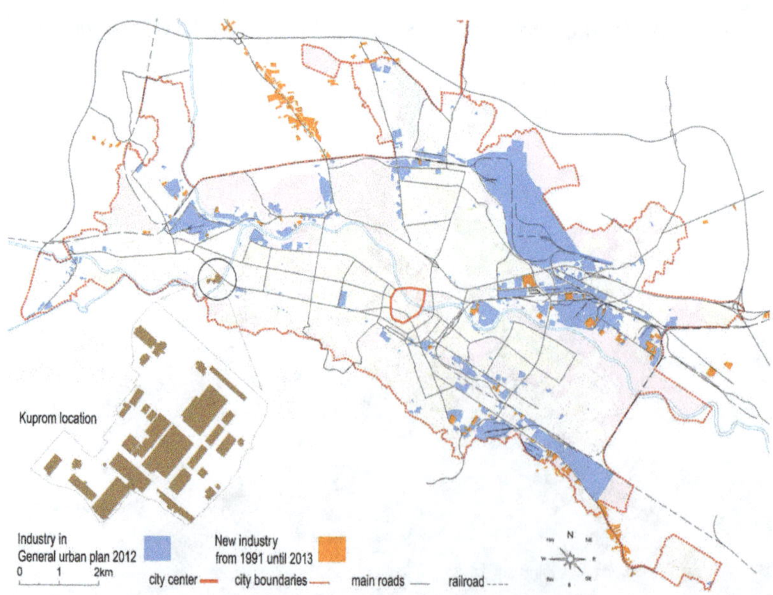

Table 2. Overview of functional transformation of socialist industry in Skopje, as at 2013 (Мицковски, 2015)

Industrial zone Transformation	West zone	Southeast zone	Rudnici and Zelezarnica	Northeast zone	Scattered industry	Total area by transformation (ha)
Urban renewal(ha) (no. of locations)	0	0	0	1,61	12,67	14,28
	(0)	(0)	(0)	(1)	(6)	(7)
Adaptive reuse (ha) (no. of locations)	1,79	3,21	0	10,47	6,36	21,83
	(2)	(5)	(0)	(7)	(4)	(18)
Total functional transformation (ha) (no. of locations)	1,79	3,21	0	12,07	19,03	36,11
	(2)	(5)	(0)	(8)	(10)	(25)

can be achieved through most profitable options for the owners and investors. The rehabilitation and re-utilization of these locations shall enable the redevelopment of urban fabric in Skopje and change the situation where abandoned and troubled industry on the borderland has turned city's edges into inactive territories.

We might resume that the spatial distribution of industry during the post-socialist period in Skopje, from mono-centric (centralized in city's urban areas), became dispersed and scatted in various specialized and fragmented production places located in and beyond urban territories. The decentralization of industry, i.e. creating new places of economic activity on urban edges and outside city borders, has the tendency of local concentration and agglomeration. The new spatial forms of economic and production activities, the commercial-economic and industrial-technological zones, appear as new poles of development (Figure 2).

The new paradigms of urban planning and the trends of transformation of urban and surrounding space of the city, due to new dominant ways of land appropriations and spatial organization of production activities verifies the fact that new industrial spaces have and shall have greater role in terms of directing the new spatial distribution of industry, known as new spatial pattern of industry.

The latest urban planning document (General Urban Plan of city of Skopje) was adopted in 2012,

and it involves new approaches in urban planning: the complexity and integrity, multidisciplinary and interactivity, as well on the principle of sustainable development. Proposed spatial organization is based on the assumption of recovery, modernization and expansion of inherited capacities and change of their industrial activity on the current location.

The Plan (Figure 2) defines city limits and its contact zone in wider area of the region. For urban planners industry is still significant for the future development of the city, although it is not the primary and only economic activity. Moreover, all heavy industries are planned to be relocated from the city and replaced by light industries or commercial activities.

The attractiveness of the location for industry development in the period of transition has threatened all activities for redevelopment and re-utilisation of the abandoned industry in the outskirts of the city. This way they have lost their position and meaning as important generator of activities and demographic exchange and interconnectivity of the urban-suburban-exurban territories. These abandoned or misused structures have affected a new edge conditions and permeability of the "membrane", changing demography, mobility and use patterns over the years.

Dispersion of new industrial activities beyond the urban boundaries of the city or on the edge, in the space between urban and rural, where most of the abandoned industrial structures are located, should enable transformation to modern service based industry, hence the industrial location attracts users from both, the city and its surroundings, re-establishing the border territories exchange and urban-rural synergy.

This determination is in accordance to the Global Sustainable Goal which our country has adapted to the national policies and programmes to implement. The United Nations, adopting the latest development policy, the Agenda 2030, at the United Nations Sustainable Development Summit 2015, held 25 - 27 September 2015, in New York (United Nations General Assembly, 2015), and the Global Sustainable Goals has established a clear political objective and platform for sustainable development and actions. In this regard we might say that the Goal 9- Build Resilient Infrastructure, promote sustainable industrialization and foster innovation (Targets 9.2 and 9.4 for promotion of inclusive and sustainable industrialisation and greater adoption of clean and environmentally sound technologies and industrial processes) is correspondent to some propositions we elaborated for changing economy modalities and reutilisation of the abandoned industrial facilities on the outskirts of the city (United Nations General Assembly, 2015).

PROJECT RESEARCH: CASE STUDIES OF FACTORY KUPROM

Case studies that are chosen to be presented are research projects/master thesis completed for obtaining master degree at University Ss. Cyril and Methodius, Faculty of Architecture. Projects presented are individual end-semester project- based master thesis supported by survey and theoretical approach elaborated. Among several project thesis we have chosen only three that share similar ideas and approach to the industrial heritage protection and to the need for revitalisation and reutilisation of the derelict and unused facilities of the old factory facilities which are positioned on the outskirts of the city of Skopje.

They all use the former factory for colored metals "Kuprom" for their research and project, which entirely belongs to the category of abandoned industry in Skopje. The factory did not survive the competition, market-oriented economy and new management methods and was forced to transformation and dismission of so-called redundant production. The envisaged approach target multiple problems simultaneously, few of them being, the brownfield issue, lack of social spaces, lack of job positions, passivation of city border territories, and off course the way the values of the inherited industrial architecture should be maintained.

They all promote the idea of positive re-evaluation and rehabilitation of the abandoned industrial facilities on the outskirts of the city, which have lost their opportunities to adapt to the political, ideological, economical system changes. Each of the case study presented offers different, but innovative approach and solution to the problem recognised, new functionality and re-programming of the unused facilities, supporting the concept of sensible city and new economies as plausible way the city should change future development of the structures alike.

The new economy and technological performances are perceived as a new opportunity for rehabilitation of the industrial fringe structures in revealing the possibilities for redefining the use and functionality of the abandoned and unused facilities, as well as new possibilities and modalities for reshaping the structures, enhancing their symbolic and social value of the old "Kuprom" factory which once had a role of a real generator of the urban-rural synergy development.

The factory "Kuprom" is located in the western part of the city of Skopje. Historically, this area represented peripheral settlement where the agriculture was developed, later developed into industrial zone and today, after the expansion of the city's border territory, is becoming highly urbanized with predominantly residential use. The factory was built during the WWII as a unique complex for processing chromium ore and for many decades it was considered as driving force for industrial and economic development of this part of the city, a stimulus for development of surrounding urban, suburban and extra urban areas and several neighborhoods such as Chrom and Vlae. In the processes of transition the factory faced ownership and property-based constraints and has since been locked up and left to decay. Today the abandoned factory is another example of unimproved potential of the industrial heritage from the socialist period.

The spacious location and remaining structures of the old factory "Kuprom" dominate the surrounding urban context. At the moment there are twelve buildings in different condition still standing on the site of the former factory 'Kuprom', but only three buildings in which the production process took place: production hall 1, production hall 2 and boiler building have preserved some of their original look to the present. The architectural expression is manifested with simplicity, while their spacious interiors are adjustable to many different programs.

Most of the projects consider program changes through adaptive reuse, taking into account the architectural heritage and apply restructuring only to the inner space of the buildings. In one case, more radical approach is applied, partial urban renewal, when only through building of completely new facilities it was possible to meet the needs of the new program, or when the devastation of the buildings was great. By this transformation the location and image of the place have changed completely.

The new interventions are supposed to act as catalyst for regeneration of whole changing border area. Such intervention become direct antithesis of the act of erasing abandoned heritage from the collective memory and image of the city. Moreover, these projects suggest that similar responsive approach should amalgamate the appalling conditions of other neglected and abandoned structures around the city. Proposed procedure of rehabilitation is expected to enable redevelopment of urban fabric of post-socialist Skopje and activate the border zones of exchange.

The first presented master thesis project, "New economies in transition society and their impact on the urban physical environment-Research on the link between small businesses and the urban development", presented in Figure 3, is by Ivana Angelova (2016) and mentored by Prof. Jasmina Siljanoska. The project aims to respond to the changing economic environment young people are faced with nowadays and to the present situation of the post-socialist city where interventions in urban development, in particular transformation of the abandoned industrial sites on the fringe, are necessitate.

The project is challenged with the real effect renewal and re-construction within the existing urban context and complex ecological environment might have on the perspective of the future development, therewithal to achieve the aspired sustainable development with minimal ecological footprint.

The re-utilization of the abandoned industry supports the idea of citizens self-empowerment through technology. In this regard the new sharing economy is seen as an offering escape route. The novel ideas are based on postulates that: the future of architecture and design lies in the power of the community and in the open participatory design; It should represent a balanced mixture of hi-tech (borderline clean technologies like artificial intelligence, virtual reality and augmented reality) and low-tech (traditional way: crafts); and the architect of tomorrow should be a mediator between the different actors taking part in the process of creating a place.

Figure 3. New economies in transition society and their impact on the urban physical environment-Research on the link between small businesses and the urban development (Angelova, 2016)

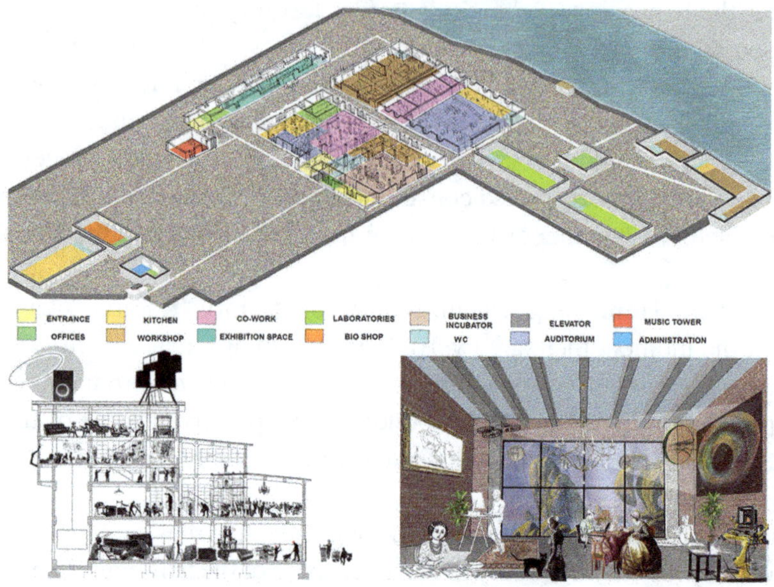

The project itself explores the transformative urban typologies and underpins innovative approach to participatory urbanism and more inclusive planning process than the existing insufficiently developed bottom-up local governance. A real challenge was to propose involvement of a different approach than the existing inflexible and nontransparent one in the urban policy decision making process, which makes long-term sustainable solutions almost impossible.

The new solution is presented as a model of mixed partnerships involving a new method of management and horizontal organization. Complementary programs and public amenity should enable mixed cultures, religions and people of social statuses, of inclusion and equality to participate in operation. It should support new types of economy and share similar "DNA" with some other places in the rural-urban fringe of the city.

Adaptive re-use and re-programing of the remaining structures is proposed to incorporate mix-uses and facilities such as: administration building, restaurant + bio shop, a vegetable garden, small artist residency, incubator and technology park, musical tower, co-work space + workshop, laboratories and workshops for small bio tech businesses and robotics, craftsmen workshops + bar and an open multi-functional stage.

Since the participatory and speculative design method was engaged as the most appropriate to the survey undertaken, proposed design is open ended. The designers are as well the users themselves, and they could gradually restore the space and shape it according to their needs, while the architects might help on the way through. The facilities should be opened to everyone and are intended to stimulate the production of creativity and goods. This way the whole space becomes a large workshop which constructs and maintains itself at the same time.

The master thesis project "Contextual reading of industrial heritage", Figure 4, by Angela Dimitrova (2018), mentored by Assoc. Prof. Aleksandar Radevski, shows strong attitude towards re-validating of the abandoned industrial structure as an important construct of the built environment and landscape.

Figure 4. Contextual reading of industrial heritage (Dimitrova, 2018)

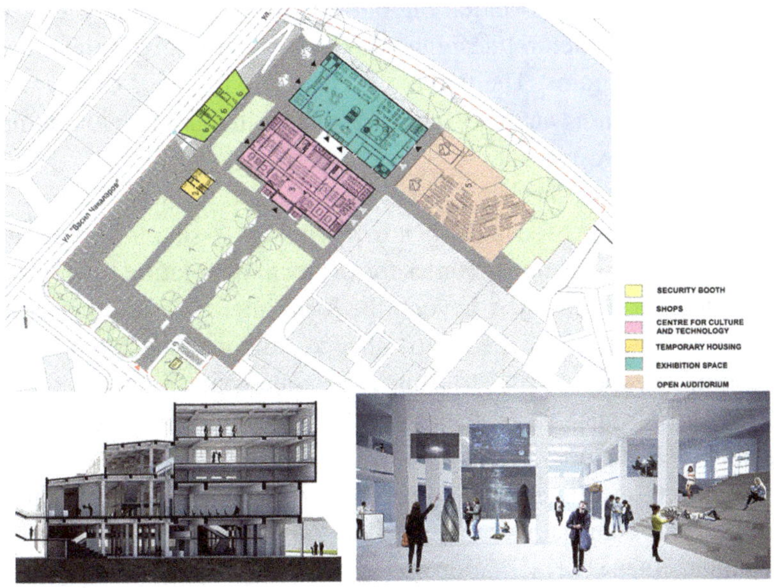

Restoration strategies are proposed primarily in terms to provide connectivity and integration of the fragments of which the local urban landscape is re-constructed. Protection stimulus is justified by introducing adaptive re-use approach to revitalization of the abandoned industrial structures and by associating them with the aspect of sustainable development, while following the modern trends and innovative industries. The main idea is to preserve the memory of the place and to use the potential of the remaining in the future urban development of the city, appropriate to the spatial, sociological and cultural context. Of particular importance to the urban design and adaptive re-use is to enrich the current predominantly housing and a small percentage of commercial use in the settlement Chrom. This way the potential of the border territory in connecting urban, suburban and extra-urban areas and citizens might be harnessed as advantageous.

By introducing a new typology aiming in emphasizing the characteristics of the inherited structures, it was decided to preserve the buildings which in the past have served for the production process in particular. Namely, the production hall 1, the building of the boiler room and the guardhouse, with their spatial definition, architectural elements and facade shaping, are treated as valuable examples of industrial buildings built of the 1940s.

Adaptive re-use method in the retained structures of the production halls is engaged accordingly to the "production" principle, and it is decided the technological park and a business incubator, an innovation center and co-operation spaces for start-up companies to be the new programs inserted.

The small startup businesses are seen as having strong potential for the process of rehabilitation and adaptive re-use, because they are part of the growing community of young professionals who are attracted to creative and innovative industries and technologies.

The preserved production hall 1 is redesigned and adapted to the new programs which incorporate spaces for courses and training, technological workshops for individual and group research with 3D printers and lasers, show rooms of innovations on the first level. Above the classrooms a platform is placed, which forms a gallery space open to the middle part, on which a co-working space develops. Due to the poor condition in which the structures of the production hall 2 were and not much of a characterized architectural expression values, it was decided the building to be removed, and rebuilt again. Certain changes in the architectural and structural design were made accordingly to the requirements and standards of the newly proposed programs. The volume itself is reminiscence of the old production hall 2. The new designed structure contains an expo space for displaying exhibits and temporary exhibitions, and on the upper floor there is a restaurant.

The last, but not the least, master thesis project to be presented is "Adaptive reuse of the former factory 'Kuprom'", Figure 5, by Sofija Stoeva (2017), mentored by Assist. Prof. Goran Mickovski, which aims to enhance the social and economic outcomes for the local population by introducing a new activities and use. Proposed changing industry is beyond the local interest, but rather a point of interest and development for the city as a whole. The idea is to enhance the bond between the place and the citizens and re-evaluate the cognitive experiences. Offering a dynamic space for a creative production: a non-material industry to re-habit the abandoned structures the project tends to restore the meaning once the place has had in the collective memory.

Architectural interventions to the existing buildings apply an adaptive re-use method primarily serving the requirements of the new program habitation, re-evaluation and re-habilitation of the industrial heritage architecture method which aimed at preservation and at subtle interfering with the present exterior of the building. However, a clear distinctiveness between 'the old' and 'the new' is instantly apparent on site.

Extensive analyzes have come to a conclusion that the city of Skopje is in deficiency of places for creative expression. Not only in the city, but in the country-wide, video as a medium and movies production studios bear unavailability or inadequate space for the expression and realization of the ideas. This was the idea behind the proposition the former factory of "Kuprom" to become an important cultural hub, the influencing city's development epicenter of cultural creation.

Figure 5. Adaptive reuse of the former factory "Kuprom" (Stoeva, 2017)

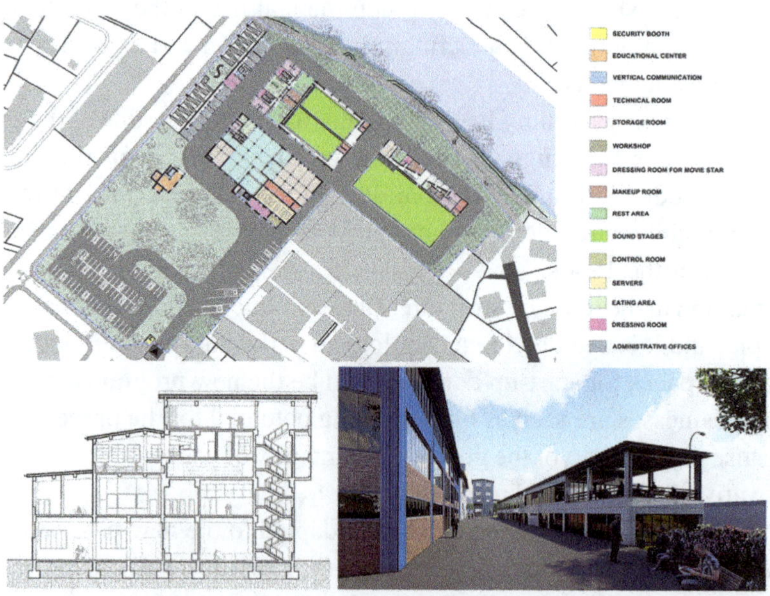

The advantages of the existing buildings on site, their vast interior spaces and structural characteristics, high ceilings etc., enabled to be re-programmed and adapted accordingly to the requirements and standards of a modern film studio. Production hall 1 is adapted to encompass a pre-production facility which incorporate a space for scene creation and video footage shooting, as well as facilities for post-production with adequately equipped video and audio processing and animation studios. Production hall 2 is adapted to a production area consisted of two sound scenes and production capacities. The boiler building is intended to turn into an educational center for the needs of the film production process.

In the end, we should not miss to mention that above presented project ideas, as well as of other master thesis projects on abandoned factory "Kuprom", which have been completed at the Faculty of Architecture-University of Ss. Cyril and Methodius, challenged a wider public debate and growing public interest for recognition and protection of its industrial architecture values. Public debate involved local community, non-governmental organizations, students and academics, professionals, cultural creators and public figures. The case encouraged non-governmental organization "Cultural Echo" and a group of young enthusiastic architects and students from our faculty, actors and film-makers, developers and lawyers to initiate introduction of the concept of industrial heritage and involvement of creative approach to its revitalisation and reutilisation, in this case into an open students' city, or into a new cultural distric. However, the real impact of these activities and initiated ideas for revitalisation and reutilisation is still missing.

CONCLUSION

We arrived at a time when responsive environments and transformation of the cities change planning paradigms. Redevelopment of abandoned buildings, and even entire city areas, as a consequence of shrinking process or economy and system changes, brings new possibilities through the methods of re-utilizing and re-programming with participatory design involved. We believe that the future of architecture is in the "power" of the community, thus inclusion of all the actors in the participative design process could revitalize and rebirth such places in the most meaningful and responsive way to the needs of both, urban and rural, population.

All these tendencies and shifting paradigms should affect the way the cities are planned and governed. In this regard, a sensible city describes contemporary paradigm shift in the way the citizens are included in the planning process for improvement of the delivered services and in urban transformation towards sustainable and efficient environments. For the future development of the city of Skopje a paradigm of sensitive city should be a real challenge for the future urban planning activities.

The emergence of new forms of production and new spatial distribution of industry influence the future city changes and development. In this process of transformation of the industrial space the role of abandoned industries and industrial activities on the city border and beyond cannot be neglected. It is expected that technology and citizens self- empowerment will play a key role in this process. The new sharing economy is offering an escape route, but it is important the micro-level projects to be protected and promoted.

Moreover, flexible organization of production, the introduction of new computer-based technologies and new model of employment relationships, have created new forms of organization of production and changes in the configuration of modern industrial spaces. Innovative approach embraces internet, nanotechnologies, telematics and bionics, among others. All these might represent "generative" points for the city outskirts redevelopment platform, based on new programs and new economy as transitioning form from manufacturing-based economy to a service-based economy, such as business incubators and clusters, or establishment of innovation centers and technology parks.

Through adaptive reuse and re-programming of the old factories whereby no radical change occurs in the urban landscape, from production of goods to production of knowledge, culture and leisure, is enabled reshaping of old into new structures which meet the needs of citizens and the new economies. The existing premises of former industrial buildings, worth as architectural heritage as well, this way might transform the sites and entire surrounding area, while re-activating the in-between urban-rural zone.

REFERENCES

Мицковски, Г. (2015). *Трансформација индустријских простора у контексту пост-социјалистичке транзиције града Скопља (Необјављени докторски рад). Универзитет у Београду - Архитектонски факултет.* Београд, Србија.

Angelova, I. (2016). *New economies in transition society and their impact on the urban physical environment-Research on the link between small businesses and the urban development* (Unpublished master thesis). Ss. Cyril and Methodius University, Skopje, Republic of North Macedonia.

Barta, G., Beluszky, P., Czirfusz, M., Gyiri, R., & Kukely, G. (2006). *Rehabilitating the brownfield zones of Budapest* (Discussion Paper No. 51, pp. 57-58). Pécs, Hungary: Centre for Regional Studies of the Hungarian Academy of Sciences Press.

Dimitrova, A. (2018). *Contextual reading of industrial heritage* (Unpublished master thesis). Ss. Cyril and Methodius University, Skopje, Republic of North Macedonia.

Drucker, P. F. (1993). *Post-capitalist society*. New York, NY: Harper Business.

Felson, M., & Spaeth, J. L. (1978). Community structure and collaborative consumption: A routine activity approach. *The American Behavioral Scientist*, *21*(4), 614–624. doi:10.1177/000276427802100411

Florida, R. (2005). *Cities and the creative class*. New York, NY: Routledge. doi:10.4324/9780203997673

Greco, I., & Bencardino, M. (2014). The paradigm of the modern city: SMART and SENSEable Cities for smart, inclusive and sustainable growth. In *Proceedings of the International Conference on Computational Science and Its Applications* (pp. 579-597). Cham, Switzerland: Springer International Publishing. 10.1007/978-3-319-09129-7_42

Hamilton, F. E. I., & Carter, W. F. (2005a). Foreign direct investment and city restructuring. In I. Hamilton F.E., K. Dimitrovska Andrews, & N. Pichler-Milanovic (Eds.), Transformation of cities in Central and Eastern Europe: Towards globalization (pp. 116-152). Tokyo, Japan: United Nations University Press.

Hamilton, F. E. I., Pichler-Milanovic, N., & Dimitrovska Andrews, K. (2005b). Introduction. In I. Hamilton F.E., K. Dimitrovska Andrews, & N. Pichler-Milanovic (Eds.), Transformation of cities in Central and Eastern Europe: Towards globalization (pp.3-21). Tokyo, Japan: United Nations University Press.

Harvey, D. (1990). *The condition of postmodernity*. Cambridge, UK: Blackwell Publishers.

Kaufman, C. A. (2015, August 10). *Stephen Hawking says we should really be scared of capitalism, not robots*. Retrieved January 14, 2019, from http://www.huffingtonpost.com/entry/stephen-hawking-capitalism-robots

Kiss, E. (2002). Restructuring in the industrial areas of Budapest in the period of transition. *Urban Studies (Edinburgh, Scotland)*, *39*(1), 69–84. doi:10.1080/00420980220099078

Kiss, E. (2007). The evolution of industrial areas in Budapest after 1989. In K. Stanilov (Ed.), *The Post-Socialist City: Urban Form and Space Transformations in Central and Eastern Europe after Socialism* (pp. 147–172). Dordrecht, The Netherlands: Springer. doi:10.1007/978-1-4020-6053-3_8

Kumar, K. (2005). *From post-industrial to post-modern society: New theories of the contemporary world, sec*. Cambridge, UK: Blackwell Publishing.

Lefebvre, H. (1991). *The production of space (D. Nicholson-Smith, Trans.)*. Oxford, UK: Blackwell Publishing. (Original work published 1974)

Lux, G. (2008). *Industrial development, public policy and spatial differentiation in Central Europe: Continuities and change* (Discussion Paper No. 62, pp. 26-28). Pécs, Hungary: Centre for Regional Studies of the Hungarian Academy of Sciences Press.

Markusen, A. (1996). Sticky places in slippery space: A Typology of industrial districts. *Economic Geography*, *72*(3), 293–313. doi:10.2307/144402

Mason, P. (2016). *Postcapitalism: A guide to our future* (1st American ed.). New York, NY: Farrar, Straus and Giroux.

Mickovski, G., Radevski, A., & Karanakov, B. (2016). Transformation of industrial space in the context of post-socialist transformation of the city of Skopje. In *Proceedings of Conference Inclusive/Exclusive Cities* (pp. 264-281). Skopje: City of Skopje

Nedović-Budić, Z., Tsenkova, S., & Marcuse, P. (2006). The urban mosaic of post-socialist Europe. In S. Tsenkova & Z. Nedović-Budić (Eds.), *The urban mosaic of post-socialist Europe. Space, institutions and policy* (pp. 3–20). Heidelberg, Germany: Physica-Verlag HD. doi:10.1007/3-7908-1727-9_1

Pratt, A. C. (2008). Creative cities: The cultural industries and the creative class. *Geografiska Annaler. Series B, Human Geography*, *90*(2), 107–117. doi:10.1111/j.1468-0467.2008.00281.x

Rifkin, J. (2014). *The zero marginal cost society: The internet of things, the collaborative commons, and the eclipse of capitalism*. New York, NY: St. Martin's Press LLC.

Scott, A. (1988). *Flexible production systems and regional development: The rise of new industrial space in North America and Western Europe. Centre for Urban and Community Studies University of Toronto* (Research Paper N° 168, pp. 1-31). Toronto, Canada: University of Toronto Press.

Scott, A., & Storper, M. (1990). *Regional Development Reconsidered. Lewis Center for Regional Policy Studies* (Research Paper N° 1, pp. 1-50). Los Angeles, CA: University of California Los Angeles Press.

Soja, E. W. (1989). *Postmodern geographies: The reassertion of space in critical social theory*. London, UK: Verso.

Soja, E. W. (2000). *Postmetropolis: Critical studies of cities and regions*. Oxford, UK: Blackwell Publishers.

Stanilov, K. (2007). The restructuring of non-residential uses in the post-socialist metropolis. In K. Stanilov (Ed.), *The Post-socialist city: Urban form and space transformations in Central and Eastern Europe after socialism* (pp. 73–100). Dordrecht, The Netherlands: Springer. doi:10.1007/978-1-4020-6053-3_5

State Statistical Office. (2011). Statistical Yearbook 2011. Skopje, Republic of North Macedonia: State Statistical Office.

Stevens, Q. (Ed.). (2015). *Creative milieux: how urban design nurtures creative clusters*. New York, NY: Routledge.

Stoeva, S. (2017). *Adaptive reuse of the former factory "Kuprom"* (Unpublished master thesis). Ss. Cyril and Methodius University, Skopje, Republic of North Macedonia.

Tsenkova, S. (2006). Beyond transitions: Understanding urban change in post-socialist cities. In S. Tsenkova & Z. Nedovic-Budic (Eds.), *The urban mosaic of post-socialist Europe. Space, institutions and policy* (pp. 21–50). Heidelberg, Germany: Physica-Verlag Heidelberg. doi:10.1007/3-7908-1727-9_2

United Nations General Assembly. (2015). *Transforming our world: the 2030 Agenda for Sustainable Development*, 21 October 2015, A/RES/70/1. Retrieved June 14, 2019, from https://www.refworld.org/docid/57b6e3e44.html

Wu, X., & Zhi, Q. (2016). Impact of shared economy on urban sustainability: From the perspective of social, economic, and environmental sustainability. *Energy Procedia*, *104*, 191–196. doi:10.1016/j.egypro.2016.12.033

KEY TERMS AND DEFINITIONS

Adaptive Reuse: Refers to implementing new functions through renovation and restoration by creating necessary conditions for performing new activities.

New Economies: Activities based on knowledge, information and high-technologies rather than traditional manufacturing economy.

Post-Socialist Development: Transformation, economically and socially, as well as spatially, in shaping the new character in the cities with legacy of the socialist development.

Re-Utilize Industry: Once abandoned, misused, underused or unused industrial facility to turn into new use or different practical use.

Rehabilitation: The action of restoring structures and facilities, which has been damaged, abandoned or unused, to its active useful condition.

Sensible City: The concept of smart city which integrates the ICT (information and communication technologies) together with the human, social and relational capital for enhancement of the environmental quality.

Urban Boundaries: The boundary between the end of the city and the beginning of nature or urban/rural fringe, a border land in which interactivity and urban-rural synergy is applicable.

Chapter 6

Urban-Rural Synergy Between Housing Spatial Patterns and Landscape:

Typological Classification of Belgrade Socialist Settlements From an Environmental Perspective

Ana Nikezić
Faculty of Architecture, University of Belgrade, Serbia

Jelena Ristić Trajković
Faculty of Architecture, University of Belgrade, Serbia

Aleksandra Milovanović
Faculty of Architecture, University of Belgrade, Serbia

ABSTRACT

The morphogenesis of the urban territory and its contact with the non-urbanized, natural environment of the wider metropolitan area distinguish issues of the synergy between landscape and spatial patterns in order to achieve their balance, optimization, and harmony. This chapter highlights the conceptual framework of landscape ecology as linking to place-based design approach for studying the synergy of landscape and housing spatial patterns in order to improve their integration in future perspective. The territory of the city of Belgrade is recognized as a specific spatial-morphological system that is a consequence of the urban-rural synergy between socialist housing settlements and environmental processes. The chapter points at the environmental and functional values of nature with a particular focus on housing typology in the process of urban planning and architectural design.

DOI: 10.4018/978-1-5225-9932-6.ch006

INTRODUCTION

The proposed thematic framework of this chapter studies the role of landscape studies in a comprehensive process of planning, programming and designing of housing spatial patterns, considering the processes of morphogenesis and environmental processes of the territorial development of the city with the recognition of border, non-urbanized zones that have achieved urban-rural synergy in the process of urbanization. Accordingly, the initial objective of this research is the improvement of relations between architectural design, urban planning and, landscape ecology, as connecting disciplines of territorial, spatial and environmental studies. In this way, the issue of environmental planning and development of housing spatial patterns is set in the specificity of the spatial framework of the wider metropolitan area of the city of Belgrade in the period from 1947 -1972. This period is recognized as relevant time point in the urban development of Belgrade and in that order, it is taken as a reference point in relation to which the development of a housing landscape is considered, and also its variability and conditionality in line to the spatial-morphological and content-functional system of the city.

The main objective of this research is to establish a perspective for future development of socialist housing settlements built between 1947 and 1972 in line with recognized developmental tendencies and axes, through (1) a critical overview of the existing morphology and the planning framework, and (2) a scientific examination of the landscape studies role in the planning of housing spatial patterns. The overall objective is to scientifically explain the position and role of the typological classification of socialist housing settlements from an environmental perspective in the process of planning, programming, and designing of the relationship between the built and the natural environment at different spatial levels of housing typology.

The contribution of this research to architectural design and urban planning, as well as urban and environmental studies, can be expressed through the function of scientific knowledge that takes place in three phases and in accordance with which the chapter structure was established. The first phase involves a scientific explanation, which takes place through the discussion of importing the landscape-ecological concepts into housing typology planning and design. The second phase involves the contextual analysis of planning and development of the Belgrade territory in the period between the two Master plans from 1950 and 1972, which significantly influenced the physiognomy of the urban landscape of Belgrade, the metabolism of the city and the identity of housing settlements in line with landscape-ecological principles: (1) comparative analysis of the adopted urban plans and the realized urban structure, and (2) mapping the professional dialogue of the housing planning of Belgrade. The third phase implies a typological classification of Belgrade housing spatial patterns from an environmental perspective through the definition of their territory, morphology, spatial form, programming and character of the housing landscape.

The importance of landscape studies for the territory of Belgrade was previously highlighted through a series of relevant studies with a particular focus on the typology of Belgrade's landscapes for the needs of application of European Landscape Convention (Cvejić, Vasiljević & Tutudžić, 2008), the role of the concept of landscape planning as an instrument of spatial development (Vasiljević, 2012), but also an overview of the transformation of the landscape structure in the process of development of the territory of Belgrade in the second half of the 20th century (Janković, 2017). These studies form the basis of landscape studies for the context of the territory of Belgrade, while the purpose of this chapter is to focus on thematic research based on the principles of landscape-ecological and environmental studies, focusing on the relationship between housing spatial patterns and landscape at different spatial levels.

Respecting the goals and principles established in the *2030 Agenda for Sustainable Development*, it is expected that the results of this research indicate the necessity of ''awareness for sustainable development and lifestyles in harmony with nature'' (United Nations, 2015: 23). It is expected that the identified housing settlements types contribute to the establishment of future planning perspectives and design principles for the housing typology in the comprehensive territory of Belgrade, and also, to point out the synergy of landscape and housing spatial patterns.

BACKGROUND

Contemporary urban studies focusing on the relationship between the landscape and the urban morphology are considering changes in the territories and urban conditions of development, which requires (1) understanding, mapping and describing the process of a urban transformation and its impact on nature, and (2) recognizing the possibilities of developing nature in the city and its impact to a certain urbanized environment in order to shift the rural into an urban landscape. The *European Landscape Convention*, as one of the leading global agenda about landscape issues, defines a landscape as ''an area, as perceived by people, whose character is the result of the action and interaction of natural and/or human factors'' (Council of Europe, 2000: 2). On the other hand, the dynamic character of the landscape has led to new perspectives of observation and study that are directed towards the constituent elements of the landscape. The landscape is thus viewed as a social and cultural record of processes that take place parallel to the existence of spatial patterns and affect their identity (Duncan, 1990; John & Mitchell, 2000). In this research, the landscape is interpreted as a holistic, hybrid, entity of natural, cultural and social values that changes and develops, or as ''a complex system of mutual relations'' (Van der Poel, 1976: 365).

The landscape is thus viewed here from the perspective of the system conceptual framework in order to understand transformation and shifting of the natural landscape to cultural. This process involves the synergy of environmental processes and social activities and flows in a certain part of the landscape viewed through different spatial levels. Within the *Second Millennium Ecosystems Report* (Millennium Ecosystem Assessment, 2005), through the consideration of factors that indirectly affect biodiversity, including technological progress and the transformation of lifestyles, a distorted balance between ecosystems and human well-being is recognized. This points to the necessity to develop new approaches to researching the relations between natural and cultural landscapes in order to achieve synergy between environmental processes and socio-economic flows within and through the landscapes. In 1992 the UNESCO World Heritage Committee adopted the cultural landscapes as a new category of Word Heritage, defining cultural landscapes in the following way:

Cultural landscapes are cultural properties and represent the "combined works of nature and of man". They are illustrative of the evolution of human society and settlement over time, under the influence of the physical constraints and/or opportunities presented by their natural environment and of successive social, economic and cultural forces, both external and internal. (UNESCO World Heritage Centre, 2005)

According to Tylor and Lennon (2011), the landscape is primarily explained as a social construct that builds on the relation between spatial and political phenomena under the influence of ideological values and four dimensions of the authenticity concept including: (1) *location and setting* – in order to create the identity of place and place attachment, (2) *form and design* – in order to create linkages with living elements and environment, (3) *use and function* – in order to produce the impacts of use, and (4) *essence* – in order to provide emotional impact of place. In order to achieve synergy and balance between the mentioned dimensions which built the cultural landscape, it is necessary to include perspectives focused on the human-environmental relationship in landscape ecological research (Wu, 2010).

Considering that the housing typology can be understood as the most complex and intense level of everyday life of the human-environmental relationship and as a complex social mechanism that is created under the influence of ideological values that are reflected at different spatial levels, the leading research perspective in this chapter is the understanding of the of a specific type of cultural landscape – housing landscape.

The general theoretical approach is reflected in the complex study of spatial-programming relations between the built environment, natural conditions and people. The research is basically transdisciplinary and examines the key points of interaction between the field of architecture and urbanism, and the field of landscape ecology including geography, ecology and social anthropology. The research includes mixed topics from these disciplines such as spatial ecology, social geography and cultural ecology, which have important influence on the development of territorial and environmental theoretical frameworks (Wu, 2012).

The research of the relationship between the built environment and nature takes place through a specific set of research methodologies: (1) morphology - a description of the structure and its elements, (2) typological classification - systematic mapping, (3) chronology - studying the variability over time, and (4) relationships - studying the relationship between all elements and processes in the landscape at different spatial levels (Zonneveld, 1990). Study of the relationship between the elements requires a description of the form and structure, as well as their classification, while on the other hand, classification and mapping require consideration of the relationships between elements, which further include studies of changes in the landscape and introduces a spatial-temporal dimension of study. Therefore, the aim of this research is to chronologically recognize the development of the morphology of the city, which is mostly developed of housing settlements from the socialist period, and to establish a typological classification from an environmental perspective in relation to the recognized characteristics.

Multi-Scale Approach

Both landscape studies and architectural practice are trying to approach a whole, a synthesis - observing different levels in space and time. Relevant theoretical paradigms of landscape-ecological studies take on a significant position in the place-based research approach - hierarchy, heterogeneity, and holism - which contribute to the understanding of different approaches in the study of the relationship between the built environment, nature and people. In order to explain the above theoretical considerations, it is necessary to define the position and role of the concept of scale by linking two dimensions: (1) spatial, horizontal dimension based on a geographical perspective, and (2) functional, vertical dimension based on an ecological perspective (Troll, 1971).

A significant chapter in the study of landscape and territory was established by the introduction of a *soft paradigm* in landscape-ecological theoretical framework which primarily focuses on the functional aspects, and then on the spatial relations, structure and patterns (Toth, 1988). Functional aspects can be identified and read by observing a man in relation to the landscape spatial dimension in order to understand (1) the adaptation of the user and environment and (2) an effort to create a sense of place, orientation and hierarchy in line to environmental dimension. In the context of the transition from industrial to post-industrial information age within which the social system is at the turning point of its relationship with nature, Naveh (1990; 1995; 2000a; 2000b) points to the synthesis of natural and social systems into a *total human ecosystem*, as the highest level of complex coevolution and ecological hierarchy. Such changes are possible by the realization of the dynamic natural, social, cultural and economic processes synthesis in the landscape, which would further enable transdisciplinary integration and cooperation with relevant types of social sciences and humanities. In this sense, it is also important to highlight the temporal perspective of the landscape ecology research, which must find the answer how a certain landscape is developed, how it sustains and how it will change in the future, viewed through a prism of ecological aspects.

Based on the tendency that environmental functions and patterns of community organization vary at different levels in space and time, Delcourt and Delcourt (1988) defined the spatial-temporal scale as an *organizational paradigm* in the landscape ecology that provide a range of spatial-temporal configurations within which each dimension is limited to a higher one, and at the same time includes all the patterns and processes that take place at the lower level of the hierarchy. The concept of scale can thus be defined as equally spatial and temporal dimension in order to explain (1) development and dynamics of spatial heterogeneity in rural areas, (2) interaction and exchange through heterogeneous landscapes, and (3) the influence of spatial heterogeneity on the social and ecological mechanisms of landscape (Turner, 1989).

According to the research conducted by Gibson et al (2000) the most important issues related to scale can be grouped into four theoretical areas: (1) identification of spatial patterns through scale, (2) conditionality of the scale and social phenomena, (3) connection of different phenomena on spatial and temporal scales, and (4) optimization of the process in certain scale.

The described standpoints of the relevant authors clearly indicate the importance of exploring the landscape by considering all spatial and temporal levels through multi-scale approach. In this regard, the intention of the chapter is to recognize the relevant spatial levels that are suitable for exploring the relationship between spatial housing patterns and landscapes.

Place-Based Approach

As previously indicated in the introduction, a place or a landscape is not only a territorially bounded spatial unit, but it is also including cultural, social, economic and political influences through spatial and temporal hierarchy (Balvanera et al, 2017). In that order, place-based approach allows a better understanding of interrelations between social changes, economic flows, and ecological processes on the one side, and spatial patterns on the other, in the pursuit of sustainability and optimization of landscape system.

One of the leading research issues of designing the identity of housing spatial patterns and strengthening social identity in such a spatial framework pointing to the increasing need to ''provide of belonging, identity and individuality to each person'' aside personal taste for pure form and designing housing for the masses (Mills, 1966: 9). His research has significantly contributed to the consideration of spatial-programming relations within high-density housing such as: (1) the domain of social urbanity - including the issue of an individual, family and level of privacy, (2) social formulation of community and neighborhood, and (3) designing social structure through recognition of social and physical needs and socioeconomic structure at appropriate spatial levels of neighborhood, community and city.

On the other hand, the central research question in the study of the place-based approach concerned the phenomenology of the place through review of space and its relationship to place and understanding the dimension of human life and experience (Relph, 1973; Norberg-Schulz, 1980). Relf's (1973) view is based on understanding the heterogeneity of space that involves many complex dimensions in everyday life by identifying several different modes of spatial experience which plays a leading role in conceptual engagement between space and place: (1) instinctive, bodily, and immediate modes of spatial experience - pragmatic space, perceptual space, and existential space, (2) cerebral, ideal, and intangible modes of spatial experience - planning space, cognitive space, and abstract space. Idea of a place can provide a concept which would act as a bridge between the various fields pointing out as the leading position of psychological, behavioral and environmental aspects and study of individuals and their comprehension of places (Canter, 1977). This can mean that place-based approach enables users to understand, experience and create places through understanding (1) the influence of place on user actions and, (2) qualitatively difference between places and objects.

HOUSING LANDSCAPE

The observational process in relation to the urban environment establishes each individual by building his own view of the built and other dimensions of the city, housing patterns and social living communities in it, which determines the identification and orientation in the environment as a constant existential interactivity in it. The urban housing structure thus represents an important segment of the complex relation to objective reality, considering the level of everyday observation that the user builds with his residential environment. The observation takes place through the recognition of morphological, ambient, anthropometric, environmental and behavioral aspects, and interrelations between natural and urban elements: (1) natural - condition the layout and hierarchy of urban physical structure, and (2) urban - act in their spatial, formative and symbolic form and its effect on spatial patterns at different levels by determining anthropological observation relation towards the overall territory of the city. Having in mind the relation between natural and built, rural and urban, as well as the relationship between architecture and landscape, a number of synergistic and dichotomous relations can be identified that can significantly contribute to the creation of a housing landscape:

By examining the relationship between landscape and architecture and emphasizing their dichotomy, we concluded that the ultimate meaning of any building goes beyond architecture. Architecture is seen as a supplement of nature facilitating our existence within it. It is a translational element that enables human survival in nature. Anselm helped us to understand that creating designs that are integral to nature lies in understanding the natural environment (topography, terrain, management of climate and energy) and involving the ecosystem by way of its interactions with building design. We agreed that designing with nature starts from a profound understanding of the place, careful management of local climate conditions and considering its characteristics (winds, orientation, sun reflections) in architectural design. (Nikezić & Marković, 2015)

At the present time, housing in a planning and design perspective no longer merely represents a category of standards and living conditions, but evolves as a complex system in space and time that includes environmental, behavioral, ambient and many other aspects and problems. Housing construction, its impulse, its content, its architectural character, undoubtedly are the needs of social development, everyday life needs, economic opportunities and culture. Housing spatial patterns are thus the determinants of spatial and temporal developmental layers at different scales.

In order to establish perspective for sustainable urban planning and architectural design through landscape-ecological study this chapter proposes a double-oriented approach which will be implemented through a case study (Figure 1):

1. **Place-based approach** for better understanding of interrelations between social changes, economic flows, and ecological processes on the one side, and spatial patterns on the other, in the pursuit of sustainability and optimization of housing landscape of Belgrade territory, and
2. **Multi-scale approach** for identification of housing patterns through spatial and temporal scale.

These two approaches are based on a cross-examination of four thematic frameworks: (1) heterogeneity - study of the transformation of urban patterns and their impact on nature in order to identify the relationship between urban and rural in contemporary conditions, (2) hierarchy - understanding the complex manifestation of spatial patterns influence at different levels and identifying their interdependence, (3) holism - an integral study of the social, cultural and natural values that shape spatial patterns, and (4) system – highlighting landscape as a complex system of mutual relations.

Figure 1. Diagram of overlapping thematic frameworks and research approaches
Source: Authors

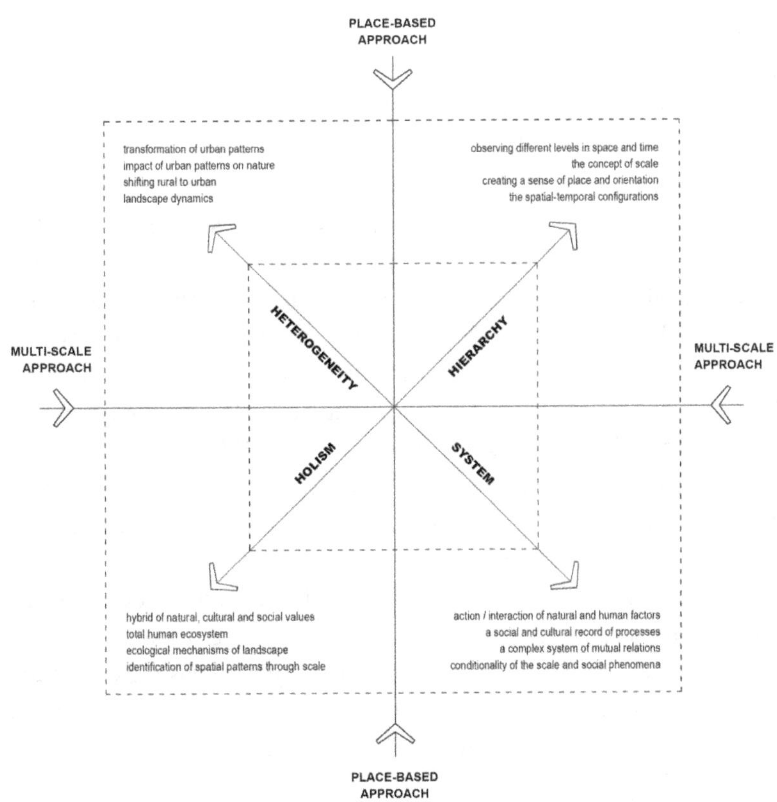

CASE STUDY: URBAN-RURAL DEVELOPMENT OF BELGRADE HOUSING LANDSCAPE 1945-1972

Among so many things that Belgrade does not have, there is something that Belgrade has and cannot be taken away, although it can be broken: it is a natural position. It is a natural plastic as created for the modelling of a larger city: the mouth of two great rivers, the hilly relief of the right riverbank and the plains of the left riverbank of the Sava and the Danube. Long-distance observation possibilities, especially from the north side, then from the riverside, place great demands in shaping the city. Such a natural situation requires a wide and powerful design. (Minić, 1955)

The prominent geographical position of Belgrade was one of the leading criteria for morphogenesis of the city territory through its long and turbulent historical development. The position of Belgrade in a very extensive agricultural area, at the crossroads of important road and waterways, and at the present time infrastructural crossings both at national and international level, has always sought to maintain a balance between infrastructure and superstructure systems on the one hand, and natural and ecological systems on the other.

After the Second World War and the period of country reconstruction, new socio-political conditions enabled the long-term planning of industrialization and economic development of the country, which opened up the perspective for planned development of the Socialist Yugoslavia capital city. The post-war period also changed the position of Belgrade from the border to the central one, which required a new reflection on the planning of the expansion of the city's territory, which spontaneously developed with a sudden increase in the population, due to which the social centers remained rather underdeveloped (Minić, 1954). Such circumstances led to the preparation and first studies for the development of a new Master Plan of Belgrade in 1945 under the leadership of architect Nikola Dobrović, considered as a pioneer of Yugoslav contemporary urbanism (Blagojević, 2007). In the middle of 1948, the *Conceptual overview of the general solution of transport and land use of future Belgrade* was completed. This study can be considered as the first urbanistic document of the comprehensive territory of Belgrade, which was given full publicity, was widely considered, served for current investments according to the first five-year plan for the development of Belgrade from 1947 to 1951, and primarily the basis for the development of the Master Plan (Stojanović & Martinović, 1978). For the urban development of Belgrade, the improvement of housing issues, the development of the social and cultural life of the city, the transformation of Belgrade according to the principle of contemporary urbanism, it was of historic importance to adopt the aforementioned *First Five-Year Plan* for the development of Belgrade within which the leading task is the wide development of housing settlements (Maksimović, 1973).

In the period between the two Master Plans of Belgrade from 1950 and 1972, a series of international events and conferences pointed out on the development of complex issues in the planning and design of residential settlements: (1) perspective housing programs, (2) industrialization of housing, (3) optimum size of housing communities, (4) housing communities, neighborhood and concept of extended neighborhood, (5) housing community structures, and (6) the necessity of complex construction and financing problems. The framework of the socialist self-governing society that was present in this period demanded the integral action of several categories from standards and conditions, social structure to productivity and general urbanization (Vujnović, 1975). In 1947, the Ministry of Construction of SFR Yugoslavia established the first norms regulating certain issues of mass housing construction. The regulations were made on the basis of an analytical procedure conducted by the Institute of the Ministry of Construction with the Republic Project Organizations and in cooperation with the Federal Planning Commission (Bajlon, 1975). The development conditions of housing issues have enabled intensive housing construction, which has become the main influencing factor in the expansion of the Belgrade territory. In the period from 1945 to 1972, between the two Master plans, residential settlements were built according to a linear principle in relation to comprehensive morphology and in an extremely expansive relationship with the rural context.

At the *Congress of C.I.B. (Conseil International du Batiment)* in Rotterdam in 1959, the opinion of the sociologist on the topic of socialization of housing and industrialization of residential architecture was especially emphasized. As one of the leading issues, the influence of the housing unit on a modern man and family is highlighted, guided by the thesis that a family housing unit is not an ordinary mechanism and pointing to the importance of the environmental relationship of the user and the structure of the housing spatial entities at different scales (Macura, 1960). At the *Seminar of the Housing Committee of the European Economic Commission in Zagreb*, held in 1961, the housing is distinguished as an integral part of comprehensive socio-economic development. One of the leading references of this seminar was housing programs, insisting on the differences between programs and plans in socialist countries and defining various types of programs - according to the time limit and territorial interaction, integration into economic or spatial programs (Stefanovic, 1961).

Within the framework of the Symposium on the topic of "Housing community as a subject of spatial planning" organized by the Federal Institute for Urbanism, Communal and Housing, held in 1962 in Belgrade, one of the basic tasks of housing settlement planning is reflected in the regulation of mutual relations of housing buildings and their spatial-programmatic additions that enhance functional, environmental, economic and aesthetic aspects (Jernjec, 1962). Housing communities as territories of self-governing communities were legally defined as a political-legal construct in planning and were considered as spatially and socially territorial in the organization of people (Bjeličić, 1962). Observing the hierarchical concept that develops in the spatial and temporal system in synergy with the natural, it is necessary to understand what is the unit of housing planning in the morphology of the city. The planning framework defined the establishment of housing communities for settlements that signify the urban unit (housing block, district) so that it is an autonomous spatial, functional and communal entity respecting the concept of boundaries in relation to spatial patterns and nature on relations new-old, high-medium-low, rural-urban, residential-non-residential. If a city is viewed as a heterogeneous totality from an urban perspective, then even the housing community as its smaller units cannot be conceived only on the basis of the norms of urban planning. Therefore, one of the basic tendencies in the planning of housing communities was that each community should include a territory that has its own specific character, as natural borders as well as synergy with the spatial context and its naturalness (Sila & Miščević, 1962).

That same year, the *26th World Congress of the International Federation for Housing, Urban Planning and Regional Planning* was held in Paris with the title "Housing and Civilization". This topic has been elaborated in three separate reports dealing with the problem of housing design and housing construction from different perspectives: (1) the evolution of housing from the point of view of man and society, that is, the development of housing along with the development of the structure, functions and life of the family of recent times, (2) technology and science in the development of urban life, and (3) decentralization of cities and independence of residential complexes, that is, the development of a network of centres that are gravitated by housing communities (Maksimović, 1962).

In order to explain in more detail, the leading role of planning and development of housing patterns below is a chronological analysis of the leading master plans for the territory of Belgrade which provide insight into the complex context of development following the perspective focused on the landscape and environmental conditions. Through this analysis, three relevant periods have been identified, which are described below in the chapter.

Phase 1: Towards Monolithic Landscape of Old and New Belgrade Territory

The *Master Plan* finalized in 1950 under the leadership of the architect Miloš Somborski, officially marks a milestone in the urban development of Belgrade, as well as the development of urban institutions and practices in general (*Figure 2A*). On the basis of economic and demographic studies, the population triple increase has been envisaged, which changed qualitatively the needs of the future million city, and recognized the necessary radical extension of residential and city territory. The largest part of the new development area is planned on the left bank of the Sava River, on the territory between the old centre of Belgrade and the oldest suburb of Belgrade, Zemun, thus forming a unique, organized, monolithic urban totality. The basic intention was that the territory of New Belgrade occupies a central rather than a peripheral character in its place and its surface. A significant contribution of this Master Plan was the contemporary understanding of the multiple importance of the greenery role in the city and the enhancement of the city landscape physiognomy (Gavrilović, 1967). Considering that Belgrade was composed

Figure 2. Morphogenesis of Belgrade territory in relation to the planned framework (Milovanović, 2018)
Source: Milovanović, 2018

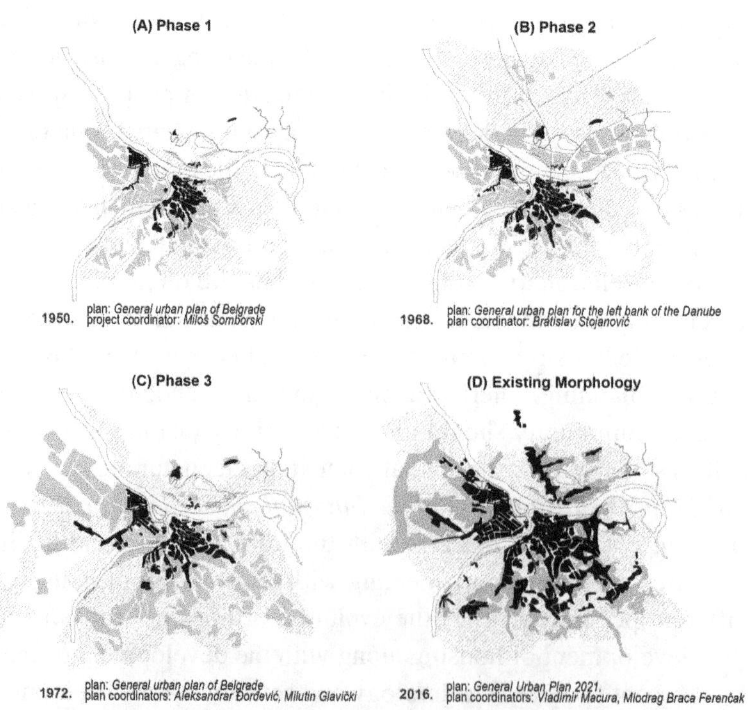

of spatial frameworks with higher topography on the right riverbanks of the Sava and Danube and on the lower plain spatial frameworks on their left riverbank, these riverbanks were, until the intensive urbanization, wetlands, deserts and steppe. In this sense, numerous criticisms are also recognized regarding the unbalanced development of the morphological image of the city, or its plastic, as well as the insufficient direction of the development axes towards the riverbanks:

But if there were attempts to form certain totality in the centre of the city, the wider, plastic design of the city, the design of its silhouette was completely ignored, although the configuration of the land on which Belgrade is located, his so often praised position, imperatively demand. It is a well-known fact that Belgrade is not river-oriented, that it is cut off by iron railroads, warehouses and unregulated areas, that the houses are turned towards the Sava River and a nice view, revealing its most rugged side view from the Sava and the Zemun side, from where, is one of the main roads in Belgrade. (Minić, 1954)

This approach to planning was to deny the natural context and geological conditions on the micro plane, which characterized this area before construction but was based on creating a new naturalness within individual blocks, which formed the basic organizational unit of the planned composition of New Belgrade. On the other hand, the Sava River, the former border and the outer periphery, is becoming a central urban motif according to the Master Plan and as such is the primary urban factor in the formation of morphology and landscape ecology of the overall urban scale.

The value of the Master Plan from 1950 from the ecological perspective is reflected in the essential respect of the Belgrade environment identity. This plan is primarily based on the fact that Belgrade is positioned at the border of two landscape totalities of a European scale - the Pannonian plain in the north and the hilly land of the Balkan Peninsula in the south. This plan indicates the tendency of the Belgrade system decentralization regarding the need to decompose the wider metropolitan area in relation to the infrastructure points (Šimić, 1985).

Phase 2: Impulses of Third Belgrade Development

As mentioned, the focus of the Master Plan from 1950 was to urbanize the left riverbank of the Sava River, while the spatial framework of the left riverbank of the Danube River is rather undefined with the marking of the green city belt and agricultural areas, as well as one residential and commercial zone (Stojanović, 1971). Since the adoption of the Master Plan and its active realization, more and more complex issues have been developed regarding the spatial framework of the Danube left riverbank, which resulted in the *Decision of the National Board of the Belgrade City on amendments to the Master plan* in 1956 (Stojanović, 1967). This decision enabled the inclusion of the Danube left riverbank into the existing urban morphology and urban structure, thus making Belgrade use the position on two large European rivers and becoming a city on all its shores (*Figure 2B*). Through its morphogenesis, the territory of Belgrade has built a relationship with the riverbanks, which can be recognized in its spatial-morphological system consisting of three subcategories: Old, New and Third Belgrade. In that sense, (1) Old Belgrade implies the compact morphology historical core of the city on the right riverbanks of Sava and Danube, (2) New Belgrade, assumes a specific spatial-morphological totality, created through application of modernistic planning on the left Sava riverbank, while (3) Third Belgrade represents a developmental framework on the left Danube riverbank of the Belgrade administrative territory (Milovanović, 2018). The developmental tendency of the new urban, morphological totality within the Third Belgrade spatial framework has been confirmed in the current planning regulation. According to the Master Plan of Belgrade 2021 (Sl. list grada Beograda, 27/03) and the Regional Spatial Plan of the administrative area of Belgrade (Sl. list grada Beograda, 10/04), the development axes of the expansion of the city territory are directed towards this territory.

Undefined planning framework of Third Belgrade is in line with the industrial development and the population concentration in urban centres, especially in metropolitan areas such as this area, caused the process of far greater engagement of landscapes in concentration zones. Such a process took place faster than planned measures and their realization, which led to the start of unplanned construction and exploitation of non-urbanized territory of the wider metropolitan area of Belgrade. On the other hand, such uncontrolled processes outperform biological potentials related to disorders of the natural balance, an ecological factor, both the landscape and the human being in it become degraded and obtain a significantly lower overall value.

In order to overcome the uncontrolled exploitation of the rural, metropolitan area in 1969, the Urban Planning Institute and the Republic Institute for the Protection of Nature of Serbia have begun drafting a *Landscape plan in the comprehensive metropolitan area of the Belgrade* (Milinković, 1969). The methodology of this plan included several phases: (1) an overview of the existing quality and morphology of the landscape, (2) an analysis of the landscape from different approaches, (3) the diagnosis of the landscape, (4) the valorization of the landscape based on the biological potential, (5) the preparation of programs for the use and conservation of the landscape in the future, and (6) the integration of the landscape plan into a master plan.

The life of a contemporary city, with its numerous economic, cultural and social functions, has an impact on its near and further environment, its content and its functions, the character of utilizing agricultural land and forest complexes, the character of settlements and the working structure of their population, on the intensity of traffic flows between the city and the broader metropolitan area. The complex organism of a city gets much more favorable conditions for the planning of individual urban structures in wider free spaces outside of the city territory, which initiates the process of urban-rural synergy. The surrounding territory of the city, on whose contents and structure, urban morphology makes visible pressure, which in its functions meets the different needs of city life and thus makes it a functional and organic complement of the city organism, is the basic spatial component of Belgrade landscape ecology.

Phase 3: The Archipelago of the Housing Patterns in the Sea of Greenery

So, we know the sea of blue water - when the Danube is flooding the high-water level and floating the plain area; we also know the yellow sea of rye cane and steppe grasses during summertime and drought periods; we know the green sea of young cereals; and the purple seas of ripe maize and grain from the time of contemporary agricultural cultivation of the city's territory on the left riverbank of the Danube. But it's not a question of those "seas". It's not a word about the Pannonian Sea, which is personified by its geographical offspring - the Danube, but the Pannonian Sea, its geological outlet: remains and witnesses at the bottom of that sea. It is about the Pannonian Sea, it remains a hundred meters below the present territory of the left riverbank of the Danube, about the depths of seawater obtained by wells at a depth of more than ninety meters (Stojanović, 1991)

The new Master Plan was adopted in 1972 under the leadership of architect Aleksandar Djordjevic (*Figure 2C*). One of the leading problems in this plan was the housing issue. Opportunities for spatial organization of the urban morphology has been explored in this plan through setting up more alternatives and through qualitative and quantitative analysis, in order to fulfil the goals of the long-term development of Belgrade (Đorđević, 1972). The methodology of this plan was developed primarily on the basis of the environmental criteria for the formation of the spatial organization of the city. The spatial organization and the environment were not observed separately but were regarded as symbiosis, as a synonym, given their indivisibility of a comprehensive quality. With this approach, the plan was harmonized program and urban matrix; and recognized that the city will not spread spontaneously in even concentric circles, but in a pronounced linear direction of expansion. The leading planning and design approach was based on the positioning of the ''archipelago of settlements in the sea of greenery'' (Đorđević, 1972: 15). In this sense, new housing spatial patterns designed in this plan are positioned in the most favourable microclimate areas in greenery, with significantly less construction densities than previously planned residential settlements.

During 1973, the *Spatial Plan of the territory on the left bank of the Danube*, which was not covered by the Master Plan itself, was adopted. This plan envisages that the largest part of the spatial coverage of Third Belgrade consists of the settlements of the Agricultural Combine of Belgrade so that the basic urban unit is a production-settlement (Stojanović, 1973). Residential settlements in this area are very specific in nature, since they belong to one large economic organization on the one hand, and on the other, they are in an extraordinary landscape-ecological system. This concept of housing development has offered new opportunities to improve the general conditions of life, housing, work, the use of leisure time on the one hand and the possibilities for establishing a unity between the city and the natural environment, for organizing life in the function of natural processes on the other (Tomić, 1980). Based on the general planning concept, residential urban settlements are central places for smaller settlements of predominantly rural character, which gravitate them. With the increase of material possibilities and with the increase of leisure time, the role of the core of Belgrade, as a social centre for the population from the metropolitan area and from the wider area, is enhanced.

TYPOLOGICAL CLASSIFICATION OF HOUSING SETTLEMENTS FROM AN ENVIRONMENTAL PERSPECTIVE

A review of the planning and research framework with particular focus on housing settlements development contributed to the establishment of a series of planning and design principles whose conceptual framework is reflected in the landscape-ecological approach and the development of environmental relations between spatial patterns and processes: (1) hierarchy - the character of the housing community in relation to the comprehensive territory of the city, the immediate environment and the place, the urban factors of the relationship between the housing community and the larger and smaller units of the city, (2) heterogeneity - natural and physiographic factors, which include relief, terrain configuration, attitude toward natural boundary zones such as parks or protected green zones, (3) orientation - distances and gravity within housing communities, (4) function - the size and optimal capacities of the accompanying objects according to the planned number of inhabitants, socio-political factors, the number of inhabitants as a factor of the size of the housing community, and (5) morphology - method, urbanistic parameters and spatial principles of construction.

In order to explain the diversity of housing patterns developed in Belgrade in the socialist period and context, the typological classification has been carried out through the establishment of four basic criteria which include (*Figure 3*):

Figure 3. Typological classification of Belgrade housing spatial patterns from an environmental perspective: nucleus, longitudinal and lamella
Source: Authors

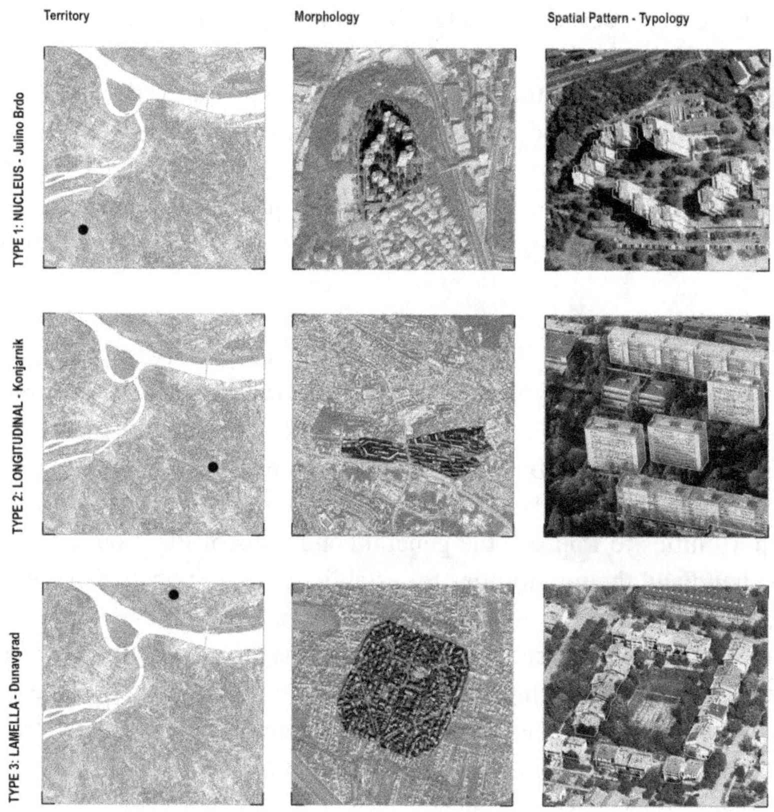

1. **the character of the territory**, in order to understand influential factors on the level of comprehensive territory of the city,
2. **the character of morphology**, in order to understand spatial process of creating and development of housing urbanity in line with landscape,
3. **the character of the housing spatial pattern**, in order to understand it as equally spatial and programmable urban construct which determinant spatial and temporal developmental layers and morphological character of housing typology at relevant levels and scales, and
4. **the approach to programming the housing pattern**, in order to understand a generic framework of housing spatial pattern for its development in the future.

In the next step, three relevant types of housing settlements are classified in relation to the defined criteria and recognized characteristics. The following table describes the most important characteristics of the types according to the criteria (*Table 1*).

Table 1. Typological classification of Belgrade housing spatial patterns from an environmental perspective

	Type 1: Nucleus	Type 2: Longitudinal	Type 3: Lamella
Character of **Territory**	• Influenced by the dynamic of nature which defines the topological contour lines.	• Characterized by natural dynamics and continuity of landscape.	• Characterized by reflection and imagination of infinite landscape.
Character of **Morphology**	• Characterized by interpenetrating of architecture and nature and enhancing by disjunctions and centralization.	• Built in line with the topography of natural landscape and represents the primary generator of landscape patterns composing and topological networking.	• Based on the principle of topography redefinition in landscape architectonic sense with respect to ecological processes.
Character of **Housing Spatial Pattern**	• Generated a balance between the autonomy of design and the irregularity of landscape context.	• With a matrix of spatial flows laid notionally over the natural landscape and its components are arranged geometrically. The architectonic structural forms are connected with elements from nature.	• Reduced, rationalized and activated in the ground plan of design in line with the fragmentation of the urban grid.
Programming approach	• Based on incorporation and transformation of themes from nature and their integration into landscape physiognomy.	• Generated in relation to patterns of movement, so it is enabled flexible composition that can grow and overlap.	• Generated in line with agricultural sense of landscape, economic and cultural interests, contact with agrarian nature and human activity in such conditions.

The first type, the nucleus is positioned in privileged locations with exceptional geographical predispositions. The spatial composition of nucleus settlements is conceptualized as centralized and enables the autonomous formation of the natural environment, which is especially pronounced in relation to the housing unit and environmental values. This type includes the settlements Knezevac-Kijevo, Julino Brdo, Banjica and Medakovic settlement.

The second type, longitudinal, is conditioned by a linearly dominant traffic infrastructure using predispositions of connectivity to a wider spatial context. The spatial composition of longitudinal settlements is heterogeneous and allows the flow of movement so that the relation between the residential building and the associated open space is particularly pronounced. The housing patterns are designed by introducing open or semi-open public spaces and activating the free ground, thus creating spatial-programming connections between people, buildings and nature. In addition to the basic urban rolls, open green spaces play the rolls of microclimate correctors in order to preserve certain natural ambient atmosphere. The settlements Gornji Zemun, Kanarevo brdo and Konjarnik belong to this type.

The third type, the lamella is designed as a system of free-standing buildings in greenery - modern spatially composed. Housing buildings are designed in a way that allow direct synergy with open public spaces and the environment. In this sense, the spatial element of design is the connecting element - covered, open or semi-open spaces for pedestrians - for the establishment of spatial and functional links between housing buildings towards the development of artistic-aesthetic synthesis. This type includes the settlements of Zeleznik, Karaburma, Sumice, Miljakovac, Dunavgrad - Kotezi and Kumodraz.

Figure 3 shows socialist housing settlement examples for each of the identified types from an environmental perspective.

In order to recognize the mutual connection between specific planning perspectives and design principles in accordance with the construction period, a chronological identification of the periods during which the socialist housing settlements were realized is established. In this sense, it is recognized that type 3 - lamella is the most represented in the comprehensive city morphology and represents the initial type that developed in the first postwar period. Type 2 was realized in the period of the 60s, which, when it comes to the comprehensive development of the city, and precisely characterized by the development of primary transport infrastructure, which was the basic constitutive motive for conceptualizing the housing spatial pattern. Type 1 was implemented during the period of preparation for the adoption of the new Master Plan of 1972, which is based on,, an archipelago of settlements in a sea of greenery'', which made it possible to urbanize the previously rural landscapes and to take advantage of the natural predispositions these sites have. The map below illustrates the distribution of types in the territory of Belgrade, which also illustrates the intensive development of the housing stock in the case study period (Figure 4).

SOLUTIONS AND RECOMMENDATIONS

Planning and design approach, in the period of socialist housing development in Belgrade, partly opened a debate on the topic of humanizing the housing environment, which, in addition to approximating urban functions, also implies the integration of nature into the environment of the housing unit. Also, this is confirmed by numerous globally established documents and agendas from this period, which among the key objectives and issues for the development of cities in the future stand for the preservation and protection of landscapes and natural resources, with the maximum reduction of negative impacts on biodiversity and ecosystems. An exceptional contribution to the global thinking of the environment and its relationship with natural conditions in the period of development of housing landscape Belgrade was achieved within the *United Nations Conference on the Human Environment (Habitat I)* in Stockholm, which brought together industrialized and developing countries to describe the rights of people in a healthy and productive environment. The result of this event is the *Stockholm Directive*, which defines the principles for the preservation and improvement of the environment and the renewal of people's relationship with nature (United Nations, 1972). The increasing impact of uncontrolled urbanization that is endangering natural conditions and the environment was recognized at the *United Nations Conference on Human Settlements* held in 1976 and confirmed by the *Vancouver Declaration of Human Settlements* (UN Nations, 1976).

The above clearly indicates the importance of synergy between the ideology of humanizing the urban environment and urbanistic achievements in order to bring nature closer to everyday life of man. In this sense, as one of the basic approaches to the structure and composition of the housing settlements as organic incorporation in the urban concept is environmentally oriented. This realization that the free space near the housing unit can become functional, the living supplement of housing in the natural environment has significantly contributed to the development of design approaches in defining the relationship between spatial patterns of residential settlements and open spaces. Based on the identified tendencies, several important principles are proposed below in the context of the recommendations in shaping the relationship between landscape and the housing spatial pattern.

Figure 4. Chronological overview of Belgrade housing spatial patterns, built between 1947-1972 with typological classification from an urban-rural synergy perspective
Source: Authors

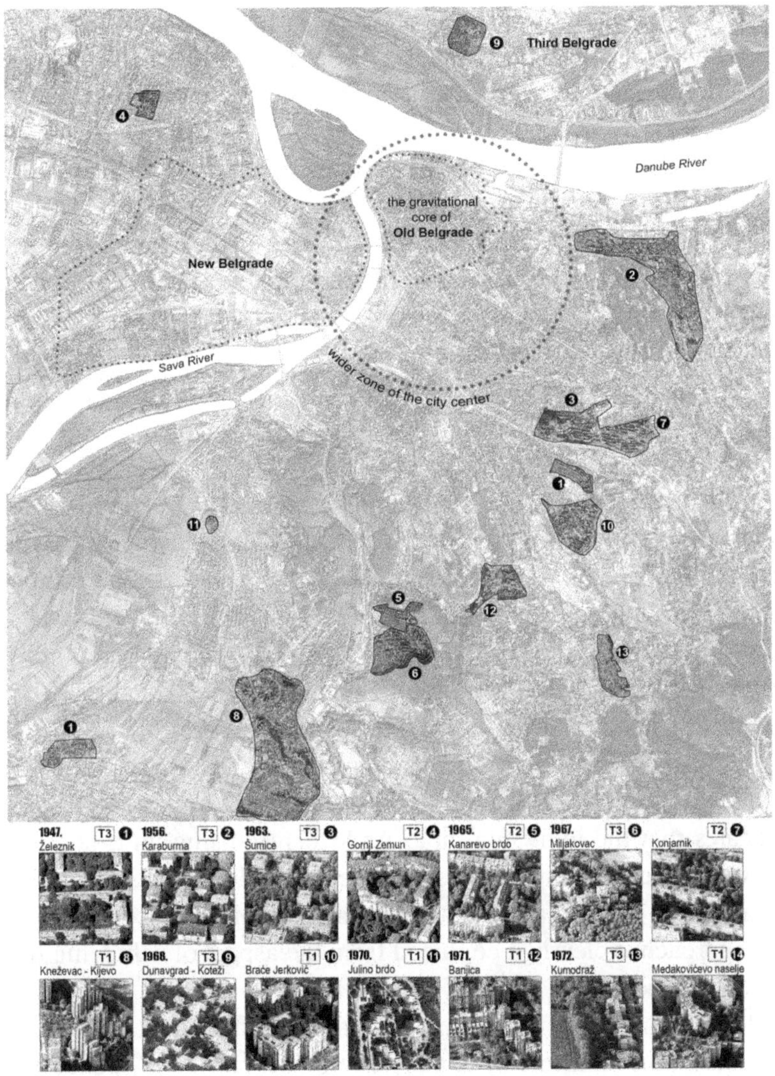

General approach for all classified types housing settlements is based on **stepping up of built and nature environment according to the system of the housing zone breakdown.** This approach implies a complementary approach to the hierarchical concept and is based on linking the organization of open spaces to the urban concept of stepping into several spatial levels of the housing group, housing settlements and housing areas. Programming the content of open spaces and the concept of their spatial structure depends on the size of the gravitational area and the need for space use.

Establishing a sense of belonging to nature in a housing community is proposed as a relevant approach for nucleus type. This approach is recognized as an important determinant in creating the identity of a housing community in accordance with nature and landscape, so that the user becomes an integral part of the environment. Due to the pronounced functional character that builds free space in synergy with the spatial housing pattern, it is possible to initiate an additional psychological and emotional characteristic of belonging within which this space makes an organic supplement and the whole of the housing.

Creating an attractive atmosphere of naturalness is proposed as a relevant approach for longitudinal type. This approach is the leading principle for the establishment of aesthetic and spatial quality of open space and the continuous use and connection of open spaces at different spatial levels established through the principle of stepping.

Determining the content and character of interrelation between nature and built environment is proposed as a relevant approach for lamella type. This approach is based on the programming of a functional breakdown of open space according to the level and size of the housing community to which the open space belongs, according to the needs of different users, according to the size of the apartments, that is, according to the prevailing family structure.

FUTURE RESEARCH DIRECTIONS

In today's conditions, the relationship between man, built environment and the natural environment can be characterized as inevitable. Strengthening the environmental perspective that looks at this phenomenon has led to more intensive research, assessment and consideration of the environment state in order to plan and design a positive urban environment in providing quality living conditions. Influential factors that shape the relationship between the built and the natural environment are numerous, ranging from spatial, functional, physical to aesthetic, psychological and design influences. The complexity of the housing typology, in spatial sense, is reflected in the fact that it is not limited to the spatial framework of the housing unit, but also engages the immediate environment, that is, the open space that includes more spatial levels of residential streets, housing blocks and common spaces.

The research indicates that the individual housing levels (housing unit, building, settlement, city) are dominantly observed in the planning and design of residential settlements within Belgrade territory, without essential overlap and consideration of the interaction between spatial levels. In this sense, the restriction of such an approach is clearly recognized from the aspect of achieving maximum results and quality of the living environment.

Each of the spatial levels has a specific role and requirements from the aspect of quality and improvement of housing. Housing unit (apartment) is the place in which a user spends most of his time and which is built by his personal characteristics and identity. It is the place of attachment and the highest level of privacy. Singular building/house is an intermediator between the housing unit and the settlement, the place for establishing relations between the individual interests and community/collective interests. Housing settlement represents a link between the singular building and the city, thus representing a compromised realization of interest groups, that is, individual and collective interests. At the end, city constitute a specific territorial and geographical entity in which satisfaction of general, specific and individual needs and interests of the population is realized. Starting from understanding the concept of the scale, the variety of scaling levels, as well as their often conflicting interests and requirements, it is essential that future research perspectives include specific techniques, methods, principles and strategies for designing and planning in order to best integrate scaling levels in space and time. The particularly important focus within future research should be dedicated to the relationship between the natural and built environment at all spatial levels. In previous studies, this relationship was dominantly explored from the aspect of city scale, while the other scaling levels were considerably less included, especially if we are talking about individual housing buildings and units.

The issues of sustainable development and environmental protection, as well as the problems of resource and biotechnological limitations of city growth, are just some of the urban issues that have become actual and crucial for the life and future of every urban environment and landscape physiognomy. However, in the extensive literature dealing with this topic, little attention is focused to urban as a process of socio-ecological exchange, so discussions of environmental problems and the possibility of a sustainable future commonly ignore the urban origin of many problems. Having in mind that the field of this chapter is based on the research of a complex, comprehensive and interdependent system of interaction between people, objects and nature, the processes of socio-ecological exchange are very important in the context of environmental-behavioral research, that is, understanding the connections between people, the natural and the built environment (Ristić Trajković, 2016).

CONCLUSION

The research has shown the characteristics of the Belgrade morphogenesis of the urban territory and its contact with the non-urbanized, natural environment of the wider metropolitan with a particular focus on socialist housing settlements. The focus of the research was achieved synergy between environmental processes and housing spatial patterns in order to achieve their balance, optimization and harmony. Accordingly, this chapter provides the conceptual framework of housing landscape as linking to place-based design approach for studying the relations between natural and social processes, including the problem of urban-rural transformation and the synergy of landscape and housing spatial patterns in order to improve their integration in future perspective. The chapter points at environmental and functional values of nature in line with different housing spatial scales, as well as the importance of overlapping and interaction between these scales in the process of urban planning and architectural design. The research also points to the importance of integral and comprehensive research of a complex and interdependent system of interaction between people, objects and nature, as well as the process of socio-ecological exchange in the context of environmental-behavioral research, and the improvement of housing at all relevant spatial levels in space and time.

ACKNOWLEDGMENT

This paper was written as a part of the projects “Research and systematization of housing development in Serbia in the context of globalization and European integrations for the purpose of improving housing quality and standards” (TR36034) and “Studying climate change and its influence on the environment: impacts, adaptation and mitigation” (43007), financed by the Ministry of Education and Science of the Republic of Serbia.

REFERENCES

Bajlon, M. (1975). Stan u Beogradu. *Arhitektura urbanizam, 74-77*, 23-42.

Balvanera, P., Caldero-Contreras, R., Castro, A. J., Felipe-Lucia, M. R., Geijzendorffer, I. R., Jacobs, S., ... Gillson, L. (2017). Interconnected place-based social–ecological research can inform global sustainability. *Current Opinion in Environmental Sustainability, 29*, 1–7. doi:10.1016/j.cosust.2017.09.005

Bjeličić, S. (1962). Stambena zajednica – Osnovna planska urbanistička jedinica grada. *Arhitektura urbanizam, 14*, 43-44.

Canter, D. (1977). *The psychology of place*. London: The Architectural Press.

Council of Europe. (2000). European Landscape Convention. *European Treaty Series, 176*.

Cvejić, J., Vasiljević, N., & Tutundzić, A. (2008). *Tipologija predela Beograda za potrebe izrade Evropske konvencije o predelima* [Typology of Belgrade's landscapes for the needs of application of European Landscape Convention]. Beograd: Univerzitet u Beogradu - Šumarski fakultet i Sekretarijat za zaštitu životne sredine Grada Beograda.

Delcourt, H. R., & Delcourt, P. A. (1988). Quaternary landscape ecology: Relevant scales in space and time. *Landscape Ecology, 2*(1), 23–45. doi:10.1007/BF00138906

Đorđević, A. (1972). Ekspoze povodom usvajanja Generalnog urbanističkog plana 1972. *Urbanizam Beograda, 19*, 13–19.

Duncan, J. (1990). *The City as Text: the Politics of Landscape Interpretation in the Kandyan Kingdom*. Cambridge, UK: Cambridge University Press.

Đurović, Đ. (1969). Sociološki pristup planiranju budućeg razvoja Beograda. *Urbanizam Beograda, 4*, 19.

Gavrilović, N. (1967). Komunalna opremljenost Beograda. *Arhitektura urbanizam, 41-42*, 27-34.

Habitat, U. N. (1976). *The Vancouver Declaration on Human Settlements. United Nations Conference on Human Settlements*, Vancouver, Canada.

Janković, N. (2017). *Transformation of Landscape Configuration in the Process of City Territory Development: Belgrade in The Second Half of XX Century* (Doctoral dissertation). Belgrade: University of Belgrade - Faculty of Architecture.

Jernejec, M. (1962). Stambena naselja danas i sutra. *Arhitektura urbanizam, 14*, 45.

John, W., & Mitchell, T. (2000). *Landscape and Power*. Chicago: University of Chicago Press.

Li, B. L. (2000). Why is the holistic approach becoming so important in landscape ecology? *Landscape and Urban Planning, 50*(1-3), 27–41. doi:10.1016/S0169-2046(00)00078-5

Macura, M. (1960). Sociologija stanovanja i industrijalizacija stambene arhitekture. *Arhitektura urbanizam, 2*, 31-33.

Maksimović, B. (1962). Stanovanje i civilizacija. *Arhitektura urbanizam, 18*, 43-46.

Maksimović, B. (1969). Funkcionalne vrednosti slobodnih prostora u blizini stana. *Arhitektura urbanizam, 56-57*, 96-104.

Maksimović, B. (1973). Kritika urbanističke perspketive Beograda iz 1948. Godine: Povodom izlaganja o planu rekonstrukcije Beograda. *Godišnjak Muzeja grada Beograda, 20*, 429-443.

McHarg, I. (1969). *Design with nature*. American Museum of Natural History.

Milinković, S. (1969). Osnovne koncepcije planiranja slobodnih površina. *Urbanizam Beograda, 4*, 27–29.

Millennium Ecosystem Assessment. (2005). *Ecosystems and Human Well-being: Synthesis*. Washington, DC: Island Press.

Mills, E. (1966). *Social identity in high-density housing* (Fifth Year Architectural Thesis). School of Design North Carolina State University.

Milovanović, A. (2018). Treći Beograd: pregled razvoja urbanističke misli i delovanja u periodu od 1921. godine do danas. *Arhitektura i urbanizam, 45,* 16-25.

Minić, O. (1954). Razvoj Beograda i njegova arhitektura između dva rata. *Godišnjak Muzeja grada Beograda, 1*, 177-188.

Minić, O. (1955). U potrazi za likom Beograda. *Godišnjak Muzeja grada Beograda, 2*, 449-460.

Naveh, Z. (1990). Landscape ecology as a bridge between bio-ecology and human ecology. In H. Svobodova (Ed.), *Cultural Aspects of Landscape* (pp. 45–58). Wageningen, The Netherlands: Pudoc.

Naveh, Z. (1995). Interactions of landscapes and cultures. *Landscape and Urban Planning, 32*(1), 43–54. doi:10.1016/0169-2046(94)00183-4

Naveh, Z. (2000a). What is holistic landscape ecology? A conceptual introduction. *Landscape and Urban Planning, 50*(1-3), 7–26. doi:10.1016/S0169-2046(00)00077-3

Naveh, Z. (2000b). The Total Human Ecosystem: Integrating Ecology and Economics. *Bioscience, 50*(4), 357–361. doi:10.1641/0006-3568(2000)050[0357:TTHEIE]2.3.CO;2

Nikezić, A., & Marković, D. (2015). Place-Based Education in the Architectural Design Studio: Agrarian Landscape as a Resource for Sustainable Urban Lifestyle. *Sustainability, 7*(7), 9711–9733. doi:10.3390u7079711

Norberg-Schulz, C. (1980). *Genius loci: toward a phenomenology of architecture*. New York: Rizzoli.

Palmer, M., Bernhardt, E., Chornesky, E., Collins, S., Dobson, A., Duke, C., ... Turner, M. (2004). Ecology for a crowded planet. *Scienece, 304*(5675), 1251–1252. doi:10.1126cience.1095780 PMID:15166349

Relph, E. C. (1976). *Place and Placelessness*. London: Pion Limited.

Ristić Trajković, J. (2016). *Architecture and nature: The application of environment-behavior theories in architectural design* (Ph.D. Thesis). University of Belgrade – Faculty of Architecture.

Sila, Z., & Miščević, R. (1962). Optimalne veličine stambenih zajednica. *Arhitektura urbanizam, 14*, 46.

Šimić, Lj. (1985). Namena površina i životna sredina. *Arhitektura urbanizam, 96*, 34-35.

Stefanović, D. (1961). Seminar Komiteta za stanovanje Evropske ekonomske komisije u Zagrebu. *Arhitektura urbanizam, 11-12*, 46.

Stojanović, B. (1967). Beograd na levoj obali Dunava. *Arhitektura urbanizam, 41-42*, 96-98.

Stojanović, B. (1971). Beograd na levoj obali Dunava. *Godišnjak Muzeja grada Beograda, 18*, 333-383.

Stojanović, B. (1973). Prostorni plan Beograda na levoj obali Dunava. *Urbanizam Beograda*, *24*, 8–18.

Stojanović, B. (1991). Beograd na Dunavu. *Godišnjak Muzeja grada Beograda, 38*, 201-215.

Taylor, K., & Lennon, J. L. (2011). Cultural landscapes: A bridge between culture and nature? *International Journal of Heritage Studies*, *17*(6), 537–554. doi:10.1080/13527258.2011.618246

Tomić, V. (1980). Humanizacija života u gradu. *Urbanizam Beograda*, *59-60*, 25–26.

Toth, R. E. (1988). Theory and language in landscape analysis, planning, and evaluation. *Landscape Ecology*, *1*(4), 193–201. doi:10.1007/BF00157692

Troll, C. (1971). Landscape ecology (geoecology) and biogeocenology–a terminology study. *Geoforum*, *2*(4), 43–46. doi:10.1016/0016-7185(71)90029-7

Turner, M. G. (1989). Landscape ecology: The effect of pattern on process. *Annual Review of Ecology Evolution and Systematics*, *20*(1), 171–197. doi:10.1146/annurev.es.20.110189.001131

UNESCO World Heritage Centre. (2005). *Basic Texts of the 1972 World Heritage Convention*. United Nations Educational, Scientific and Cultural Organization.

United Nations. (1972). *Report of the United Nations Conference on the Human Environment*. United Nations Publication.

United Nations. (2015). *Transforming our world: the 2030 Agenda for Sustainable Development*. A/RES/70/1.

Van der Poel, A. J. (1976). From landscape care to research in landscape planning. *Landscape Planning*, *3*(4), 363–370. doi:10.1016/0304-3924(76)90080-0

Vasiljević, N. (2012). *Landscape planning as an instrument in spatial development of Serbia* (Doctoral dissertation). Belgrade: University of Belgrade - Faculty of Forestry.

Vujnović, R. (1975). Prognoze i tendencije u stambenoj izgradnji u narednom desetogodišnjem periodu. *Arhitektura urbanizam, 74-77*, 9-10.

Wiens, J. A., & Milne, B. T. (1989). Scaling of "landscape" in landscape ecology, or, landscape ecology from a beetle's perspective. *Landscape Planning*, *3*(2), 87–96.

Wu, J. (2008). Making the case for landscape ecology: An effective approach to urban sustainability. *Landscape Journal*, *27*(1), 41–50. doi:10.3368/lj.27.1.41

Wu, J. (2012). Landscape Ecology. In R. Leemans (Ed.), *Ecological Systems: Selected Entries from the Encyclopedia of Sustainability Science and Technology* (pp. 179–200). Springer New York. doi:10.1007/978-1-4419-0851-3_575

Zonneveld, I. S. (1989). The land unit – a fundamental concept in landscape ecology, and its applications. *Landscape Ecology*, *3*(2), 67–86. doi:10.1007/BF00131171

Zonneveld, I. S. (1990). Scope and concepts of landscape ecology as an emerging science. In I. S. Zonneveld & R. T. T. Forman (Eds.), *Changing landscapes: An ecological perspective* (pp. 3–2). New York: Springer-Verlag. doi:10.1007/978-1-4612-3304-6_1

KEY TERMS AND DEFINITIONS

Housing Landscape: A complex social mechanism that is created under the influence of everydayness and ideological values that are reflected at different spatial levels and scales in line with nature and environmentalism.

Housing Physiognomy: Generated spatial framework and character of housing in accordance with morphological, typological and program aspects.

Housing Spatial Patterns: An equally spatial and programmable urban construct which determinant spatial-temporal developmental layers and morphological character of housing typology at relevant levels and scales.

Morphogenesis: Spatial-temporal process of urbanity that defines the morphological image of the city, urban units or spatial units.

Multi-Scale Approach: Based on the concept of scale as equally spatial and temporal dimension in order to understand spatial heterogeneity and the influence of spatial heterogeneity on the social and ecological mechanisms.

Programming: A complex analytical method which involves complex operations in classifying design aspects in order to define the problem which becomes the focus of solving through design or planning process.

Section 2

Urban–Rural Cooperation in Tourism

Chapter 7
Information System Supporting Heterogeneous Sources for Promoting Destination of Rural Areas

Miroslava Raspopovic Milic
Belgrade Metropolitan University, Serbia

Milena Vukmirovic
https://orcid.org/0000-0002-0239-3157
Department of Landscape Architecture and Horticulture, University of Belgrade, Serbia

Svetlana Cvetanovic
Belgrade Metropolitan University, Serbia

ABSTRACT

Tourism in rural areas is considered small-scale tourism, which represents a strong potential for growth of areas traditionally characterized as agricultural areas. Rural tourism tends to be heterogeneous with many different and scattered stakeholders, making the efficiency and effectivity hard to achieve. Many ICT technologies have found their way into smart tourism. Even though there are plentiful user generated data and smart tourism applications, they represent a very heterogeneous sources that are challenging to integrate in one scalable system. The aim of this research is to propose a model for information system that will increase efficiency of the rural small-scale tourism by using both internal and external systems, such as social networks, local services, geotagged resources, sentiment analysis, and data- and text-based mining systems. The goal of this information system is to gather a rich database that will allow users to identify their next destination and to identify most valuable assets for each location in the region.

DOI: 10.4018/978-1-5225-9932-6.ch007

INTRODUCTION

Information and communications technology (ICT) tools and applications have allowed increase of the performance and effectiveness in tourism. ICT has influenced ways tourism services are accessed and consumed. Tourists typically use their ICT devices to search for information to help them form decisions in regard with their trip (Buhalis & Amaranggana, 2015). This is one of the reasons why technology and its applications have had instrumental role in tourism marketing and many businesses had to redefine their marketing model. Usage of new technologies provided benefits not only as the increase in the supply of information, but it also provided a channel for the information sharing to larger audience, while reducing the cost of production.

While there are various benefits to using different technologies and application, there are also drawbacks of their usage. The abundance of information on the Web poses a challenge when trying to filter useful information on various destinations. This is especially true for tourism industry where information is distributed over various websites and stored using heterogeneous formats (Haller, Pröll, Retschitzegger, Tjoa, & Wagner, 2000).

Destination image formation no longer relies on brick-and-mortar travel agencies and glossy travel brochures, but rather on the user created content, usually based on various shared experiences, advices, comments and photographs (Hunter, Chung, Gretzel, & Koo, 2015). Currently, on TripAdvisor there are 465 million reviews and opinions and 390 million monthly average unique visitors (TripAdvisor, 2017). Furthermore, usage of mobile devices has changed tourist services and offerings. Mobile technologies offer information to tourists not only while planning a trip, but also at the destination. GoodWorkLabs reports that 85% use smartphones to plan their travel when on leisure tours, 72% people will post photos about their travel on social platforms like Facebook, 30% use mobile apps to find the best hotel deals, 29% use mobile apps to find the best flight deals, 55.8% tourists use mobile apps to check weather, 49.1% tourists use mobile apps to use mapping features, 62.1% tourists use mobile apps to search for nearby restaurants, 48.1% tourists use mobile GPS to get travel directions, 46% tourists use apps to find hotels (GoodWorkLabs, 2015).

While the industrial tourism is popular and ICT usage in its promotion is visible, rural tourism is largely individual oriented and may or may not have a mass appeal (Shanker, 2008). Many of the assets in rural tourism are not very well known. Using ICT tools and applications, these assets can be brought into mainstream tourism. In order for this to happen, it is essential to create a model for a user-friendly platform to promote local regions. Tourist services in rural areas are typically not well established. Moreover, they need to be continuously innovated and diversified so that tourist can maintain their interest in the region and to increase in the influx of tourist visits to rural areas. Rural areas typically hold their income based on the agricultural activities. Hence, developing and diversifying rural tourism can counteract the income from the agricultural sources, as tourists bring income and purchasing power to rural areas.

Rural area tourism provides perspective and new economic possibilities to the younger generations. Keeping younger generations in rural areas can preserve villages and strengthen the communities by not only strengthening them economically but by preserving local culture and heritage. Tourism areas may offer a dynamic environment for young generations given good social and job opportunities and a generally smooth entry into the labor market (Möller, 2015).

The main objective of introducing a novel information system (IS) within rural areas is to establish the practice that will generate positive impact on tourism and contribute to competitiveness in promotion of regional assets. The idea behind information rich information system is to promote not only very well know regional assets, but also a less known "hidden jewels" of the region and some less known activities and accommodations that are hard to find on the web. Information rich IS utilizes different technologies, while gathering different format of digital media that represent the region through digitalization of its assets. Digitalization of assets allows one to present valuable information about the region, which is not easily accessible on a daily basis.

BACKGROUND

We are living in the world dominated by the process of urbanization. According to the UN projections it is expected that 68% of world's population will live in cities by 2050. "Projections show that urbanization, the gradual shift in residence of the human population from rural to urban areas, combined with the overall growth of the world's population could add another 2.5 billion people to urban areas by 2050, with close to 90% of this increase taking place in Asia and Africa" (United Nations Department of Economic and Social Affairs, 2018). The mentioned situation has a great impact on rural areas and their development.

Following the general statistics for EU, 28% of the EU-28 population lived in rural areas in 2015. Considering the period from 2010 to 2015 there was a gradual increase in the number of people living in rural areas across the EU-28, their relative share of the total number of inhabitants rising by 1.7 percentage points (EUROSTAT, 2018). Observed migrations to rural areas is recognized as a new phenomenon in the European context, having significant and growing impact on peripheral and rural areas (Kasismis, 2010).This situation was explained focusing on a number of real and perceived advantages that could attract the people to live in rural areas such as lower housing and living costs, more space, a better social fabric, less pollution, closer proximity to nature and less stressful lifestyle. Also, a number of factors are identified as possible explanations for this situation divided into two groups. The first looks at the restructuring of agriculture as a reason that create significant demands for labor which could not be satisfied because of the unfavorable demographic changes in rural areas related to rural exodus and ageing of the population. The other explanations is seen in development of other non-agricultural activities such as "manufacturing, tourism, housing expansion, new consumption patterns connected to leisure and recreation that have increased demand for labor" (Kasismis, 2010).

However, at the same time, there are a number of serious shortcomings of rural areas that do not support this positive trend. They include fewer local education or job opportunities and choices, difficulties in accessing public or transport services and a lack of cultural and social venues for leisure activities requiring infrastructure. Also, considering the spatial distribution of the mentioned positive trends, it could be concluded that over the past decades "a shrinking population has become the normal trajectory for many rural regions as agriculture has been restructured and population and employment have become increasingly concentrated in urban centers" (ESPON, 2017). The policy recommendation to deal with these challenges is in finding a multidimensional and innovative solutions to address shrinkage. On the national and the sub-national level the Policy (ESPON, 2017) recognized six solutions of which one relates to the development of alternative economies and smart diversification, where changes in tourism and residential patterns is seen as one of the opportunities. In view of this, regional innovation potentials can be further exploited and "capitalized upon through targeted investment in ICT in the context of joined-up regional adaptation strategies" (ESPON, 2017, p. 12).

Smart tourism deals with products and services that can help identify and increase efficiency in usage of accommodation, transportation, events, restaurants, local heritage sites, which can all be stored as data in ICT applications, which provide easy and efficient access of information to travelers and service providers. Usage of different ICT platforms and devices such as smart phones have provided better flexibility in establishing the connection between humans and smart applications, hence, providing a sustainable model and improving the quality of life (Gretzel et al. 2016; Rivera et al. 2016). The widespread of smartphone applications raise more interest in establishing user friendly systems that provide more mobility to their users, "smart, connected products... new functionality, greater reliability, higher product utilization" (Porter & Heppelmann, 2014, p.4). However, even though tourist are likely to use mobile applications, the abundance of different applications are making it harder to narrow down the search, as tourist are looking for optimized search that can fit their time and budget. Personalized services can be a strong predictor when booking and using the accommodation (Ozturk, Nusair, Okumus, & Singh, 2017).

Rural Areas, Rural Tourism, and Digital Technologies

Three theoretical approaches are used to define term rurality: negative, sociological and economic. For the first approach, everything that is not urban is rural, i.e. low density, with little artificial character, dispersion of activities and communities. In terms of socio-cultural criteria, rurality is related to specific social relationships, the value system, and lifestyle and consumption patterns. While the economic approach observes rural as areas where economic activities are little diversified and where agricultural activity is dominant in terms of its share of jobs and income. At the international level, the most frequently used approach is that proposed by the OECD. They have established a regional typology according to which region has been classified as predominantly urban, predominantly rural and intermediate.

Accordingly to the mentioned approaches, rural areas are dominant for the EU territory (Figure 1), and therefore, development of rural areas represents a significant challenge as it has a great relevance for the economic and social context of EU[1] (ESPON and UHI Millennium Institute, 2011). EU rural areas are facing multiple problems in the domains of demography (aging, gender disparities); remoteness, infrastructure, and access to basic services (housing, access to transport, commuting, ICT, access to health care and social services); education and human capital; problems of labor market and vicious circles (European Commision, 2008). In regards to this, any potential for rural development is seen as "a vector of geographic cohesion, driving the competitiveness of rural areas and supporting the sustainability of human activities and rational management" (CIHEAM, 2008).

Observing the weakness of rural areas, tourism has attracted growing interest of academics, tourism professionals, investors and politicians alike, due to its apparent potential as a development tool. This was also confirmed by an indication of the various types of secondary gainful activities (EUROSTAT, 2016, p. 13) that were practiced by agricultural holdings in 2010 with 12.5% of agricultural holdings that also offered tourism services in the EU-27.

Figure 1. a) EU Degree of urbanization for LAU2 (Eurostat - GISCO, 2013) b) EU Structural typology map (ESPON and UHI Millennium Institute, 2011)

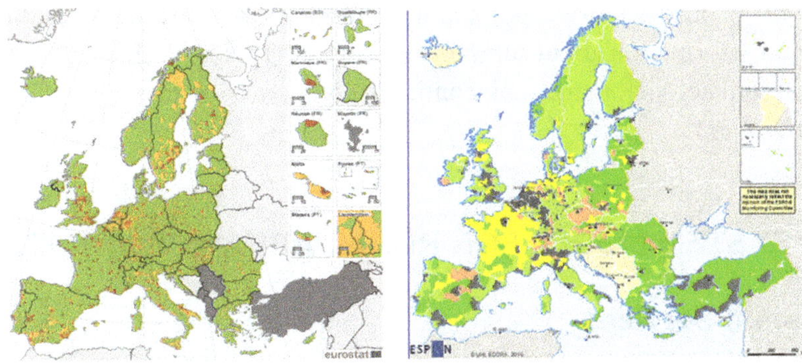

There are different views of what is rural toursim. The OECD (1994) define rural tourism as all types of tourism taking place in rural areas, while some authors under the rural tourism considered a quite specific tourism product, with some requiring the presence of agriculture as a core element (Cavaco, 1999). Following the negative approaches in defining term "rural", rural tourism could be as well defined as something opposed to mass and resort/urban forms of tourism. In such way, rural tourism is characterized by "features such as small scale, personalized contacts, the traditional character of service elements and environments, the presence of nature and agriculture and the existence of traditional social structures, reflected in a specific way of life that tourists wish to discover and participate in" (Eusébio, Kastenholz, & Breda, 2014, p. 15). This characterization is very important since there are different kinds of rural areas with diverse opportunities for development, which dictate the character of the tourism product.

The main goal of the Europe 2020 strategy to create the conditions for smart, sustainable and inclusive growth, poses a complex and difficult challenge for rural areas. This challenge in particular relates to the general characteristics of the rural areas reflected in "low accessibility, negative migratory balance, and low education levels" (Naldi, Nilsson, Westlund, & Wixe, 2015). Having in mind that smart growth of rural areas comprises of exploitation of local amenities and development of creative economies, rural tourism is emerging as a very important activity. However, reaching of embeddedness, relatedness, and connectivity as key areas in order to reach smart growth, as well as expected tourist demand can be considered difficult for the achievement.

In accordance with the presented situation problems can be overcome by using digital technologies, i.e. ICT. Implementation of ICT tools and applications can achieve positive outcomes in three areas: connectivity, entrepreneurship and human capital. In the first domain, telecommunications technologies and the Internet diminish, if they do not entirely erase, the tyranny of space and distance. From the second flexible manufacturing and smaller optimal sizes of products provide greater possibilities for rural firms against giant competitors. The third is related to continuing population growth in rural areas, both from new migrants attracted by rural amenities and by return migrants, promises an upgrade of skills demanded for the new economy (Malecki, 2003).

Furthermore, ICT could also contribute to more sophisticated ways of finding, as well for offering of diverse tourist products and services. This is very important characteristic because of the "shift from standardized mass tourism to more individualistic patterns, in which greater flexibility and a more meaningful experience have gained prominence" (Briedenhann & Wickens, 2004, p. 73). Because of the different motivations that might include ecological uniqueness, special adventure opportunities, cultural attractions, or the peace and quiet of the countryside there was a unique opportunity for rural operators to manage in terms of 'economies of scope' and a need for creation of the special digital tool for the integration of different information about rural tourism offers. This is particularly important because of "a complex, multi-faceted activity, marked by continuously increasing diversity" (Eusébio, Kastenholz, & Breda, 2014, p. 16).

INFORMATION SYSTEM MODEL FOR RURAL AREAS DESTINATIONS

Information System Services

The primary aim of the rural area information system (RAIS) is to integrate both existing systems that provide information about the rural areas, as well as to use internal systems that will provide one well rounded and self-populated system. It is important not to rely only on the information that is provided by RAIS users, but also by external sources such as local tourism enterprises, weather forecast services, rent a car services, other tourism web pages, forums newsgroups, and most importantly social networks. Using different sources of information can provide accurate and up-to-date information about rural areas' cultural, historical, geographical, ethno sites, as well as the relevant facilities they offer such as accommodation, restaurants and different agricultural products. The aim of the RAIS is to attract tourists to visit certain rural areas, and increase visibility to local events, restaurants, accommodation, products, produce, cultural heritage, etc. By using such a diverse sources of information, RAIS will enable access to large amount of other relevant tourism information already distributed all over the web such as information about weather and traffic conditions, schedules of trains, buses, low cost airlines or shuttle transportation.

In order to support those objectives, RAIS services can roughly be categorized in three different ways:

- Services that are accessible to the public, whereby the most important ones are: (a) services for retrieval of all types of tourism information in order to find relevant tourist products, (b) booking services for accommodation that are based previous and current user search, (c) services for recording digital content generated directly by users that are relevant for the information system.
- Services that allow authorized stakeholders (travel agencies/organizations, local tourist enterprises or local community) to create and update information about their local assets of interest.
- Configuration services which are accessible only to stakeholders with administrator privileges and which allow system configuration in various ways over the Internet.

RAIS defines an integrated system with both internal and external systems. The goal of this information system is to reduce extensive research that potential tourists usually go through in order to identify their next destination. With the information system with rich database, the objective is to provide easy approach to identify most valuable assets for each location in the region.

Internal systems have several important functions:

- RAIS relies on the very rich database, which is updated by the shared activity of various public and private entities

- Database allows data collection through web platform and portable devices such as mobile phones and tablets. Data collection through mobile phones and tablets allow for tourist to upload their photos, information, opinions and experiences for a given asset.
- Mobile application provides easier access to the database, allowing Mobile tag system to sort and store digital assets by users. Mobile tag system can also provide variety of information through the tag reading software including displaying text, images, video, audio, hyperlinks, contact information, etc.
- Using the data in the RAIS's database, RAIS has the functionality to automatically generate the guidebook from the selected spots of interest. As such, RAIS can serve as a virtual and portable visitor information center.
- Unlike most tourist ICT platforms which allow search by destination only, RAIS provides opportunity to also select assets by type and interests. Type of destination is used as a primary search category: cultural site, historical site, geographical/nature site, ethno site, events, accommodation, restaurants, etc.
- RAIS allows to input different forms of digital assets such as virtual tours, augmented reality tours, videos, text/description of destination, contact information, maps, photographs imported by owners of hotels, restaurants, tourist organizations, etc, photographs imported by tourists who visited the place, interesting happenings and information (i.e. festivals).
- Flexible booking system for accommodation, which is compatible with external booking systems.
- Internal systems have to be integrated with information provided by external systems and resources. Internal systems are using information provided both by its users and by external systems, for the goal of recommending certain location, event or cite. These information are used to:
    - geotag resources for particular location,
    - conduct sentiment analysis for particular location
    - data and text mining from social networks for providing more insights for the recommender system.

Integration with external systems, such as Google maps, weather forecast, car rentals, social networks and public transportation, provides a complete solution for all key stakeholders (Figure 2). There are

Figure 2. Collection and distribution of data

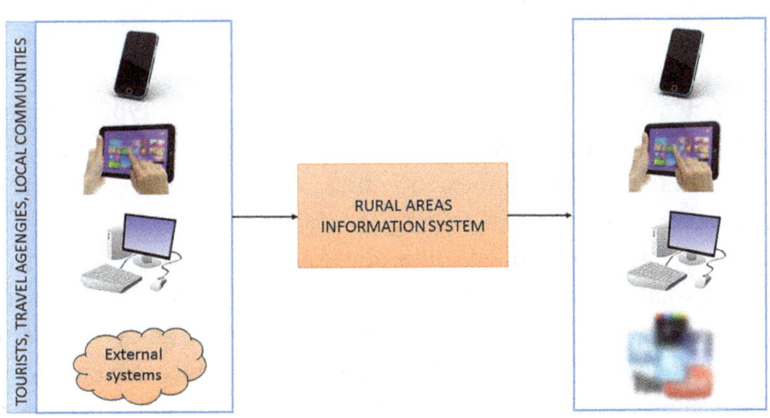

three critical key stakeholders: travel agencies, local community promoting their local assets (restaurants, accommodation, etc.) and potential tourists. System is not excluding other stakeholders, but they are not the scope of this paper and are left to be considered for future research.

Heterogeneous Sources

Tourists have become inexhaustible source of information, which can help in personalizing services for each tourist using smart tourism applications. Different systems have different approach in structuring and processing their data. Trying to integrate different systems, in order to create one homogeneous and easy to use system for users, poses many challenges. One of the challenges is collecting useful data, and once collected, how to process data from all these diverse systems. Data from external systems can be obtained as structured, semi-structured and unstructured data. This is the reason why RAIS has multiple input sources. While RAIS can control format of data stored from its users, data formats from external resources cannot be controlled and have to extra processed and formatted. Ultimate goal of RAIS is to provide recommendation based on all of the findings and data processing of its internal and used externals systems. RAIS integrates and uses data from all these systems. The Recommender systems collects and presents rural areas information through and is based on the combination of text and data mining, sentiment analysis, geotagged resources, as well as additional information provided by car rentals systems, public transportation services, tourist office systems, weather forecast systems. Based on all these systems, ranking is done for each location of interest. Recommender system uses this ranking together with the ranking from the internally stored information from RAIS users, as well as the user preferences. Recommender system provides users with top ranked locations, their photos, current events and points of interest for that location (accommodation, restaurants, etc.).

There are several systems that should be carefully considered in this integration: social networks, weather forecast systems, local tourist information system, car rental services, google maps, and local heritage application. The biggest challenge come from collection and mining of social networks data, as they come in both unstructured and semi-structured formats. However, social networks data carry very useful information that can uncover the localities based on geotagged resources from social media for cultural heritage tourism. Given a particular geolocation (such as latitude and longitude), RAIS can extract from social network data local resources and information that can be provided to end users.

RAIS collects data from social networks, and at the same time it filters it from the existing noise. Social network noise can be generate in different ways, but nevertheless it is important to try to use de-noising techniques in order to reduce influence of noise, and try to avoid invalid and unreliable conclusions and recommendations. Noise and bias can be generate by social bots (Allem & Ferrara, 2016), data collection itself (Raspopovic & Vukmirovic, 2018) and mostly by users itself whose posts consist of spelling errors (intentional or unintentional), included images, links or any other type of additional information, other than text. This type of noisy data requires advanced text mining, natural language processing and data mining techniques to extract useful information that can be related to different behavior (Rashidi, Abbasi, Maghrebi, Hasan, & Waller, 2017).

In RAIS social network data represent an important source for most of its internal systems. Based on the geotag data in the social network data pool, data can be classified and structured for particular location. RAIS uses geotagged resources (photos, comments and tags), analyzes these resources and uses these information to guide tourists in either deciding their next travel location or guides them during their traveling (Nguyen, Camacho, & Jung 2017). In order to determine relation between geotagged resources and location itself, method proposed by Nguyen et al. is used. In this approach combination of semantic tags and media data such as image data bring out much more useful information for users during their rural area travels. Furthermore, geotagging provides more insightful information for all other internal systems, as all of their finding and information will essentially have to be geotagged. For instance, geotagging is essential in the combination with conclusions and opinions that come from sentiment analysis.

Sentiment analysis provide subjective impressions, attitudes, stands and opinions about certain topic, which are extracted using many different techniques from data and text mining, natural language processing, machine learning, etc (Mäntylä, Graziotin, & Kuutila 2018). Sentiment analysis tools and methods classify opinions as positive, negative or neutral (Dave, Lawrence, & Pennock, 2003). Based on sample of data from social networks RAIS's goal is to determine how people are responding to certain rural areas and its events, locations, accommodations, etc. Focusing on textual contents including emoticons in relation with hashtags and geotags, in the context of sentiment analysis, significant data regarding the quality of rural area location can be obtained (Raspopovic & Vukmirovic, 2018). RAIS considers the quality criteria of each location by additionally introducing interpretation of emoticons as an additional factor for opinion polarity determination. Outputs of the sentiment analysis represent one of the inputs for the Recommender system.

The idea of RAIS's text mining system is to build a database of local events, restaurants, and accommodations for location of interests in rural areas, and to build pool of information which will allow to study travelers' preferences. In a sense, text mining works similar to the data mining. However, text mining deals mostly with unstructured or semi-structured data (web pages, forums, newsgroups, etc.) (Berson, Smith, & Thearling, 2000). This text mining tool turns unstructured data into structured data, keeping only the relevant information for Rural Area Information System. The goal is to jointly analyze quantitative and qualitative data with data processing system and feed it to the Recommender system.

The data mining system is another internal system within RAIS, which plays a crucial factor in providing critical information to the Recommender system. Due to the large amount of collected data analytics are needed to make sense out of them. The objective of the data mining system is to improve decision-making, opportunities, and overall performance of RAIS. Big data can be used to interconnect the dispersed information from different systems and then improve decision making capability (Song & Liu, 2017).

Software Architecture

RAIS's software architecture is designed to be as flexible as possible, in order to serve existing internal and external systems, as well as to leave room to add more additional systems in the future with ease. Figure 3 represents RAIS's software architecture, which is service-based framework that has both Web services (WSs) and mediator elements (proxy and adapter). The integration of the *Recommender system* with other internal components in the architecture is based on the mediator elements: *Data mining system mediator, Sentiment analysis system mediator, Geotag processing system mediator, Text mining system mediator* and the *Internal data processing system mediator*. The role of each presented mediator element is to facilitate communication between the *Recommender system* and the correspondent internal systems. This communication is enabled using output data generated by each of these internal systems and is fed to the *Recommender system*. RAIS is integrated with social networks and other external systems by using the pre-existing SOAP based or RESTful WSs. WSs based on SOAP use a single standardized SOAP

Figure 3. RAIS's software architecture

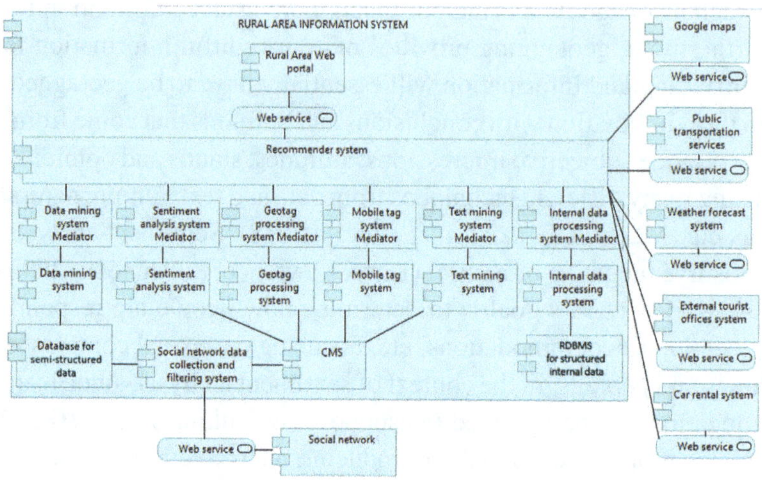

message format, while RESTful WSs use a variety of MIME document types, which include SOAP and JavaScript Object Notation (JSON).

Direct access to all services is offered through common Web portal dedicated to all stakeholders. In order to enhance its accessibility, Web portal should be implemented by employing the Web technologies recommend by the latest W3C standards and novel technologies in modern smart phones. Web portal deals with structured and semi-structured tourism data generated by the Recommender *systeml*. In order for all these data to be presented to users in an understandable format, various visualization techniques are used. One of the most commonly used interface designs is the visual dashboard, which aggregates and displays information from various sources.

RAIS and its functionalities are strongly relying on the extensive usage of social media platforms, such as Facebook, Twitter, LinkedIn, as well as the smaller and more specialized sources such as Internet forums, blogs, Wikis etc. All these platforms represent big data environments, and their data are used as a source of structured and unstructured data that needs to be processed and utilized for RAIS purposes. All textual and multimedia data like pictures, videos, sounds and geolocations can be treated as unstructured data. On the other hand, data about RAIS users' profiles and friend/follower relationship are the example of structured data.

RAIS's component *Social network data collection and filtering system* has multiple roles. On one hand, this components main objective is to collect relevant data across popular social media platforms using WSs. On the other, this component does initial processing of the data, with the aim to reduce noise that came with data collection. Besides the noise cancelation techniques, the idea is to filter relevant from the irrelevant data. As RAIS's *Recommender system* needs to make recommendations to users based on all of the findings, unstructured data needs to be formatted in the form that RAIS's internal systems need it, either as semi-structured or structured data. This functionality is also conducted within the *Social network data collection and filtering system.*

Unstructured data are stored in the content management systems (CMS). CMS provides preservation, organization and dissemination services for digital collections. CMS is the place where geotagged resources and other multimedia resources like virtual tours, videos, maps, photographs of hotels, and restaurants are stored for a particular location. All of these data can be imported from external systems or can be placed in the system by RAIS's users through Web portal. This the reason why CMS is presented as the base for *Mobile tag* and *Geotag processing systems*. Also, CMS is the system reserved for storage of: (i) textual information extracted from external tourists blogs and tweets, and (ii) opinions and textual descriptions about visited location from RAIS's users. CMS can be treated as the input component of the *Text mining* and *Sentiment analysis systems*.

The *Text mining system* uses unstructured data from CMS in order to structure them, derives appropriate patterns from data and finally evaluates and interprets the results. These results are passed on to the *Recommender system*. The *Text mining system* organizes its data into relevant subgroups (clusters) for further analysis. A significant challenge in the clustering process is to form meaningful clusters from the unstructured textual data which can be used for data distribution and classification.

Another way to understand textual information from CMS is to apply sentiment analysis which is done by RAIS's internal component the *Sentiment Analysis system*. *Sentiment analysis system* scores textual data about location of interest based on the text and emoticon analysis. Text-sentiment scoring is based on the sentiment-based and emoticon-based lexicons. Sentiment score for the relevant location or its amenities is calculated as the average of all sentiment-scores and is passed to the *Recommender system*.

Component *Data base for semi structure data* is intended for the storage and manipulation of semi-structured data. Semi-structured data is a form of structured data that does not conform with the formal structure of data models associated with relational databases, but contains tags or other markers to separate semantic elements and enforces hierarchical structure within the data. Dealing with XML semi-structured data formats requires RDBMS to be enhanced with new data types for representing XML documents inside databases and new capabilities for querying and managing the XML documents (Oracle 8i and IBM DB2 support such characteristics). But in the case of JSON data format usage, document-based NoSQL database (MongoDB) can be used.

Data mining system is an important component of RAIS architecture. This component obtains data from popular travel destination, places of interests and popular cultural attractions. Mined information can help travelers with the best travel routes and different recommendations. *Data mining system* uses different machine learning and statistical and visualization techniques.

All information on tourism products can be treated as structured information. This data is stored within a central *Relational database management system (RDBMS)*, which is constructed on the basis of conceptual data model that incorporates all conceptual entities gathered during the requirements definition with numerous tourism information providers. *RDBMS* is the base of *Internal data processing system* which main role is to store and retrieve structured data of RAIS's users: their personal data, gender, age, accounts, previously visited destinations etc. In order to generate relevant and personalized information for each particular user, this data is related to multiple structured and semi-structured data.

Pros and Cons of the RAIS System

There are many advantages of RAIS system usage. RAIS provides efficiency in integrating different types and forms of information for certain localities of rural tourism. Furthermore, RAIS provides services that allow for this information to be consumed and shared to a larger audience: travel agencies which can offer their tourist packages, potential tourists who can retrieve information in order to find and book relevant destinations, local community that can promote their local assets. Because of the fact that RAIS integrates many different systems, it resolves the problems of filtering different kinds of information distributed over various websites at different location. Rural tourism information are not relied only on travel agencies brochures which can often be incomplete and unreliable but are also based on various shared experiences, advices, comments and photographs of tourists that really visited specific location. As RAIS is structured to support different platforms and devices, such as smart phones, users can get relevant information not only while planning a trip, but also at the moment when they are at the actual destination. This allows users to search for "hidden jewels" of the region, meet some less known activities and find accommodations that are hard to find on the web. Unlike systems that support standardized mass tourism, RAIS allows their users to input different forms of digital assets (virtual tours, videos, textual description of destination, contact information, maps, photographs, interesting happenings) and in that manner, it shows greater flexibility and presents more individualistic patterns.

One of the advantage of RAIS system is the possibility to use different sources of information provided by internal RAIS systems and external systems such as travel agencies' information systems, weather forecast services, rent a car services, other tourism web pages and the most importantly social networks. The integration of different homogeneous system presents the advantage for users who can easily retrieve relevant information, but also it can pose many challenges. Internal systems are updated by the shared activity of various private entities and their main goal is to gather information about geotag resources and conduct sentiment analysis for particular location. On the other hand, entering information into the system is not mandatory for users and very often up-to-date information about rural areas' cultural, historical, geographical site and relevant facilities cannot be provided. On the other side, the problem is how to ensure that data from external systems can be retrieved and automatically entered into RAIS when they are updated or entered into each of those external systems. The biggest challenge presents collection and mining data from social networks. There is also an issue related to format of external systems' data. While RAIS can control format of data stored in its internal systems, data formats from external resources cannot be controlled and have to be extra processed and formatted.

Since the number of web sites and social media pages presenting the rural tourism and complementing offer is increasing, the RAIS system seem to be very efficient platform for information exchange regardless the geographical distance. In this way many rural locations and their offer will be more visible and accessible to a large number of usual as well as sophisticated seekers of new experiences.

FUTURE RESEARCH DIRECTIONS

Exploitation of data from social networks poses a lot of challenges and limitations. As the data collection from social networks should be automated or semi-automated and triggered by user queries, it still remains to improve noise cancellation techniques that will give a more reliable data set for the data processing and the recommender system. More importantly, RAIS system relies a lot on geotagging which is collected from the social networks, however, geotagging is usually not a mandatory data that comes along with data from social resources. Only handful of posts will have geotag incorporated. Something to consider for the future improvements should include localization from mobile devices and GPSs, along with sensors included in sensor networks or emerging trends such as Internet of Things (IoT). This will of course bring more challenges with new data formatting, which will lead to improvement of methods for effective transformation of unstructured data into structured and semi-structured data. This will also lead to the new research of how to more efficiently utilize ranks and scores from social networks and incorporate it with RAIS opinions and scores.

Even though RAIS uses diverse set of external system, it still remains to explore additional systems that can improve user experience and provide any necessary information that is currently missing. For instance, not all social networks were integrated in RAIS up to this point, and the future work will explore integration with wider group of social networks in order to utilize more geotagged resources. Also, another beneficial feature that rural areas tourism can benefit from is highlighting more of cultural heritage of the particular area, so it should be analyzed how to incorporate cultural heritage application in RAIS, either as an internal or external system.

Finally, but not the least important, each internal system remains to be improved with more effective techniques for data processing. This is especially considering data and text mining systems. All in all, the use of big data in the tourism is still in its infancy, and it remains to explore more data analysis techniques. Fortunately, more data analysis techniques and tools are emerging, so smart tourism applications will be able to benefit from them both on macro and micro levels. On the other hand, sentiment analysis system at this stage of RAIS development uses data processing for English, but it will be analyzed in the future how these techniques can be implemented to other languages of interest.

CONCLUSION

Inclusion of public and private entities in developing rich and up to date database of assets of rural regions can pose to be challenging. In order to involve local community and other stakeholders, local enterprises and local actors should be informed about the aims and expected benefits of using such system for the promotion of their local region. Furthermore, all stakeholders should be aware of benefits of using ICT services, which represent an important tool and marketing-channel for the entire local community and their service offerings to tourists. Furthermore, as ICT provides strategic importance for the marketing of the rural region tourist organizations and associations should be prepared to invest time and resources into development of new ICT tools and applications.

Allowing system access to different types of stakeholders will require cleansing of information in order to avoid repetitions and errors. Information cleansing will take time and funds. In general, it also contributes to achievement of a successful rural tourism that could play a significant role in rural development. From one side it has an economic impact and potentially large multiplier effects, and from the other positive social and cultural impact seen as interaction between inhabitants and tourist, as well as new residents.

Tourist offer and maintain of its quality due to its authenticity could contribute to the overall preservation of rural ambience and its culture. In this way, by the responsible use of all its potentials and resources, rural areas could achieve smart, sustainable and inclusive growth which is the main goal of the Europe 2020 strategy.

REFERENCES

Allem, J. P., & Ferrara, E. (2016). The importance of debiasing social media data to better understand e-cigarette-related attitudes and behaviors. *Journal of Medical Internet Research, 18*(8), e219. doi:10.2196/jmir.6185 PMID:27507563

Berson, A., Smith, S., & Thearling, K. (2000). *Building data mining applications for CRM*. New York: McGraw-Hill.

Briedenhann, J., & Wickens, E. (2004). Tourism routes as a tool for the economic development of rural areas—Vibrant hope or impossible dream? *Tourism Management, 25*(1), 71–79. doi:10.1016/S0261-5177(03)00063-3

Buhalis, D., & Amaranggana, A. (2015). Smart tourism destinations enhancing tourism experience through personalisation of services. In *Information and Communication Technologies in Tourism 2015* (pp. 377–389). Springer International Publishing. doi:10.1007/978-3-319-14343-9_28

Cavaco, C. (1999). O turismo rural nas políticas de desenvolvimento do turismo em Portugal. In C. Cavaco (Ed.), *Desenvolvimento Rural: Desafio e Utopia* (pp. 281–292). Lisboa: CEG.

CIHEAM. (2008). Development strategies for rural areas. In *Mediterra. The Future of Agriculture in Mediterranean Countries* (pp. 275–301). Montpelier: SciencesPo. Les Presse.

Dave, K., Lawrence, S., & Pennock, D. M. (2003, May). Mining the peanut gallery: Opinion extraction and semantic classification of product reviews. In *Proceedings of the 12th international conference on World Wide Web* (pp. 519-528). ACM.

ESPON. (2017). *Policy Brief // Shrinking Rural Regions in Europe*. Retrieved from https://www.espon.eu/rural-shrinking

ESPON and UHI Millennium Institute. (2011). *EDORA: European Development Opportunities for Rural Areas*. Luxembourg: ESPON & UHI Millennium Institute.

European Commission. (2008). *Poverty and Social Exclusion in Rural Areas. Directorate General for Employment, Social Affairs and Equal Opportunities, Unit E2*. Rome: European Commision.

EUROSTAT. (2016). *Rural development statistics by urban-rural typology*. EUROSTAT.

EUROSTAT. (2018). *Statistics on rural areas in the EU*. Retrieved from https://ec.europa.eu/eurostat/statistics-explained/index.php/Statistics_on_rural_areas_in_the_EU

Eurostat - GISCO. (2013). *Eurostat. Statistical Atlas Eurostat regional yearbook 2013*. Retrieved March 10, 2017, from http://ec.europa.eu/eurostat/statistical-atlas/gis/viewer/?year=2013&chapter=01&mids=2,116&o=1,1&center=50.0958,20.00071,3&ch=113,114&

Eusébio, C., Kastenholz, E., & Breda, Z. (2014). Tourism and susteinable development of rural destinations: a stakeholders` view. *Revista Portuguesa de Estudos Regionais, 2. Quatrimestre,* (36), 12-21.

GoodWorkLabs. (2015). Retrieved from http://www.goodworklabs.com/how-mobile-app-benefits-travel-and-tourism-industry/

Gretzel, U., Zhong, L., & Koo, C. (2016). Application of smart tourism to cities. *International Journal of Tourism Cities, 2*(2). doi:10.1108/IJTC-04-2016-0007

Haller, M., Pröll, B., Retschitzegger, W., Tjoa, A. M., & Wagner, R. R. (2000). Integrating heterogeneous tourism information in TIScover: the MIRO-Web approach. *ENTER 2000: 7th. International Congress on Tourism and Communications Technologies in Tourism*. 10.1007/978-3-7091-6291-0_7

Hunter, W. C., Chung, N., Gretzel, U., & Koo, C. (2015). Constructivist research in smart tourism. *Asia Pacific Journal of Information Systems, 25*(1), 105–120. doi:10.14329/apjis.2015.25.1.105

Kasismis, C. (2010). Demographic trends in rural Europe and international migration to rural areas. *Agriregionieuropa (Ancona), 6*(21). Retrieved from https://agriregionieuropa.univpm.it/en/content/article/31/21/demographic-trends-rural-europe-and-international-migration-rural-areas

Malecki, E. J. (2003). Digital development in rural areas: Potentials and pitfalls. *Journal of Rural Studies, 19*(2), 201–2014. doi:10.1016/S0743-0167(02)00068-2

Mäntylä, M. V., Graziotin, D., & Kuutila, M. (2018). The evolution of sentiment analysis—A review of research topics, venues, and top cited papers. *Computer Science Review, 27*, 16–32. doi:10.1016/j.cosrev.2017.10.002

Möller, P. (2015). Young adults' perceptions of and affective bonds to a rural tourism community. *Fennia-International Journal of Geography, 194*(1), 32–45.

Naldi, L., Nilsson, P., Westlund, H., & Wixe, S. (2015). What is smart rural development? *Journal of Rural Studies, 40*, 90–101. doi:10.1016/j.jrurstud.2015.06.006

Nguyen, T. T., Camacho, D., & Jung, J. E. (2017). Identifying and ranking cultural heritage resources on geotagged social media for smart cultural tourism services. *Personal and Ubiquitous Computing, 21*(2), 267–279. doi:10.100700779-016-0992-y

OECD. (1994). *Tourism Strategies and Rural Development*. Paris: OECD.

Ozturk, A. B., Nusair, K., Okumus, F., & Singh, D. (2017). Understanding mobile hotel booking loyalty: An integration of privacy calculus theory and trust-risk framework. *Information Systems Frontiers, 19*(4), 753–767. doi:10.100710796-017-9736-4

Porter, M. E., & Heppelmann, J. E. (2014). How smart, connected products are transforming competition. *Harvard Business Review*, 1–23.

Rashidi, T. H., Abbasi, A., Maghrebi, M., Hasan, S., & Waller, T. S. (2017). Exploring the capacity of social media data for modelling travel behaviour: Opportunities and challenges. *Transportation Research Part C, Emerging Technologies, 75*, 197–211. doi:10.1016/j.trc.2016.12.008

Raspopovic, M., & Vukmirovic, M. (2018). Sentiment Analysis of Twitter Data for Exploration of Public Space Sentiments. *6th International Academic Conference on Places and Technologies*, 212-216.

Rivera, M., Croes, R., & Zhong, Y. (2016). Developing mobile services: A look at first-time and repeat visitors in a small island destination. *International Journal of Contemporary Hospitality Management, 28*(12), 2721–2747. doi:10.1108/IJCHM-02-2015-0052

Shanker, D. (2008, May). ICT and Tourism: challenges and opportunities. In *Conference on Tourism in India–Challenges Ahead* (*Vol. 15*, p. 17). Academic Press.

Song, H., & Liu, H. (2017). Predicting tourist demand using big data. In *Analytics in smart tourism design* (pp. 13–29). Cham: Springer. doi:10.1007/978-3-319-44263-1_2

TripAdvisor. (2017). Retrieved from https://tripadvisor.mediaroom.com/US-about-us

United Nations Department of Economic and Social Affairs. (2018). *68% of the world population projected to live in urban areas by 2050*. Retrieved from https://www.un.org/development/desa/en/news/population/2018-revision-of-world-urbanization-prospects.html

ENDNOTE

1 PR and IR regions account for 91% of EU territory and 56% of the population of EU27. These regions generate 43% of Gross Value Added (GVA) in the EU and provide 55% of employment (European Commision, 2008, p. 49).

Chapter 8
Evaluation of Technologies That Help to Identify Hazards for Cyclists in Cities

Christine Chaloupka
Independent Researcher, Austria

Ralf Risser
Palacky University Olomouc, Czechia

Elisabeth Füssl
https://orcid.org/0000-0003-3596-5489
Apptec Ventures, Austria

ABSTRACT

How do people look at sites and places and perceive details? Studies are referred dealing with the looking behaviour of cyclists, under the assumption that the method used could also be applied for assessing looking behaviour of tourists and to learn more about risks for cyclists. In the frame of a naturalistic cycling study in Austria, among others, a method should be developed that would help to find out where bicyclists direct their visual attention on their ways. Use was made of mobile eye tracking glasses. Results will be shown of the development work that demonstrate that points of interest can be identified via the detection of gaze plots of samples of cyclists. No technology helps to register what is perceived by peripheral vision which would give a more complete picture of reality. Any technological method that today registers where cyclists, or customers, or visitors direct their attention to has to be completed with verbal data from interviews, questionnaires, etc. in order to assure themselves of what really has been perceived.

DOI: 10.4018/978-1-5225-9932-6.ch008

INTRODUCTION

It belongs to the topics of the congress series Places and Technologies to ask how people look at sites and places. How do they perceive the whole picture, how do they perceive details? There is no research known to the authors of this paper that deals with methods concerning looking behaviour with respect to architectonic or design characteristics of the public space. The idea therefore was to have a look at studies from other areas that cover the question how to analyse looking behaviour. The authors therefore chose to refer to studies dealing with the looking behaviour of cyclists, under the assumption that the method used there could also be applied for assessing looking behaviour of users of the public space – like tourists for example.

The bicycle is an appropriate means to explore the city. The speed is slower than that of a car and this allows a better perception of details, it is smooth in the sense that many places can be accessed more easily than by car. Not least because the bicycle, in the worst case, can be pushed. From a touristic point of view, the bicycle seems to be a very attractive „old technology": Many city administrations all over Europe already offer bicycle renting systems to visitors who use it for different touristic transport purposes within city limits. The problem, however, is that cyclists are facing considerable traffic safety risks. We can read in the statistics of many European cities that cycling is an increasingly popular alternative transport mode. But there were far too many cycling fatalities over the past decade (European Transport Safety Council ETSC, 2015). 138,400 pedestrians and cyclists were killed in the EU in the time period from 2001 to 2013. Vulnerable road users (VRU) – i.e. pedestrians and cyclists together - account for 29% of all road deaths across the EU; pedestrians for 21% and cyclists for 8%.

According to a TRL report (Knowles et al., 2008) more than a quarter of all cycling deaths in 2005-07 happened when a vehicle ran into the rear of a bike. This occurred to more than one-third in rural areas and to 40% in collisions that took place away from junctions. According to the same report it seemed that one cause for cyclists´ accidents was that car drivers as well as cyclists "failed to look properly", as was the interpretation in the TRL report. Therefore, it appears to be important to have a closer look at the looking resp. searching behaviour of road users, hoping to understand what in this area and why obviously important aspects in interaction settings between cyclists and car drivers are not perceived appropriately.

Target in General

In order to learn more about possible risky interactions that bicyclists get involved in the authors of this paper, i.e. FACTUM Chaloupka & Risser OG in Vienna, together with colleagues from Salzburg Research, Prisma Solutions EDV Dienstleistungen GmbH, Plansinn GmbH, and the University of Salzburg, Institute of Geoinformatics – Z_GIS, worked together in the project BIKEALYZE. This first Austrian Naturalistic cycling study (project BIKEALYZE, funded by the Austrian Ministry for Transport, Innovation and Technology, funding programme "Mobility of the Future") was carried out from April 2015 to March 2017. The main target of this project was to evaluate different methods for data acquisition concerning cyclists' behaviour, namely (1) mobile eye tracking, (2) GPS and sensor-based motion data acquisition and (3) video-based data acquisition for analysing the interaction of cyclists with their environment in the context of a naturalistic cycling study. The focus of this paper is on the analysis of the eye-tracking data, and on the strengths and weaknesses of this method.

In total 24 participants have been asked to complete standardized as well as free-choice routes in Salzburg and Vienna. The participants collected data for 1 week on their daily trips. Altogether approx. 130 hours of data were collected. A sample out of these data was analysed in detail. The category system used was based on literature about traffic safety and comfort (Ovstedal et al., 2002, Otte et al., 2009) and the operationalization of these concepts (Kraay & v. Horst, 1985, Hydén, 1987). In addition mobile eye tracking data were collected from 12 participants on standardized routes and analysed with regard to points of interest and frequency and movement of gazes. Collected data were analysed with respect to safety-critical events and (semi-)automatically referenced to the digitalised cycling infrastructure. The project team (1) evaluated the different methods concerning feasibility and added value, (2) identified and measured safety-critical events in the context of the given cycling infrastructure, (3) proposed measures to increase bicycle safety, and (4) defined standards for conducting naturalistic cycling studies in Austria.

The scientific focus of the FACTUM team lay on developing a method that would help to find out, among other issues, where bicyclists direct their visual attention on their ways thus identifying elements in the traffic system (infrastructure, other traffic participants) which hinder a safe and comfortable ride through the city. The hypothesis was that bicycle riders will pay attention to road users or infrastructure elements that are relevant for themselves, regarding critical events. To study visual behaviour, mobil eye tracking glasses were used that usually is utilized in marketing, with the goal to find out where customer look in shops or on websites. Recently, the method is also used in the frame of geographic studies. Brychtová and Coltekin (2014), for example, ran a study on the readability of map labels by utilizing an eye marker system. They studied eye tracking metrics fixation frequency, fixation duration and scan-path speed, and they conducted an area-of-interest (AOI) analysis to understand the performance and strategy issues that may be influenced by colour, distance, and font size during map reading. This user experiment – as most of others - was carried out in a controlled setting, but it gave results that allowed to assume that the method could be applied in different contexts in the field, as well. In the study presented here a video system (GO PRO hero session) was used, which were fixed on the test persons helmets. The eye tracking data was collected with Tobii Pro Glasses 2 (further information about this system see later). The study was carried out in two Austrian cities – Vienna and Salzburg -, providing data related to bicycle riding in dense city traffic. According to our assumption, it should of course be fully possible to also use the system to analyse where the focus of the attention of visitors in a city is. It should as well be possible to use the technical system when being in motion, focussing on subjects or objects that are in motion themselves, like other road users, viz. users of the public space.

In this paper some results of the development work will be shown that clearly tell that points of interest can be identified via the detection of gaze plots produced by samples of cyclists. It is important to add, though, that knowing where the focus of the vision of persons is does not necessarily mean that we also know what has really been perceived by any subject. Studies with car drivers showed that the searching and scanning strategies vary if drivers of cars are experienced or not. Experienced drivers have far fewer fixations and their fixation durations are much shorter than those of new-comers, meaning that they cover a lot of the necessary information intake with the help of peripheral vision (Chapman et al., 2001). According to perception psychology this is to be expected. We perceive many things with the help of our peripheral vision. The problem is that there is no technology, yet, that helps us to register mechanically what is perceived with the help of this peripheral vision. Gaze plots, at least, reflect some variation in the fixation of people and, hopefully, one may assume that part of the people fixate spots represented by the plots while others cover those areas with their peripheral vision. The fact that we can register a substantial amount of fixations, there, allows us to assume that the spot covers a point in the street that is relevant (for individuals). Still, to assure oneself that this assumption is correct, any method that helps to register where cyclists – or customers, or visitors – direct their attention to has to be completed with verbal data from interviews, questionnaires etc.

The general target of the study BIKEALYZE was to evaluate different registration methods for analysing the interaction of cyclists with their environment in the context of a naturalistic cycling study. The aim further on was to develop a category system which helps to synopsize the GPS and sensor-based measured motion data and the video data as well as the data from mobile eye tracking (MET). This procedure was used to identify probable explanations for the measured values and to classify these events into categories, which inform about safety criticality as well as limitation of comfort. The following sections of this paper will focus on the pros and cons of an existing high-tech vision-investigation tool – the mobile eye tracking - and will discuss it, with respect to the registration of points of interest (POI) in general and, more specifically, to critical incidents. The theoretical framework refers to communication in traffic and to infrastructure elements, both contributing to the generation – or to the avoidance – of safety critical events. The results of the study, which is an explorative one, are mainly qualitative as the number of the observed riders has been 24 both in Salzburg and Vienna. And those with the eye tracking system in addition have only been 12. The data does not cover a representative sample of riders, thus we will not present statistical data analysis results.

Theory 1: Safety Critical Events

As crash data show (see introduction), cycling safety is a pressing issue. In order to prevent cyclists from getting involved into crashes it is necessary to focus on problems in traffic that occur before crashes happen (Hydén, 1987). These problems usually arise in the frame of communication processes, and there are many different methods in different countries to identify critical events (Zheng et al. 2014). Within the Austrian research project BIKEALYZE a special methodology was developed. In the following the theoretical background of the developed methodology is presented, encompassing 1) psychophysical aspects in critical events, 2) the concept of communication, 3) contributing factors for critical events and 4) a systematic for identifying accident causation elements especially for bicycles.

Safety-Critical Events in Traffic

Safety-critical events may arise both, when being alone on the street, and in the context of interaction between persons. In order to decide whether any event is safety critical, apart from the objective safety risk one also have to consider the subjective perception of the situation. Does the involved individual feel any risk, or is he or she uneasy concerning the safety situation? On the other hand, usability issues related to infrastructure have to be considered appropriately. Crash data only contain objective data but do usually not include subjective safety elements. However, the latter are often responsible for the misinterpretation of situations or of infrastructure conditions. Therefore both have to be considered and examined separately with different qualitative and quantitative methods.

But what do we mean by objective or subjective safety, and by safety relevant situations?

We applied the following generic definitions:

Objective safety: The probability of a crash, or the frequency of crashes in a special situation within a defined timespan.

Subjective safety: The personal assessment of the risk that a crash might be imminent, reflected by both cognitions and feelings.

Safety Relevant Situations

A so-called safety relevant situation during a cycle ride is a situation where objective and/or subjective safety deteriorates in relation to at least one of the following five aspects in road traffic

1. optimal roadway arrangement
2. optimal weather conditions and related quality of the roadway
3. optimal infrastructure
4. optimal encounter with other traffic participants
5. optimal driver operation characteristics

To make sure that those aspects are clear here some explanations are provided:

Optimal roadway arrangement

In some Austrian routing guides normally there exists advise, regarding the safety and/or comfort of specific cycle routes. If such routes are not chosen by the rider objective safety or subjective safety could deteriorate. An example: A cycle path ends and the cyclist has to continue his ride on a car lane → safety relevant situations could result from the choice of such a roadway.

Optimal weather conditions and maintenance quality of the roadway

Optimal conditions in this respect would be dry cycle path, free from puddles, free from loose chippings, foliage, items etc. Further good general condition (no lane grooves, potholes etc.), no snow, rain or fog.

Optimal infrastructure

Many hazard areas arise because of shortcomings in planning and design because of small edges, adverse manhole covers or trees and roots, too narrow cycle lanes, abrupt end of cycle path, routing on main streets or alongside parked cars, lacking of, or unclear signposting.

Optimal encounters with other road users

During a ride, cyclists often meet car drivers, pedestrians or other road users. This is called an "encounter". An optimal encounter would be any situation where no traffic participant is on a collision course towards another traffic participant. On the other hand one is on a "collision course" if one accepts the possibility of a collision within three seconds if further pursuing one's driving direction without taking action. Such an action is called evasive action (by braking, swerving, or accelerating) in order to avoid a collision. The timespan between the rise of a collision course and the point of time of a potential collision is called "time to collision" (TTC) (Hydén, 1987). To understand and register an encounter as "not optimal" several actions have to be taken into consideration. According to literature (Zheng et al., 2014) those actions could be the following:

Traffic Conflict: A traffic conflict is a situation in which two or more road users are on a collision course in a way that a collision would happen within 3 seconds (TTC < 3s) if no evasive action is taken. This means that a collision is imminent and that there is a realistic probability that a person could be harmed or that there could be material damage if there is no evasive action. When TTC >= 3s one can assume that traffic participants adjust to each other, give way to each other, or arrange things in any other safe way. From a TTC of < 3s on road users find themselves in a slight traffic-conflict situation, and from TTC < 1s on in a serious one. The shorter the timespan between a necessary evasive action and a potential collision, the more serious is the degree of the traffic conflict.

Other relevant factors for the assessment of the seriousness of a conflict are the expected consequences of a collision and the expected collision energy, both depending on speed differences, and the vulnerability of the traffic participants. Thus, influencing factors are driven speeds, available space, angle of approach, and what kind of traffic participants are involved.

Concerning the space which is provided for the road users, situations could be safety relevant in different ways. For instance, a safety relevant situation could arise if there is not sufficient space for an evasive action.

Evasive Action: If two or more road users are on a collision course, any action which prevents a collision from happening is an evasive action. Usually this could be: braking, swerving or accelerating.

In our case an evasive action is necessary if a cyclists and at least one other traffic participant are on a collision course; at least one of the involved traffic participants will have to set an evasive action in order to avoid a collision.

Near Miss: Two or more traffic participants could pass each other at a very narrow margin, without setting an evasive action. Those situations are safety relevant because any small disturbance in such an interaction could cause a collision.

Post Encroachment Time (PET): Time span between the moment the first traffic participant leaves the course of the second traffic participant and the second traffic participant reaches the course of the first one (Kraay, 1986). The PET value shows how close the involved road users „missed" each other. In a city, a PET value <1s is defined as safety relevant.

Optimal driver operation characteristics

This could be situations where speed is well adapted to the respective situation or the distance to other road users (e.g. by overtaking cyclists) is in an appropriate way. In general, all driver operations which are according to the traffic law are included in this section.

Theory 2: Main Factors Causing Safety-Critical Events

To sum up the last chapter: Safety relevant situations have two main sources: **Road users** and their behaviour especially their communication or the lack of it, and the **environment**.

Environment

"Environment" has been explained before under the safety relevant situations by the points 1) to 3). Accident black spots show where the interplay of human and infrastructure and/or the interaction between road users do not function in an optimal way. Environmental conditions are rarely the only reason for crashes or traffic conflicts. On the contrary, safety relevant events occur often within the process of interaction between different road users, in many cases facilitated by the given environmental conditions. Interaction problems arise especially at intersections where special attention is required when negotiating a limited space between all traffic participants.

Road User

According to many studies human failure is responsible for accidents in traffic in approx. 90% of all cases (NHTSA, 2008, Singh, 2015 and many others). Failure is often due to the neglecting of rules, to hindrances of the communication between road users, e.g. because of the given infrastructure, to the psychophysical condition of the road users, or to misinterpretation of actions of other road users.

The role of the **road user** as the one who interacts with the environment is illustrated in the ACAS model by Otte et al. (2009), where road users could ask themselves the following questions, the latter always based on the former one:

- How do I perceive the environment?
- Did I recognize the situation correctly?
- Did I assess the situation correctly?
- How did I decide?
- How did I act?

The following paragraph provides a short introduction to the ACAS (Otte et al., 2009), which was chosen to serve as a model for the investigation and explanation of critical events for cyclists. As ACAS shows, within the communication process various tasks have to be completed, and failure might lead to conflicts or crashes.

ACAS Categories

ACAS, the *Accident Causation Analysis System (ACAS) by Otte et al (2009)* is a tool for collecting human error causation data, which describes the human information processing factors in a chronological sequence. It differentiates between phases of information processing where errors may arise, as well as human misbehaviour, conflicts or crashes.

The five subsequent categories: information access, information admission, information evaluation and finally the resulting action can be easily deciphered by asking respective questions:

Information access (Could all relevant information be detected?): The availability and the access to information is especially important regarding the design of infrastructure. Markings on the pavement and other signs, which are necessary for orientation belong to this category.

Information admission (Has all relevant information been recognized? Was the driver able to identify subjectively every information which is relevant for the traffic situation)): Lack of information admission is the most important source for errors and thus for safety relevant events. The in depth investigation of Otte (2009) showed that about 58% of all failures which lead to accidents could be derived from problems in the information admission area. Problems may arise if it is not possible (e.g. because of distraction, wrong focus of attention etc.) to identify all relevant information in order to be able to have a frictionless encounter with other traffic participants. But also wrong expectations (e.g. "at this time of the year there are no cyclists on the road because of bad weather") can lead to a delayed or not at all recognition of this road user group.

Information evaluation and action: has all information been assessed correctly? Was there an appropriate choice of action and was the action operation correct?

Communication

Communication is necessary in traffic in order to understand what other road users do or will do, to inform other road users about one's own intentions and needs, to coordinate mutual actions and to give feedback if a situation/interaction tends to be complicated, uncomfortable or unsafe. In communication, apart from formal rules, informal rules play an important role in road traffic, as they guide road users´ actions and interactions. Communication of course requires that one sees the other road users, is conscious of what they do and are going to do, and understands what the actions of the other road users imply for oneself.

According to findings of accident analysis studies, it appears that the lack of appropriate ability to recognize and interpret such information correctly proved to be the main source of errors in traffic (Otte et al., 2009). This corresponds with findings of the NHTSA (Singh, 2015) which show that inattention, internal and external distractions, and inadequate surveillance lie behind such recognition errors. According to Singh they are the most frequently recorded reasons for problems (41% ±2.2% of all accidents).

As communication is a process of information exchange, the "Accident Causation Analysis System" (ACAS)-tool was considered as a valuable hint at which levels of information processing problems may occur, which then lead to critical events.

Theory 3: Main Factors Causing Comfort Relevant Situations

Even if this has not been discussed extensively, so far, comfort is a relevant issue in connection with safety, as the demarcation between lack of comfort and the rise of hazards is diffuse. Situations which have a relevance for comfort are mainly related to infrastructure: Comfort is influenced by construction measures, but is also highly related to the behaviour of (other) road users. Relevant aspects of comfort (or the lack thereof) are for example:

- Complexity of the route
- Smoothness of the ride, traffic fluidity
- Aesthetic aspects (ugly environment)
- Frictions with other road users (e.g. perceived lack of respect)

Uneven pavements, potholes, steep curbs, manhole covers, stairs, greenery, narrow angles, little space to pass each other, slopes etc., they all could produce comfort problems, but also safety impairments.

Operationalisation: Results

The following section will give an overview of the operationalisation of 'critical events' related to information admission and will illustrate how interaction data is registered by the different methods in the BIKEALYZE project. As said before 'information admission' is the most problematic part within the process of information processing. Therefore this category has been used to illustrate the development of the methodology for identifying critical events.

The main question related to this category is, "could all relevant information be recognized objectively and subjectively?". The baseline hypothesis (H0) therefore is: Necessary information is available for the bicyclist at least objectively and is possible to be recognized by the cyclist objectively (according to Otte et al., 2009).

The following overview (Table 1) will give information about the assumed possibilities to operationalise some aspects of the category "information admission" by different technological tools as described above.

Video Observation

When it comes to bicycle studies in natural environment, video recordings are quite popular and therefore frequently used because they are cost-efficient and an easy way of using a method which is able to provide highly precise information of the driving situation. In the current study helmet cameras (GO PRO HERO session) were used for data collection (see Johnson et al., 2010). The cameras, pointing in the driving direction, were fitted on the helmets of the test persons. Thus the cyclists' line of sight and the surrounding situational context like traffic density, physical barriers or weather conditions could be captured. Former studies discussed the advantages and disadvantages of certain positions for fitting of the camera (e.g. camera mounted on handlebar see Gustafsson & Archer 2013, Dozza et al,. 2014), or the usage of helmet camera (see Johnson et al., 2010). An advantage of helmet cameras is that one is able to register head movements (turning of head) which are an important way for cyclists to either communicate with other road users (look at them, show them that they are active in searching for information) and/or to collect information from the surrounding.

Table 1. Methods for operationalisation of behaviour, related to lack of information admission in road infrastructure and to elements obstructing information admission (Chaloupka-Risser & Füssl, 2017)

Elements in road infrastructure and behaviour that affect information admission	Appropriate methods
Barriers for orientation: (traffic) signs, trees etc.	Video, interrogation, eye tracking
Lacking orientation points	Video, interrogation, eye tracking
Deceleration before intersection or encounter	Video
Distraction because of special traffic sounds, mobile phone, shop window etc.	Video, eye tracking, interrogation
Distraction because of maintenance of bicycle (e.g. mud guard broken)	Video, eye tracking, interrogation

Elicitation Interviews

During the video observation, the researchers interviewed the participants about their subjective feeling of safety while riding on this specific route and their feeling of comfort. Although the videos showed the behaviour of the driver and the other road users as well as the environment/infrastructure there was no information about the subjective feelings of the cyclists, viz. whether they would categorise a situation or an infrastructure-setting as problematic or not. E.g., the participants informed the researchers that they had experienced unsafe, uncomfortable feelings whenever they had been overtaken by car-drivers without decent side distance and they interpreted this as a bad communication act which sometimes led to a conflict and an evasive action (= swerving of the cyclist). They also very often had perceived a lack of information about further route guidance. Both – careless car drivers and lack of route information - were classified as critical events. Safety–critical events were reported by the bicyclists mainly due to interactions with other road users – but also due to poorly maintained infrastructure. Self-reported pedestrian–bicycle events primarily consisted of pedestrians starting to cross the bicycle path without looking which sometimes led to braking by the cyclist and is therefore classified as a conflict – and here again a sign for bad communication.

Related to the ACAS category (lack of) "information admission" three main results could be derived from the interviews: 1) Cyclists reported missing guiding markings on the road – or on the other hand if available – the helpfulness of such infrastructure elements. 2) As a function of the density of traffic more or less infrastructure elements – helpful or not – could be registered by the cyclists on the same route. 3) The density of traffic itself gave an influence on the feeling of safety and/or comfort.

One result of the study therefore is that video data can be relevant to hypothesise what could be a critical event, but it is insufficient for learning about subjective feelings concerning safety and/or comfort of riders. The necessity of knowing more about subjective feelings of safety or comfort is overt when it comes, as a consequence of those feelings, to an avoidance of roads as well as special cycle lanes or parts of the lanes. The explicitly mentioned lack of communication between road users as a source for unsafe or uncomfortable feeling could only be derived from interviews.

Mobile Eye-Tracking

Mobile Eye Tracking (MET) data is used in order to analyse gazes of persons in any specific environment. In naturalistic cycling studies these data provide information about the areas bikers fixate, the direction and duration of their gazes. So called areas of interest (AOI) can be defined in order to investigate if persons may be assumed to have noticed specific elements in their environment. Recently eye tracking has been used in geo-informatics studies (e.g. Brychtová & Coltekin, 2014, Popelka, 2016) but scarcely in naturalistic cycling studies. In the BIKEALYZE project, the eye tracking data has been collected with Tobii Pro Glasses 2. The glasses record gaze data which include fixations as well as saccades. In addition, various metrics about the gaze data, e.g. interval duration and area of interest (AOI) can be deduced from the data with the help of the analysis software Tobii Pro Glass Analyser. The use of MET for naturalistic cycling studies is an almost uncharted field in traffic science, except for very few studies (see Vansteenkiste et al., 2013). Thus, there is hardly any reference available.

Data Quality

A relevant aspect related to data import is the quality of the collected data. The target is to register a high amount of eye movements during the whole test route. Different variables may influence the recording: blinks, condition of light, head and/or body movements of the test riders etc. The maximum amount of registered eye movements recorded by Tobii Glasses Pro 2 would be 50/sec.

External variables during the test rides could not be influenced, thus some test riders who had a very low gaze sample rate where asked to take a second ride along the same route after a renewed calibration. Nonetheless, the different gaze sample rates indicate that eye movement data can vary considerably between the test riders along a test route.

When it comes to interpretation of eye movement data during a naturalistic riding study it is also necessary to know that the producer of Tobii Glasses Pro 2 emphasizes, that a 100% gaze sample rate is not possible. The reason for this, among others, is that during a blink 5 to 10% of the gaze sample gets lost. In the frame of naturalistic riding studies this is a handicap of the device in its actual technological standard. This would make the use of a second device (a video camera carried by the testee or an observation by an accompanying person) necessary, in order to understand the lack of gaze rates. This lack also impedes the comparison between different test riders because of incomplete data sets of part of the tested persons.

Heat Maps

On basis of the eye movement data “heat maps” are generated. At some selected locations they serve as examples to illustrate fixations in a graphic way. These heat maps are of a higher relevance when a lot of data points could be registered or when the analysed intervals at points of interest (POI) lasted longer so that more gaze data could be mapped. Heat maps represent an ideal base material for graphic presentation of duration and intensity of attention of riders in special situations. The colour spectrum varies from red (highest amount of fixations or longest duration of fixation) to green (lower amount of fixations or shorter duration).

Gaze Plot

Beside Heat Maps “gaze plots” are available. They give an overview about the chronological order of the fixations and their respective position. Gaze plots serve as strict filter in situations where there are only little data points. They reduce data only to fixations without intake of saccades. The more complex a situation is, the better is the interpretation of the amount of information which is given by the sequence of the gazes. In traffic scenarios of comparatively low complexity (e.g. no crossings, no moving parts etc.) the analysis may provide helpful indications to which areas on the road, along the road or outside the road the attention of the testees has been directed. The only problem with gaze plots is that if the amount of gazes is very high this reduces the clarity of the representation – it is not obvious any more where people actually look, or what they look at.

Some Examples

The analysis of a complex traffic scenario like a crossing, including dynamic elements there, makes sense especially if the test riders do not move much themselves (no swerving or abrupt braking) and if their head is rather static, so that the test scenario does not change.

The scenario of the crossing serves as a good example because during the chosen analysis interval (10,1 sec) cars scarcely moved. The Heat Map shows that the attention of the test rider was focussed on two areas: the crossing traffic and the direction the test rider had to move on after crossing the road (on the picture to the left).

Taking into account the Gaze Plots it becomes overt that the condition of the pavement played a special role in this situation, regardless of the high amount of traffic and the orientation and navigation activities which would be necessary for the rider. The first fixations go to the surface immediately in front of the bicycle and further on to the possible unevenness of at the manhole near the crossing

Whilst the visualisation of the data via Heat Maps does not provide a differentiation between riders, the cumulation of data of all riders can provide an information about potential problems of riders with special infrastructure or traffic elements. If they all together lead their attention to similar areas one could filter aspects of distraction and identify orientation necessities and thereto connected problems.

The actual example shows clearly that the focus of the riders lays on the further riding direction if the lane is straight without recognizable unevenness of the roadway. Otherwise, manholes and other static elements of the infrastructure attract the attention of the rider. Accordingly, rubbish containers as well as parking cars also on the left hand side are fixated by the rider.

Moreover there are some fixations alongside an advertising pillar, which may be a distraction according to the navigation target. This leads to the research question in how far advertisement can draw attention and thus function as a distracting element in traffic.

The Difficulty of Interpretation

The interpretation of a specific detected behaviour, e.g. an uncertain lane keeping, could be that it might either indicate lack of information access or erroneous information processing, like poor communication or coordination with other traffic participants. This is relevant, because information (e.g. for orientation) has to be provided in due time so that it can be processed appropriately, e.g. if the cycle lane ends and if there is no information via the infrastructure how the route for the bicyclists will go on (ACAS, Otte, 2009). So if there is lack of information, the time to coordinate one´s behaviour with other road users will be limited or there will be none at all. In the worst case this may result in a crash.

The gaze plot gives an impression about data obtained from the eye tracker (Tobii glasses) showing all gazes during a time span of 1,92 sec of approach of a cyclist to an intersection. The cyclist had to take decisions after his searching activities within this very short time span (less than 2 sec).

One can see that the focussed elements are mainly other road users (car drivers), who might be relevant – and possibly safety critical - communication partners. But infrastructure elements are hardly focussed by the riders, like traffic lights or street markings. But does this mean that these elements have not been registered?

Peripheral Vision

This example shows that it cannot be made sure that the cyclist – although gazing at a special item – has definitely been able to recognize this item consciously, viz. if he/she had recognized it anyway even without gazing at it, by the peripheral vision. We have to be aware of the fact that addition to focal vision, recognisable with the help of registered fixations, we also perceive stimuli outside of the fovea. Peripheral vision for instance functions quite well for identifying movements or brightness or figural contrasts in our periphery. This means that variables which are not fixated can be perceived and thus affect behaviour.

In fact, peripheral vision is relevant for everyday situations in traffic where movements and contrasts have an influence on perception, but it is not possible to register it with the help of eye-tracking hardware. This causes relevant restrictions for the interpretation of eye-tracking data. There is common knowledge that the human field of vision is nearly 220°, while only a central area of 1° to 2° (= foveal field of vision) can be focussed. The rest is covered by the peripheral view (Lachenmayr, 1995). Experiments show that the eye may perceive elements by the peripheral view and the human being acts accordingly although the eye focusses a target laying in the foveal area (Fei-Fei, VanRullen, Koch & Perona, 2002). In dynamic scenarios in which cyclists move within a complex traffic system the peripheric view plays an important role, especially concerning the registration of movements and brightness and figural contrasts, as said. Accordingly, in the periphery there may be elements - relevant or non-relevant for the given situation - which take in the attention and distract it from the actual riding task but are not detected by eye trackers. When riding along a test scenery in a naturalistic cycling study an experienced bicycle rider catches the scene and relevant elements easily, thereby also making use of peripheral vision. If the person moves quickly with the help of this capability the setting of the scene is also captured quickly, often without much fixating of concrete objects (Parker, 1978; Henderson, McClure, Pierce & Schrock, 1997).

Thus, the central point of critique concerning the use of eye-tracking hardware concentrates on the fact that the range of what we perceive visually is larger than the foveal view within which elements are focussed and perceived. Especially in the traffic context perception is distinctly broader than registered by MET. We have to accept that visual attention and perception are cognitive processes which cannot be recorded only by fixations.

As a way out, Schepers et al. (2013) suggest that the combination of a collection of eye movements and subsequent interrogation may give an extensive picture about the factual reaction to the visual stimuli. In accordance with this suggestion and as already mentioned above with regard to the Video method, interviews after the riding with MET would have been necessary in order to understand more about the riders' information admission. However, such interviews have not been carried out because they were not planned at the beginning of the project. When planning, the focus lay on the evaluation of the MET data, thus missing an important portion of interpretation possibilities in the study.

CONCLUSION

This study has presented the development of a methodology for identifying critical events as an important indicator for improving infrastructure and prevent cyclists from crash. Behaviour data, infrastructural conditions and gaze behaviour data can be interrelated and could finally provide an improved understanding of their reciprocal influence on each other. Why did the authors of this study assume that it would be interesting for a congress like Places and Technologies? The reason is that as psychologists we are certain that the subjective view on sites and places, viz. how individuals see the city, is a relevant question. But there are no studies available concerning the analysis of visual behaviour in connection with the perception of the city. Thus we resorted to a study dealing with the visual behaviour of cyclists, hoping that what we learned in this study could be made use of in future analyses of the subjective perception of sites and places in the city.

The Naturalistic cycling data is a valuable source for pointing out existing shortcomings in infrastructure. Within the BIKEALYZE project a demonstrator of a planning tool was presented to policy makers and planners, which shows at which locations safety and comfort relevant situations could be met. The potential of such an evidence based planning tool with real-time data was considered as enormous. Data about cycle traffic efficiency (average travelling speed, number of stops per trip etc.) is considered to be an important source for defining areas and locations where the need for action is high, either by improving the infrastructure or the traffic management. Further research is necessary in order to advance the automatic processing of a bigger amount of motion data, e. g. by identifying unambiguous fingerprints of safety and comfort relevant situations. This would be an essential benefit for raising awareness where need for action exists and for defining measures of priority.

The importance of appropriately considering interpersonal communication in naturalistic cycling studies in the city has to be underlined. For this, we need a more complete picture of events which are relevant for comfort and safety. It could be shown that quantitative investigation methods need additional qualitative methods in order to understand what really is happening, objectively and subjectively. Count and measuring events is not sufficient, the most important short-coming being that information taken in with the help of peripheral vision cannot be identified with help of any eye-tracking hardware by the experimenter. Verbal data are needed in order to correctly interpret what has been registered with the help of technical tools.

At this moment, the obtained results can be adapted and used for drivers' and cyclists´ training, with the target to improve communication between different road users. In addition, recommendations for the adaptation and/or improvement of cycling infrastructure especially regarding route guidance measures can be given on a more objective basis than just by feedback given by riders by chance using a mobile phone platform. A challenging issue with regard to analysing communication and critical events based on different methods is the integration of the resulting data (video data, eye tracking data and interview data) into a joint category system. This is a time consuming task, as concepts and theoretical approaches have to be linked considering their epistemological foundation. Especially the use of eye tracking data in the field of traffic is a rather new issue - and seems to be a fancy tool - but much further research is necessary to clarify what the instrument Mobile eye tracking is exactly able to measure. Coming back to the original question, whether the tools and procedures presented in this paper can be of any use for studying issues related to the subjective perception of the city, the answer is short and simple: Tools and procedures can be used on tourists, walkers, sojourners, at any place in the city, thereby considering the mentioned short-coming by completing technical data with interviews. The technical problems that we have met due to the fact that the technology is still "young" will probably be overcome very soon.

REFERENCES

Bell, D., & Leitinger, S. (2016). *Results of the applicability of the technology Mobile Eye Tracking in naturalistic cycling studies*. Bikealyze, Deliverable 2, unpublished, Salzburg Research.

Brychtová, A., & Coltekin, A. (2014). An Empirical User Study for Measuring the Influence of Colour Distance and Font Size in Map Reading Using Eye Tracking. *The Cartographic Journal Maney Publishing*.

Chaloupka-Risser, Ch., & Füssl, E. (2017). The importance of communication between cyclists and other traffic participants and its potential in reducing traffic safety-critical events. Transactions on Transport Sciences 2017, 8(1), 24-30. doi:10.5507/tots.2017.004

Chaloupka-Risser, C., Risser, R., & Zuzan, W. (2011). *Verkehrspsychologie: Grundlagen und Anwendungen*. Wien: Facultas.

Chapman, P., King, S., & Underwood, G. (2001). *The Role of Experience in Searching Scenes, Accident research Unit*. School of Psychology, University of Notthingham.

Dozza, M., & Fernandez, A. (2014). Understanding Bicycle Dynamics and Cyclist Behavior From Naturalistic Field Data. *IEEE Transactions on Intelligent Transportation Systems, 15*(1), 376–384. doi:10.1109/TITS.2013.2279687

European Transport Safety Council 2015: Europe must boost safety for cyclists and walkers. (n.d.). Assessed 27.6.2016. http://etsc.eu/europe-must-boost-safety-for-cyclists-and-walkers/

Fei-Fei, L., VanRullen, R., Koch, C., & Perona, P. (2005). Why does natural scene categorization require little attention? Exploring attentional requirements for natural and synthetic stimuli. *Visual Cognition, 12*(6), 893–924. doi:10.1080/13506280444000571

European Transport Safety Council 2015: Flash, P. I. N. 29 Making walking and cycling on Europe's roads safer. (n.d.). Assessed 27.6.2016. http://etsc.eu/wp-content/uploads/ etsc_pin_flash_29_walking_cycling_safer.pdf

Gustafsson, L., & Archer, J. (2013). A naturalistic study of commuter cyclists in the greater Stockholm area. *Accident; Analysis and Prevention, 58*, 286–98. Retrieved from http://www.sciencedirect.com/science/ article/pii/S0001457512002266

Henderson, J. M., McClure, K., Pierce, S., & Schrock, G. (1997). Object identification without foveal vision: Evidence from an artificial scotoma paradigm. *Perception & Psychophysics, 59*(3), 323–346. doi:10.3758/BF03211901 PMID:9136264

Hydén, Ch. (1987). *The development of a method for traffic safety evaluation: The Swedish Traffic Conflicts Technique*. Department of Traffic Engineering, Lund University.

Johnson, M., Charlton, J., Oxley, J., & Newstead, S. (2010). Naturalistic cycling study: identifying risk factors for on-road commuter cyclists. *Annals of Advances in Automotive Medicine/Annual Scientific Conference, 54*, 275–83. Retrieved from http://www.pubmedcentral.nih.gov/articlerender.fcgi?artid=3242541&tool=pmcentrez&rendertype=abstract

Johnson, M., Charlton, J., Oxley, J., & Newstead, S. (2010). Naturalistic cycling study: identifying risk factors for on-road commuter cyclists. *Annals of Advances in Automotive Medicine/Annual Scientific Conference, 54*, 275–83. Retrieved from http://www.pubmedcentral.nih.gov/articlerender.fcgi? artid=3242541&tool=pmcentrez&rendertype=abstract

Knowles, J., Adams, R., Cuerden, R., Savill, T., Reid, S. & Tight M. (2008). *Collisions Involving Cyclists on Britain's Roads: Establishing the Causes*. TRL (PPR 445).

Kraay, J. H., & van der Horst, R. (1985). *The Trautenfels study. A diagnosis of road safety using the Dutch conflict observation technique DOCTOR*. Report 85-53, The Netherlands: SWOV.

Lachenmayr, B. J. (1995). *Sehen und gesehen werden: Sicher unterwegs im Straßenverkehr*. Aachen: Shaker.

NHTSA. (2008). *National Motor Vehicle Crash Causation Survey, Report to congress No. DOT HS 811 059*. Washington, DC: National Highway Traffic Safety Administration.

Otte, D., Pund, B., Jänsch, M. (2009). Unfallursachen-Analyse ACAS für Erhebungen am Unfallort. *Zeitschrift für Verkehrssicherheit*, 122–128.

Øvstedal, L., & Ryeng, E. O. (2002). *Understanding Pedestrian Comfort in European Cities: How to improve walking conditions?* Paper presented at the European Transport Conference 2002, Homerton College, Cambridge, UK.

Parker, R. E. (1978). Picture processing during recognition. *Journal of Experimental Psychology. Human Perception and Performance*, *4*(2), 284–293. doi:10.1037/0096-1523.4.2.284 PMID:660101

Popelka, St. (2016). *Eye-tracking with dynamic stimuli*. Olomouc: Dep. of Geoinformatics, University of Olomouc.

Sanders, R. L. (2015). Perceived traffic risk for cyclists: The impact of near miss and collision experiences. *Accident; Analysis and Prevention*, *75*, 26–34. doi:10.1016/j.aap.2014.11.004 PMID:25460088

Schepers, J. P., Den, B. B. P. L. M., De, W. D., Twisk, D. A. M., Schwab, A. L., & Smeets, J. B. J. (2013). Studying the role of vision in cycling: Critique on restricting research to fixation behaviour. *Accident; Analysis and Prevention*, *59*, 466–468. doi:10.1016/j.aap.2013.06.027 PMID:23911618

Singh, S. (2015). *Critical reasons for crashes investigated in the National Motor Vehicle Crash Causation Survey. Traffic Safety Facts Crash•Stats. Report No. DOT HS 812 115*. Washington, DC: National Highway Traffic Safety Administration.

Vansteenkiste, P., Cardon, G., & Lenoir, M. (2013). Dealing with head-mounted eye-tracking data. *Proceedings of the 2013 Conference on Eye Tracking South Africa – ETSA*, *13*, 55–57. 10.1145/2509315.2509325

Zheng, L., Ismail, K., & Meng, X. (2014). Traffic conflict techniques for road safety analysis: Open questions and some insights. *Canadian Journal of Civil Engineering*, *41*(7), 633–641. doi:10.1139/cjce-2013-0558

KEY TERMS AND DEFINITIONS

Area of Interest (AOI): A tool to select regions of a displayed stimulus, and to extract metrics specifically for those regions.

Elicitation Interview: A method for gathering detailed accounts of human experience. It can be applied to help understand how people experience and interpret visualizations as part of data analysis processes.

Evasive Action: If two or more road users are on a collision course, any action which prevents a collision from happening, like braking, swerving or accelerating, is an evasive action.

Eye Tracking: The recording and study of the movements of the eyes when following a moving object, lines of printed text, or other visual stimuli, used as a diagnostic procedure or a means of evaluating the visual presentation of information.

Gaze Plot: Give an overview about the chronological order of gaze fixations and their respective position.

Gaze Points: Show what the eyes are looking at. If an eye tracker collects data with a sampling rate of, for example, 60 Hz, one will have 60 individual gaze points per second.

Heat Maps: A representation of data in the form of a map or diagram in which data values are represented as colours.

Near Miss: Two or more traffic participants pass each other at a very narrow margin, without evasive action (braking, swerving, accelerating).

Peripheral Vision: All that is visible to the eye outside the central area of focus; the ability to see objects and movement outside of the direct line of vision.

Points of Interest (POI): A specific point location that someone may find useful or interesting. The term is used when referring to hotels, cafeterias, museums etc. or any other categories used in modern (automotive) navigation systems.

Post-Encroachment Time (PET): Time span between the moment the first traffic participant leaves the course of the second traffic participant and the second traffic participant enters the previous course of the first one.

Traffic Conflict: A situation in which two or more road users are on a collision course in a way that a collision would happen within 3 seconds (TTC < 3s) if no evasive action is taken.

Chapter 9
Rural Tourism Development as a Catalyst of Urban–Rural Synergy in the Region of Southern and Eastern Serbia

Milica Igić
Faculty of Civil Engineering and Architecture, University of Niš, Serbia

Milena Dinić-Branković
Faculty of Civil Engineering and Architecture, University of Niš, Serbia

Mihailo Mitković
Faculty of Civil Engineering and Architecture, University of Niš, Serbia

Jelena Đekić
Faculty of Civil Engineering and Architecture, University of Niš, Serbia

Petar Mitković
Faculty of Civil Engineering and Architecture, University of Niš, Serbia

Ivana Bogdanović Protić
Faculty of Civil Engineering and Architecture, University of Niš, Serbia

Milica Ljubenović
Faculty of Civil Engineering and Architecture, University of Niš, Serbia

ABSTRACT

Rural areas in Serbia occupy about 3/4 of total country territory and they are home for almost half of total population. These areas are characterized by very heterogeneous environment with not expressed anthropogenic negative influence. Because of a few types of the settlements and very rich rural and natural landscapes, there are great opportunities for development of diverse economies, multifunctional use, and a variety of different activities. This chapter discusses rural tourism development in the Region of the Southern and Eastern Serbia and its impact on the urban-rural synergy. The main goal is to investigate the impact of the tourism development on the connections on the relation urban to rural on the example of this region, and to point out importance of this synergy for future balanced rural and regional development. Rural tourism represents drive for urban-rural synergy and it is one of the key factors of rural development.

DOI: 10.4018/978-1-5225-9932-6.ch009

INTRODUCTION

Rural tourism in Serbia started to develop in the second half of the XX century, but because of the political events in the last decade of the XX century its further development stagnated. In the first decade of XXI century, rural tourism became popular again and different national strategies and laws, regarding rural tourism, were adopted. Unfortunately, today rural tourism is still not adequately developed as in other European countries, though there is huge potential, because about 75,1% country territory is defined as rural area and within which lives approximately 49,9% of total population (Ministry of Agriculture, Forestry and Water Management, 2014). In Serbia, according data from last census, there are 6,158 settlements in total, and only 193 are classified as urban (3.1%), while 5,965 are classified as "other settlements" (Ministry of Agriculture, Forestry and Water management, 2014).

This paper discusses rural tourism development in the least developed region in Serbia - Region of the Southern and Eastern Serbia, and its impact on urban – rural synergies within this Region. Aim of this paper is to analyze current situation regarding rural tourism in the Region of the Southern and Eastern Serbia, to investigate possible directions for development and its impact on the connections urban to rural regarding rural tourism. Rural tourism can be defined as a tool for more balanced regional development, and a main hub for rural development in general. The research is based on analytical multidisciplinary approach in analyzing current state and possibilities of rural tourism, which relies on methods of case study, SWOT and comparative analysis based on spatial, functional, economic and social criteria, in order to identify possible setbacks and discuss guidelines for rural tourism development. In this paper approach to the problems and possibilities is from different points of view - not only economic. Rural tourism cannot be observed only as a potential for economic growth, urban, ecological and social development should be also considered. By highlighting potentials for rural tourism development, it is possible to point out importance of adequate and effective planning for improving this process. Critical analyses of existing planning, legal and strategic documents, show main threats for further tourism expansion. In order to develop rural tourism and promote rural development in general, it is necessary to have good connections with urban areas because of their good strategic locations and better connections with other regions. Rural tourism represents catalyst – essential link in urban – rural synergy which is a base for improvement of life quality in rural settlements and even regional development.

BACKGROUND: RURAL AREAS IN SERBIA – DEFINITION, CLASSIFICATION AND CURRENT ISSUES

In order to comply with European division of territory according NUTS nomenclature, Serbia was divided into five statistical Regions in 2010. Unfortunately, disparities between different regions are more pronounced, and today the Region of Southern and Eastern Serbia is the least developed. Even there are some EU concepts for better regional development such are system of polycentric development and organization of settlements, decentralization of public services and development of main technical infrastructure, still insufficiently developed spatial planning and unsuccessful implementation process contribute to more pronounced regional disparities (Maksin Mićić et al, 2009).

Rural areas are not strictly defined because of their specific characteristics, these areas differ from urban areas in terms of population density and built settlement area. As main activity in these settlements is land use for agriculture, husbandry and forestry. Traditional structure, heritage and customs are also important marks of these settlements. Most of the countries accepted Organization for Economic Co-operation and Development (OECD) definition according which rural areas have population density lower than 150 inh/km^2 (OECD, 1994). From the economic point of view, rural areas can be defined as areas where economy is oriented towards agriculture and food production, poverty is on a high level and resident's incomes are lower than in the other areas. From the social point of view, rural areas are characterized by expressed stagnancy in terms of technological and cultural development. If we observe demography, rural areas can be characterized by negative demographic growth, sparsely populated settlements and with very low population density. From spatial point of view, territory with mainly agriculture land and different land use than urban represents rural (Milić, 2011).

Statistical classification of settlements in Serbia determines only two categories: urban and „other settlements", and all the settlements that are not urban in character are determined as other. In order to preserve rural settlements that are facing huge depopulation problems, it is obligatory to determine new classification. Spatial Plan of the Republic Serbia identifies three types of rural settlements depending on their location and urban connections: *Successfully integrated areas* – settlements with favorable socio-economic and spatial characteristics and developed economy; *Central rural areas* – settlements that are near urban centers and beside agriculture, there is an "industry in development" as well as basic service activities; *Remote rural areas* – mostly in the Region of the Southern and Eastern Serbia, located on the periphery – edge territory of the regions which are mostly underdeveloped (Republic Agency for Spatial Planning, 2010). According population dynamics between last few censuses in Serbia, four types of rural settlements can be defined (Antić et al, 2017): *Progressive type* – settlements along main urban areas and centers and with huge influence of adjacent urban areas; *Stagnant type*: mainly settlements in Vojvodina, settlements next to the main corridors and those near settlements of progressive type. *Regressive type* –settlements are mostly in Central Serbia and east Banat and they are characterized by negative demographic growth and unfavorable age structure; and *Dominant regressive type* – these settlements are mainly in the Region of the Southern and Eastern Serbia and in southwest Serbia. There are approximately 32% of total number of settlements within this type, these settlements are underdeveloped and are facing huge demographic decline.

In the period of the industrialization, there were huge migrations on relation rural to urban, because working population always chooses to live near their jobs. Lately, there is an opposite situation and we have frequent migrations from urban to rural areas because of reduced life quality in cities and high values of the properties. From the social point of view, these migrations cause „urbanization" of rural areas because urban population brings their habits into rural areas and opposite – "ruralization" of urban areas. Because of that, rural areas and their development cannot be foreseen only through economy, social aspect – especially urban – rural sociology, must be taken into account (Todorović, 1964). Urban – rural connections are very complex, not only in spatial terms, but also in terms of economy and social and ethnic structure. This is especially pronounced in case of Serbia, and mutual connections - urban – rural synergy can be considered as one of the drivers for more balanced regional development and "revival" of devastated rural areas. One of the huge problems encountered by the rural population is social exclusion in the economic aspect and in the context of the market of the agricultural production and employment opportunities in non-agricultural sectors, access to social services, as well as social and cultural participation of the rural population (Cvejić et al, 2010). Rural areas, represent very healthy environment for living because of intact nature landscapes and rich biodiversity, and due to "less anthropogenic influences", harmful emissions and changes in land use and climate changes impact are not the same as in urban areas (Zelenakova et al, 2015). On the other hand, because of the mountainous and hilly topography, there is a greater risk of natural disasters caused by the negative impact of climate change. If considering activating rural areas for tourism, due to global warming there is a problem with loss of snow cover on the mountains and hilly areas, degradation of landscape, floods and torrential floods due to heavy rainfalls and landslides (European network for Rural development, 2010). Due to growing negative impact of climate change, adequate agro-ecological measures must be also taken in to account. These measures must be comprehensive in order to improve soil and water quality and fertility, keep existing biodiversity, develop organic agricultural production as a way of sustainable production (Sarudi et al, 2003).

Rural Areas in the Region of the Southern and Eastern Serbia

Within the Region of the Southern and Eastern Serbia lives approximately 1/4 of total country population. This Region occupies surface of 26.248 km^2 which is about 33% of total country surface. In terms of administrative division, this Region includes nine districts within which are 1.973 settlements with 1.521.081 residents and population density is 58 inh/km^2. This Region is characterized by negative demographic growth and in period from censuses in 1948 and 2011 there is a decrease in population for approximately -180.000 residents and annual population growth is – 12.371 (Statistical office of the Republic of Serbia, 2014 and 2018). Lack of workplaces and educational institutions network are causing migrations of residents, and average age structure is approximately 44 years. According to the Law on territorial organization of the Republic of Serbia, in Serbia there are only 29 settlements that are characterized as cities which represent economic, administrative, geographical and cultural centers. From this number, 9 cities are within this Region – Bor, Vranje, Zaječar, Leskovac, Niš, Pirot, Požarevac, Prokuplje and Smederevo, which are also centers of administrative districts that are 9 in total in this Region. Except cities of Bor and Niš, all other municipalities and settlements are underdeveloped. From total number of 1.973 settlements in the Region of Southern and Eastern Serbia, only 46 are classified as „urban" and 1.927 are classified as „other" settlements which is presented on the map on the Figure 1. This area is characterized with very diverse topography, rich hydrographic network, natural landscape, flora and fauna and very fertile land. Due to specific topography – dominantly high hilly and mountainous terrain, large surface of the territory is poorly populated and most of the settlements in the eastern part, especially on the border line, are declared as "weak – remote".

According to the Plan of Strategy for Rural Development for period 2009-2013, Region of the Southern and Eastern Serbia is mainly classified as Region III, except urban areas of the cities Niš and Leskovac, and area around city of Vranje which is Region II because of intensive agricultural production. Region III is characterized by mostly mountainous areas where agriculture and mining are developed, and the focus is on the exploitation of natural resources. The main problems this Region is facing, because of its heterogeneous structure, are depopulation, unemployment and underdeveloped economy. Each municipality has the network of settlements concentrated around main cities. Rural areas are in dependence of urban in terms of administration, education, health care, trade, and on the other side urban areas are depending on rural areas in terms of food production and natural resources. There is unbreakable link between these two types of settlements which is a good base for developing urban – rural synergy within this Region. Functional links between them are different and depend on the spatial disposition and economic development, but because of uneven development it can be said that they are limited and underdeveloped. (Ministry of Agriculture, Forestry and Water Management, 2009.). Rural settlements in the Region of the Southern and Eastern Serbia usually emerged spontaneously, and their territory

Figure 1. Map of spatial distribution of districts, statistical data regarding demography and administrative territory for the Region of Southern and Eastern Serbia (source: main map http://www.lokalnirazvoj.org/upload/Book/Document/2012_03/Southern_and_Eastern_Serbia_Region.jpg; data source: Statistical office of the Republic of Serbia, 2014)

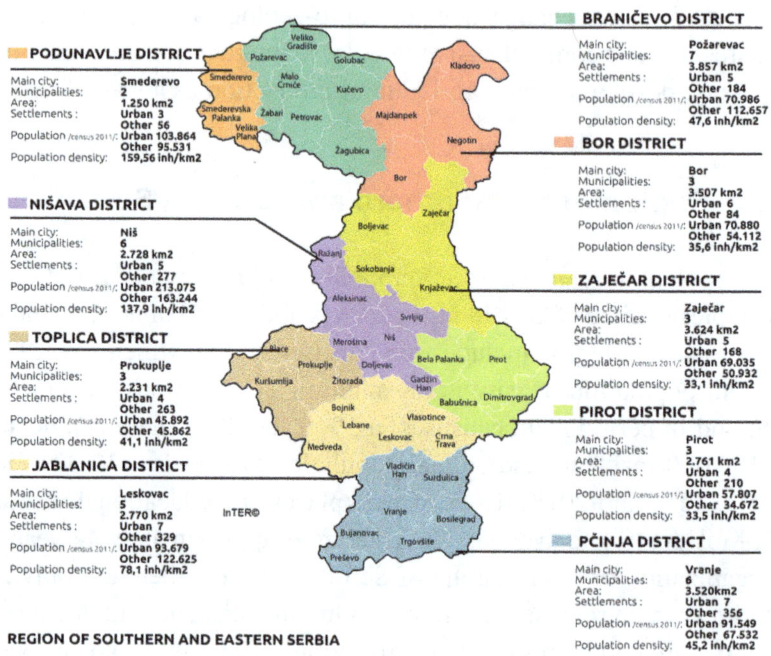

spreading is uncontrolled. Lack of infrastructure, negative demographic growth and insufficient connections with urban areas made these areas less attractive for both tourists and investors.

RURAL TOURISM

Rural tourism has long history and there are many theories discussing evolution of rural tourism. According Butler's cyclical evolution theory of rural tourism, there are 7 stages of evolution: *exploration, involvement, development, consolidation, stagnation, decline and rejuvenation.* Many factors must be taken into account – social, political, environmental, because of their influence on development process which are not always resulting in positive way (Butler, 1980). This theory was more developed by Miossec in 1993, where author defined four characteristics – drivers for tourism development - resort, transport, tourist behavior, and attitudes as well as attractive areas for tourism (Streimikiene & Bilan, 2015). Rural tourism is one of the developing rural industries in Serbia. Even there is no precise definition of rural tourism, tourism in rural areas can be defined as diversification process from traditional dominant agricultural activity to non – agricultural activities. Because rural tourism represents very complex activity which involves large number of people, simple definitions where rural tourism is recognized as tourism that is happening in village – rural areas, cannot be applied. There are many problems which are enabling to have clear definition of rural tourism that could be used for all rural areas. According to Lane, there are several problems – rural areas are not strictly defined, and their classification varies depending on the country; rural areas worldwide are going under great changes and tourism within them is not only farm based; there is a huge influence of urban area on rural area because tourism can take part in rural area and be urban in "form"; tourism started to develop in urban areas and most of the tourists are from urban territory which is could result in "urbanisation" of rural areas (Lane, 1994). Difference between rural and conventional tourism is that it takes place in rural areas with very low population density, natural environment is one of the main characteristics of these areas, it is subordinated to the individual experience in nature environment and it connects different types of heritage: cultural, historical, natural, traditional with different events while protecting local customs and tradition (Nagaraju & Chandrashekara, 2014).

Regarding tourism in rural areas, three main concepts can be recognized: *rural tourism* which is also the most common term for tourism and different activities in rural areas. It implies all activities and it is not necessarily connected with local households; *village tourism* is more specific, connected to the village and its surrounding, where local farms are involved and activities in their surrounding; *agritourism* is oriented to farm or households with primary agriculture activity where tourist service is additional activity. These farms offer stay for tourist and they are engaged in agriculture offering their agriculture products to the visitors (Đenadić et al, 2016). Because of the complexity of this kind of tourism, it is hard to define all the types of rural tourism. Because rural areas are recognized as agriculture oriented, one of the basic and most common type of rural tourism is *agritourism*. This type of tourism implies that tourists actively take part into agriculture activities or they visit farms in order to buy local agriculture products. *Ecotourism* is one of the types of rural tourism where tourists are traveling to nature landscapes and they are using local resources without harming nature. This way of tourism is mostly in mountainous areas. Another type is *Culinary – gastro tourism* where tourists are visiting certain areas and testing their local foods and products. In Serbia, and also in other surrounding countries, wine routes are very popular and people are visiting wineries and domestic distilleries in order to test different beverages made from local fruits. *Creative tourism* which represents traveling for learning new customs, crafts and tradition. This type of tourism offers visitor to have new experience and improve its skills. Many workshops, art and graphic colonies are organized where local residents are sharing their knowledge and skills with the visitors. *Ethnic tourism*, represents specific type of tourism which offers a tourists to visit specific ambient units and learn about way of life of minority of population is living or has lived by traditional way (Sasu and Epuran, 2016).

Tourism can be also defined as one of the most working intensive industries which is base for new jobs and economic development through activating local resources in rural areas, and it is also recognized as one of the elements in rural development strategies (Mitchell & Hall, 2005). According to the World Tourism Organization (WTO, 2004), rural tourism represents type of tourism where rural culture is main characteristic of the touristic offer, and where users – tourists are allowed to enjoy and participate in everyday rural lifestyle and to take part in all activities within households. Implementation of strategies for tourism development must have multidisciplinary approach in order to observe and control not only economic, but also urban, social and ecological impact (George, et al 2009). Rural tourism is a part of *"travel – tourism and leisure chain"* where main focus is orientated towards development and exploitation of local resources in the most efficient way. According to Greffe, there are three ways of understanding term "rural tourism": first where rural tourism is defined as "cheap" way of tourism where tourists is staying with a local resident in rural area; second way is where beside accommodation there are activities that tourists can take part in; third way where tourist is staying on the farm where he is included in activities organized around farms (Greffe, 1994). According to Aronsson, sustainable tourism is based on a system of the tourist demand and supply – local environment offer, and the connection between them is meeting place. This concepts has few categories, and the meeting place represents encounter between the visitors-tourists and local environment (Aronsson, 1994). There is also concept of integrated rural tourism where multidimensional sustainability, strengthening of the involvement of local residents, use of local resources and local ownership of the resources, cooperation with stakeholder (Cawley & Gillmor,2008). In order to make better and authentic presentation, "identity" of place is also important in order to have unique presentation of the local resources and customs and also to encourage local residents to promote their lifestyle and tradition. Cultural and spatial identity can be one of the essential catalyst for rural development and improvement of touristic offer by promoting local values (Crouch, 1994). Nowadays, as a new concepts, creativity and authenticity can be also determined within rural tourism (Sasu and Epuran, 2016).

As one of the greatest impacts of rural tourism on economy is poverty reduction – these areas are facing huge financial problems but they have outstanding natural resources, which is one of their major strengths (Jing, 2006). In most of the European countries rural tourism is recognized as main activity for economic growth on the one side, but also as a mechanism that can prevent depopulation process and bring young people in these ambiences. Most of the strategies and national plans in Europe recognize rural tourism as a possibility for entrepreneurship and for creating local businesses that could provide constant incomes for the residents. Also, this kind of tourism is customer oriented tourism, which relies on natural and local heritage and resources (Nylander & Hall, 2005). European commission has adopted Rural Development Programme for period 2014-2020, where rural tourism, as one of the major sectors in rural areas, is recognized as a drive for rural development within all the defined priorities (European Commission, 2014). According to the regulations of European Parliament regarding support for rural development through EAFRD, establishing of small and medium enterprises (SMEs) is crucial for economic development of rural areas and also it can prevent depopulation trend (European Parliament, 2013). Support of EAFRD to 2014-2020 programming period regarding rural tourism is focused on farm and business development, providing basic services and village renewal in rural areas, co-operation between different partners, knowledge transfer and advisory services (Noev, 2013). This policy also covers the issue of quality of life in rural areas in order to reduce the occurrence of depopulation and to create high quality and attractive places for the life of different age categories and for the development of tourism (Brauer, 2014; European Commission, 2016). Tourism can be developed through multiple activities, and agriculture, as one of the main „marks" of rural areas can be complementary developed with tourism through various activities. Concept of multifunctionality of agriculture can be used to improve economy and develop agricultural tourism that can contribute to diversification of rural tourism offer (Vuković et al, 2012).

RURAL TOURISM IN SERBIA

As above mentioned, in last twenty years agricultural employment stagnated and also entire industry related to it, which left many rural residents without any income. New job opportunities and agricultural industry development started in last decade, but only in the areas that are in the vicinity of urban areas. At the same time, rural areas on the periphery are endangered because of negative demographic growth, lack of infrastructure and traffic network. Upon defining rural areas, it is necessary to identify various types of rural settlements because in Serbia they are not homogeneous. There are differences among them and these may be due to various factors such as geographical characteristics, accessibility, population changes and migration, infrastructure, different environmental conditions, favorable or disadvantaged agricultural structures and diversified local economies (Bogdanov, 2007; Efstratoglou et al, 2007) Some ambient sets in Serbia possess such developmental values that with adequate valorization may take part in international development processes. Their ecological connections and influences are very strong. Isolation of these areas from main traffic routes and development is precisely what gives them their autochthonous quality with a rich natural and historical heritage. (Todorović & Bjeljac, 2009)

Real rural tourism development in Serbia started with economic and political change in the very beginning of the XXI century. Rural tourism started also in the village settlements and tourism became one of the main factors for integrated rural development and balanced regional development. (Todorović & Bjeljac, 2007). The main problem is that nobody considers different rural areas separately because of their specific positions and characteristics. All the plans are unitary for all and specific problems were not solved on local level - everything was always planned on national level.

Current Legislation, Planning and Strategic Framework Regarding Rural Tourism

In 2006. Strategy for tourism development of the Republic of Serbia was adopted (Ministry of trade, tourism and telecommunications, 2006). This strategy recognizes rural tourism as great potential for tourism development in general, with greatest potential in mountain areas - national parks or protected areas. The main idea is to promote rural areas as an affordable way of tourism. There are no suggestions to overcome infrastructure problems and to improve protection of traditional ambient that these settlements have. Promoting only the protected areas on the other side has caused expansion of illegal construction which endangers the natural landscape.

In 2011. Master plan for sustainable development of rural tourism was published. This program is based on research of the current state in rural areas and as the main issue are recognized funds, and because of that improvement public-private partnership is suggested. (Ministry of Finance and Economy, 2011).

In 2016. new Strategy for tourism development of the Republic of Serbia for period 2016-2025 was adopted (Ministry of trade, tourism and telecommunications, 2015). This new strategy recognizes rural tourism as one of the main activities for developing the tourist offer of Serbia. Rural tourism is not presented only as an affordable way of tourism, but also as a developing potential for every region in Serbia. As one of the weaknesses for tourism development, this Strategy recognizes the lack of infrastructure in rural areas, and not so good connections between rural and cultural tourism. Start-up investments in non-agriculture activities and SME have been suggested as measures that could help the improvement of rural tourism.

In 2018 draft version of new Law of tourism was presented by Ministry of trade, tourism and telecommunications. As main problems for rural tourism development were high values of taxes that householder had to pay and standards that households had to comply, which led to the expansion of illegal tourist offers and many rural households were "invisible" to official offer. Rural tourism is recognized as "special product" that should be developed and because of that, new Law allows householders to rent their properties without agents and they can advertise their tourist offers directly online. In rural areas, for householders it will be obligatory that if they offer catering they must sell their local food, in order to promote tourism and food production (Ministry of Trade, Tourism and Telecommunications, 2018).

In the period 2010-2015, Republic Agency for spatial planning developed Regional spatial plans for all administrative districts within this Region. These plans discuss rural tourism development in all nine districts, establishment of competitive and innovative urban areas that are functionally connected with underdeveloped rural areas, which comprise majority of the Regions territory. Because of numerous spas, good hydrographic network and hilly topography, main tourist activities are towards activation of these resources. Tourism and complementary activities are recognized as significant factors for development of deprived rural areas –periphery of the districts. As main threats, beside negative demographic growth, are lack of main technical infrastructure, absence of main public services in these settlements and poor functional connections urban to rural (Republic Agency for Spatial planning, 2010, 2011, 2013, 2015).

RURAL TOURISM IN THE REGION OF THE SOUTHERN AND EASTERN SERBIA- CURRENT STATE AND CHALLENGES

Region of the Southern and Eastern Serbia represents the main hub of strategic roads that connect Europe with middle east, central Europe with south Europe and it is located in the center of the Balkan peninsula. This area was subject of interests for different rulers during centuries and there are many sites located both in urban and rural areas. Because of activation of airport in city of Niš, there is large influx of tourists. According to author's research and available online data, there are approximately 150 villages engaged to rural tourism which is presented on the map on the Figure 2.

Figure 2. Map of distribution of main potentials for rural tourism development within different districts in the Region of Southern and Eastern Serbia
(source: authors)

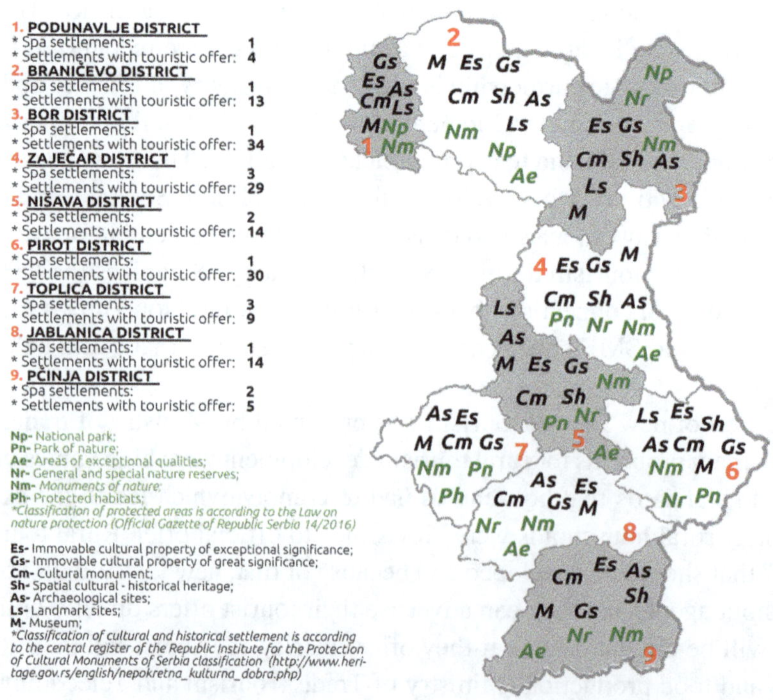

Because of the problems of illegal tourist offer, it is hard to determine exact number of households that are offering accommodation. Offers are different from those with only rooms within host's house to ethno villages and small aparthotels – especially on Stara Planina Mountain. Within this Region there are numerous of settlements with thermo mineral water springs, but according data from the field, there are 15 spas with offer of accommodation and health services (Figure 2). Census of the Agriculture in 2012 presented number of registered agricultural households, and in 2012 there were 187.744 registered agriculture households in Serbia. This census gave data regarding households that beside agricultural production, also have some nonagricultural activity. Results obtained within this census show that there are 514 registered households that beside agriculture are also involved in tourism, and from that number, 112 are within the Region of Southern and Eastern Serbia. The largest number of these households is in Zaječar district 27, and after that in Braničevo, Bor and Pirot district 18-23. National tourism organization of Serbia published Catalogue of village touristic households in 2017. This publication does not cover entire country and has only data regarding registered households within this organization. For the Region of Southern and Eastern Serbia, there is only data for the villages on Stara planina, Sokobanja spa, Braničevo and Timočka Krajina districts and only 38 households were presented.

According data from the field, data from the internet and from local tourist offices, in the Region of the Southern and Eastern Serbia there are approximately 300 different types of household and ethno villages offering accommodation. This number is not exact, and because of the problem of illegal offers, new Law provides better solutions in the struggle with the "grey economy". Figures 3 and 4 shows official data regarding total number of tourists within districts in period 2015-2017. This data is for both urban and other settlements, and it can be seen that the highest number of tourists are visiting Nišava district (city of Niš) and Zaječar district (spa Sokobanja).

Increasing number of tourists – approximately 250.000 in this period, shows that this Region has potential for developing tourism and that tourism can be one of the key factors for further Region and districts development. Unfortunately, because of the absence of the unique database or online platform, it is impossible to have precise data regarding number of accommodations, tourists, overnight stays and their distribution separately for urban and rural settlements. In order to create complete and diverse offer,

Figure 3. Statistical data regarding tourist visit in period 2015-2017 in the Region of Southern and Eastern Serbia
(source: Statistical office of the Republic of Serbia 2016,2017,2018)

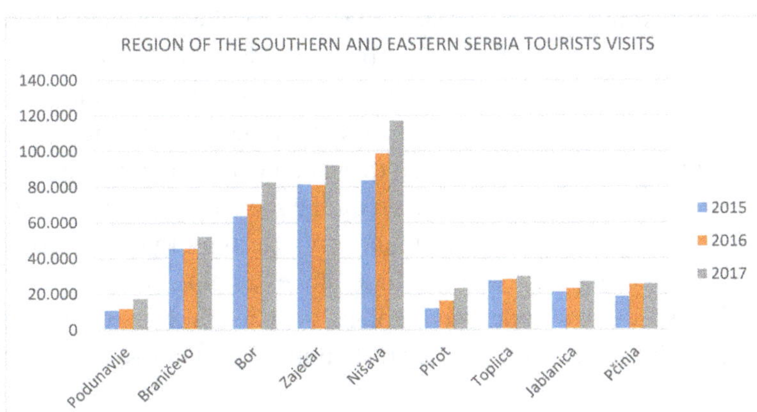

Figure 4. Statistical data regarding tourist overnight stays in period 2015-2017 in the Region of Southern and Eastern Serbia
(source: Statistical office of the Republic of Serbia 2016,2017,2018)

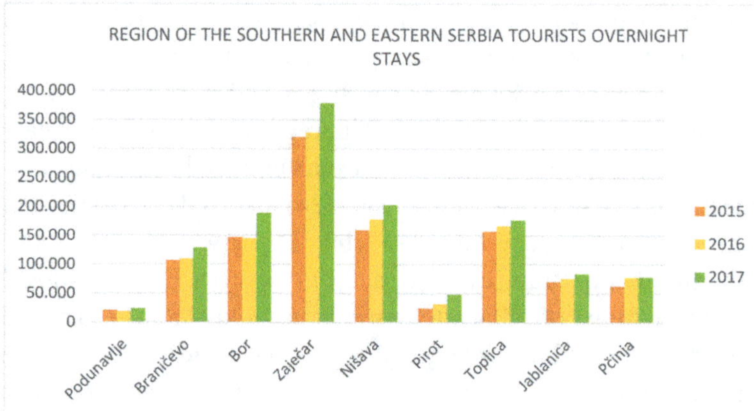

but also to consider all resources on local and regional level, it is possible to have local – micro tourist destinations that are based on local resources of one municipality or to have complete regional – national – macro offer that can combine more settlements and sites. Tourism management must be comprehensive and it must have defined precisely all aims and future solutions must rely on adequate strategies

There are different activities that could be offered to the visitors in this Region, and based on that, we can determine few categories of rural tourism that could be developed. **Agitourism** offers tourists possibility to actively participate in agricultural work or on farm. Many settlements offer active rural vacation with outdoor activities that are not related to agriculture and it can be classified as **recreational tourism**. **Cultural tourism** can be developed since these areas have numerous of cultural manifestations and cultural and historical heritage. One type of cultural tourism can be also **event tourism** which is related to the numerous of out and indoor manifestations in rural areas. There is also **eco-tourism** which is based on natural resources protection. Intact and unpolluted natural landscape provides excellent opportunities for development of **educational tourism**, teaching – school in nature and recreational teaching can be organized for pupils of all ages. Food – **gastronomic tourism** is also a way of tourism where different types of food and drinks from local surrounding can be presented and also can be used for area branding. Within this Region there are 15 spas, so it is possible to develop **health- spa tourism**. As a common way of visiting the rural areas is classical rural experience where tourists can walk through the settlement and experience traditional ambience and way of living.

All these types of tourism, even happening in rural areas, have very strong connection on relation urban – rural. Lately, because of the rising soil and air contamination, organic farming is more popular and organic products are more and more buying and they are mostly produced in rural areas. Agriturism is a kind of tourism where people from urban area go on the local farms and buy these products, sometimes even participate in production. Many restaurants are offering organic foods from specific farms and wineries. Because of increasing cities densification process, rising values of air pollution, and huge problem with green public spaces "disappearing in plans", urban residents are spending their free time in surrounding rural areas with intact nature. Many restaurants, sport clubs and resorts are in rural areas in vicinity of the cities offering daily or weekend escape for urban population from polluted neighbourhoods. It is obvious that there is very close connection between urban and rural areas and population and that urban – rural synergy must be more improved in order to create quality and efficient settlement network and promote sustainable development.

Development of tourism is often analyzed together with sustainable development. There are many definitions regarding sustainable development but it is certainly that it combines development theory and environmental sustainability (Lélé, 1991). Even tourism is orientated towards "product" touristic offer, principles that are used for it development are in accordance with environment protection and sustainability. Developing rural tourism in this Region would have great impact on both regional and sustainable development because it would bot preserve natural landscape from illegal construction and devastation and also would promote abandoned natural values. In order to have successful and sustainable touristic offer, it is important to have permanent residents in rural areas so that they could be "coordinators" for its further development.

On figure 3 are shown characteristic traditional houses which are abandoned in the village Rsovci in Pirot district. This village is located on the slopes of Stara Mountain and it is distanced just 20km from the city of Pirot but there is only one road to the village. Village is located near Zavoj Lake, waterfall Tupavica, Gostuša village, church in the cave, and last few years some households started with tourism activities. This is not the only example in Pirot district, around this village there are few more – Gostuša, Dojkinci, Velika Lukanja, Visoka Ržana that are offering touristic activities and accommodation where people from the nearby cities are spending their time. This settlements are very close to the city and they can be used as one day trip or weekend travel.

SWOT ANALYSIS

In order to analyze properly rural areas within the Region of Southern and Eastern Serbia, SWOT analysis was conducted and presented within Table 1. Analysis was done by authors for purposes of this research and it represents important element for defining future strategic development and aim is to highlight main potentials, and at the same time setbacks that could endanger future development.

Development of this kind of toursim cannot rely only on potentials of these sites, it must also have good connections with urban areas because synergy of these two areas can create healthy environment and better life quality. Rural areas depend on urban areas in terms of infrastructure and economy but in this connection they are important part of tourist offer and regional development (Van Leeuwen, 2015). Due to very diverse topography and possibility of organizing different type of activities, rural tourism in this Region can have offers for entire year activities regardless season. One of the main benefits of rural tourism is prevention of conspicuous depopulation process through activation of deprived local economy and creation of new jobs. Development of rural tourism has positive economic and financial effects, it decreases negative population migrations and also stimulates migration on relation urban to rural – especially in remote areas, and it promotes transfer of knowledge and new ideas from urban areas. Development of economy is not based only on agricultural activities, accent is on diversification of economy by activating local potentials. All the benefits, unfortunately, have other negative side which is great threat to sustainable development. With expansion of rural tourism and investors, wellbeing and intact nature are endangered and also quality of life can be significantly reduced.

RURAL TOURISM FOR URBAN-RURAL SYNERGY: SOLUTIONS AND RECOMMENDATIONS

Urban – rural synergy can be defined as mutual connection of urban and rural area in terms of distribution of urban functions and economy, creating spatial connections and improvement of social cohesion of population. Development of rural tourism and its impact on urban – rural synergy can have many positive effects for further development. Economic prosperity in terms of direct income from tourism development: impact of tourism development can be visible on multi levels – not only local, and it can improve regional disparities. Economic prosperity from the accompanying activities of rural tourism development in terms of constructing infrastructure network, opening of new construction sites, development of agricultural and organic production, better marketing and web promotion, can also ensure improvement of urban-rural synergy. Local context and resources have great impact on possible entrepreneurial activities development because quality of entrepreneurship depends on a diversity of local offers. On the other side, quality of the entrepreneurship affects economic and social-demographic development which directly affects regional rural development. All these factors are very connected and they are interdependent and successful impact can be achieved only with their strong mutual connections and also with the connections with the urban area which is main support for rural tourism. Successful and developed rural tourism is also one of the key elements for even regional development. Purpose of urban rural synergy is not only economic prosperity and mutual gross product growth, it is also development of spatial connections of settlements and locations, connections of population from both areas and also connections between urban and rural goods.

Within Table 2. main economic, social, functional and spatial characteristics of urban and rural areas in this Region are shown based on the research data. It is obvious that urban and rural areas are complementary within this Region and that their mutual connection is obligatory for further development.

Urban areas cannot develop independently from rural areas in terms of food production, high percentage of population in these areas and also traditional values which represent origin of one nation. On the other hand, rural settlements are depending on urban in terms of economy, urban functions that are within cities and strategic connections – airport and main roads. This mutual synergy would ensure new business ideas and possibilities, especially for small and medium enterprises, it would ensure better social inclusion and services, more secure food system – production and distribution, it would develop cultural awareness of residents from both areas and would ensure better cultural connections. With this connections and future development it is possible to: achieve better demographic and age image of the rural settlements; to control migrations towards urban areas and improve life quality in rural areas that are satisfactory to all residents; different distribution of educated people because there will be higher percentage of educated residents in rural areas thanks to the different seminars and workshops; diversification of industries can improve inclusion of women and more balanced gender structure in rural areas by opening new job opportunities. Rural areas, that gravitate to urban areas and that are functionally connected with urban areas represent "resilient" settlements in terms of economy. In order to become more responsible and resilient settlements in terms of climate change impact, rural areas need coordination with urban areas in case of natural disasters and long term processes of mitigation. Better organization of polycentric settlements network can be achieved with better connection urban – rural and also planning system and tools will be improved through common planning for both areas. Development of rural tourism unequivocally impacts both urban and rural areas and also has impact on Regional development. Currently there are many links and connections between urban and rural but not efficiently developed

Table 1. SWOT analysis (by authors)

Strengths	Weaknesses
Vicinity to urban areas - extraordinary geographical and strategic location; **Intact natural landscape** and biodiversity; **Developed agricultural production**, "domestic" production with different types of households; **Rich architectural fund** in traditional spirit from different time epochs; **Hospitalit**y of local residents; **Wide range of local food and crafts** specific for this Region; **Unpolluted environment** with high level of life quality; **Favorable geomorphological characteristics** of the terrain, large water resources, moderate climate; **Large number of abandoned public building**s - wine cellars, culture, mills, churches and monasteries; **Existing transport network;** **Thermal mineral water springs;**	**Depopulation and unfavorable age structure** in large number of settlements; **Absence of basic infrastructure** in settlements; **Lack of entrepreneurship**– small agriculture producers that are not marketcompetitive market; **Lack of funds and finances;** **Low capacity of accommodation** and insufficient utilization of them; **Offer of accommodation is not divers**e –only budget friendly offers without higher categorization; **Mainly accommodation withou**t catering and offer for more activities than just accommodation; **Lack of cooperation with tourist offices** on all levels – local, regional, national; **Authentic households without investments to comply with needs of the visitors and regulations;** **Awareness and educational level** regarding tourism of hosts is inadequate; **Lack of web and national presentation;**
Opportunities	**Threats**
Private investments in different activities than agriculture; **Improvement of neglected traditional crafts & customs;** **Protection of natural landscape;** **Equal rights and inclusion of all residents** with different age, gender and educational structure; **Better valorization of inherited and natural resources;** **Available foreign funds** for improvement of built structures, organization of cultural events etc.; Technical and transport **infrastructure improvement**; **Development of different clusters on relation urban –** rural in order to involve residents from "both sides"; **Rehabilitation of existing cultural centers** in order to create space for the gathering of young people; **Improving health care** with the introduction of additional shifts and capacity expansion; **Education about use of modern technologies;** **Organizing courses and workshops for traditional crafts;** **Renovation from local materials;** **Creating urban-rural touristic offers;**	**Very pronounced depopulation**, and increase in the number of elderly; **Problems with revitalization** of municipal owned infrastructure; **Income from rural tourism** can easily become part of the city's budget and not of local community; **Unequal distribution of finances**; **Distance between settlem**ents and poor public transport connections; **Complex administrative procedures** and legislation in the field of food production; **Abandonment and marginalizatio**n of certain activities of the local population and old crafts; **Endangering authentic "rural pattern"** and social and cultural diversity; **Lack of planning documents;** **Climate change impact – lack of strategies for preventing natural disasters;**

and lack of functional connections is one of the greatest threats for rural and balanced regional development. Rural tourism can be defined as one of the catalysts for this synergy because both areas can have excellent benefits from its further development and expansion. Better social inclusion, prevention of further negative demographic growth, better life quality and connections between built and natural landscape, easier and faster knowledge transfer on relation urban-rural, promotion of traditional values as one of the important characteristics of one nation, responsible management of natural resources. Figure 6 shows mutual connections and dependence between urban and rural, and rural tourism as a drive for their synergy.

Development of rural tourism and urban – rural synergy will help decision making on local level and improve connections of local resources and promote local values. Local economy will be boosted and also transfer of knowledge can be improved. With this together and local decision making, strategies and

Table 2. Main Economic, Social, Functional and Spatial characteristics of urban and rural areas within this Region (conducted by authors)

REGION - URBAN	REGION-RURAL
ECONOMIC AND FUNCTIONAL CHARACTERISTICS	
9 cities which are underdeveloped in terms of economy and industrial production. Only 2 cities – Bor and Niš with Republican average, and Niš is the third largest city in the country	Large number of settlements with potentials for development of diverse industries, especially those along Danube, Nišava and Toplica river, on the slopes of mountains in Southern and Eastern Serbia, also there are more than 10 lakes on this territory.
Diverse possibilities for new industries - great potential for tourism development and connections with rural areas in vicinity. Good traffic connections within the Region, newly constructed motorway towards Bulgaria goes through this Region	Possibilities for agriculture development with modern technologies and development of organic farming and production, creation of brands from protected geographic origin. Large market within surrounding urban areas
International airport in the city of Niš with more than 20 European destinations	Programs of cross-border cooperation in rural areas IPA with Bulgaria and Romania, Danube projects, and different non-governmental institutions programs that could enable better connections to foreign market and "inflow" of different type of tourists.
Eligibility for application for many foreign funds because of the ongoing cross border cooperation with Bulgaria and Romania – borders with this two EU countries are within this Region	Developed only housing, agriculture and husbandry. Only basic education, high schools in Municipalities and Faculties in cities. There is no administration beside local offices for basic documents.
SOCIAL STRUCTURE / DEMOGRAPHY	
Negative demographic trend, except in the city of Niš, according census 2011 results	Decreasing number of population and negative age structure in almost all village settlements in the Region.
Different age, educational and gender structure. There is University within this Region.	Low percentage of educated people in area of tourism and use of modern technologies for touristic promotion.
High level of social inclusion of all ethnic groups.	Social inclusion on low level – many people are living on the margins of the existence
Large number of tourist from different parts of the world and from different cultures visiting this Region which could be seen on figures 4 and 5 that this number is increasing last few years,	Heterogeneous structure of different traditions of every district and multicultural structure, especially in border area
SPATIAL ORGANIZATION	
Almost all the cities are located on some river flow	Biggest river in the country- Danube goes mainly through rural landscape in the Region of Southern and Eastern Serbia
Spatial distribution of population and different urban functions is on satisfactory level	High hilly and mountainous topography with mountains and great potential for tourism
Built area with huge air pollution - Niš, Bor and Leskovac especially	Unfavorable spatial distribution of settlements and underdeveloped polycentric system of settlements
Good connections and spatial organization of infrastructure	Poor spatial connection between settlements and lack of main technical infrastructure in most of them

financial plans can be shaped in the way local context dictates which will assure future development. This means that local residents will take active part in decision making and local entrepreneurships will be established. This synergy can help in providing sustainable and smart growth not only in terms of economy, but also in social, technological and spatial terms, and it can create network of responsible and resilient settlements to negative impacts of climate change.

Figure 5. Traditional abandoned houses and natural landscape in village Rsovci – Pirot district (photograph taken by authors in September 2018.)

Figure 6. Urban – Rural synergy (by authors)

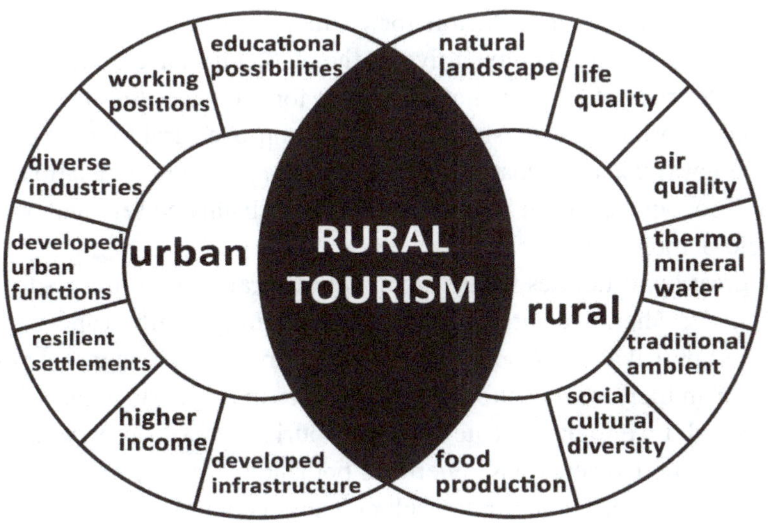

FUTURE RESEARCH DIRECTIONS

There are many possibilities for future research within this topic. It could be further investigated which type of tourism is best drive for urban rural synergy and for further rural development. Also, exact number, as well as unified register of all households for rural tourism within this Region can be established.

Since there are four more regions, this research can be done for some "more developed" region in order to investigate relations between rural tourism and urban and rural connections. In order to improve urban – rural synergy and investigate all possible setbacks for its development, future research could be oriented towards micro solution by analyzing small scaled areas – micro level case studies. This Region has very heterogeneous structure, so would be useful to conduct researches on different levels and within different districts as micro case studies.

CONCLUSION

Analyzing rural areas and their potentials, it is obvious that the Region of Southern and Eastern Serbia, and entire territory of Serbia in general, have great potential for this kind of tourism. As main problems there are inadequate infrastructure lines and waste management, which are endangering quality of life and quality of built environment, there are poor and incomplete official offers for tourist tours for these areas. Awareness of the residents is on a very low level and many people from city don't even know about sites that are distanced from them in just few kilometers. Settlements are not connected well, not only in terms of road network, but also in terms of mutual cooperation and economic development. Budgets of the cities are limited regarding rural areas because there are lot of investments within the urban territory.

Strategies that were defined by government are not implemented in adequate manner and there was no monitoring or evaluation period. There are many foreign funds that could bring investments in rural areas, but no local initiatives because residents are living in very difficult conditions, low rate of employment and very low incomes are main factors for young people and working population migrations. On the other side, rural areas were not competitive on the market because of their state but still promoting them as affordable tourist potentials is not great help for their development. Master plan for rural tourism considered public-private partnerships in order to improve rural tourism development because local communities are not financial prepared for all challenges. Within legislation and strategic documents, rural tourism is recognized as an important element in balanced regional development in terms of economic and socio-demographic growth.

Rural areas have great opportunities for development of rural tourism thanks to rich natural, architectural, cultural and historic heritage but also thanks to the vicinity of urban areas. Connection between urban and rural areas is unbreakable not only in economic terms, but also in terms of spatial distribution of functions and also in ambience connections. Heritage sites in the urban areas are connected with the sites in the rural areas and if we want to create complete touristic offer it is necessary to include sites in both areas. Also, urban – rural synergy has benefit for both sides because vicinity of the urban areas to the rural helps faster and more efficient rural development in economic and educational terms.

REFERENCES

Antić, M., Šantić, D., Kašanin-Grubin, M., & Malić, A. (2017). Sustainable rural development in Serbia – relationship between population dynamics and environment. *Journal of Environmental Protection and Ecology*, *18*(1), 323–331.

Aronsson, L. (1994). Sustainable tourism systems: The example of sustainable rural tourism in Sweden. *Journal of Sustainable Tourism*, *2*(1-2), 77–92. doi:10.1080/09669589409510685

Bogdanov, N. (2007). *Mala ruralna domaćinstva u Srbiji i ruralna nepoljoprivredna ekonomija*. United Nation Development Program.

Brauer, R., & Dymitrow, M. (2014). Quality of life in rural areas: A topic for the Rural Development policy? *Bulletin of Geography. Socio-Economic Series*, *25*(25), 25–54. doi:10.2478/bog-2014-0028

Butler, R. W. (1980). The concept of a tourist area cycle of evolution: Implications for management of resources. *The Canadian Geographer. Geographe Canadien*, *24*(1), 5–12. doi:10.1111/j.1541-0064.1980.tb00970.x

Cawley, M., & Gillmor, D. A. (2008). Integrated rural tourism. *Annals of Tourism Research*, *35*(2), 316–337. doi:10.1016/j.annals.2007.07.011

Crouch, D. (1994). Home, escape and identity: Rural cultures and sustainable tourism. *Journal of Sustainable Tourism*, *2*(1-2), 93–101. doi:10.1080/09669589409510686

Cvejić, S., Babović, M., Petrović, M., Bogdanov, N. & Vuković, O. (2010). *Socijalna isključenost u ruralnim oblastima Srbije*. UNDP- Sector for inclusive development.

Đenadić, M., Muhi, B., & Jovanović, D. (2016). Rural tourism – Serbia's missed chance. *Economics of Agriculture*, *2/2016*, 515–529.

Efstratoglou, S., Bogdanov, N., & Meredith, D. (2007). Defining Rural Areas in Serbia and Their Typology. *Proceedings of the 100th Seminar Development of Agriculture and rural areas in Central and Eastern Europe*, 553-562.

European Commission. (2014). *Rural Development Programme 2014-2020*. European Commission.

European Commission. (2016). Cork 2.0 Declaration: A Better Life in Rural Areas. *Conference on Rural Development*.

European network for Rural development. (2010). *Climate Action. EU Rural Review No 4*. European Commission.

European Parliament. (2013). *Regulation 1305/2013 of 17 December 2013 on the European Agricultural Fund for Rural Development, Article 18*. Author.

George, W., Heather, M., & Reid, D. (2009). *Rural tourism Development Localism and Cultural Change*. Channel View Publications.

Greffe, X. (1994). Is rural tourism a lever for economic and social development? *Journal of Sustainable Tourism*, *2*(1-2), 22–40. doi:10.1080/09669589409510681

Jing, X. (2006). Rural Tourism and Sustainable Community Development. *Proceedings of International forum on rural tourism, final report*, 73-80

Lane, B. (1994). What is rural tourism? *Journal of SustainableTourism, 2*(1-2), 7-21

Law on territorial organization of the Republic of Serbia. *Official Gazette RS, No. 129/2007, 18/2016 and 47/2018*. pp. 237-325.

Lélé, S. M. (1991). Sustainable development: A critical review. *World Development, 19*(6), 607–621. doi:10.1016/0305-750X(91)90197-P

Maksin Mićić, M., Milijić, S., & Nenković Riznić, M. (2009). Spatial and environmental planning of sustainable regional development in Serbia. SPATIUM International Review, 21, 39-52.

Milić, B. (2011). *Ruralni razvoj - praktikum za lokalne aktere*. Stalna konferencija gradova i opština SKGO.

Ministry of Agriculture, Forestry and Water Management. (2009). *Plan for the rural development strategy for the period 2009-2013*. Retrieved January 17, 2019. http://www.agrif.bg.ac.rs/files/events/38/Plan%20strategije%20ruralnog%20razvoja%202009-20013.pdf

Ministry of Agriculture, Forestry and Water Management. (2014). Strategy for Agriculture and Rural Development of the Republic of Serbia 2014-2024. Official Gazette RS, No. 85/2014. pp. 36.

Ministry of Finance and Economy. (2011). *Master plan for sustainable development of rural tourism in Serbia*. Retrieved January 15, 2019. https://futurehospitalityleaders.files.wordpress.com/2012/11/master-plan-odrzivog-razvoja-ruralnog-turizma-u-srbiji.pdf

Ministry of Trade, Tourism and Telecommunications. (2016). Strategy for tourism development of the Republic of Serbia for the period 2016-2025. *Official Gazette of Republic Serbia No. 98/2016.*

Ministry of Trade, Tourism and Telecommunications. (2018). *Law on tourism – draft version*. Retrieved January 15, 2019. http://www.parlament.gov.rs/upload/archive/files/lat/pdf/predlozi_zakona/2018/3153-18%20-%20Lat.pdf

Ministry of Trade, Tourism and Telecommunications. (2006). Strategy for tourism development of the Republic of Serbia. *Official Gazette of Republic Serbia No. 91/2006.*

Mitchell, M., & Hall, D. (2005). Rural Tourism as Sustainable Business: Key Themes and Issues. In D. Hall, I. Kirkpatrick, & M. Mitchell (Eds.), *Rural Tourism and Sustainable Business* (pp. 3–14). Channel View Publications.

Nagaraju, L. G., & Chandrashekara, B. (2014). Rural Tourism and Rural Development in India. *International Journal of Interdisciplinary and Multidisciplinary Studies, 1*(6), 42–48.

National Tourism Organization of Serbia. (2017). *Catalogue of village touristic households*. National tourism organization of Serbia.

Noev, N. (2013, December). *EAFRD support for sustainable rural tourism 2014 – 2020*. Paper presented at "Eurometing 2013," Brussels, Belgium.

Nylander, M., & Hall, D. (2005). Rural Tourism Policy: European Perspectives. In D. Hall, I. Kirkpatrick, & M. Mitchell (Eds.), *Rural Tourism and Sustainable Business* (pp. 17–40). Channel View Publications.

Organization for Economic Co-Operation and Development. (1994). Tourism Strategies and Rural Development. *OCDE/ GD (94)49.*

Organization for Economic Co-Operation and Development Republic Agency for Spatial Planning. (2010). Spatial plan of the Republic of Serbia for period 2010-2020. *Official Gazette of the Republic of Serbia, No. 88/2010.*

Republic Agency for Spatial Planning. (2010). Regional spatial plan for Southern Pomoravlje Municipalities. *Official Gazette of Republic Serbia No. 83/2010.*

Republic Agency for Spatial Planning. (2011). Regional spatial plan for Timočka Krajina – Bor and Zaječar districts. *Official Gazette of Republic Serbia No. 51/2011.*

Republic Agency for Spatial Planning. (2013). Regional spatial plan for Nišava, Toplica and Pirot districts. *Official Gazette of Republic Serbia No. 1/2013.*

Republic Agency for Spatial Planning. (2015). Regional spatial plan for Podunavlje and Braničevo districts. *Official Gazette of Republic Serbia No. 8/2015.*

Sarudi, C., Szakaly, Z., Mathe, A., & Szente, V. (2003). The Role of Organic Agriculture in Rural Development. *Agriculturae Conspectus Scientificus, 68*(3), 197–202.

Sasu, K.A., Epuran, G. (2016). An overview of the new trends in rural tourism. *Bulletin of the Transilvania University of Brasov, Series V: Economic Sciences, 9*(58), 119-126.

Statistical office of the Republic of Serbia. (2013). *Census of Agriculture 2012: Agriculture in Serbia* (vol. 2). Statistical office of the Republic of Serbia.

Statistical office of the Republic of Serbia. (2014). *Comparative overview of the number of the population in 1948, 1953, 1961, 1971, 1981, 1991, 2002 and 2011.* Statistical office of the Republic of Serbia.

Statistical office of the Republic of Serbia. (2016). *Municipalities and regions in the Republic of Serbia in 2015.* Statistical office of the Republic of Serbia.

Statistical office of the Republic of Serbia. (2017). *Municipalities and regions in the Republic of Serbia in 2016.* Statistical office of the Republic of Serbia.

Statistical office of the Republic of Serbia. (2018). *Municipalities and regions in the Republic of Serbia in 2017.* Statistical office of the Republic of Serbia.

Statistical office of the Republic of Serbia. (2018). *Profile of the Region of the Southern and Eastern Serbia. Statistical office of the Republic of Serbia.* Retrieved January 25, 2019.: http://devinfo.stat.gov.rs/SerbiaProfileLauncher/files/profiles/sr/1/DI_Profil_Region%20Juzne%20i%20Istocne%20Srbije_EURSRB002002.pdf

Streimikiene, D., Bilan, Y. (2015). Review of Rural Tourism Development Theories. *Transformations in Business & Economics, 14*(2), 21-34.

Todorović, A. (1964). Ruralna i urbana sociologija. *Sociologija i prostor,* (3), 37-43.

Todorović, M., & Bjeljac, Ž. (2007). Osnove razvoja ruralnog turizma u Srbiji. Bulletin of the Serbian Geographical Society, 87(1), 135-148.

Todorović, M., & Bjeljac, Ž. (2009). Rural tourism in Serbia as a concept of development in undeveloped regions. Bulletin of the Serbian Geographical Society, Acta geographica Slovenica-geografski zbornik, 49(2), 453-465.

Van Leeuwen, E. (2015). Urban-Rural Synergies: An Explorative Study at the NUTS3 Level. *Applied Spatial Analysis and Policy*, *8*(3), 273–289. doi:10.100712061-015-9167-x

Vuković, P., Kljajić, N., & Arsić, S. (2012). Multifunctional agriculture as an assumption and a condition for rural development in Serbia: Special turn to rural tourism. *International Journal of Sustainable Economies Management*, *1*(2), 24–32. doi:10.4018/ijsem.2012040103

World Tourism Organization (UNWTO). (2004). *Rural Tourism in Europe – Experiences, Development and Perspectives. World Tourism Organization.* UNWTO.

Zelenakova, M., Purcz, P., Hlavata, H., & Blištan, P. (2015). Climate Change in Urban Versus Rural Areas. *Procedia Engineering*, *119*, 1171–1180. doi:10.1016/j.proeng.2015.08.968

KEY TERMS AND DEFINITIONS

Balanced Regional Development: Equal development of all regions within one country through decentralization of economy and better spatial and functional distribution and social cohesion of all residents.

Ethno Village: Small scaled village with traditional architecture that consists of accommodation in cottages, small restaurant where local food is served, small farm with domestic animals, shops where local products are sold, and other objects that represent local traditional ambient.

Negative Depopulation Trend: Demographic occurrence where there is constant decrease in population number caused by negative birth rate, high percentage of mortality or due to expressed migrations on relation rural to urban settlements.

Region of Southern and Eastern Serbia: Least developed administrative region in Serbia characterized by large depopulation process and unfavorable age structure.

Rural Areas: Underdeveloped areas where main activities are related to the land use for the purposes of agriculture and husbandry and with underdeveloped economy, depopulation trend and low population density.

Rural Tourism: Tourism which is oriented towards promoting rural tradition and rural areas where accommodation is in traditional spirit, where local food is served and all the services are based on local resources.

Urban-Rural Synergy: Mutual connection of urban and rural area in terms of distribution of urban functions and economy, creating spatial connections and improvement of social cohesion of population.

Chapter 10
Visitor Centres as a Contribution to Improving Urban-Rural Relations:
From Primary Needs to Landmarks

Aleksandar Videnovic
Faculty of Architecture, University of Belgrade, Serbia

Milos Arandjelovic
Faculty of Architecture, University of Belgrade, Serbia

ABSTRACT

The architecture of public buildings in rural areas, through the advancement of skills, knowledge, technologies, and materials, has become increasingly representative in recent decades, especially considering tourism as a global theme. The work is related to the planning of visitor centers in rural areas and the main task is defined by the effort to improve the quality of life in such areas, that is, highlight the major advantages in space, such as the natural environment. The aim of the research is defined by establishing certain elements in the planning of the visitor centers within the idea of promoting local values and cultural heritage. The first part of the chapter has been defined as an analysis of the theoretical views. The second part of the study has been defined as an analysis of the visitor centers. Through a case study, in the third part of the chapter, the work presents a comparative overview of the process to achieve two individual similar investments in Serbia and Bosnia-Herzegovina.

DOI: 10.4018/978-1-5225-9932-6.ch010

INTRODUCTION

The shift in cultural patterns impacted contemporary architecture by making it subordinate to strive for spatial attractions, occasionally overshadowing even the very functionality of the specific facilities. The arrival of new, overlarge, geometrically aggressive forms, usually unsuited to the relevant context, was largely caused by the technological and technical progress which introduced a new way of working and thinking. In that sense, the need for originality and recognisability at any cost can be named as one of the reasons for inferiority of certain architectural designs.

Visitor centres are mandatorily info-promotional facilities, often focused on service and accommodation, and less often on research and education, and which should satisfy different needs of a specific natural or artificially created area by means of a narrower or wider range of different activities. They are very important as special purpose points for development of a tourist offer in a specific wider area, but also as marketing and promotional markers for popularization and attractiveness, i.e. "new recognisability" of individual locations in a more narrow spatial sense. This paper overview of needs for their individual aspects in rural spaces and areas of Serbia, while taking into consideration the environment and experience they are to present to users.

The analysis of urban-rural relations in this paper, as well as the placing of those relations in the traditional and contemporary context, aims to provide understanding of the elementary principles and approaches to design in these communities. The relationship between urban and rural architecture is defined by unbreakable interaction between traditional values of construction heritage. It is precisely the traditional construction values that have been defined in the continuation of this paper as one of the main criteria and the starting point for comparison of examples. In the form of an overview of a modern approach to design in the rural context, this analysis of specific examples should provide a verification of theoretical assumptions on the relationship between past rural construction and modern architectural trends.

In the framework of the presented theoretical background, this paper should provide an analysis of the buildings in Goc and East Drvar for the purposes of reviewing new trends in design and organizing of rural spaces. The approach to design has been highlighted as the main contribution to the reflection on introduction of new facilities whose visual reflection and character aim to enhance the qualities of natural and man-made spaces, free from desire for attractiveness at any cost.

MODERN ARCHITECTURE AS A PREREQUISITE FOR IMPROVEMENT OF URBAN-RURAL RELATIONS

Rural-urban relations, when viewed, should be observed primarily through the lens of the living, housing and working styles. Life in the country is characterized by individual housing that implies free space surrounding the house - a backyard, a building (the house) and auxiliary, agriculture- related buildings. As far as the economic structure of rural communities is concerned, it should be noted that the number of non-agricultural households has been increasing in recent years, significantly impacting the changes in architecture of specific buildings.

Pointing out the similarities and differences of architecture in urban and rural areas in the research is important because it indicates the possibility of a certain cultural and thus economic cooperation between the city and the village. The result of this cooperation, on the other hand, can be a framework for improving the degree of rural development, but certainly defining the starting points in the contemporary design sense. More specifically, in which direction it is necessary to consider and develop the architecture of the Visitor Centers as tourist oriented objects. A theoretical analysis of planning and construction in the city or countryside in this regard contributes to the understanding of a way of life that largely determines the criteria for building such content. Establishing certain criteria on the other hand is important, because besides the program typology, it indicates the possibility of cooperation of similar centers in the regional - interstate framework.

The city as a spatial group is characterized by high concentration of people and as such is capable of absorbing different cultural modality models, thus creating an architecture that is considerably detached from the traditional framework. Unlike individual housing in the country, collective housing presents the dominant system in urban communities. The organization of life and work is completely different, which is reflected in the structure of proper buildings. The links and co-dependency between living and work spaces so characteristic for the rural environments are not present in the city (Kojić,1973). Modern cities are the hubs for flow of capital, people and commerce, which is why architecture in such dynamic environments most often strives to create representative and attractive spaces.

The city as organized space is developed in line with predefined rules. In the urban architecture's mass production individual buildings largely lose their specific personalities. As such, they become subordinate to the whole, norms, rules and regulations, thus being important only as one segment of the city (Tafuri, 1979). Buildings in natural surroundings are seen as works of architecture, while grouped in one whole they belong to a different type of art (Cullen, 1990).

The circumstances of the social and cultural context of the rural and urban environment initiate diametrically opposed architectural concepts. Branislav Kojić (1973) considered that the cities, unlike villages maximally used the benefits of modern technical and technological achievements, while rural settlements and the buildings in them preserved the "traditional" form and design framework.

Villages are characterized by spontaneous origin and architecture adapted to the conditions of the given region, i.e. natural environment. Even though organized/planned rural settlements do exist in practice, visual identity of rural buildings continues to follow the basic trends of traditional construction as the inherited postulates of the context in which they originated.

Traditional - Modern

Architecture basically initiates a dialogue with the tradition of the particular social and historical context, thus actually striving to create new forms of artistic action and aesthetical norms. The understanding of tradition, as well as regional specificities and culture, is required in the domain of modern construction practice (Petrović & Rašković, 2011). Contemporary architecture, regardless of whether it is located in the cities or villages is mainly based on certain values of cultural heritage which provides a stand for the creation of new design forms.

Architecture is defined by several other determinants apart from functionality. These are first and foremost the conditions in the immediate location, conditions in the wider environment (climate, natural surroundings) and the characteristics of the materials used. It would be completely illusory to ignore these design parameters, as well as the (social, spatial, cultural etc.) context. It would mean that identical buildings could be built at the seaside and on a mountain peak, in a village or in a city. And by saying "identical" we refer to identical architectural expression (design, form), use of materials… It would also be illusory to keep building the way it was done in the previous periods, since architecture stems from specific social and cultural circumstances, so it should accordingly follow the contemporary development trends.

Architecture in rural areas has always been representative of the culture of its respective region. Construction and design specificities reflected regional characteristics of a given area and tradition, as well as the elements of ethnographic heritage and cultural and historical context. Special attention was paid to a rational approach to construction, combining the values of everyday needs of the population for which the buildings were intended, thus making the said architecture contextual.

Traditional rural architecture was characterized by the use of local materials whose characteristics and structures were essentially appropriate for the climate conditions in the area. The same material was sometimes used for construction elements, sometimes for purposes of roofing or facade cladding, and sometimes in the building interior, for construction of furniture. Such practice is supported by a wide range of different options for processing of natural materials which formed the basis of architectural aesthetics in rural areas. Due to this, village buildings were an integrated part of the rural landscape.

Rural architecture heritage was always a source of inspiration for architects desiring to achieve a special relationship with the tradition they wanted to transform and reinterpret in a modern manner. International practice indicates the existence of efforts to translate traditional values, spirit of a place, and cultural context into a specific architectural expression. 'Critical regionalism' as an approach that has defined the postulates of such architecture, represented an opposite of modern architecture which started to degrade the values of cultural context with its unified design (Frempton, 2004).

Aesthetical and functional principles of traditional architecture represent an answer to man's fundamental needs. The existence of universal construction principles based on anthropological principles guarantees the creation of architecture tailored to human beings (Krier, 2008). The efforts to overcome these postulates in order to achieve purposeless attractiveness results in alienated architecture unsuited to humanity's needs. Architect and theorist Robert Krier (2008) emphasizes that aesthetical uniformity of contemporary architecture has actually stemmed from excessive development and progress of the society, i.e. uncontrolled entropy. He considered the omnipresent strive for simplicity of form as unnecessary.

DEVELOPMENT OF VISITORS' CENTRES: SERBIAN RURAL AREAS IMPROVEMENT

In the circumstances of altered natural living conditions (where the climate factor is only one consequential but important factor, and which takes the role of a generator of further environment development), as well as a more significantly altered social-economic environment framework, almost all aspects of life are constantly reviewed, and not solely in Serbia. Almost daily, dynamic and diverse sociological changes in the rhythm of life, routine behavioural habits and human relations (between people and between people and the environment) irrevocably lead to deliberations on the quality of established or spontaneously evolved attitudes towards potential and possible resources of future social development.

Tourism, i.e. all its applicable traditional and contemporary forms, represents one of the more important established potentials of further development of Serbian socio-spatial framework. Although publicly accepted as such, it still remains in the framework of routine understanding and doctrine of tourist services, confined in an ancient, out-dated view of current issues, changed demand and needs of users. If we also take into consideration the rural character of the majority of Serbia, where the essential items of potential tourist offer are located, exist and are manifested, it becomes clear that the organization of modern use of these potentials primarily lacks the informational, and then all other (reception, accommodation, presentation, education, culture, recreation, and accompanying) new or updated suprastructures.

Apart from their rural character, the areas of special natural and artificial features in Serbia are also characterized (with well-known exceptions) by a specific measure of preservation, not only in the sense of nature and environment quality, but also from the aspect of autochthonous visual and ambient-related values of the environment. Such spaces, through advancing tourism knowledge, can be understood not only as a development resource but as a core capital of rural areas. (Garrod *et al.* 2011) The reason for such a favourable environment for future action and development can be found primarily in the social-ideological inferiority and disregard for rural territories in the second half of the last century, but also in the lack of social circumstances, initiatives, investments and general (not solely material) poverty during the last decades of the transition between centuries.

In the transition social-economic circumstances, in the atmosphere of rare investments with often uncertain profitability, with pioneer investments in the initiation of the development of various under-developed areas, the flexibility of new programmes and rationalization, i.e. multifunctionality, become crucial planning factors. Apart from the aforementioned, in the conditions of the permanent lack of resources required for the construction of new content, resourcefulness and initiative-taking in reconstruction, adaptation, refurbishment, innovation, repurposing and updating, hence maintenance, of the full range of existing facilities (a large number of which exists in the field) becomes a social behaviour imperative, and a basic aspect of rational action with regards to attempts to develop certain environments.

Problem With Tourism in Rural Areas of Serbia

Although exceptionally rich with potential tourist locations, rural Serbia, barring rare exceptions, is organizationally inaccessible to consumers with different needs and interests. In order to clarify the need for Visitors' centres in the aforementioned areas, we will provide a brief review of the basic reasons for such tendencies.

Reason One: In the circumstances of an increasingly faster and more stressful way of life, the tourist offer consumers are less likely to go vacationing once or twice a year in the framework of usual summer (seaside) and winter (skiing) packages. The need for more frequent but shorter vacations, which should possess an active recreational and learning component, is becoming increasingly pronounced. Changes in the concept of recreation and tourism are the result of numerous changes at the global level, taking into account, above all, the term globalization as a model of accelerated social, cultural and economic development. (Hall *et al.* 2006). Such packages require a different philosophy and manner of organization in tourism. Namely, this is based on the principle of continuous 24-hour work of tour operators who accompany tourists on their journey, and which excludes the usual organization based on change of tourist shifts after a several-day-long stay at a single destination, but is instead based on a dynamic change of trip locations and overnight stays, with visits to important sights in the framework of a predetermined tour. The lack of modern tourist offers and modern concept of the use of potentials is but one of the reasons for the poor development of tourism in the rural areas of Serbia.

Reason Two: We can look for the second reason in the lack of attention paid to tourist activities organizing based on contemporary principles of thematic units – clusters. Although top-quality conditions for organization of specialized trips (quite often called routes, hence the Roman Emperor Routes, Wine Routes; or tours – hence the Monastery Tour, Danube Tour, Fruska Gora Tour…) exist, few of them are actually implemented in practice. What seems to be becoming an axiom in Serbian tourism is the organization of different annual events (lasting usually one, less often several days) in different towns and villages, and which are rarely dedicated to cultural content, and more often limited to different culinary fairs and which cannot possibly present a foundation for serious tourist activities, since nothing else happens all year long at these locations. Tourism based on various, differently motivated events taking place throughout the year can survive in bigger cities, since by rule of thumb, some kind of event is always taking place in them. However, a continuous annual tourist offer in smaller towns and rural settlements and areas has to be differently founded and organized.

Reason Three: The third important reason for unsuccessful tourism in rural regions is the issue with authority over and coordination of tourist activities and facilities in a specific area. Natural attractions, entities and related content representing a single tourist brand usually cover the territories belonging to several municipalities, and sometimes to more than one District. In such cases the main issue with the organization of top-quality, well-designed tourism most often lies in the lack of initiative in relation to maintenance of facilities and design of programmes in their proper territory for certain administrations, as well as the unwillingness to cooperate with neighbouring administrations and jointly use the relevant resources.

The aforementioned reasons, which represent only some of the unfavourable circumstances impacting the development of high-quality tourism in rural areas, paint a clear picture of the situation in the rural Serbia tourist offer. These circumstances inspire us to consider attempts aimed at improvement of the tourist offer, which would be based on the establishing of new coordinates, networked locations, habitat-centred facilities and activities or memory places. If we would like for such destinations to be at least minimally functional (e.g. only an overlook, a place with a beautiful view), or to possess a more complex multifunctional structure (appropriate for longer stays, seminars, congresses, recreation, research, thematic, presentation, commercial or marketing events…), that would largely depend on the position in the hierarchy of related locations, but also on the specific features of individual locations, requirements for a longer stay, potential special protection duties, i.e. the size of the territory and the site, geographical unit or area which we wish to promote. It is implied that such review takes into consideration both the construction of new ones and use and rehabilitation of existing facilities.

The planning of facilities such as visitor centers in order to promote rural tourism requires the involvement of numerous stakeholders from local communities. The reason for this is that by engaging in the process of promoting local values, they contribute to establishing better relationships between service providers and users-tourists. Some studies point to the need for their mediation in the field of planning tourism activities in order to ensure the sustainability of content (Panyik *et al.* 2011). Not by any members of the local community but by those selected on the basis of carefully defined socio-demographic and economic criteria (Muresan *et al.* 2019:2-4).

Visitors' Centres as Elements of Tourist Route Network

Visitors' centres are primarily educational facilities, which endeavour to provide information to its visitors, as well as new insights on the specific traits of a particular area they are located in, and thus provide a significant contribution to development and promotion of rural areas, i.e. the dissemination of different, largely positive views of them. Contemporary understanding of heritage requires modern ways and forms of presentation that sometimes go beyond certain elements of protection. (Radosavljević *et al.* 2019)

Development of Visitors' centres is encouraged by the interest of visitors and the need for different services within a tourist offer in a certain area. Tourism as a separate branch of economy is being increasingly forced to take initiative in the generation, popularization and presentation of less obvious specific characteristics of specific areas in the global market, i.e. in the deliberate creation of an image or brand of specific sites, with the aim of increasing the number of visitors.

In this aspect, rural areas possess a huge potential, primarily due to their natural and cultural diversity. The term "rural" can encompass non-urban or even suburban areas, whose basic physical and geographical characteristics are determined in the framework of the use of land for agricultural production and forestry. The considerable interest of the international market for a rural tourist experience has initiated the modernization of the said offer, i.e. facilities, which, again, have to correlate with modern lifestyles.

The crucial factor in the planning of future Visitors' centre facilities should be linked to the choice of the proper location and content. The surrounding area has to possess specific features attractive to future visitors. Tourist activities and establishment of such facilities in rural areas encourage their direct economic development. The interest of visitors also can be focused on individual households (a parallel can be drawn to the expansion of ethno-farms during the last few decades) within a settlement, so they can indirectly become elements of "visitor" composition. Households are in this manner promoted for the purposes of tourist activities and transformed into accommodation facilities, illustrating the international trend of translating ex-agricultural entities to different types of tourist products. Apart from those, the rare historical vernacular architecture heritage sites (such as Staro Selo in Sirogojno) can also assume a new function of tourism route stops. In the geographical sense, these examples can play an important role in the establishment of Visitors' centre network.

It would also be pertinent to mention large number of facilities, either in rural settlements or outside them, which have lost their initial importance and purpose, and which could – post-rehabilitation – become a synergy points for tourism actions in Serbia (cooperative buildings, trade cooperatives, depopulated schools, valuable and rehabilitation-available vernacular facilities with traditional construction elements, abandoned barracks, and many other existing forms in rural areas who can be used owing to their preserved intrinsic value). This would achieve at least two positive effects: rescue of buildings from falling into disrepair, and placement of new functions important for survival of villages through tourism development.

The phenomenon where whole new settlements, such as Drvengrad at Mokra Gora and Kamengrad in Visegrad (Republika Srpska), are constructed, only confirm the exceptional need for this type of content. (Figure 1) These two examples present typical cases of creation of entities which can be called and treated as Visitors' centres in the real sense of the word. Apart from possessing urban planning and architectural solutions which fully acknowledge the tradition and region in which they were established, they also, due to their functions, represent a magnet to visitors who come to the region with the specific purpose of visiting exactly these locations, while taking the opportunity to get familiar with the wider area and neighbouring facilities and amenities. Visegrad, with its famous old bridge was no longer attrac-

Figure 1. New ethno-style settlements, as a specific type of Visitors' centres with complete packages of events and activities (Drvengrad Mecavnik in Mokra gora on Tara Mountain and Kamengrad, Andricgrad on the river Drina – Visegrad / RS)
Source: From the authors' archive

tive to visitors travelling the region since they would already have had the opportunity to see the famous bridge on river Drina. A new town – Visitors' centre – provides a new motivation for second visit and improvement of the wider area. It can satisfy a whole range of need, starting from the basic ones such as short or long stay, overnight stay, catering, trade, information, but also much more complex demands related to culture, education or knowledge.

The same usefulness-based philosophy applies to Mecavnik, where visitors come not only for basic services, but also in order to participate in now traditional festivals, concerts, events, and the wider region also provides interesting sights, such as a rehabilitated narrow-gauge heritage railway, Sargan Eight, and other attractions in the direction of Uzice and Visegrad, or Bajina Basta across the Tara mountain.

Although perhaps under a different name and with a different intent, the authors of these new towns have achieved that which is proposed in this paper, all in order to improve the tourist offer in the rural Serbia.

Current Forms, Structure, and Content of Visitors' Centres

Visitors' Centres have become a synonym for multi-purpose, flexible structures. The complexity of their services reflects the complexity of modern social needs. The multi-purpose role provides an advantage in organization of different programmes enabling the active functioning of the centre throughout the year, which is a primary prerequisite if we desire for such services to survive, be cost effective and able to promote a specific area, which all results in the commercialization as the final goal.

We can notice two basic types of Visitors' centres. The first represents individual info-spatial facilities (info-points), which serve exclusively to provide necessary information and additional knowledge about the relevant site to the visitors. The second are multipurpose flexible structures which, unlike the first type, provide various additional options, ranging from education, accommodation, relaxation, museum and gallery exhibits, organization of conferences, to the presentation of production processes, tasting of products, and the commercial aspect related to their sale and placement (as is the case, e.g. with vineyards and Visitors' centres located in them).

Figure 2. Primary visitor spots: A vista of lake Perucac from "The Cross" on Tara (Source: from "Drina" monograph – Institute for textbooks and teaching materials – Belgrade), and Lookout Timmelsjoch – Austria/Italy (Architect WernerTscholl) on the cliff side lay-by
(Source: Journal C3 no. 359/2013.)

Figure 3. Memorial Home in Ravna Gora (designed by architect Spasoje Krunic) by the structure of its facilities represents a Visitors' centre at a historical site, with basic (info-presentation) and minimal accommodation facilities
Source: From the authors' archive

It is important to emphasize that visitor centers do not always have to be newly built facilities. It is also possible to reconstruct and adapt the existing facilities. The Republic of Serbia, for example, belongs to countries of a post-socialist context, and accordingly, in many rural settlements there are co-operative homes that were built en masse after the Second World War (Aranđelović and Videnović, 2016: 952). Most of these homes are decaying and serving no purpose today - they are out of function, so the potential for conversion to visitor centers needs to be considered.

Newly formed spatial structures do not necessarily have to be physically connected. They can be integrated into a system of pavilions with several independent facilities. In that way the facilities can be constructed in stages, i.e. the offer can be incrementally improved, depending on the interest of potential consumers. Of course in such cases, especial care needs to be paid to the aesthetics of the complex being developed during a longer time period, i.e. the form of the whole structure needs to be defined while the idea is still being formulated.

If we set aside the primary (basic) Visitors' centre type - a physically and design-wise marked point in space, which can possess a bared function, but which still has a specific purpose and intent (e.g. an overlook) - and move from philosophical observation of content phenomenon to programme in a more detail, then the basic content of Visitors' centres should consist of an information/presentation (exhibit-oriented and commercial) section, with a small cafe-bar for refreshments, toilets, and, potentially, an appropriate storage place.

The centre can then be amended with the next related step – space dedicated to education and research, such as all-purpose halls for lectures, screenings, plays and performances, exterior and interior amphitheatres, classrooms (workshops), internet-rooms, media libraries, libraries …

All these facilities make a centre more useful, but also cause a need for several-days-long visits, the visitors thus becoming users, which in turn requires more ambitious programmes, as well as the construction of rooms for overnight stays, common space for social gatherings, and a restaurant with a kitchen.

If we were to develop the programme further from the aspect of rest and relaxation, it could contain spa and wellness facilities, potentially a gym, a smaller swimming pool and exterior sports facilities, and in the case the centre is sports-focused, also interior halls for some of the sports.

On the other hand, if the Visitors' centre was to be located next to some specific facilities typical of the surrounding landscape, such as vineyards, it could contain a modest production line (winery) aimed at presentation of the production process and wine preservation, as well as the premises for presentations and tastings, i.e. the premises for commercial sales. This, in turn, implies the need for special purpose restaurants, such as taverns.

Hence, depending on the ambition, investment cost-effectiveness, and for purposes of attractiveness and exclusivity, such facilities can always be amended and enriched with different additional content depending on the specific features of the environment they are supposed to promote, and the decision of the investor on the level of investment into amenities.

Visitors' Centres should be defined as attraction spaces and active elements of the natural environment with which they are actively interacting. The compatibility of facility and environment is the goal to strive for during the planning of such structures. They are designed as exhibition, pavilion, spatial compositions representing the autochthonous culture and protected natural riches of a specific area in a completely rational manner. This is precisely why the rural areas represent ideal training grounds which can also provide the application and verification of ideas on sustainable development and use of renewable energy sources (e.g. sunlight, water, and wind power).

The expansion of Visitors' centres during the last few years depicts the general consumer society trend where natural and cultural resources are charged and completely exploited. The tourist offer diversity can have a positive reflection on the diversity of Visitors' Centre facilities and activities. Their multipurpose character contributes to the attracting users of different profiles. The function and form authenticity is defined by a new and different manner of promoting natural and cultural heritage of rural areas to the modern society.

We were correct to believe that the networking of central entities and new functional nuclei in villages and rural areas could encourage a new attitude of hesitating visitors towards their habits in relation to aforementioned spaces. New attractive content, which must not be standardized or homogeneous, but always original and adapted to the environment in which the intervention is taking place, could become the engine of revival of village spaces in Serbia, having in mind that the long-lasting trends of move from cities to preserved environments with lower stress levels and tempo of life are soon to be generally accepted in our society as well.

Visitors' Centres as an Evolving Model

The importance of establishment of facilities such as these is reflected primarily in the framework of presenting natural and cultural heritage to public, then in the economic, infrastructural and other forms of general rural areas. Visitors' centres represent an evolving model which redefines the existing knowledge on planning and design of vacation facilities and hotel structures by integrating different facilities (reception, information, education, accommodation, culture, services, recreation…) in a compact unit. Tourism in seaside rural regions is not limited to simple rest, but includes active introduction into activities, nature, tradition, heritage, and other features differing from the city life. This is why the visitor centres can be characterized as a response to the process of global "tourist-mania" and the growing desire of the society to visit unknown and neglected spaces outside of the boundaries of urban centres.

The development of tourism activates and engages the wider context of an area, so that different (natural, cultural, industrial etc.) heritage factors can be integrated in the cognitive and interactive offer in the goal of the best possible promotion and presentation of new spaces in front of the public. Multipurpose role of Visitors' centres absorbs characteristics and authentic features of a site, which essentially determine its future purpose. Such principle enables development and planning of different activities encouraging different aspects of tourism (e.g. rural, village, medical, cultural, hunting, sports and recreation, educational wine-oenological, eco-tourism, ethno-tourism etc.).

Visitors' centres architecture is defined by a synergy of simplicity and rationality of form and capacity. The consistency of such a structure must possess a balance of psychological and physical phenomena, in order not to disrupt the natural environment's spatial assembly which actually generates and shapes it. Favouring traditional planning, design and construction principles can be unjustified, since it can lead to a complete slighting of rich cultural and historical values and deviation from modern society achievements.

VISITOR CENTRES AS MULTIPURPOSE FACILITIES: STUDY OF TWO CASES IN SERBIA AND BIH

Through case studies, third part of this chapter presents a comparative overview of the implementation of two similar individual investments in Serbia and BiH. A comparative overview of specific activities indicates the options for improvement of overall process efficiency through transfer of best practice examples from one country or another, i.e. through delivery of specific conclusion which can improve similar endeavours. More precisely, we focus on secondary educational facilities, scientific and teaching bases or field laboratories which in both cases represent branches- field offices or stations for practical training provided by Faculties of Forestry in both countries. Apart from the academic educational activities, these spaces are also intended to provide accommodation and potential tourist-oriented facilities for a wider circle of users. From the aspect of programming, these investments required designing of multifunctional spaces which were supposed to provide not only practical training and accommodation for students, but also accommodation and animation services for tourists in the mountainous tourist regions in which they are located. This paper also reviews the issue of rehabilitation, reconstruction, adaptation and expansion of already existing facilities, as well as the design and implementation of completely new functions.

From the aspect of location, we consider the issues of adaptable designing of open and closed spaces at higher altitudes (≥ 1,000 meters above the sea level) in mountainous regions with thick forests and severe winter weather conditions. These stations are primarily intended for the required/mandatory activities, or their specific basic purpose, but they can also be especially functional from the aspect of tourist network elements development in a specific wider area, and as marketing and promotional markers for popularization and attractiveness, i.e. "new recognisability" of specific sites in a more narrow spatial sense. As modern landmarks, they could use their functionality and symbolism of their form to generate considerable development and increase the number of visits to rural areas, where the sites which interest the wider circle of users are usually located.

Through case studies, third part of this chapter presents a comparative overview of the implementation of two similar individual investments in Serbia and BiH - from initial ideas, through programming, urban and architectural design, to preparation and construction of specific facilities at specific locations. A comparative overview of specific activities indicates the options for improvement of overall process efficiency through transfer of best practice examples from one country or another, i.e. through delivery of specific conclusion which can improve similar endeavours. More precisely, we focus on secondary educational facilities, scientific and teaching bases or field laboratories which in both cases represent branches- field offices or stations for practical training provided by Faculties of Forestry in both countries. Apart from the academic educational activities, these spaces are also intended to provide accommodation and potential tourist-oriented facilities for a wider circle of users. From the aspect of programming, these investments required designing of multifunctional spaces which were supposed to provide not only practical training and accommodation for students, but also accommodation and animation services for tourists in the mountainous tourist regions in which they are located. This paper also reviews the issue of rehabilitation, reconstruction, adaptation and expansion of already existing facilities, as well as the design and implementation of completely new functions.

From the aspect of location, we consider the issues of adaptable designing of open and closed spaces at higher altitudes (≥ 1,000 meters above the sea level) in mountainous regions with thick forests and severe winter weather conditions. These stations are primarily intended for the required/mandatory activities, or their specific basic purpose, but they can also be especially functional from the aspect of tourist network elements development in a specific wider area, and as marketing and promotional markers for popularization and attractiveness, i.e. "new recognisability" of specific sites in a more narrow spatial sense. As modern landmarks, they could use their functionality and symbolism of their form to generate considerable development and increase the number of visits to rural areas, where the sites which interest the wider circle of users are usually located.

Specific Characteristics: The Developmental Potential

The majority of Balkan Peninsula is covered by hills and mountains and the territories of Serbia and Bosnia and Herzegovina are no exception. Over 2/3 of Serbia's terrain are located at higher altitudes, and in Bosnia and Herzegovina such areas are even more predominant. The logical further social and spatial development of these territories indicates the need for reaffirmation and use of those resources that they objectively have at their disposal. Due to synergy of general historical, social, sociological, ethnological and other factors, as well as specific territorial and mostly peripheral traffic conditions, it is precisely the mountainous regions that have been neglected from the aspect of development in the past, or they have at best stagnated. Due to their developmental advantages, which arose consequentially, spontaneously or intentionally, they should be the focus of social interest, and to a greater degree than in the past (Ploeg and Renting, 2000).

Among other specific characteristics, the developmental potential of rural spaces are reflected in unquestionable opportunities for tourism development (Hall and Richards, 2002). In the current circumstances of untapped potential of mountainous areas, we have the fortune that those characteristics and requirements for tourism development which have been popular, appreciated and in demand during the last couple of decades have been preserved in precisely these areas. These are a part of the modern view of this branch of economy, which focuses on attractiveness of regions, abundance of natural and manmade values, specificity of certain regions, level of preservation of natural environment and potential for environmental approach to development (Garrod *et al.* 2004). Apart from all of the aforementioned, we should also take into consideration the constantly present physical structures dating from different periods, their service properties and the options for additional use. Objective social and economic circumstances of social climate in Balkan regions cannot afford to ignore what they have at their disposal and neglect the previously invested developmental efforts. Rationality and adaptability in treatment and use of already existing facilities, but also the creation of potential for multifunctional and universal use during planning and design of new investments are becoming the prerequisite for infrastructure construction policy, offer design, and character and brand creation in the tourist development of mountainous regions. This paper, based on the comparison of specific traits of proposed models – two case studies – aims to shed a light on specific aspects of such processes from the viewpoint of their reality and experience.

Two Projects: Sites in Their Contexts – Basic Coordinates

Case 1: The first analysed project (the existing - adapted, reconstructed and expanded buildings), is located at Gvozdac on Serbian Goc Mountain, in the territory of the Municipality of Vrnjacka Banja. The teaching base camp of the Belgrade Faculty of Forestry ("Pyramid" hotel) is located at the altitude of 900 meters. It is intended for practical training of students in an exceptional natural environment – a forested area, appropriate for the intended empirical and education purposes, as well as vacation and rest in nature. The buildings were constructed in different time periods (immediately after Second World War, in the case of student accommodation and the mountain lodge, and a little later in the case of the newer part of the pyramidal reception desk and the annex for accommodation of professors and guests). Apart from the mentioned facilities, the complex also includes two older accommodation facilities-lodges located near the relevant site, and the complex comprising of a farm, saw mill and a heating plant in the wider area, as well as several additional buildings located around a small lake with elements of outdoors furniture (canopy, bridges over the rivers and streams, desks and benches, hard landscaping – all made from forest timber), and a construction intended for an oven in the vicinity of the lake.

Case 2: The second project, is planned in BiH in the Republika Srpska entity, more precisely East Drvar municipality which was created from small villages Potoci, Uvala and Smetica. The centre of this municipality is Potoci village at Klekovaca Mountain, located at the altitude of more than 1000 meters and numbering only 66 inhabitants according to the 2013 census (the 1991 census recorded 27 inhabitants). (Figure 4) Such an example of transforming small settlements into administrative centres (where in the case of Potoci it was founded in early XX century as a workers and railway settlement with 600 inhabitants and started dying out with the elimination of narrow gauge railways in 1967) is the usual practice with previously large cities where the majority of urban settlements now belongs to BiH, and only a few smaller ones to Republika Srpska. At entity border areas, municipalities were formed out of settlements which do not possess the appropriate characteristics required for their role of centres. This is precisely why there is a desire in this Case to provide additional facilities (apart from the existing Post office building, municipal administration, medical centre East Drvar, one residential building and the central office of one of the organizational units of Republika Srpska Forests – forest estate "Klekovaca – Potoci") in order to define and empower the municipal centre. Thus, the design and construction of a teaching base of Banjaluka Faculty of Forestry (modelled on the base on Goc Mountain in Serbia, i.e. Case 1) was initiated here, with the aim of developing an outpost in this forested region, which also holds a protected nature reserve – a 300 ha large Lom virgin forest (Google, 2018).

Relations: Case 1 is a typical socialist relict, originating from "enthusiastic" 1950s and riding on the wave of university teaching development, when the forest wealth was made available to the government and the "new" society, and thus to "Workers'" University due to the shift to social property structure. The teaching base and forest estate facilities, intended for student training and accommodation, were later used for commercial and tourism purposes in the wider sense, owing to relaxation of authority and competences. In the situation where such privilege is innovated and prepared for successful business operations in the circumstances of transitional market, this approach can become a role model for design of Case 2 investments in BiH due to its specific combination of educational and purely commercial activities. Along with forest wealth and further development of university level forestry education in Banjaluka, the teaching base in Potoci-East Drvar could be important not only from the commercial aspect, but also as a factor of village urbanization and elevation of its significance as the new municipal centre.

Figure 4. Banjaluka Faculty of Forestry teaching base in Potoci - East Drvar - BiH. On the left: Settlement centre urban plan. On the right: Composition plan.
(Design: 2014. Authors (source): A. Videnovic, D. Radonjic, architects)

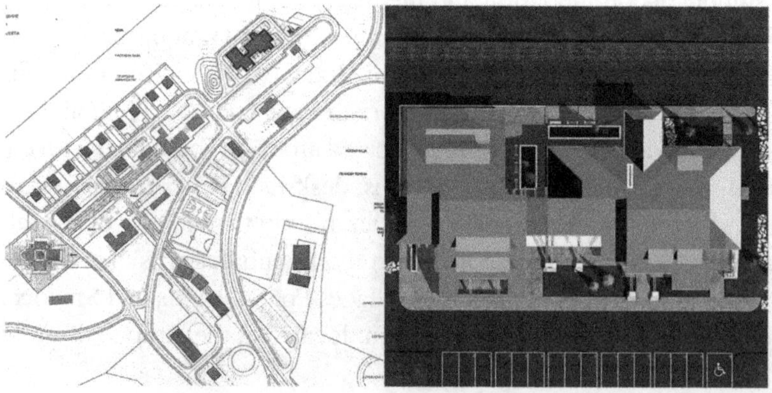

Comparison 1: Position, Context, Traffic Communications, and Relations

Case 1: The site of Goc Teaching base is, as usual in mountainous areas, not easily accessible. The main connection with bigger roads in Serbia is by an over 20 km long side road which diverges from Kraljevo-Krusevac main road. The same road continues towards Aleksandrovac and Brus, and it has a problematic, 9 kilometres long poor quality section. Another, even worse road connects the site with Vrnjacka Banja, in whose administrative territory it is actually located. There are no other facilities in the immediate vicinity of the Teaching base. Several hundred meters further we can find the planned tourist centre of Goc Mountain with several accommodation buildings, a hotel and a children resort. Two ski tracks are located next to these facilities. The entire complex is rarely visited by tourists and recognized only in the local community of Kraljevo.

Case 2: While the site in Case 1, although rarely used, is still accessible, Potoci settlement in BiH (Case 2) is almost totally inaccessible. The settlement can be reached only by local, non-paved forest roads, intended for forest exploitation purposes and appropriate for off-road vehicles. There is a lack of communications with its immediate surroundings and wider environment.

Relations One: In both cases, the basic quality of the relevant location can be found precisely in its exceptional natural environment and high level of nature preservation, which provide the best foundations for development of any form of tourism. In spite of the position of these rural centres and even the context of remoteness and non-urbanized surroundings, the efforts aimed at their rehabilitation demand a considerable improvement of road infrastructure, as well as planning of additional facilities in Case 2 (Videnovic, 2017).

While Case 1 represents an autonomous complex of self-sufficient facilities, comprised of an accommodation base, teaching station, a hotel located in natural surroundings and additional facilities in relative vicinity, the design in Case 2 aims to provide immediate contact with the nearby forest, but also to help the formation of Potoci street plan, i.e. to contribute to the design of the rural region centre in order to provide the same centre with characteristics of a planned, designed and functional space.

In this sense, the project of the Potoci Teaching base/ tourist accommodation facility, which is awaiting implementation, was accompanied by the Urban Planning Project for regulation of settlement centre in line with the fact that this centre was awarded an important regional administrative function.

Comparison 2: Functional and Organizational Concepts and Facilities

Case 1: As far as the preservation level of Goc old teaching base is concerned, the reception desk, hallway and gallery in the pyramidal object, the nearby guest accommodation building and the annex connecting them, i.e. the hall leading to student accommodation building, are still standing. All other facilities were removed for being either derelict or inappropriate. Even though these buildings (except some less valuable forms and facilities) are representatives of autochthonous, mountain architecture with elements of vernacularism, the student accommodation buildings – old villas – and dining halls with massive stone fire places were simply removed. (Figure 5) Without any sentimentality or rational reasons for keeping them, these building were replaced by new ones with the same function, but modern design and equipment needed so they could satisfy the requested program in the best way possible.

Figure 5. Belgrade faculty of Forestry teaching base at location (Gvozdac – Goc – Serbia) On the left: Ground floor and first floor layout. On the right: Model of the complex.
(Design: 2013. Authors (source): A. Videnovic, D. Radonjic, D. Danilovic, architects)

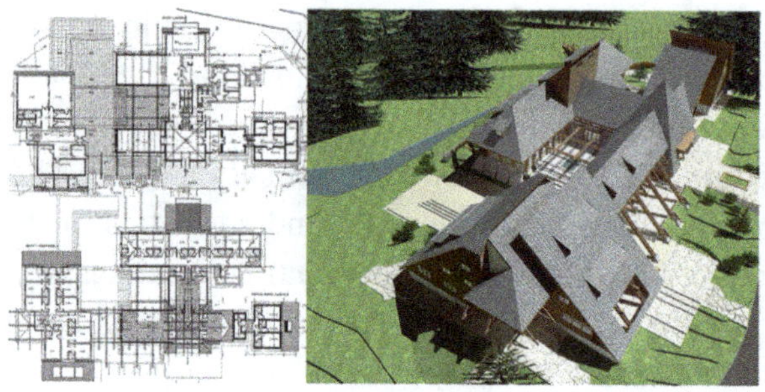

Case 2: Potoci facility (BiH), although inspired by Goc project, has followed programming requirements of a completely new complex with a more modest programme, so it could be aligned with the regulatory plan of the growing settlement and, although formed in the character of urban freestanding buildings, could contribute to the origins of urban development in the current ambient circumstances. Thus the design has adopted a line concept, with a clearly differentiated public and accommodation facilities, and the central reception desk with a hallway connecting them. Cut into unlevelled terrain next to a forest complex, the building can be accessed from several sides and since it doesn't have to be permanently linked to the surrounding landscape (as the facility in Case 1 does), it is conceptually focused on internal facilities and their as rational as possible relationship, with the possibility for multifunctional use of specific facilities.

Relations Two: Although quite similar from the aspect of purpose and concept, the projects (buildings) on Goc and in Potoci are still considerably different in their approach to investments and with regards to users' expectations. While in Case 1 the focus is on innovation, updating, improvement of teaching and other tourist activities, in Potoci we can see the very origin of specific facilities and functions, i.e. the intention to resolve many burning issues and needs of that settlement with one stroke. Hence, in Case 2 it was necessary to consider the needs of the investors related to the Banjaluka Faculty of Forestry teaching base, the need for accommodation of guests, the organization of specific forms of tourism, introduction of additional activities to the basic economic activity in the community – maintenance and exploitation of forests, but also the improvement of central settlement functions from the aspect of facilities and regulatory planning.

Comparison 3: Materialization and Form in Different Circumstances and in Relation to Different Priorities

Case 1: The architects used the opportunity to design the elements of roof construction in the form of prolonged struts which create the impression that the buildings are adhering to the ground, i.e. growing out of that ground. Due to different circumstances, as it often happens during the construction phase, this idea was not realized. This has considerably reduced the aesthetics of the buildings and created an impression of ordinary and relatively average architecture.

Case 2: In the Potoci project such elements weren't even attempted, due to limited space available for construction of facilities. Also, the need for rationality has overshadowed the desire for diverse forms, so this project can also be placed in the category of creations void of any explicit aspirations from the aspect of form.

However, all of the aforementioned does not mean that these are not concrete projects which do not possess high quality elements of transposed vernacularism and regional characteristics, i.e. that they are not or will not be interesting from the aspect of their architectural form.

Relations Three: We can definitely say that the designing of any facilities in the both cases analysed here would be at least inappropriate if timber, predominant in these forest environments, was not used as main or at least important material in construction. But, on the other hand, any excess in this sense would lead to lack of measure in the use of shaping elements.

In both cases pitched roofs were used as the usual design in mountainous regions with abundance of snow. In Case 1, the challenge laid in integration of new and existing facilities. A vertical element in the form of wooden lookout tower was used in Case 2 as an attempt to establish specific visual identity and a symbol for the building and related facilities, but also as a symbol for forest related activities, hunting and the need to move away from the ground which is not always safe or comfortable.

Contemporary investments require multifunctionality as the most important determining factor of the space that is being designed. The speed of changes in the modern society, behaviours, priorities, values. customs, habits and trends, requires the option of adapting every space, especially those which are not standard, mass produced and easily exchangeable, to the needs of real time events in a painless, rational and efficient manner. In this sense no design can be approached unilaterally without any options for combinations of functional elements, changes in purpose of space, or adaptability and adjustability to newly arisen circumstances. If intelligence is defined as the ability to navigate new circumstances, the same concept can be applied to architectural design. In other words, the designed space is of good quality if it can be relatively easily adapted to different types of activities. The both of these reviewed and compared cases had to contain educational, accommodation and leisure facilities intended for students as well as a wider circle of users. In case 1, the design and the building possess a sound link with the environment in the form of several interspaces intended for external use.

In the Case 2, there are no such spaces since the terrain and urban regulation plan did not allow for it, which, again, does not mean that they cannot be realized after the construction through further design of surrounding space. The quality of both designs lies in the possibility of continuous multifunctional use, regardless of the rhythm of specific education or climate calendars.

The role of new designs and buildings as potential landmarks in the wider regions or areas in which they are designed and constructed also plays an important, strategic role in the fields of architecture, construction, but also environment aesthetics, organization of tourist events, and finally the creation of distinct brands for specific areas (Videnović and Aranđelović, 2016:194). It does matter how an architectural work will look like and whether it will attract the visitors to stay in it or visit it when passing through the region. Entire new strategies of tourism, and not only tourism, are based on the phenomenon of attractiveness, i.e. exclusivity and importance of the seen and the experienced. In that sense, both of these analysed cases can represent "true" architecture in the relevant locations from at least two aspects, the aspect of location and the aspect of an aesthetic and functional symbol.

If we focus on the aspect of location, Case 1 (the complex on Goc Mountain) represents an improvement of an already established and well founded facility intended primarily for faculty students with the option of accepting a wider circle of interested persons. So it is possible for this facility, together with other old villas in the complex, to represent one of the stops on the dynamic road of regional tourism development. Outside of the usual standards of stationary tourism based on a long stay at a single location, there is a contemporary form of dynamic vacation, where tourists spend every night at a different location and go to day-long tours, or spend several nights at the same location taking field trips to closer or more distant destinations. From the aspect of location, Potoci project, or Case 2, should represent a modern regional element of newly formed administrative centre urbanization. From the aspect of university education, it would signify a new era of higher quality content and teaching. The effect of such a facility's offer in such an unexpected location could result in a modernization of the attitude towards the economic activity in the area, where it would no longer focus only on the sphere of forestry and administration but instead on development of tourism that can be expected in such a natural environment.

Having in mind the aesthetic and functional aspect of these two analysed cases, the appropriate suggested materialization, application of forms transposed from the traditional forms of Serbian and Bosnian mountainous region architecture, then the respect of the rules of appropriate and logical spatial relations, as well as maximum flexibility regarding the potential for use of suggested elements and programmes during the design process indicate that currently, after the construction of one of the projects and prior to the construction of the other, we can certainly expect results deserving of the concepts and good intentions of their investors and authors.

Whether we are discussing the reconstruction and expansion of buildings, as in Case 1, or the construction of completely new facilities as in Case 2, these are positive trends in the shaping and planning of rural spaces and ambient habitats i.e. in the behaviour in spaces characterized by their natural surroundings. It is extremely important to apply a careful approach to the design of public spaces in rural communities, especially having in mind that in the future it will become even more relevant and frequent. In the circumstances of less attractive surroundings and specific spatial situations, it is necessary to create a space free from the impact of fashionable, transitory and exclusively commercial trends.

CONCLUSION

The production of new living culture and society development influence the establishment of different values of function and form. Cultural diversity, as one of the modern society characteristics, can impact the establishment of future facility features, in preference to modern uniformity and unified design. Architecture cannot survive in the modern society if it defines only basic construction principles. The necessity of sensation and creativity presents an important requirement in the process of designing multipurpose facilities, such as Visitors' centres. Educational dimension of these facilities establishes new modalities for construction of multipurpose structures aimed at preservation of natural and cultural identity of rural areas.

It is logical to plan visitor centers in areas that have certain natural or cultural specificities, that is, elements of tourist and educational attractiveness, in order to contribute to stronger social and economic development. Interstate or regional co-operation of similar content would bring together and enhance their functioning in every sense, contributing essentially to the development of a sustainable concept.

Social development and promotion would entail the existence of certain social services, above all necessary for the local population. More extensive economic development would refer to the possibility of marketing agricultural products and local specifics within the range of visitor centers. It does not exclude the possibility of networking both individuals and entire local communities in order to create the basis for tourism and economic sustainability.

But still, development of Visitors' centres network, regardless of whether these are new or refurbished and adapted facilities, is still not the only and sufficient condition for renewal and activation of Serbian rural spaces. In the set of necessary actions, or steps to make, this is probably only the first but important requirement for the forming of a critical mass of positive energy, knowledge, desires and opportunities around these centres, needed in order to fundamentally change negative trends in the management and administration of rural spaces of Serbia which are rich with resources bur poor from the aspect of ideas and willingness.

The recommendations for setting up visitor centers mainly relate to the decentralization of the rural area administration, as well as a set of other relevant state measures that could engage entrepreneurial companies and tourist agencies which would activate the inert rural space by paying state taxes while simultaneously improving the social and economic situation in rural regions of Serbia.

Finally, it should be noted that unifying and defining tourist areas would represent a further crucial step in improving the village-city relations. It is important to note that this should not be reduced to another of the many categorizations of villages or statistical atlases, but that it should be based on the establishment of new microregions, regardless of current administrative boundaries, based on common natural and created values, tradition and mentality and where networks of tourist circles and routes would be formed; ie. a set of possible tourist offers that could take different forms, duration, activities and rhythm and which could attract a wide range of users due to their diversity.

REFERENCES

МаксинМ.ПуцарМ.МилијићС.КорaћМ. (2011). Одрживи развој туризма у Европској унији и Србији. *Београд / Иаус, Посебна издања, 67.*

Зборник радова стручно – научног скупа. (2004). *Планирање и уређење села и руралних подручја.* Бања Врујци: Удружење урбаниста Србије.

Aranđelović, M., Videnović, A. (2016). Društveni značaj zadružnih domova u postsocijalističkom kontekstu. *Zbornik Matice srpske za društvene nauke*, *4*(159-160), 947-960.

Cullen, G. (1990). *Gradski pejzaz.* Beograd, Srbija: Gradjevinska knjiga.

Frempton, K. (2004). *Moderna arhitektura, kritička istorija.* Beograd, Srbija: Orion art.

Garrod, B., Wornell, R., & Youell, R. (2006, January). Re-conceptualising rural resources as countryside capital: The case of rural tourism. *Journal of Rural Studies*, *22*(1), 117–128. doi:10.1016/j.jrurstud.2005.08.001

Garrod, B., Youell, R., & Wornell, R. (2004). *Links Between Rural Tourism and Countryside capital.* Cheltenham, UK: Countryside Agency.

Hall, D., & Richards, G. (2002). *Tourism and sustainable community development*. London: Routledge.

Hall, M. C., & Page, S. J. (2006). *The Geography of Tourism and Recreation: Environment, place and space*. London: Routledge. doi:10.4324/9780203420249

Kojic, B. (1973). *Seoska arhitektura i rurizam*. Beograd, Srbija: Gradjevinska knjiga.

Krier, L. (2008). Settlements of the future. In T. Haas (Ed.), *Beyond Urbanism and the Future of Cities*. Milan, Italia: Rizzoli.

Muresan, I., Harun, R., Arion, F., Oroian, C., Dumitras, D., Mihai, V., ... Chiciudean, G. (2019). Residents' Perception of Destination Quality: Key Factors for Sustainable Rural Development. *Sustainability*, *11*(9), 2594. doi:10.3390u11092594

Nikolić, M. (2011). Visitors' centers at archaeological sites in Serbia as an input for sustainable development of the country. *Journal of Applied Engineering Science*, *9*(1), 253–258.

Panyik, E., Carlos Costa, C., & Rátz, T. (2011, December). Implementing integrated rural tourism: An event-based approach. *Tourism Management*, *32*(6), 1352–1363. doi:10.1016/j.tourman.2011.01.009

Petrovic, B., Raskovic, I. (2011). *Tradicija – tranzicija: upotreba nasleđa u arhitekturi*. Beograd, Srbija: Arhitektonski fakultet, Orion Art, Institut za arhitekturu i urbanizam Srbije.

Pride, R., (2015): *Дестинација као бренд*, Београд, Srbija: Klio. Пуцар, М., Јоксимовић, Б., (2010). *Одрживи развој бањских и туристичких насеља у Србији*. Зборник радова НИ пројекта. Београд, Србија: Иаус - Посебна издања бр.64.

Radosavljevic, U., Đorđević, A., Lalovic, K., Živkovic, J., & Đukanović, Z. (2019). Nodes and Networks: The Generative Role of Cultural Heritage for Urban Revival in Kikinda. *Sustainability*, *11*(9), 2509. doi:10.3390u11092509

Stovel, H. (2003). *Cultural landscape: The new approach to the preservation of cultural heritage*. Beograd, Srbija: Glasnik Društva konzervatora.

Tafuri, M. (1979). Radical Architecture and the City. In M. Tafuri (Ed.), Architecture and Utopia: Design and Capitalist Development. Cambridge, MA: MIT Press.

Van Der Ploeg, D., Renting, J., & Renting, H. (2000). Impact and Potential: A Comparative Review of European Rural Development Practices. *Sociologia Ruralis*, *40*(4), 529–543. doi:10.1111/1467-9523.00165

Videnovic, A. (2017). *Obnova centara u ruralnim podrucjima - Novi zapisi*. Beograd, Srbija: Arhitektonski fakultet.

Videnović, A., Aranđelović, M. (2016). Visitors' centres – New Coordinates of Serbian rural areas improvement. *Architecture and Civil Engineering, 14*(2), 191 – 200. doi:10.2298/FUACE1602191V

KEY TERMS AND DEFINITIONS

Cooperative Homes: Buildings constructed mostly according to standardized design in the Yugoslav villages after World War 2 as multipurpose social and ideological creations (they served as culture events halls, agricultural produce storages, general stores…) which were tailored to the characteristics and size of their respective settlements. They also possessed a utilitarian character and ideological task to direct the population (especially the youth) towards new (socialist) social circumstances through intensive propaganda.

Critical Regionalism: This is an approach in architecture that strives to oppose the universality and lack of meaning common in modern architecture by using the elements and influences of the environment which contribute to the feeling of belonging and meaning. It is considered a reaction to universality and standardization brought by the Modern architecture.

Entity Republika Srpska: According to the Dayton accord (i.e., the peace accords that ended the war [1992 – 1995]), Bosnia and Herzegovina was divided into two entities: Republika Srpska and the Federation of Bosnia and Herzegovina.

Istočni Drvar: A newly formed municipality in Republika Srpska (BiH), with the centre in the village of Potoci. Bigger cities are generally located at the borders between the two entities, so they have been divided into main urban areas in the territory of the Federation and smaller suburban or rural areas which settlements that do not meet the requirements of centricity and which are located in Republika Srpska.

New Recognisability: Constructing architectural creations in immediate surroundings or in synergy – symbiosis with the existing natural or man-made sites, and in the goal of reaffirming those sites (i.e., promoting the interest in increased number of tours or re-visits), in order to raise the profits from tourism.

Non-Agricultural Households: Households in rural settlements whose members do not engage in agriculture but in administrative occupations (in education, healthcare, administration, government agencies, services or crafts), so they do not work on their land or keep livestock. If, however, they do also perform these activities on their farms, these households are defined as mixed.

"The Šargan Eight": A heritage railway created on an old narrow-gauge main line between Belgrade and Dubrovnik, i.e. on the section between Užice (Serbia) and Višegrad (BiH). The railway, due to a big difference in altitude that it has to overcome during a very short section in the surroundings of Šargan village, forms a double serpentine turn by passing through a number of tunnels. Viewed from above, this section forms an eight. For the last 15 years, during the period between April and October, multiple round-trips from the village of Mokra Gora have been organized on daily basis. On this ride, the passengers can enjoy mountain landscapes with short stops in small stations in places with interesting surroundings.

Species Range: Part of the territory or aquatic surface in which a certain population performs its life processes. The shape and form of the territory covered by this range are impacted by the interaction between ecological factors and the specific population (species) and the history of it development (phylogenesis). The borders of this territory are the borders of a species distribution.

"Staro Selo" Museum: An exhibit consisting of a reconstruction of a traditional old Serbian village in Sirogojno settlement at Zlatibor Mountain in West Serbia. Old houses and auxiliary buildings have been transplanted form different original rural settings, and thus saved from natural destruction (erosion) so they could testify on customs, vernacular, sociological and other traditional characteristics and values of life in the past in the form of a designed museological open-space exhibit. Village also provides accommodation, services and activities for the tourists.

Vernacular Architecture: Authentic, folk architecture that belongs to the indigenous population and which stemmed from the experience of self-taught naïve builders.

Section 3

Urban–Rural Cooperation Through Technologies

Chapter 11
Upgrading the Building Facades in Low-Density Residential Areas:
Implications on the Overall Building Performance

Katerina Tsikaloudaki
Aristotle University of Thessaloniki, Greece

Stella Tsoka
Aristotle University of Thessaloniki, Greece

Dimitra Tsirigoti
https://orcid.org/0000-0001-6995-5102
Aristotle University of Thessaloniki, Greece

Theodore Theodosiou
https://orcid.org/0000-0002-3385-119X
Aristotle University of Thessaloniki, Greece

ABSTRACT

The most common action for the buildings' energy upgrade across Europe is the addition of thermal insulation on the external walls. Such interventions, although simple on their construction, cause significant changes on the building's behavior, not only on its energy needs, but also on the hygrothermal and visual performance. The effects are not always positive; for example, thicker insulation may result in lower thermal transmittance and better thermal energy performance, but on the other hand the thermal bridging effect is amplified, and the daylight levels are decreased. This research intends to quantify these impacts by analyzing the relevant parameters for different regions of Europe. The analysis aims at explaining the complicated interrelationships on the building physics' aspects encountered through interventions on the building envelope, but also at identifying appropriate measures that could counterbalance the negative impacts and enhance the overall building performance.

DOI: 10.4018/978-1-5225-9932-6.ch011

INTRODUCTION

The building sector contributes significantly to the final energy consumption, given that in 2011 its share was equal to 38% for the EU-28. For residential buildings in EU-28 countries, which correspond to 75% of the building stock, the share of consumed energy is equal to 25% and concerns mainly gas (36%) and electricity (25%) (Directorate-General for Energy-European Commission, 2012). At the same time, a substantial part of the building stock in Europe is older than 50 years, while many buildings in use today are hundreds of years old. More than 40% of the European residential buildings have been constructed before the 1960's, when energy building regulations were very limited.

Yet, it has to be mentioned that the buildings' energy demand is strongly related not only to the envelope's thermal properties and the HVAC systems' efficiency, but also to its surrounding environment, characterized by the surface density, the opening to the sky, the presence of pervious surfaces etc. Buildings on low-density urban areas are generally characterized by lower heating and cooling energy needs compared to the compact high-rise areas due to the more favorable climatic conditions, induced by the respective morphological characteristics (Stewart and Oke, 2012). On the other hand, most sub-urban areas of European cities suffer from urban sprawl resulting in dispersed building development often with a low construction quality which leads in large energy demand. Either in the peri-urban areas which suffer from degradation and low quality of life or in low rise developments of detached housing areas built during the past decades, the building shell retrofit in these low density areas could contribute significantly in the decrease of the consumed energy of buildings and at the same time in the improvement of the quality of life of the inhabitants (Tsirigoti et al, 2018).

To continue, since the construction rates remain currently at low levels across Europe, the perspectives of applying effective energy renovation measures on existing buildings seems very promising. The retrofit actions in low-density areas, where mostly low-rise, detached buildings are met, appear to be even more effective, as the restrictions in applying interventions on the building envelope are very limited, at least with respect to the building planning and urbanism rules, i.e. regarding the expansion of the existing building's area or volume by increasing the width of the walls or extending the façade beyond the building line, due to the existence of open space around the buildings (Bikas et al, 2017).

Till today, the external insulation of facades is the easiest way to improve the thermal insulation of building envelopes and one of the most common measures used. Recently, the positioning of thermal insulation is often accompanied with the installation of an external cladding, offering architectural and aesthetical upgrade along with the enhancement of the building energy performance.

Either in its simpler form with plaster finishes or in its more complicated construction with the addition of external cladding -often creating a ventilated façade- the external insulation addition in vertical building elements has impact on several aspects of the building performance: it certainly improves the thermal performance of the building elements, but at the same time it is responsible for the increase of the thermal bridging effect, while at the same time the increase of the wall thickness alters the indoor visual environment due to the minimization of the sky view and the consequent decrease of daylighting (Vanhoutteghem et al, 2015; Unver et al, 2003; Reinhart et al, 2013).

However, the majority of research concerning the assessment of the upgrading interventions on the building performance focuses mainly on the energy performance and neglects their effects on other aspects of building physics. In this work, a holistic approach is attempted, which aims firstly at highlighting the positive aspects and the potential problems stemming from the addition of thermal insulation on an existing building envelope and secondly at proposing measures of counterbalancing these effects.

BACKGROUND

The most common retrofit actions that are met during the renovation of the existing European buildings concern interventions on the HVAC systems and on the building envelope. As regards the former, the interventions involve either their modernization or their total replacement with new, more efficient systems, on the basis of the condition and the performance of the existing HVAC equipment.

For the latter, the interventions are focused on the replacement of the windows, and/or the thermal protection of the opaque building elements through the addition of thermal insulation on the internal or the external surface of the building elements. The latter option offers more advantages, as among others the thermal bridging effect is reduced, the exploitation of thermal mass is not prevented, the external building elements are protected from thermal stress, the condensation risk is minimized and the building operation is not disrupted (Kisilewicz, 2019).

Several refurbishing concepts have been defined and analyzed with regard to their ability to upgrade the energy performance of buildings (Hakkinen, 2012). Moreover, there is a great number of literature analyzing optimum insulation thickness of walls for minimizing heating and cooling demand (Ucar et al, 2010; Ozel et al, Kurekci et al, 2016; Fang et al, 2014) and materials' appropriateness in terms of environmental or /and economic cost (Andeweg et al, 2007).

The external insulation of facades usually referred as ETICS (External Thermal Insulation Composite system) is the most common strategy used to improve the thermal insulation of building envelopes (Andeweg et al, 2007; Braganca et al, 2007). The thermal insulation material (usually extruded polysterine, expanded polysterine or mineral wool panels) is adhered to the wall with a special adhesive and anchors and it is covered with an organic mortar layer, reinforced by a glass fibre mesh (Figure 1). In some countries dry techniques for thermal insulation (ETIDS) are also applied (Gimaraes et al, 2015).

In other cases, the positioning of thermal insulation is accompanied with the installation of an external cladding, offering architectural and aesthetical upgrade along with the enhancement of the building energy performance (Nizovtsev, 2014). However, the last years, Ventilated Façade Systems (VFS) have gained much attention. They constitute of the insulation layer, positioned on the external surface of the wall, an air cavity and the outer cladding material. The air gap provides infiltration and, depending on the design, may provide pressure equalization across the outer skin. Any water penetrating the cavity

Figure 1. Detail of the two examined thermal protection systems, ETICS and VFS

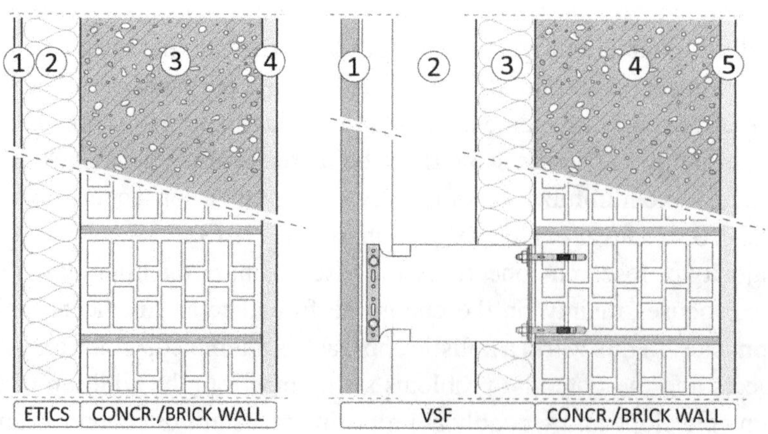

is drained away. To fix the cladding, a support system designed to carry both the mass of the system along with the projected wind and earthquake loading is installed (Bikas et al, 2017). The support system usually comprises of vertical or vertical and horizontal fixing rails fixed back to the structure using brackets (Figure 1).

The main advantage of the ventilated façade system is the wide range of architectural expression that it offers, as there are various possibilities for the cladding material. In refurbishment projects in particular, the renewed appearance with the use of ventilated façade systems is often a decision-making argument.

A limitation of the system is the added structural weight on the existing façade, particularly in case that a heavier cladding material, such as concrete panels or natural stone, is applied. Moreover, depending again on the cladding material and the fixing system, the construction cost can be significant.

While both ETICS and ventilated façade systems (VFS or VSF) aim to improve the overall energy efficiency of the envelope, their implementation may lead to different results, due to other thermal insulation related matters like thermal bridges of any type, linear of point ones (Theodosiou et al, 2019).

Comparing these technologies, the main differences can be summarized as follows:

- The ventilated air gap alters the way the façade exchanges heat with the environment. While actual energy phenomena is such systems are quite complicated, due to fluid dynamics, an approximation of heat flow can be achieved following international standards like ISO 6946 (ISO, 2017).
- Facades implementing ETICS are practically free of point (or 3D) thermal bridges and, depending on construction and morphological characteristics of the building, face only linear thermal bridges.
- Ventilated facades have a similar behavior as ETICS regarding linear thermal bridge effects, but their structural configuration, based usually on an aluminum supporting sub-structure, results in point thermal bridge problems. Although such thermal bridges are not treated in many national regulations, their magnitude might be significant according to recent studies (Theodosiou et al, 2015).

METHODOLOGY

The objective of this research is to indicate the overall influence of external thermal protection on each aspect of building physics that is associated directly or indirectly with the addition of the thermal insulation layer, such as the hygrothermal and the energy performance, as well as the indoor daylight conditions.

This holistic approach will assess the performance of the two main wall retrofit actions presented above (ETICS and VFS). More specifically, it will attempt to identify potential conflicting impacts and propose measures in order to counterbalance any negative effects stemming from their application.

The study concerns a residential building, which is located in a low density area. It is a two-storey, single-family detached house with a basement (Figure 2). More precisely, following the classification of Stewart and Oke, the building is considered to be located in an open low-rise area, placed in the suburbs and characterized by low surface density and considerable plant cover. The heated area of the building is 177m^2 while the unheated area of the basement 90m^2.

Figure 2. Axonometric plan of the examined building, which is assumed to be located in three different low density areas near Oslo, Vienna and Athens

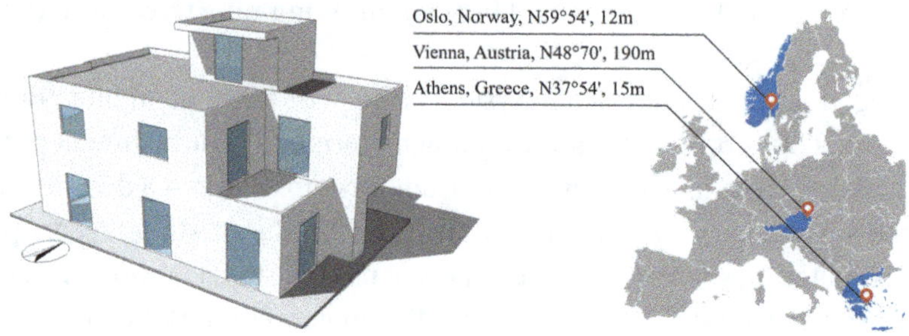

The energy needs for heating and cooling were calculated for three climatic European regions (Figure 2) across the north-south axis, to investigate the examined effects under different prevailing climatic conditions, assuming that the architectural and morphological characteristics will not be dramatically changed among the geographical areas.

The calculated parameters that will be used for the assessment of the two thermal protection alternatives are:

- The change of the U-value of the building element with regard to the wall insulation's thickness
- The thermal bridging effect calculated both for the linear and the point thermal bridges, with regard to the wall insulation's thickness
- The energy demand for heating and cooling of the examined building, with respect to the insulation's thickness
- The daylight conditions prevailing indoors, with respect to the wall insulation thickness

More specifically:

In the framework of this study, in all examined thermal protection alternatives, the thermal insulation layer thickness is gradually increased from 10 cm to 30 cm with a step of 10 cm, resulting in a range of thermal transmittance values of the façade's elements that currently represents many European national regulation requirements.

The layering and the thermal transmittance (U-value) of all examined scenarios are presented in Table 1 and Table 2 respectively for the two façade energy retrofit alternatives, i.e. ETICS and VFS.

The calculation of the U-value for both alternatives was based on EN ISO 6946. The air gap of the ventilated façade is considered as well ventilated. According to EN ISO 6946, the total thermal resistance of the wall encompassing a well-ventilated air layer can be obtained by disregarding the thermal resistance of the air layer and all other layers between the air layer and external environment, and including an external surface resistance corresponding to the internal surface for horizontal heat flow,

For all examined cases, the average U-value of the windows was regarded equal to ca. 2.60 W/(m^2 K), which corresponds to most frame types (wood, synthetic, aluminium with a conventional thermal break) and a double glazing.

Table 1. The layers forming the examined wall assemblies. and the U-values of the cases studied for each intervention alternative

ETICS				VSF			
	d m	λ W/(m·K)	d/λ W/K		d m	λ W/(m·K)	d/λ W/K
Brick-wall: 1 Plaster 2 Thermal insulation 3 Brick 4 Plaster	0.025 - 0.020	0.870 0.035 0.580 0.870	0.0287 0.345 0.029	Brick-wall: 1 Cladding 2 Ventilated air gap 3 Thermal insulation 4 Brick 5 Plaster	0.03 0.03 0.2 0.25	0.035 0.580 0.870	- - 0.345 0.029
Concrete: 1 Plaster 2 Thermal insulation 3 Concrete 4 Plaster	0.025 0.200 0.025	0.870 0.035 0.580 0.870	0.0287 0.345 0.029	Concrete: 1 Cladding 2 Ventilated air gap 3 Thermal insulation 4 Concrete 5 Plaster	0.03 0.03 0.2 0.25	0.035 0.580 0.870	- - 0.345 0.029

Table 2. The U-values of the examined wall assemblies for different thicknesses of thermal insulation

	ETICS				VFS			
Thermal insulation thickness (m)	0.000	0.100	0.200	0.300	0.000	0.100	0.200	0.300
Total wall thickness (m)	0.250	0.350	0.450	0.550	0.285	0.385	0.485	0.585
U value of brick-wall (W/m^2 K)	1.747	0.292	0.159	0.109	1.578	0.286	0.158	0.109
U-value of concrete (W/m^2 K)	3.252	0.316	0.166	0.113	2.712	0.310	0.164	0.112

For the Calculation of the Thermal Bridging Effect

According to EN ISO 20211, a thermal bridge is defined as the "part of the building envelope where the otherwise uniform thermal resistance is significantly changed by full or partial penetration of the building envelope by materials with a different thermal conductivity, and/or a change in thickness of the fabric, and/or a difference between internal and external areas, such as occur at wall/floor/ceiling junctions" (ISO, 2007).

Within the context of this study, both linear and point bridges were taken into account. According to ISO 10211, thermal bridges occur at the parts of the building where the otherwise uniform thermal resistance is significantly changed due to the full or partial penetration of the building envelope by materials with a different thermal conductivity, change in thickness of the building fabric and differences between the internal and external areas (ISO, 2007).

Thermal bridges can fall into two categories:

- Linear thermal bridges are the ones found at the junction of building elements or across edges where the geometry significantly differs from the one-dimensional assumption followed in heat flow calculations in buildings. The main characteristic of this type of thermal bridge is that the temperature distribution is uniform across a cross section of the junction along one of the three orthogonal axes.

- Point thermal bridges are localized thermal bridges, often met at the junction of three planes or at points of localized change in heat flow,

Linear thermal bridges are encountered when both thermal protection alternatives are installed, while the point thermal bridges are calculated only for the VFS at the points of the façade's anchorage. No thermal bridging effect is considered in the case of no thermal insulation.

For Linear Thermal Bridges

The magnitude of the thermal bridging effect is characterized by the value of the linear thermal transmittance Ψ, which is calculated with the use of special software for 2-D heat flow analysis from (ISO, 2007):

$$\Psi = L_{2D} - \sum_{j=1}^{n} U_j \times l_j$$

where Ψ is the linear thermal transmittance of the thermal bridge, L_{2D} is the thermal coupling coefficient obtained from a 2-D calculation of the component separating the two environments being considered, U_j is the thermal transmittance of the 1-D component j separating the two environments being considered and lj is the length over which the value Uj appears.

For quantifying the influence of the two thermal protection alternatives on the thermal bridging effect, the linear thermal transmittance values were calculated for the junction of the vertical with the horizontal building elements, i.e. the roof, the intermediate floors and the floor. The calculation was conducted with a 2-D energy analysis software, ANSYS 13.0, assuming a 20°K temperature difference between the inner and the outer environment. For the initial case, where no thermal insulation exists, no thermal bridge is taken into account.

Within the context of this paper, the junctions between the horizontal building elements (i.e. roof, floor) with a pillar (line a in Figure 3) and a beam (with a brick masonry below when needed, line b in Figure 3), both made of reinforced concrete, were examined. The junction between the walls and the intermediate floor was also examined, but it is not presented here analytically as it involves only a limited part of the building.

The flat roof (column i in Figure 3) is made of reinforced concrete and its internal surface is covered with lime-cement plaster. Above the slab the vapour barrier and the thermal insulation are positioned, followed by a layer of lightweight concrete for providing a slope for rain water drainage and the water proofing membrane. Cement tiles, fixed with cement mortar, constitute the outer surface of the building element.

The floor (column ii in Figure 3) is also structured with a reinforced concrete slab with thermal insulation below. Above the slab there is the infill layer, made of light weight and gravel concrete, followed by the flooring.

The Ψ-values were calculated both for the ETICS and the VFS alternatives, taking into account the three different insulation thicknesses (10 cm, 20 cm, 30cm).

Figure 3. Schematic plan of the examined junctions between the pillar (a) and the beam with the masonry below (b) with the flat roof (i) and the floor (ii)

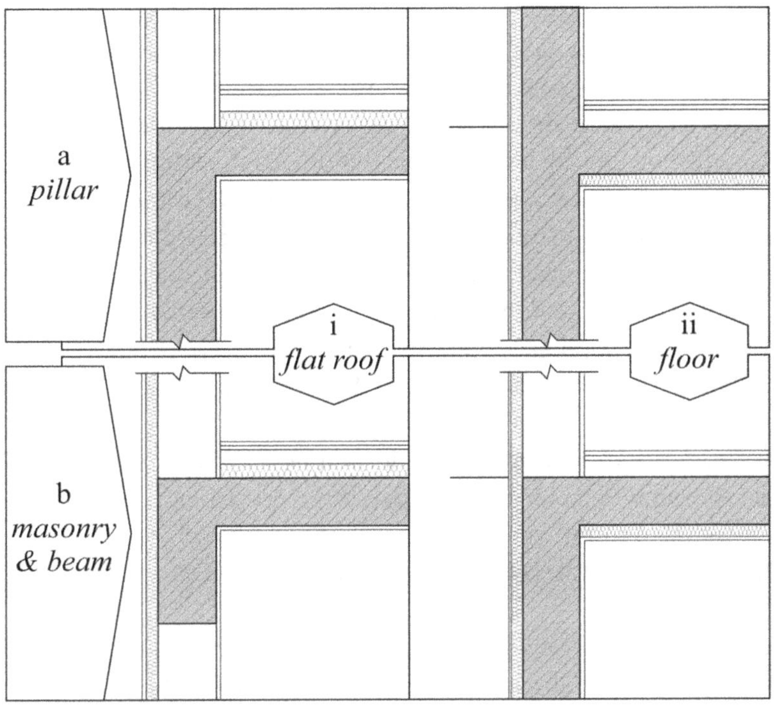

For Point Thermal Bridges

According to ISO 10211, the point thermal bridge is calculated by:

$$Ç = L_{3D} - \sum_{i=1}^{N_i} U_i \cdot A_i$$

where x is the point thermal bridge transmittance, L_{3D} is the thermal coupling coefficient, U_i is the thermal transmittance of the building element i reaching the junction of the building elements, A_i is the area of the building element i reaching the junction of the building elements.

The three dimensional nature of the thermal bridge effect on cladding systems requires a detailed calculation approach in order to take into account the complex geometry and the great differences in the thermophysical properties of adjacent materials. According to ISO 10211, in order to achieve reliable results regarding the magnitude of thermal bridging in such complicated structures, a detailed finite element analysis is needed. In the case of this study, the calculation platform that was used to calculate the thermal coupling coefficient was ANSYS Workbench. As in the case of linear thermal bridges the calculations were made assuming a temperature difference equal to 20°K.

For the Calculation of the Building's Energy Performance

The energy performance of the building was investigated by the use of dynamic whole building energy analysis using EnergyPlus simulation tool. Within the framework of the study, the energy performance is expressed by the heating and cooling needs, in order to ensure proper indoor thermal comfort conditions.

Operational schedules were set according to ISO 18523-2:2016 (ISO, 2016). The definition of heating and cooling season was different for each site and was automatically adjusted by the simulation software in order to keep the interior climatic conditions in acceptable levels throughout the year. Thermostat temperature for heating and cooling was set to 21°C and 26°C accordingly, with 16°C night-set back temperature for heating. It was assumed that cooling was not operating during night. An occupation density of 5 people per 100m^2, and ventilation for indoor air quality of 15l/s•person was considered as reasonable. Internal gains from appliances (2W/m^2) and lighting for ensuring 200lux in heated zoned during operation hours was also considered. Finally, the airtightness of the building was expressed by an infiltration rate of 0.35 air changes per hour.

The evaluation parameters include the heating and cooling energy needs of the building on an annual base for both upgrade alternatives located in the three examined cities.

For the Calculation of the Visual Performance

For evaluating the changes in indoor daylight conditions caused by the addition of thermal insulation and the following increase of the wall width, a room of the examined building was selected as a reference unit, encompassing typical geometric characteristics of main use spaces of residential buildings.

The examined room is of square plan, with 4.50 m side length and 2.80 m height (Figure 4). It is side-lit by one window, 1.70 m wide to 1.20 m high, with a sill elevated by 1.00 m above the floor. The area of the window corresponds to 10% of the floor area, a requirement that is often found in national legislations all over Europe. The window is double glazed with a visual transmittance equal to 0.75. The reflectivity of the walls is equal to 0.80, representing a light coloured smooth finish, while the reflectivity of the floor was considered equal to 0.50, representing common floor finishes, such as wood, carpet, tiles, etc. The reflectivity of the ceiling was regarded equal to 0.85, representing common light coloured plastered surfaces.

Moreover, it is assumed that the examined room has an unobstructed view to the sky vault and no shading devices. The window orientation does not play a significant role in this analysis, especially regarding daylight factor, given that the conditions examined refer to overcast skies.

The daylight levels were estimated before and after the thermal protection. For the existing conditions, the thickness of the wall, which actually refers to the thickness of the window casing, is 0.25 m. As regards the case after the thermal protection, in the daylighting study the worst scenario is assumed, i.e. the case that a very low U-value is anticipated and an approximately 30 cm thick layer of thermal insulation is positioned. That leads to a wall thickness equal to 0.550 m for the case of ETICS and 0.585 m in the case of the ventilated façade system (Figure 1, Figure 4).

It is also worth mentioning that preliminary analyses showed that the thicker the wall the worst daylight conditions prevail on the working plane of the examined room. However, there is a significant difference on the daylight levels prevailing before and after the thermal protection of the wall. For this reason, it is decided to present the results for the worst case, i.e. when a ventilated façade system with

Figure 4. Simplified axonometric view of the examined building before (left) and after (right) the thermal protection of the building unit

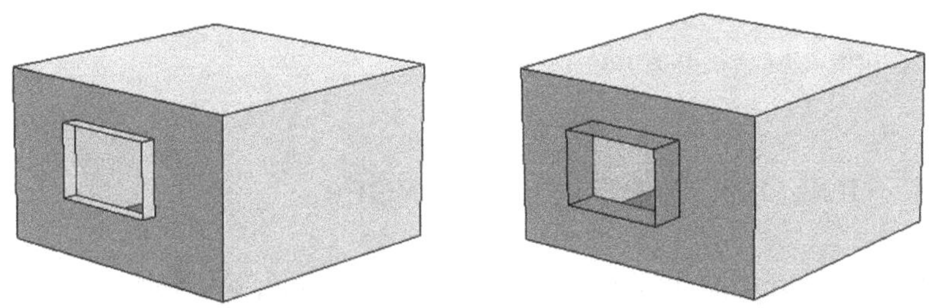

a. b.

a 30 cm –thick thermal insulation is installed. The reflectivity value of the casing around the window, which is a part of the ventilated facade system, is considered equal to 0.60.

The daylight levels were estimated before and after the retrofit system integration. Given that the quantity and quality of available daylight varies significantly with the hour of the day, time of year and meteorological conditions, the indoor daylight conditions are very often assessed through the estimation of daylight factor. The daylight factor represents the illuminance at a specified point indoors, expressed as a percentage of the simultaneous horizontal illuminance outdoors under an overcast sky. The orientation of the opening does not play an important role on the determination of the daylight factor, since the standardized overcast sky presents only altitudinal asymmetry, featured by illuminance at the zenith three times greater than on the horizon (Baker et al, 1993). On the contrary, differentiations on the angular height of the visible sky patch and on the glazing transmittance influence the daylight factor patterns significantly.

The minimum, maximum and average value of daylight factor was calculated, in order to provide the general picture of the visual performance of the room. Values of daylight factor above 2% are considered as acceptable for most of the national regulations of the EU countries, especially when artificial lighting is complementary used. Especially for residential buildings, even lower values of daylight factor are allowed; for example, UK BS 8026-2 recommends that in dwellings the average daylight factor should exceed 1.50%, 2.00% and 1,00% in the living room, the kitchen and the bedroom respectively (BS, 2008). However, it has to be highlighted than when the daylight factor values of an indoor space are reduced below 2% and the activity taking place into the room requires optical acuity, the use of artificial light is needed almost throughout the day.

However, when studying the adequacy of daylight or the energy consumption for lighting, it is very important to examine the illumination levels as well. In this study it was attempted to study whether the minimum daylight requirements are fulfilled after the integration of thermal insulation. Within this context, the worst case scenario was assumed, i.e. the estimation of daylight levels under an overcast sky. Within this context, the illumination levels of all examined cases were plotted in isolux lines with a standard scale ranging from 40 lx to 15000 lx to allow for comparison.

Additionally, the uniformity of daylight distribution was estimated, as one of the qualitative criteria of visual comfort. It is expressed as the ratio of minimum to average daylight factor and usually the acceptable values range above 0.3 (BREEAM, 2016).

All daylight metrics are estimated on the working plane, i.e. at a horizontal plane elevated by 0.80 m above the floor level. For the calculation of the daylight metrics Autodesk Ecotect analysis 2011 software was used, which employs the split-flux method. The design sky illuminance value and was calculated through Tregenza formula for each examined location.

ASSESSING THE BUILDING'S PERFORMANCE

The impact of the thermal protection alternatives on the thermal, energy and visual performance of the examined building are presented below analytically. It is worth mentioning that the results of the analysis can be generally applied to similar single family buildings located in low-rise areas with similar climatic conditions.

Thermal Bridges

The assessment of the thermal bridging effect is presented separately for the linear and the point thermal bridges.

Linear Thermal Bridges

The temperature profiles that have derived from the analytical calculations of the energy flow rate across the examined building elements are presented in Figure 5. The figure is organized as a matrix, in which the lines refer to the type of the vertical elements and the columns the type of the horizontal elements. The pictures included in this matrix refer to ETICS, with a thickness of thermal insulation equal to 0.10 m and are presented here in order to assist towards the justification of the numerical results.

The Ψ values that have been calculated for the examined junctions, with varying thickness of thermal insulation and the two different upgrading alternatives, are presented diagrammatically in Figure 6 for the junction between the wall and the roof and Figure 7 for the junction between the wall and the floor. The Ψ-values are plotted on the y-axis against the thermal thickness of the horizontal element (roof, floor). Each line appearing on the diagram refers to the different thickness of the vertical element (wall).

As regards the junction between the flat roof and the vertical elements, it is indicated that the isothermal contours in Figure 5 have a smoother distribution in the case that the vertical component is a reinforced concrete element (pillar). In the case of the beam, there is a double thermal bridging effect, one at the junction between the flat roof and the beam and one at the junction between the beam and the masonry, which creates lower surface temperatures around the beam area. This indication is also portrayed on the Ψ-values, which generally range between 0.10 W/(m K) and 0.20 W/(m K). Actually, the values are slightly higher for the case of beam-roof junction (Figure 6, down). It is interesting to notice that the slight differences between the examined junctions (pillar and the beam with the masonry below) are observed for a 10-cm-thick layer of thermal insulation on the roof. As the thickness of the roof insulation increases, the differences on the Ψ-values among the three curves (different wall insulation thickness) are eliminated. Moreover, the Ψ-values appear to be higher in the case of ETICS, but this is a fact only when a 10cm thick layer of thermal insulation is used the roof. It is worth mentioning, that for a a medium thermal insulation thickness on the wall (10 cm) the Ψ-values can be significantly decreased when a thick insulation layer is applied on the roof (i.e. 30 cm). The specific Ψ-value is even

Figure 5. The temperature profiles of the examined junctions between the pillar and the beam with the masonry below with the flat roof (left) and the floor (right)

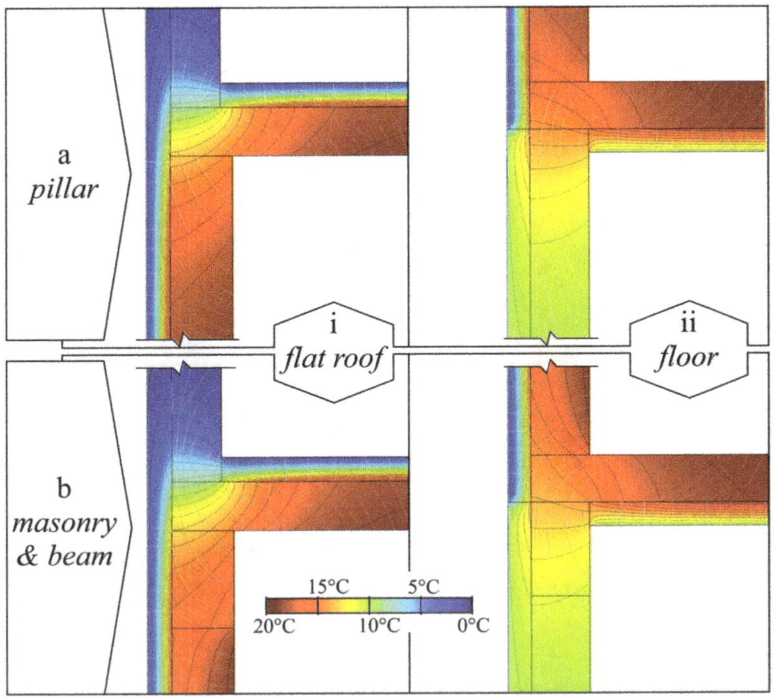

lower for the VFS when compared to ETICS. For the other thicknesses of wall insulation the Ψ-values are almost identical. This can be attributed to the geometry of the building elements forming the junction, given that when the path of heat flow is longer when thick layers of thermal insulation are used for the roof, leading to decreased Ψ-values on such junctions.

For the thermal bridge between the floor and the vertical elements, the temperature distribution appears to be less even in the case of pillar (Figure 5). This is also logical as the change of the material conductivity between the masonry and the reinforced concrete causes temperature fluctuations and differentiations on the heat flow.

The Ψ-values are higher in the case of the floor when compared to the roof junctions and range between 0.50 W/(m K) and 0.80 W/(m K). Moreover, the decrease on the Ψ-values is very noticeable as the thickness of thermal insulation on the floor increases. The difference is higher for the thinner layer of thermal insulation on the vertical element (10cm) and is mostly discerned in the case of ETICS at the junction of the pillar with the floor. In general, the two curves showing the Ψ-values for the cases of 20 cm and 30 cm thermal insulation on the vertical element are very close to each other, indicating that the thermal bridge is of similar magnitude in each case.

Figure 6. The Ψ-values derived for the junction between the pillar and the flat roof (up) and the beam with the masonry below and the flat roof (down) for different thicknesses of thermal insulation on the roof and the wall, for the two thermal protection alternatives (ETICS and VFS)

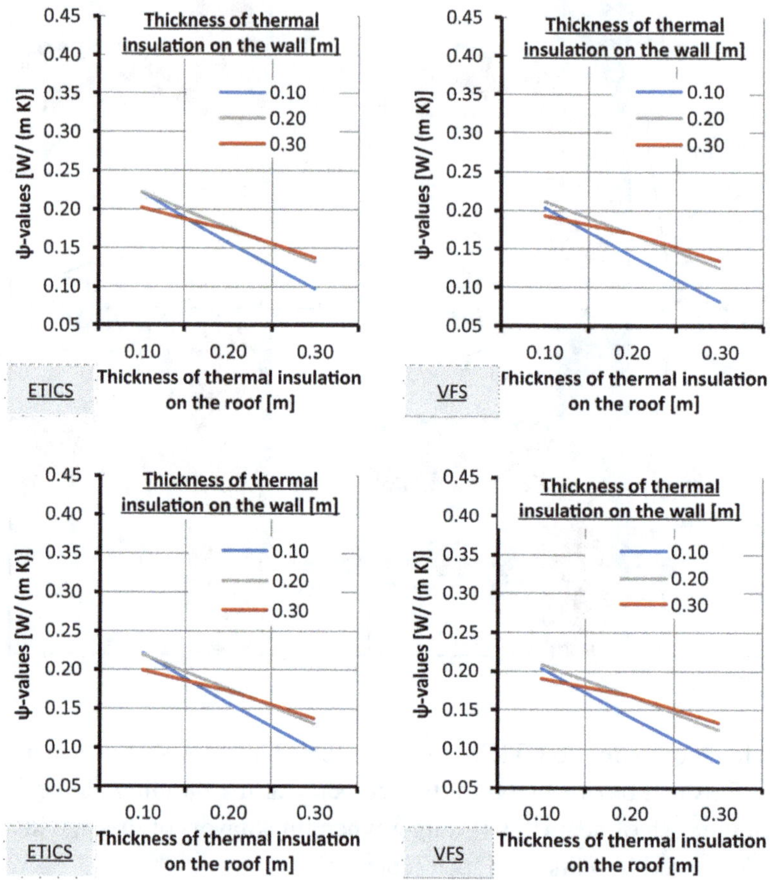

Point Thermal Bridges

In the case of cladding systems, the external envelope is anchored on the internal by penetrating the thermal insulation layer at various points, usually through materials with high thermal transmittance, like steel and aluminium. Although the size of the penetration is very small compared to the area of the façade, the transmissivity can be more than 2000 times higher than the thermal insulation material, creating intense thermal bridge heat flow at these penetration areas. Figure 8 shows the local isothermal change due to the existence of an aluminium bracket penetrating the thermal insulation layer.

The problem of point thermal bridges is a multi-parametric problem, highly affected by the geometry under study and the thermophysical properties of all materials involved within the geometry. Any change in one of the parameters of the effect can lead to changes of the thermal bridge magnitude that do not follow a linear relation and, most important, it is not always clear if the change in thermal bridge magnitude would be analogous or reverse analogous.

Figure 7. The Ψ-values derived for the junction between the pillar and the floor (up) and the beam with the masonry below and the floor (down) for different thicknesses of thermal insulation on the roof and the wall, for the two thermal protection alternatives (ETICS and VFS)

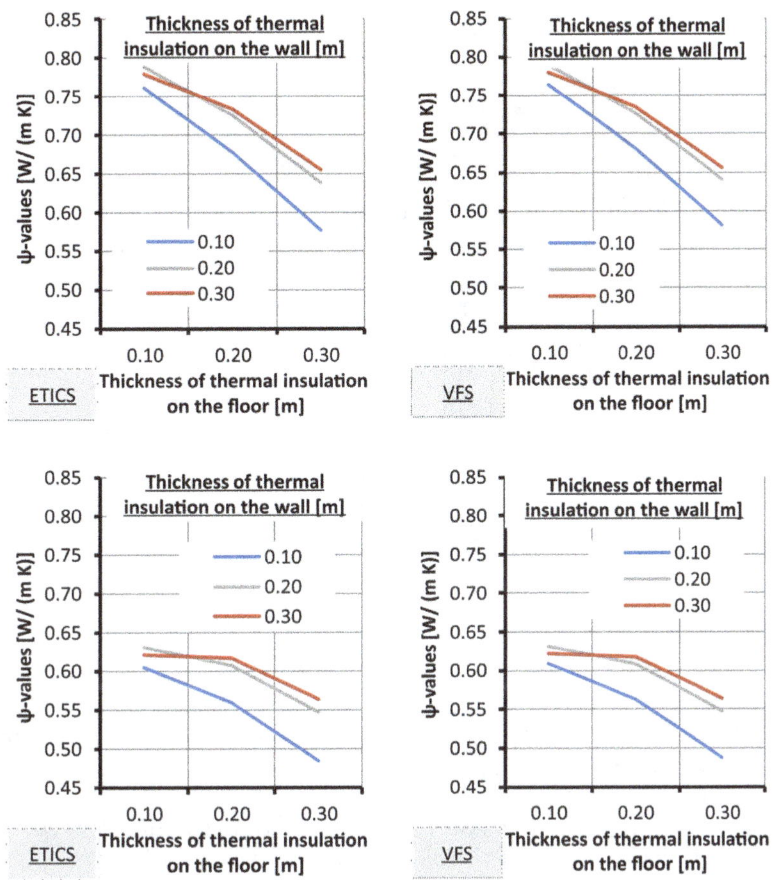

The x-values estimated for various thicknesses of thermal insulation are presented in Figure 9. It is obvious that the thermal conductivity of the substrate wall plays an important role. In cases where the wall has some thermal insulation capabilities (low thermal conductivity), the thicker the thermal insulation layer, the higher is the point thermal transmittance. On the contrary, on more thermally conductive materials, the increase in thermal insulation thickness leads to a reduction of the point thermal bridge magnitude. This behaviour represents the complexity of the subject under investigation, since the portion of the phenomenon attributed to geometric factors and the part attributed to heat flow factors are highly depended of the examined parameters.

In every case, since existing wall materials are expected to have thermal conductance more than 0.5 W/(m·K), one can conclude that the increase in thermal insulation thickness can contribute to a small decrease of point thermal transmittance in the brackets, most probably because the thickness of the insulation layer is analogous to the length that heat has to travel through the highly conductive bracket.

Figure 8. 3D sketch of the temperature distribution around the bracket of the ventilated façade system

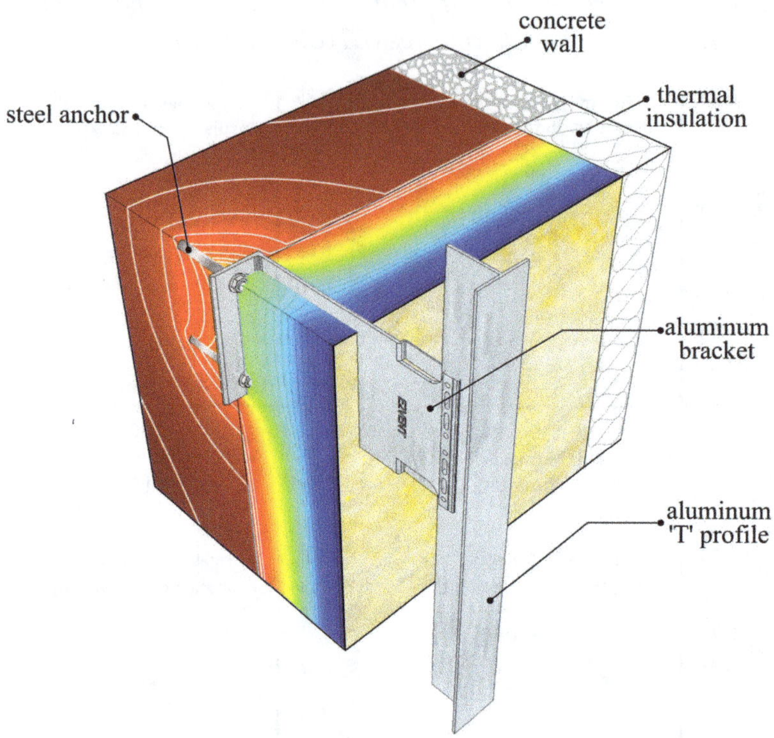

Figure 9. The effect of the thickness of the thermal insulation on the point thermal transmittance, for a variety of substrate wall thermal conductance

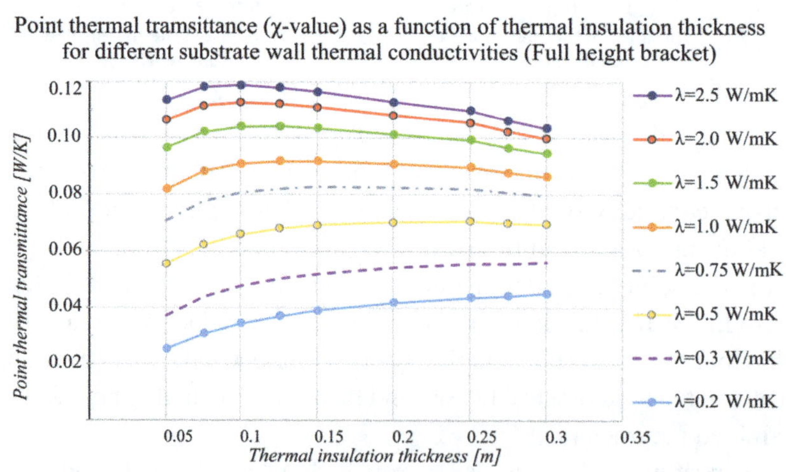

Energy Performance

The estimated annual heating and cooling energy needs of all the analyzed scenarios, as a function of the thermal transmittance of the examined buildings' façades are presented in Figure 10 and Figure 11 respectively. More precisely, the simulated energy needs are presented in relation to the thermal transmittance of the weakest part of the façades, i.e. of the concrete elements, due to their higher U-value, compared to the brick walls. Simulation results for the examined buildings in all three climatic European regions are given in pairs, with the solid and dashed lines referring to the application of the ETICS system and the ventilated façade system respectively.

Figure 10. Simulated annual heating energy needs of all the analyzed scenarios, as a function of the thermal transmittance of the facades' concrete elements

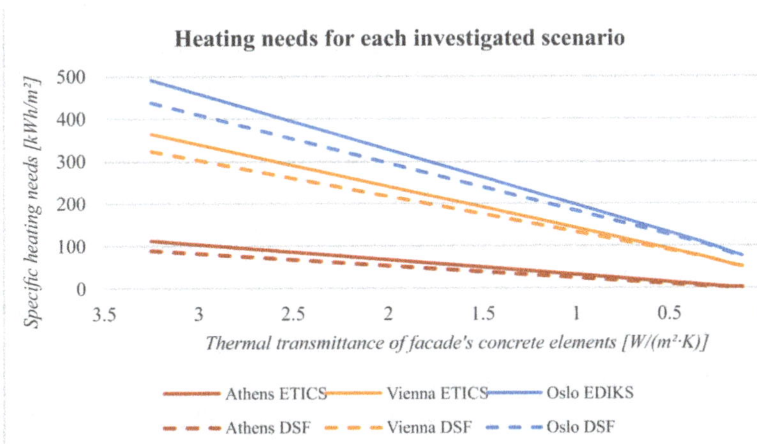

Figure 11. Simulated annual cooling energy needs of all the analyzed scenarios, as a function of the thermal transmittance of the facades' concrete elements

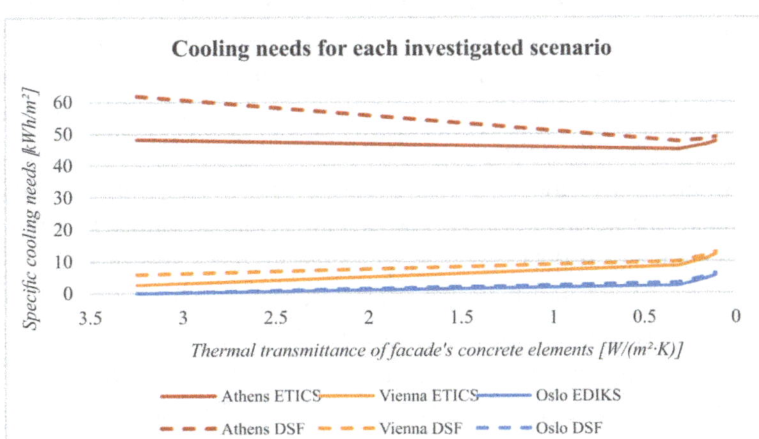

In Terms of the Annual Heating Energy Needs

In all the examined scenarios, the application of a thermal insulation layer, either with the form of the ETICS or the VFS, has contributed to an important decrease of the estimated buildings' annual heating energy needs. However, the extent of the achieved reduction strongly depends on the thermal protection level, but also on the prevailing climatic conditions. As depicted in Figure 10, the highest annual heating energy needs among the three climatic European regions have been found for Oslo, reaching even 490 KWh/m^2/year in the case of a thermally uninsulated building envelope, whereas in Athens, the respective energy needs did not exceed 115 KWh/m^2/year, given the milder climatic conditions during winter period.

The implementation of an ETICS system with an insulation width of 10 cm in both the examined buildings of Oslo and Vienna resulted in a lower annual heating energy demand by almost 80%, compared to the respective non-insulated scenario, while the achieved energy savings have been found even more pronounced in the case of 20 cm and 30 cm insulation width, corresponding to a thermal transmittance of the façade's concrete elements of 0.166 W/m^2/K and 0.112 W/m^2/K respectively. Similarly, for the building located in Athens, the application of an ETICS of 10 cm contributed to a substantial decrease of the annual heating energy demand of almost 90% compared to the non-insulated scenario.

An interesting observation is that in all cases, the estimated heating energy needs for the buildings implementing double skin facades are always lower compared to the ETICS alternative. Even in the case of the non-insulated scenario, the application of a ventilated façade in the examined buildings of all three regions has been proven beneficial in terms of the heating energy needs, compared to the conventional system. The latter effect lies on the air layer of the air cavity, being treated as a surplus on surface thermal resistance, as described in ISO 6946. Thus, for the same width of thermal protection, the overall thermal resistance of the building façade is always higher in the case of the double skin ventilated façade, leading to a lower heating energy demand. Indicatively, the application of a VFS system with 10 cm instead of the corresponding ETICS system in the building of Athens, has led to a reduction of the estimated heating energy demand by 12%, while the respective savings to Vienna and Oslo were rather minor and did not exceed 2%. Yet, it has to be emphasized, that both for the ETICS and the VFS systems, applying a higher insulation width than 20cm did not provide any considerable further energy savings.

In Terms of the Annual Cooling Energy Needs

Concerning the estimated annual cooling energy needs, there is a noticeable difference among the buildings of the 3 regions, due to the considerable different climatic conditions (Figure 11). As anticipated, the lowest cooling energy needs have been found for the buildings in Oslo and Vienna, corresponding to heating dominated climatic regions whereas the highest demand has been estimated for the building in Athens, representing mild, cooling-dominated climatic conditions (Tsikaloudaki et al, 2012).

In the latter case, simulation results revealed a moderate reduction of the annual cooling energy needs, not exceeding 7% in the case of an ETICS of 10 cm (corresponding to a thermal transmittance of the facades' concrete elements of 0.315W/m^2/K), compared to the non-insulated scenario, while any additional increase of the building's envelope thermal resistance (ETICS 20cm and ETICS 30cm) only resulted in marginal energy savings that did not exceed 1.5%. In fact, the additional increase of the thermal insulation would reduce the rate that the stored heat can be dissipated towards the outdoor environment, leading thus to a rise of the cooling energy demand. Therefore, contrary to the performance of the building during the heating period, the provision of thermal insulation layers higher than 10cm would not be the optimal solution, while similar results have been also reported in the existing literature (Pan et al, 2012; Shariah et al, 1997). Regarding the performance of a VFS system in Athens, the beneficial effect of the additional thermal resistance of the air cavity on the reduction of the heating energy needs is counterbalanced during the cooling period, since the estimated annual cooling energy demand has been found higher. Especially in the case of the non-insulated scenario, the application of a VFS system would lead to 28% higher annual cooling energy needs, compared to the ETICS alternative, whereas the discrepancy is attenuated as the thickness of the insulation layer increases.

In terms of the buildings' energy performance in Olso and Vienna, the annual cooling energy needs for the non-insulated scenarios were close to 0 KWh/m^2/year, whereas the application of an ETICS system of 10 cm has led to a higher energy demand by 2.2 KWh/m^2/year and 5.9 KWh/m^2/year respectively. Moreover, in both regions, as the insulation thickness rises, the annual cooling energy needs are also increased. Again, this effect is attributed the slower dissipation of the internal heat gains towards the outdoor environment and through external walls during summer, resulting thus in an increase in cooling energy use. Finally, as with the case of Athens, implementing a VFS system in the buildings of Oslo and Vienna would result in generally higher annual cooling energy needs compared to the respective ETICS solution, reaching 15% and 7% in Oslo and Vienna respectively, in case that a VFS with 20 cm insulation is applied, instead of the respective ETICS 20cm.

Daylighting

The impact of adding thermal insulation to an existing wall on the indoor visual conditions is depicting by comparing the illumination levels prevailing before and after the intervention.

Figure 12 presents the illumination levels prevailing on the working plane of the examined room (a.) before and (b.) after the wall's thermal protection. For the initial case, daylight levels range from 77 lx to 760 lx with an average value of nearly 200 lx. For the areas that are close to the window (i.e. at a distance of 1.5 m from the window wall –one third of the room's depth-) the illumination exceeds 300 lx, while at the rear areas of the room the illumination is poor and artificial lighting is needed.

Figure 12. The daylight levels prevailing on the work plane of the examined building unit before (a.) and after (b.) the installation of the ventilated façade system

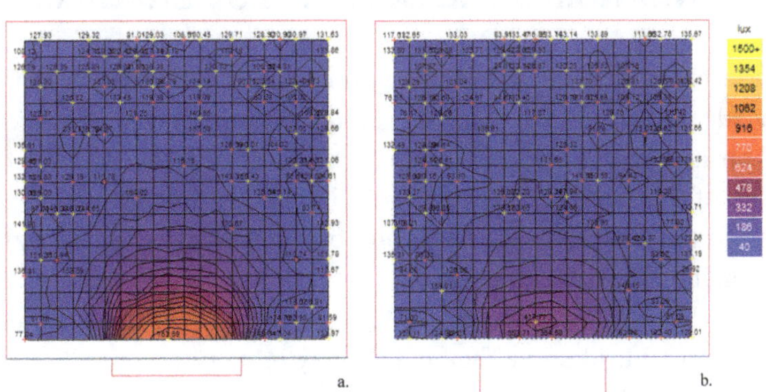

The average daylight factor is equal to 2.59%, indicating that although the room is not bright, there is adequate daylight. For specific tasks, support with artificial lighting is needed, though. Additionally, the ratio of minimum to average illumination is equal to 0.39, indicating that the daylight distribution can be regarded as uniform.

When the ventilated façade system is installed, daylight levels range from 67 lx to 417 lx with an average value of nearly 155 lx. This accounts for a 45% reduction of maximum daylight levels and a 22.5% reduction on average illumination levels. This indicates that the installation of a thick layer of thermal insulation influences mostly the areas that are close to the window, since the view to the sky vault from these points is more significantly obstructed. The zone with adequate daylight (i.e. above 300 lx) is limited to a depth of 1.0m from the window wall, comparing to 1.5m in the existing conditions.

The average daylight factor is equal to 2.05%, just slightly above the general acceptable value for the most "demanding" area of the dwelling, which is the kitchen. However, illumination values are low for most of the working plane area, which indicates that if specific tasks are going to be performed, artificial lighting may be needed.

The ratio of minimum to average illumination is equal to 0.42, indicating that the distribution is now a little more uniform than before (changed from 0.38 to 0.42), given that the average value is significantly reduced, while the minimum value is practically unchanged.

The daylight conditions presented in this section referred to Vienna conditions. It must be mentioned that the daylight factor is not influenced by the geographical location of the building. It is expressed as a ratio of indoor to ambient illuminance on a horizontal plane under overcast sky conditions and it shows mostly how the room configuration (size, window area and transmittance, reflectance, obstructions, etc.) influence the visual appearance of the room. For this reason, the daylight factors calculated and presented for Vienna are the same for the case of Athens and Oslo, or for any other geographical location.

On the contrary, the illuminance levels are not the same, since the design sky illuminance differentiates among different locations. For example, the design sky luminance was considered equal to 8028 lx for Athens, 6974 lx for Vienna and 5513 lx for Oslo. That practically means that if the illumination levels under overcast sky conditions are needed, the daylight factor values could be multiplied by the design sky illuminance value. The lower sky illuminance observed in Oslo results in lower illumination values on the working plane for all models by ca. 21%, while the higher sky illuminance of Athens will result in higher illumination levels by ca. 15% with respect to the illumination values calculated for Vienna, presented in this section.

MEASURES FOR ENHANCING THE BUILDING'S PERFORMANCE

The analysis of the thermal protection alternatives showed that the addition of thermal insulation on an existing building façade has a significant impact on the building performance. More importantly, this impact is not always positive and in some cases it may lead to the deterioration of the indoor environment. However, the benefits of thermal protection are still important and cannot be easily achieved through other measures; therefore it is very essential to propose measures for minimizing the negative effects with the aim of optimizing the indoor conditions.

The negative effects that have been indicated through the analysis concern mainly the cooling building performance, the thermal bridging effect and the daylight conditions.

Improving the Cooling Building's Performance

The analysis of the thermal protection alternatives showed that the addition of thermal insulation has a positive effect on the control of heating needs for all locations, while for the warmer climates, the increase of thermal protection may lead to the increase of cooling loads, as the dissipation of heat towards the ambient environment is hindered by the low thermal transmittance of the building elements. This can be efficiently counterbalanced by using passive cooling techniques, such as night ventilation. More specifically, the heat that has been accumulated during the daytime can be removed at night-time through natural ventilation, as the intake of outdoor cooler air reduces the indoor air temperature and the surface temperature of the building fabric. The consequent cooled structure slows the rate at which the internal temperature rises during the day [thereby decreasing the cooling energy demand (Solgi et al, 2018).

Minimizing the Thermal Bridging Effect

The addition of thermal insulation on an existing facade results in differentiated heat flows through the building elements, especially at the junctions between the vertical elements and between the horizontal and the vertical building elements. The phenomenon is more evident when the thermal insulation layer is disrupted, i.e. in the case of a balcony.

In the case of external thermal insulation, the linear thermal bridges participate in the overall heat flow towards the ambient environment and should be taken into account, not only when calculating the energy flows, but also for investigating the possibility of vapour surface condensation. Special attention should be given to the installation of thermal insulation, in order to cover the building envelope and no irregularities or cavities are created. It has to be highlighted, however that it is very difficult to eliminate the thermal bridging effect, especially along slab projections. However, the linear thermal bridges can be significantly reduced in the case of the window to wall junctions, by adding a thin thermal insulation layer around the window casing. Such an intervention can reduce the Ψ-value by at least 50% (Ben-Nakhi, 2002).

In the same vein, the magnitude of the point thermal bridges appearing in the ventilated façade systems can be minimized through the use of thermal break layers between the bracket and the wall. Such materials have a low thermal conductivity and at the same time they should have the required strength to efficiently deliver the forces from the bracket to the wall. With interrupting the direct contact of the metallic bracket and the substrate wall, the heat flow is significantly decreased.

This is clearly presented in Figure 13, where the heat flow is mainly concentrated around the anchor in the case of a bracket with a thermal break (right picture), while the absence of the thermal break pad results in a less uniformly distributed temperature profile.

It has to be highlighted that the effect is minimum in the cases of wall materials with lower thermal conductivity, since the thermal resistance is also provided by the substrate wall, and thus the contribution of the thermal break is minimized.

Figure 13. Isothermal surfaces showing temperature distribution on a plane vertical to the anchor for the scenario without thermal break (left) and with thermal break pad (right)

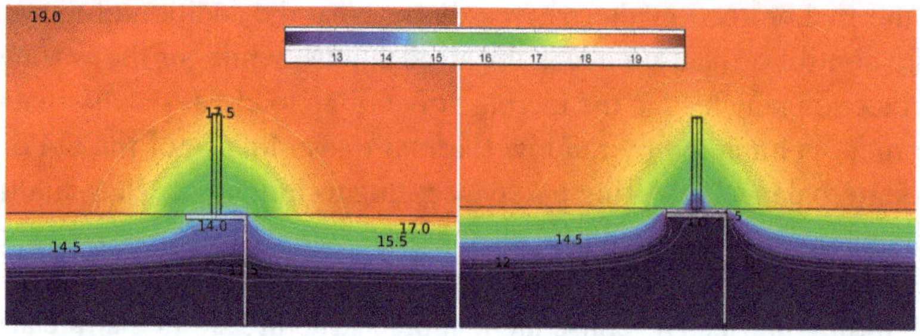

Optimizing The Daylight Conditions

It was proved that the integration of the ventilated façade system has a direct influence on the admittance of daylight in indoor spaces. This was expected, given that the examined thermal protection systems embrace the windows of the building as an external casing, providing solar protection on one hand and reduction of the visible sky view on the other. What was not known, but was revealed through the analysis, is the quantitative assessment of this reduction (approximately 22% on the average values). Although the impact can be regarded as moderate and only in certain cases is considerable, the resulting visual performance of the indoor spaces may be poor, especially in rooms where daylighting is limited. For these reasons it was decided to study and assess measures that could mitigate the impact of such interventions on daylighting.

Two measures were examined as suitable for counterbalancing this effect: the increase of reflectivity on the casing around the window and the formation of inclined lateral and upper sides of the casing (Figure 14).

Figure 14. Axonometric view of the examined building unit with the integrated ventilated façade system, in the case of inclined lateral and upper sides of the window casing and construction detail of the wall sections

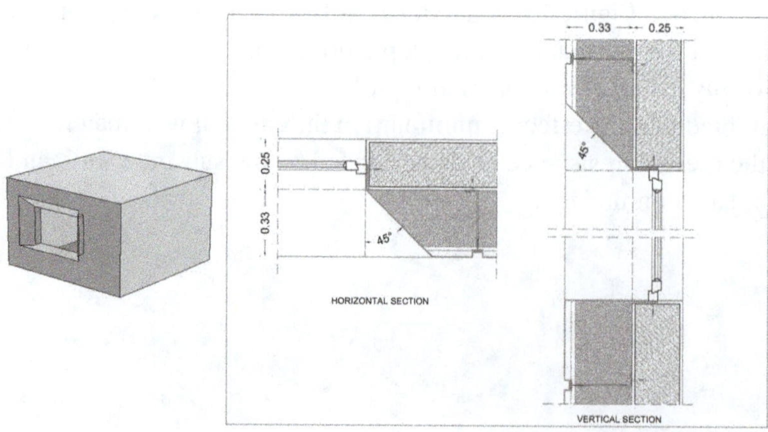

Figure 15 shows the illumination levels prevailing on the working plane of the examined building unit when these two measures are implemented. With the higher reflectivity panels (Figure 15.a), the average illumination ranges at approximately the same levels as before. There is only a 1% increase on illumination levels observed on the areas near the window, but the average values remain practically unchanged. So is the daylight factor. These findings reveal that the increase of reflectivity of the window casing is not effective in terms of mitigating the impact of the ventilated façade system on the daylight conditions of indoor spaces.

Figure 15.b presents the illumination prevailing on the working plane of the examined building unit, when the lateral and upper sides of the casing are inclined. With this specific configuration of the window casing, the illumination levels range from 64 lx to 660 lx with an average value of 187 lx. This corresponds to an increase of daylight levels by 21% with regard to the conditions formed with the standard systems. In fact, the illumination values are very close to, yet lower than, the initial daylight conditions, i.e. before the thermal protection of the facade.

It is worth mentioning, however, that the increase of daylight levels concerns mostly the front half area of the room, which results to the non-uniform increase of daylight on the working plane. The new uniformity ratio equals 0.34, 20% lower than the standard situation.

The distribution of daylight across the examined room is more vividly shown in Figure 16. It shows the distribution of daylight factor along the depth of the room, for the cases of no thermal insulation, thermal insulation without taking any measures of enhancing daylighting, as well as the two examined scenario for daylight enhancement, i.e. the increased reflectivity of the window casing as well as the configuration of the lateral and the upper parts of the window casing as inclined.

The diagram shows that the values of daylight factor differentiate significantly at the front of the room, near the window wall. It is very characteristic that the configuration of inclined window casing sides has a very positive impact on the daylight conditions, as the curve is very close to the initial condition. On the contrary, the increased reflectivity of the window casing has a small impact on the indoor illumination.

Figure 15. The illumination levels prevailing on the working plane of the examined room when (a) panels of high reflectivity are used for the window casing and (b) when the lateral and the upper sides of the window are inclined

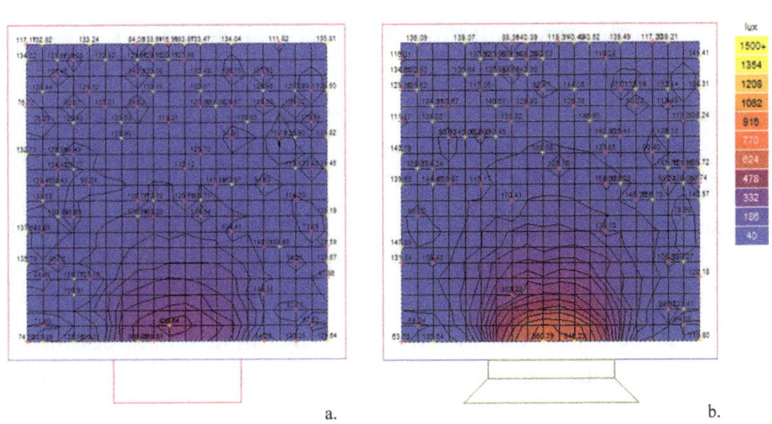

Figure 16. The distribution of daylight factor on the working plane of the examined building unit with the integrated ventilated façade, when (a.) reflectivity of the window casing is increased and (b.) the lateral and upper window casings are inclined

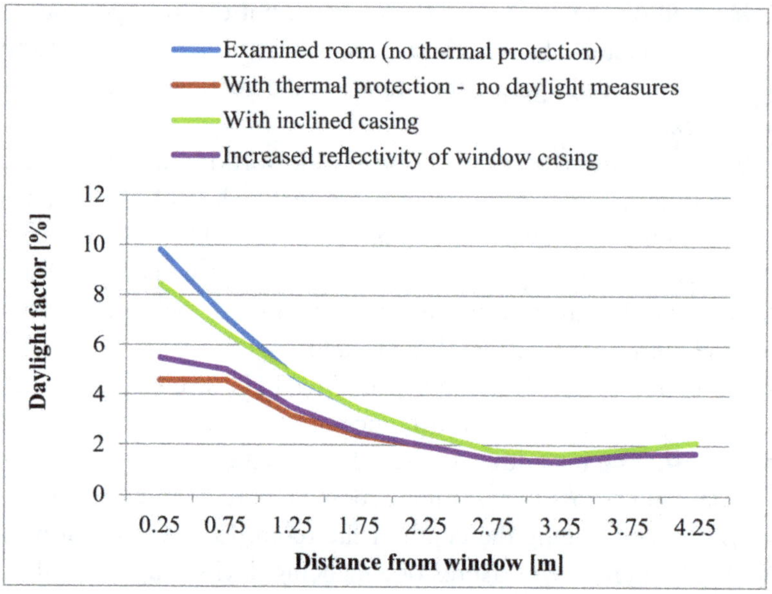

It has to be highlighted that the special configurations of the window casing examined in this section can be realized mainly in the implementation of ventilated façade systems, where the cladding can be handled more easily.

However, such a solution interferes with the architectural configuration of the building and changes its façade. Although the installation of the ventilated façade system inevitably alters the façade, the formation of inclined sides on the window causes additional changes, by giving emphasis on the windows. In fact, the inclination of the casing around the windows will appear as a frame around their lateral and vertical sides, the width of which may change with regard to the inclination angle.

FUTURE RESEARCH DIRECTIONS

The last decades the enhancement of the building energy performance has been one of the prior targets for the EU and within this context several strategies have been adopted both at European and at a national level. One common strategy among all European countries is the periodic recast of the building energy performance regulations, which in most cases impose stricter requirements for the building elements and the services.

However, this work showed that the formation of a very well thermally protected envelope does not always result to an overall upgrade of the building performance. A careful study has to be made, in order to ensure that the negative effects are minimized and the well-being of the occupants is ensured.

However, we still have to bear in mind that in the future the climatic conditions that are directly related to the building's performance will gradually change. Firstly, the continuous rise of urbanization will transform many low density areas to high density ones, which may gradually acquire the characteristics of urban areas, in terms of the building materials (increased thermal mass and decreased solar reflectance), reduced evapotranspiration due to plant deficiency, anthropogenic production of heat and atmospheric pollution. The ultimate impact of these parameters is the gradual increase of temperature in the urban fabric. Secondly, the contribution of climate change on the already high temperatures of urban environment affects significantly the building performance, which in turn adds to the urban environment additional heat and emissions of pollutants, leading to population discomfort.

Up to now, several studies have been published, dealing with the influence of climate chance on the buildings' energy behavior. These studies underline that the future estimated air temperature growth can lead to significantly increased cooling loads that outweigh the reduction of thermal loads. In the contrary, for the very cool climatic zones, the reduction of the thermal loads is much more significant compared to the rather limited increase in cooling loads.

At this point it is worth studying whether the general trend for additional thermal protection is in line with the anticipated future climate change or whether it is more appropriate to study in depth those requirements in conjunction with the future estimated conditions of each area with regard to the occurring climate change.

CONCLUSION

This work aimed at studying in depth and in a holistic way two common alternatives for the thermal protection of building facades, the ETICS and the VFS, by assessing their impact on the hygrothermal, energy and visual performance of a single family building located in three different zones of Europe. There were no significant differences found among the efficacy of the two alternatives when the overall performance of the systems is assessed. For example, although the thermal bridging effect is more noticeable in the case of the VFS, due to the point thermal bridges, the energy performance is better for the VFS as regards heating and slightly worse as regards cooling. For both alternatives the analysis showed that the installation of a thermal protection system influences the indoor daylight conditions at a maximum rate of about 20% as regards the average values. The reduction of daylight levels is observed mostly at the areas that are near the window wall, therefore the uniformity ratio is increased, producing a more even distribution of daylight on the working plane. Although this impact can be regarded as moderate and only in certain cases is considerable, the resulting visual performance of the indoor spaces may be poor, especially in rooms where daylighting is limited.

The holistic approach followed in this work enables the identification of potential and proper measures to mitigate the negative impacts. The adoption of such measures constitute the optimal way for integrating thermal insulation systems on existing buildings with the purpose of obtaining advanced building energy performance and visual comfort conditions at the same time. Furthermore, showing the benefits of building energy retrofit and indicating that potential negative effects can be easily counterbalanced through a comprehensive study, is one step towards raising the public awareness and the social acceptance of the systems, both of which are very essential for the wider diffusion of energy upgrade actions in Europe (Tsoka et al, 2018).

ACKNOWLEDGMENT

Part of this work has been developed within the project E2VENT: Energy Efficient Ventilated Facades for Optimal Adaptability and Heat Exchange enabling low energy architectural concepts for the refurbishment of existing buildings http://www.e2vent.eu/. The project has received funding from the European Union's Horizon 2020 research and innovation program under grant agreement No 637261.

REFERENCES

Andeweg, M. T., Brunoro, S., & Verhoef, L. G. W. C. (2007). *COST C16 Improving the Quality of Existing Urban Building Envelopes, state of the art* (Vol. 2). Research in Architectural Engineering Series.

Baker, N., Fanchiotti, A., & Steemers, K. (1993). *Daylighting in Architecture: a European reference book*. Brussels: James and James.

Ben-Nakhi, A. E. (2002). Minimizing thermal bridging through window systems in buildings of hot regions. *Applied Thermal Engineering*, *22*(9), 989–998. doi:10.1016/S1359-4311(01)00121-1

Bikas, D., Tsikaloudaki, K., Kontoleon, K. J., Giarma, C., Tsoka, S., & Tsirigoti, D. (2017). Ventilated Facades: Requirements and Specifications Across Europe. *Procedia Environmental Sciences*, *38*, 148–154. doi:10.1016/j.proenv.2017.03.096

Bragança, L., Wetzel, C., Buhagiar, V., & Verhoef, L. G. W. (2007). *COST C16 Improving the Quality of Existing Urban Building Envelopes, facades and roofs* (Vol. 5). Research in Architectural Engineering Series.

BREEAM. (2016). *UK New Construction technical manual*. BRE.

BS. (2008). *BS 8206-2:2008 Lighting for buildings. Code of practice for daylighting*. BSI.

Directorate-General for Energy –European Commission. (2012). *EU energy in figures – statistical pocketbook 2012*. Brussels: European Union.

Fang, Z., Li, N., Li, B., Luo, G., & Huang, Y. (2014). The effect of building envelope insulation on cooling energy consumption in summer. *Energy and Building*, *77*, 197–205. doi:10.1016/j.enbuild.2014.03.030

Guimarães, E. T., Bragança, L., Almeida, M., & Mateus, R. (2015). Analysis of Portuguese Residential Buildings' Needs and Proposed Solutions, Connecting People and Ideas. *Proceedings of EURO ELECS 2015*.

Häkkinen, T. (2012). Systematic method for the sustainability analysis of refurbishment concepts of exterior walls. *Construction & Building Materials*, *37*, 783–790. doi:10.1016/j.conbuildmat.2012.07.084

ISO. (2007). *EN ISO10211 Thermal bridges in building construction - Heat flows and surface temperatures - Detailed calculations*. Geneva: ISO.

ISO. (2016). *EN ISO 18523-2:2016, Energy performance of buildings - Schedule and condition of building, zone and space usage for energy calculation - Part 2: Residential buildings*. Geneva: ISO.

ISO. (2017). *Building components and building elements - Thermal resistance and thermal transmittance - Calculation method*. International Organization for Standardization.

Kisilewicz, T. (2019). On the role of external walls in the reduction of energy demand and the mitigation of human thermal discomfort. *Sustainability (Switzerland)*, *11*(4), 1061. doi:10.3390u11041061

Kurekci, N. A. (2016). Determination of optimum insulation thickness for building walls by using heating and cooling degree-day values of all turkey's provincial centers. *Energy and Building*, *118*, 197–213. doi:10.1016/j.enbuild.2016.03.004

Nizovtsev, M. I., Belyi, V. T., & Sterlygov, A. N. (2014). The facade system with ventilated channels for thermal insulation of newly constructed and renovated buildings. *Energy and Building*, *75*, 60–69. doi:10.1016/j.enbuild.2014.02.003

Özel, G., Açikkalp, E., Görgün, B., Yamik, H., & Caner, N. (2015). Optimum insulation thickness determination using the environmental and life cycle cost analyses based entransy approach. *Sustainable Energy Technologies and Assessments*, *11*, 87–91. doi:10.1016/j.seta.2015.06.004

Pan, D., Chan, M., Deng, S., & Lin, Z. (2012). The effects of external wall insulation thickness on annual cooling and heating energy uses under different climates. *Applied Energy*, *97*, 313–318. doi:10.1016/j.apenergy.2011.12.009

Reinhart, C., Jakubiec, A., & Ibarra, D. (2013). Definition of a reference office for standardized evaluations of dynamic façade and lighting technologies. *Proceedings of the 13th Conference of International Building Performance Simulation Association*, 26-28.

Shariah, A., Tashtoush, B., & Rousan, A. (1997). Cooling and heating loads in residential buildings in jordan. *Energy and Building*, *26*(2), 137–143. doi:10.1016/S0378-7788(96)01027-4

Solgi, E., Hamedani, Z., Fernando, R., Skates, H., & Orji, N. E. (2018). A literature review of night ventilation strategies in buildings. *Energy and Building*, *173*, 337–352. doi:10.1016/j.enbuild.2018.05.052

Stewart, I. D., & Oke, T. R. (2012). Local climate zones for urban temperature studies. *Bulletin of the American Meteorological Society*, *93*(12), 1879–1900. doi:10.1175/BAMS-D-11-00019.1

Theodosiou, T., Tsikaloudaki, K., Tsoka, S., & Chastas, P. (2019). Thermal bridging problems on advanced cladding systems and smart building facades. *Journal of Cleaner Production*, *214*, 62–69. doi:10.1016/j.jclepro.2018.12.286

Theodosiou, T. G., Tsikaloudaki, A. G., Kontoleon, K. J., & Bikas, D. K. (2015). Thermal bridging analysis on cladding systems for building facades. *Energy and Building*, *109*, 377–384. doi:10.1016/j.enbuild.2015.10.037

Tsikaloudaki, K., Laskos, K., & Bikas, D. (2012). On the establishment of climatic zones in Europe with regard to the energy performance of buildings. *Energies*, *5*(1), 32–44. doi:10.3390/en5010032

Tsirigoti, D., & Tsikaloudaki, K. (2018). The effect of climate conditions on the relation between energy efficiency and urban form. *Energies*, *11*(3), 582. doi:10.3390/en11030582

Tsoka, S., Tsikaloudaki, K., Theodosiou, T., & Dugue, A. (2018). Rethinking user based innovation: Assessing public and professional perceptions of energy efficient building facades in Greece, Italy and Spain. *Energy Research & Social Science*, *38*, 165–177. doi:10.1016/j.erss.2018.02.009

Ucar, A., & Balo, F. (2010). Determination of the energy savings and the optimum insulation thickness in the four different insulated exterior walls. *Renewable Energy*, *35*(1), 88–94. doi:10.1016/j.renene.2009.07.009

Ünver, R., Öztürk, L., Adigüzel, S., & Çelik, O. (2003). Effect of the facade alternatives on the daylight illuminance in offices. *Energy and Building*, *35*(8), 737–746. doi:10.1016/S0378-7788(02)00227-X

Vanhoutteghem, L., Skarning, G. C. J., Hviid, C. A., & Svendsen, S. (2015). Impact of façade window design on energy, daylighting and thermal comfort in nearly zero-energy houses. *Energy and Building*, *102*, 149–156. doi:10.1016/j.enbuild.2015.05.018

ADDITIONAL READING

Ballarini, I., Corgnati, S. P., & Corrado, V. (2014). Use of reference buildings to assess the energy saving potentials of the residential building stock: The experience of TABULA project. *Energy Policy*, *68*, 273–284. doi:10.1016/j.enpol.2014.01.027

Brownstone, D., & Golob, T. F. (2009). The impact of residential density on vehicle usage and energy consumption. *Journal of Urban Economics*, *65*(1), 91–98. doi:10.1016/j.jue.2008.09.002

De Boeck, L., Verbeke, S., Audenaert, A., & De Mesmaeker, L. (2015). Improving the energy performance of residential buildings: A literature review. *Renewable & Sustainable Energy Reviews*, *52*, 960–975. doi:10.1016/j.rser.2015.07.037

Frontczak, M., & Wargocki, P. (2011). Literature survey on how different factors influence human comfort in indoor environments. *Building and Environment*, *46*(4), 922–937. doi:10.1016/j.buildenv.2010.10.021

Oral, G. K., Yener, A. K., & Bayazit, N. T. (2004). Building envelope design with the objective to ensure thermal, visual and acoustic comfort conditions. *Building and Environment*, *39*(3), 281–287. doi:10.1016/S0360-1323(03)00141-0

Poel, B., van Cruchten, G., & Balaras, C. A. (2007). Energy performance assessment of existing dwellings. *Energy and Building*, *39*(4), 393–403. doi:10.1016/j.enbuild.2006.08.008

Shameri, M. A., Alghoul, M. A., Sopian, K., Zain, M. F. M., & Elayeb, O. (2011). Perspectives of double skin façade systems in buildings and energy saving. *Renewable & Sustainable Energy Reviews*, *15*(3), 1468–1475. doi:10.1016/j.rser.2010.10.016

Sierra-Pérez, J., Boschmonart-Rives, J., & Gabarrell, X. (2016). Environmental assessment of façade-building systems and thermal insulation materials for different climatic conditions. *Journal of Cleaner Production*, *113*, 102–113. doi:10.1016/j.jclepro.2015.11.090

Theodosiou, T. G., Tsikaloudaki, A. G., Kontoleon, K. J., & Bikas, D. K. (2015). Thermal bridging analysis on cladding systems for building facades. *Energy and Building*, *109*, 377–384. doi:10.1016/j.enbuild.2015.10.037

Chapter 12
Estimate of Energy Performance Indicator of Existing Single-Family Houses in Bosnia and Herzegovina

Darija Gajić
https://orcid.org/0000-0003-1679-5111
Faculty of Architecture, Civil Engineering, and Geodesy, University of Banja Luka, Bosnia and Herzegovina

Erdin Salihović
Faculty of Architecture, University of Sarajevo, Bosnia and Herzegovina

Nermina Zagora
Faculty of Architecture, University of Sarajevo, Bosnia and Herzegovina

ABSTRACT

Yielding from an overall quantitative study of the residential sector of Bosnia and Herzegovina (B&H), this chapter concentrates on the ratio between single-family and collective housing, as well as on the urban-rural ratio of the single-family housing. Based on the data from the existing building stock (buildings built by 2014) and the statistical estimates, 23% of the buildings belong to the urban areas and 77% belong to the rural areas. The main goal was to study the correlation between the characteristics of the building envelope, the shape factor (A/V ratio) and the energy savings potential for the application of conventional measures of refurbishment of the building envelope of the single-family houses (type of buildings, which dominate in rural and urban areas). The chapter wraps up with recommendations for the adequate level of the energy performance indicator (energy need for heating) for the approved energy class for single-family houses located in the climate zone of the northern B&H.

DOI: 10.4018/978-1-5225-9932-6.ch012

INTRODUCTION

The key approach for climate change mitigation is the reduction of the level of energy consumption, that is, reduction of energy intensity by preserving the same or enabling greater comfort that such energy effect offers. This approach represents the foundation for all activities being undertaken in the field of energy efficiency increase. According to the key strategic documents in Bosnia and Herzegovina (B&H), at the level of entities, region and local communities, the largest amount of energy - cca 58.44% of the total energy in B&H is used in the residential sector (NEEAP, 2012). The aforementioned assessments were not extrapolated from energy efficiency indicators studies nor from the energy need for heating calculations (Gajić, 2014), they were rather based on statistical data, energy balance and energetics development plans, at the level of final energy and energy generating products (electric, heat, natural gas, oil and oil derivatives, coal and biomass). Scientific-research project "Typology of Residential Buildings in Bosnia and Herzegovina" (Aksić-Arnautović et al, 2016) delivers the key data[1] on the quantity, type and the manner of construction, as well as the energy need for heating of the existing residential buildings in urban and rural areas of B&H. The methodological framework of typology of residential buildings research of B&H was based on European international research project „TABULA" (*Typology Approach for Building Stock Energy Assessment*) harmonized with the directives 2002/91/EC and 2006/32/EC and co-financed from the European Commission IEE (*Intelligent Energy Europe*) program. Results of the scientific-research project highlighted the fact that there is a significantly greater proportion of existing individual/single-family housing (97.6%) compared to the existing building of collective/multi-family housing (2.4%). The term *single-family housing* (or individual housing) refers to a segment of the building stock, which includes both detached single-family houses and terraced houses. The term *collective housing* refers to a segment of the building stock, which includes multi-family houses, attached apartment building in urban blocks, apartment blocks and high-rise residential towers. The present article demonstrates the potential of the segment of single-family housing and their respective sub-typologies within the entire building stock in B&H, in terms of energy efficiency and their level of the energy performance indicator (energy need for heating).

BACKGROUND

In accordance with Directive 2012/27/EU and overall economic and energy capacity, all countries are required to define an appropriate set of minimum requirements for the energy performance of buildings based on parameters that determine the *energy efficiency indicator*. In addition, when it comes to final energy consumption, Bosnia and Herzegovina's building stock is the greatest consumer (58.44%) (NEEAP, 2012), of which 64% is used for heating and the remaining 36% for lighting, electrical appliances, cooking (of which 66% using electricity) and water heating (89% using electricity[2]) (Energy statistic of B&H, 2015). B&H is not the only country that uses most of its energy to heat buildings. The situation is similar in most European countries, including those with the highest and lowest energy intensity (Switzerland). Energy statistical data show that the energy used for heating accounts for 68% and 71% of end-use energy consumption in Europe and Switzerland, respectively, while lighting and electrical appliances account for 15%, water heating for 12% and cooking for 4% (Pampuri et al, 2017).

Finding the best combination of design strategies to tackle the problem of energy performance of buildings represents a true challenge. There is a debate between different authors as to whether it is better to design a super-insulated building with minimum heating requirements, or to provide less insulation but use a variety of renewable energy sources (Moran et al, 2017). Recent studies have shown that the cost of a poor indoor environment for the employer, building owner and society as a whole is often considerably higher than the cost of the energy used in the same building (Olesen, 2012). In B&H, in the past, buildings did not need to be energy-efficient, but only serve the functional needs of society/communities. This had an impact on the design of the country's old building stock, which ensured sufficient space for the intended use of buildings. It was only in 2016 that building energy efficiency rulebooks has adopted across B&H. The aforementioned legislation, which had been in effect in two entities of the B&H (the Federation of B&H and the Republic of Srpska) since 2010 and 2016, respectively, lays a great emphasis on improving the energy performance of building envelopes, by setting forth the energy certification of buildings based on their energy need for heating.

Doctoral thesis of the first author (Gajić, 2014) of this article, presents the analysis of energy savings and cost efficiency of measures applied to the energy optimization of the entire envelope of specified examples residential buildings, and at the same time it provides adequate proposals for measurement settings for the north climatic conditions of Bosnia and Herzegovina (maximum/limit of U-value [W/(m^2K)] - thermal transmittance for the elements of the building envelope and adequate limit of energy need for heating for residential building of collective housing in Rulebooks of the Entity of Republic of Srpska). The scientific-research project "Typology of Residential Buildings in Bosnia and Herzegovina" (2016) was realized by a B&H team of experts (among others, comprising of the authors of this article), stemmed with several important conclusions, including that the the limit value of the permitted energy class C in the rulebook on energy certification for the dominant typology of individual housing - single-family houses types- was not adequately set. According to EN 15217 Annex B, the energy performance (EP) indicator for the permitted energy class C is between the reference value $Q_{H,nd}$, obtained on the basis of the lowest energy requirements, defined by energy performance regulations (R_r) and the average of the reference value R_r and the reference value of the considered building type (R_s), i.e. $R_r \leq EP < 0.5\,(R_r+R_s)$.

The „TABULA" project, whose initiators come from the IWU (*Institut Wohnen und Umwelt GmbH*) from Darmstadt, defines a single framework of classification of typology of residential buildings in Europe, with a defined methodology of calculation of energy characteristics of buildings. Starting from the common TABULA concept the project partners developed national building typologies representing the residential building stock of their countries. The typologies consist of the following elements:

- a classification concept for the existing residential buildings according to age, size and further parameters,
- a set of example buildings which represent specific building types of the national stocks,
- typical energy consumption values for the example buildings,
- showcase calculations of the possible energy savings,
- statistical data for buildings and supply systems.

The "TABULA" project was implemented in the following countries: Austria, Bosnia and Herzegovina, Belgium, Bulgaria, Cyprus, Czech Republic, Germany, Denmark, Spain, France, Great Britain/England, Greece, Hungary, Ireland, Italy, The Netherlands, Norway, Poland, Serbia, Sweden and Slovenia.

Most of the European research teams, participants of the "TABULA", typology buildings perform on the basis of available statistical data, a team of experts from the University of Belgrade, for the specific needs of the research project "National Typology of Residential Buildings in Serbia "(Jovanović -Popović et al., 2013) conducted a dedicated inventory of residential buildings with statistical data processing.

Similarly as in the case of research in the residential typology of Serbia and due to the absence of data on the existing building stock, the team of experts in B&H decided to conduct the building surveys and statistical analysis, which served as a basis for a reliable classification of building according to their common general and specific characteristics.

RESIDENTIAL SECTOR IN BOSNIA AND HERZEGOVINA: QUANTITATIVE ANALYSIS

The first part of this paper discusses the quantitative ratio of individual and collective housing in the total building stock, and the relationship between the occurrence of the category of individual housing in urban and rural areas, and in particular underscores their quantitative presence by period of construction. In the period between June and September 2015, the statistics agency IPSOS Bosnia and Herzegovina processed, by computer-assisted personal interviewing (CAPI), 13,044 buildings of individual and collective housing in Bosnia and Herzegovina with an estimated sampling error of +/- 0.92%. The model of the Agency for Statistics of Bosnia and Herzegovina and preliminary results of 2013 *Census of Population, Households and Dwellings* were used with the objective of obtaining data based on the number of dwelling units per municipalities and optimal zone mapping, starting points and enumerators' movements as possible. According to the statistical survey by the total number of buildings, the dominant share in the total number of buildings is accounted for individual/single family housing typologies (97.63%). Quantitative overview of the number of dwelling units shows that the difference between the individual and collective housing is less pronounced (66.5% versus 33.5%), but still indicates that the individual housing typologies represent the dominant category, accounting for the two-thirds of the entire building stock (Figure 1).

Throughout the history, urban areas have undergone further development. Based on the data from the existing building stock (buildings built by 2014) and the statistical estimates, 23% of the buildings belong to the urban areas and 77% belong to the rural areas (Figure 2).

Figure 1. The relation of individual/single family housing and collective/multi-family housing by total number of buildings (left) and total number of dwelling units (right)
Source: Typology of Residential Buildings in B&H, 2016

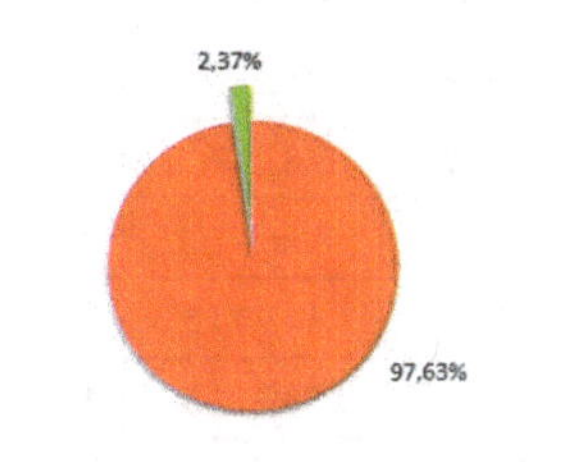

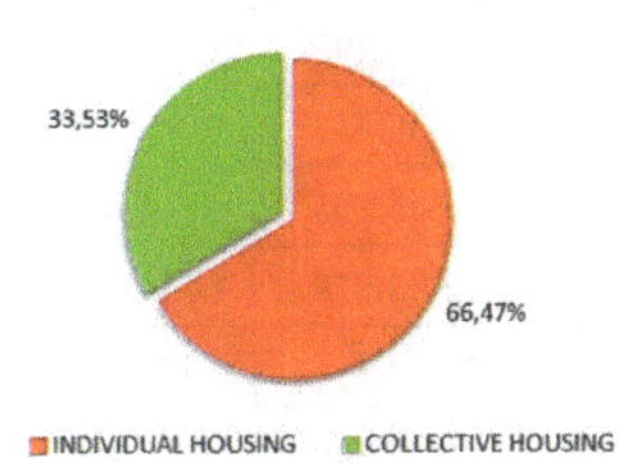

Figure 2. The correlation between the individual/single-family housing in urban and rural areas (left) and quantitative overview of individual/single-family housing typologies by the period of construction (right)
Source: Authors analyzed raw data from the project Typology of Residential Buildings in Bosnia and Herzegovina, 2014-2016

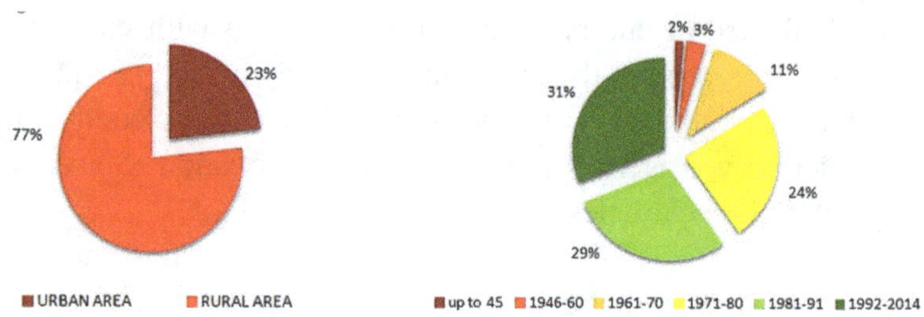

Development of building stock in urban areas was due to the requirements of regulations and legislation, which directly influenced on the energy performance of buildings. Most of the buildings of individual housing (60%) were constructed after 1981. (Figure 2, right). In urban areas, there are about 95% of the individual housing - single-family houses, as opposed to the 5% of the terraced houses. Most of the individual housing typologies were built within the period 1992-2014 (31%), but due to the length of the period, the most productive period was from 1981 to 1991. (29%, Figure 2, right and Table 1) It is necessary to point out the fact that 40% of the building stock of individual housing typologies (single-family houses and terraced houses) are older than 40 years, which would require the refurbishment of the envelope according to the practice in the European Union (Giebeler et al, 2009).

The quantitative analysis indicates that there is, in the entire residential building stock, a predominantly represented typology of individual housing (97.61%, by number of buildings, or 66.5% by number of dwelling units), and that 77% of existing buildings of individual housing are located in rural areas of Bosnia and Herzegovina, out of which 40% are over 40 years old.

Table 1. Gross surface of residential buildings per type in B&H (m^2)

	INDIVIDUAL HOUSING		**COLLECTIVE HOUSING**					
Construction period	**SFH**	**TH**	**MFH**	**AB1**	**AB2**	**H**	**TOTAL**	**TOTAL**
	Single-family house	Terraced house	Multy-family house	Attached apartment building	Apartment block	High-rise building	All types	
up to 1945	970.333	116.098	168.460	193.882		5.200	**1.453.973**	**0,89%**
1946-60	2.356.517	149.959	1.639.913	697.074	353.420	49.777	**5.246.660**	**3,22%**
1961 -70	9.132.251	888.085	3.184.346	1.192.969	3.122.201	485.596	**18.005.448**	**11,05%**
1971-80	26.704.411	1.248.793	3.587.596	1.101.982	8.758.838	489.698	**41.891.319**	**25,71%**
1981-91	38.282.654	1.035.435	1.048.335	56.310	3.280.277	76.832	**43.779.843**	**26,87%**
1992-2014	42.653.964	1.145.208	4.666.418	846.958	3.200.171	38.669	**52.551.387**	**32,25%**
TOTAL	**120.100.130**	**4.583.578**	**14.295.068**	**4.089.175**	**18.714.906**	**1.145.772**	**162.928.630**	**100,00%**
TOTAL	**73,71%**	**2,81%**	**8,77%**	**2,51%**	**11,49%**	**0,70%**	**100,00%**	

Source: Typology of Residential Buildings in B&H, 2016

ASSESSMENT OF ENERGY EFFICIENCY CHARACTERISTICS OF THE EXISTING SINGLE-FAMILY HOUSES IN BOSNIA AND HERZEGOVINA

This following section of the article shows key parameters for determining the energy indicator of building in B&H (energy need for heating), the energy performance of envelope and compactness (shape factor or A/V ratio) of existing buildings, and the potential of the building stock of individual housing in saving the energy needed for heating.

Compactness Ratio and the Building Envelope of Residential Building

According to the standard EN 15217 parameters affecting determination of the energy performance indicator (EP) for a permitted energy class, are: climate, building type and function, energy carrier, shape factor of the building, ventilation - air change rate and illumination level. To highlight the normative discrepancy between the local and European standards, it is important to note the standard EN 15217 differentiates the shape factor from the compactness ratio (Figure 3), while in the Rulebook in B&H and in the Rulebook in neighboring countries (Croatia and Serbia), the variable for assessing compactness is the building shape factor.

In rulebooks in B&H the shape factor is the ratio of thermal envelope area and the conditioned volume like compactness ratio in the standard EN 15217.

Values of shape factor have been analyzed for all types of residential buildings in order to determine most represented value in B&H, or in other words, to obtain the average shape factor for the existing houses and buildings. When it comes to designing new residential buildings, the shape factor value setup should allow freedom in designing compactness, while at the same time they should comply with the energy-efficient compactness design. The typology of existing residential houses and buildings, in general, indicates the tradition of housing construction on the territory of B&H.

Figure 3. Definition of the building shape factors and compactness ratio
Source: Standard EN 15217

6.3.2 Impact of building shape

The building shape is characterised by the building shape factor:

$$f = A_E / A_C$$

or the compactness ratio

$$c = A_E / V_C$$

where

A_E is the thermal envelope area, in m^2;

A_C is the conditioned area, in m^2;

V_C is the conditioned volume, in m^3.

Adequate shape factor for existing houses and buildings are typically determined by particular factors depending on the construction periods; however, since the rulebooks guidelines and standards should be applied to the representative building types, due to the setting of reference energy classes, leading towards the defining of the representative or the average shape factors (Table 2).

It important to note that the definition of the representative average shape factors can be partially affected by the building renovation. In the case of new residential buildings, representative shape factor should correspond the lowest recorded value shape factor, as indicated in the relevant studies.

Architectural design parameters which affect the heat loss comprise of the selection of materials which compose the building envelope as well as the composition and the ratio between the transparent and opaque surfaces, affecting the U-value - quality (α-coefficient) and quantity (thickness) of insulating materials – U_{wall}, U_{floor}, U_{roof}, the quality of windows-U_w. Average U-values of the building envelope of the representative residential buildings categorized by types and periods of construction, were estimated by the authors, as members of the B&H expert team (Table 3).

Energy Need for Heating of Residential Buildings

Calculation of the annual energy need for heating of the building, i.e. heat losses and gains, is influenced by various parameters that originate from all planning and design phases of residential buildings. In the case of new buildings, these parameters are encompassed in the project documentation; while in the existing buildings, due to sustainability and maintenance issues and due to alterations that occur during the lifetime of the building, these parameters may be registered and assessed in situ by the auditors, determining the field, and the standards specified in the regulations.

Energy efficiency indicators (energy need for heating of the building) are generally influenced by the parameters derived from urbanism and architecture, and more specifically, by the parameters that affect the heat losses and heat gain in the building, as specified by the calculation process according to EN 13790.

Table 2. Review of the representative shape factor for residential buildings in B&H

	INDIVIDUAL HOUSING		COLLECTIVE HOUSING			
Construction period	**SFH**	**TH**	**MFH**	**AB1**	**AB2**	**H**
	Single-family house	Terraced house	Multy-family house	Attached apartment building	Apartment block	High-rise building
up to 1945	0.91	0.71	0.59	0.52		
1946-60	1.49	1.36	0.69	0.45	0.61	
1961 -70	1.36	0.69	0.71	0.50	0.48	0.41
1971-80	1.04	0.91	0.53	0.50	0.45	0.36
1981-91	0.83	0.95	0.63	0.47	0.57	
1992-2014	0.92		0.55	0.59	0.40	
Average values for the representative sample	**0.85**	**0.80**	**0.62**	**0.50**	**0.46**	**0.38**
The lowest shape factor	0.70	0.50	0.40	0.40	0.40	0.36

Source: Analysis of data collected for the purpose of the research project Typology of Residential Buildings in B&H, 2016

Table 3. Review of the representative energy performance – U value of elements of the building envelope of residential buildings in B&H

TYPES	PERIODS	External wall	External wall	Partition wall to unheated staircase	Wall to the adjecent building	Windows	Ground floor or ceiling under unheated basement	Flat roof or ceiling to unheated attic
building category	construction period	W/m²K	W/m²K	W/m²K	W/m²K	W/m²K	W/m²K	W/m²K
SH	up to 1945	1.1	1.48			3.50	2.14	2.52
SH	1946-1960	1.7				3.09	4.74	2.52
SH	1961-1970	1.7				3.00	2.45	2.52
SH	1971-1980	1.64	3.07			3.00	3.63	1.75
SH	1981-1991	0.5	0.64			2.93	0.43	0.34
SH	1992-2014	0.52	0.65			2.85	4.74	0.34
TH	do 1945	0.93	1.58		1.56	5.20		0.55
TH	1946-1960	1.87			1.53	3.50		1.3
TH	1961-1970	1.58	2.86		0.89	2.90	2.89	1.82
TH	1971-1980	1.29	3.34		1.29	2.91	2.59	0.97
TH	1981-1991	1.3	1.28		1.68	3.00	0.86	2.36
MH	do 1945	1.05		0.96		3.50	1.38	1.22
MH	1946-1960	1.04		0.96		5.27	1.65	2.21
MH	1961-1970	1.88		2.34		4.50	1.30	0.68
MH	1971-1980	1.04	2.54	1.27		2.92	0.75	0.91
MH	1981-1990	1.45	3.17	1.27		2.94	0.55	0.87
MH	1992-2014	0.66		0.9		2.20	0.59	0.51
AB1	do 1945	1.3	1.28		1.68	3.00	0.97	0.49
AB1	1946-1960	1.5		1.28	1.72	3.50	1.78	0.41
AB1	1961-1970	1.79		1.62	1.68	3.50	1.64	6.5
AB1	1981-1991	0.32		0.53	0.74	3.50	2.39	6.67
AB1	1992-2014	3.47		0.36	0.35	2.20	0.89	0.62
AB2	1946-1960	1.37		1.24		3.60	1.63	0.39
AB2	1961-1970	2.31	1.81	1.76		4.50	1.11	1.06
AB2	1971-1980	1.13		0.88	0.88	3.00	1.46	0.59
AB2	1981-1991	0.49	1.83	2.75		3.00	0.51	0.51
AB2	1991-2014	0.55		0.62	0.34	2.59	0.56	0.36
H	1961-1970	1.17	3.86	2.45		3.09	0.87	1.53
H	1971-1980	0.94	1.26	2.41		3.14	1.17	0.59

Source: Analysis of data collected for the purpose of the research project Typology of Residential Buildings in B&H, 2016

Architectural parameters which impact energy efficiency indicators:

- Building shape factor ($f_o = A / V$), orientation of the building, thermal zoning, presence and number of users in the building (impact of building design, dispositions and purposes of premises in the building);
- Selection of materials which compose the building envelope as well as the composition and the ratio between the transparent and opaque surfaces, affecting the U-value - quality (α-coefficient) and quantity (thickness) of insulating materials – U_{wall}, U_{floor}, U_{roof}, the quality of windows-U_w;
- Thermal bridges ΔU_{TM}, infiltration (n) - construction details of the building envelope.
- *Factors of urban planning* additionally influence energy efficiency indicators, emphasizing the relevance of the building location and thus differentiating seemingly alike buildings, by their common architectural parameters:
- The microclimate / environmental conditions (intensity of solar radiation, duration of sunshine, temperature and humidity of the air, speed and direction of the wind);
- Topography of the terrain (flat or inclined terrain);
- Urban context (adjacent buildings, the distance of the neighboring building, trees etc.).
- Calculation of the thermal losses of the building (transmission and ventilation losses), takes into account the following factors:
- Thermal zoning of the building - depending on the purpose of the building, the internal design temperature, the type and the regime of the heating system, the building will be monitored fully or will be divided into heat zones requiring separate calculations;
- Building shape factor ($f_o = A / V$) - the surface of the building envelope to the volume of the room to be heated ratio, or, the thermal envelope area to the conditioned volume ratio;
- The size and distribution of the non-transparent surfaces within the building envelope, in relation to the outer gross dimensions, depending of the U-value and the position of the element in relation to the outside air, the heated and unheated space or in contact with the soil;
- The size and distribution of the transparent surfaces within the building envelope, depending of the U-value and the g-coefficient (for thermal gain) and according to the orientation (cardinal directions N, E, S and W);
- Thermal bridges in the building envelope (ΔU_{TM}) - linear heat losses;
- Infiltration of air - number of air changes per hour (n);
- The thermal capacity of the building (C_m) - depends on the type of construction and exterior finishes the building envelope - detailed calculation, through a catalog of thermal bridges and thermal characteristics of the material or approximate surfaces surrounding the space and influences the factor of exploitation of heat gain for the heating period (η).
- Parameters that affect the thermal gain of the building (internal gains from users, lighting and electrical appliances and the gain of solar radiation through transparent surfaces of the envelope) are:
    - Metabolic heat caused by the presence of users (occupancy) in relation to the useful heated area of the building:
        - Heat input from lighting and electrical appliances to the heated area of the building;
        - Openings / windows, involve the consideration of the frame to the glazed surface ratio, as well as the quality, the dirty/contamination and the cover/shelter of the glazing unit

- the g-coefficient and F_{sh}-factor of shading in relation to the amount of solar radiation according to this orientation.

- Values of the energy performance of the envelope and compactness were derived from the representative samples of residential buildings, the so-called *typical buildings*, which are characterized by their average physical features indicated by the survey. Assessment of the existing conditions of the envelope was based on the technical requirements and technological possibilities of the relevant period and the experts' evaluation of the degradation of building envelope through time periods (Table 4).

Energy need for heating of the typical buildings is the basis for the determination of total energy need for heating for residential buildings in the area of Bosnia and Herzegovina (Table 5). Comparison between the Table 1 and Table 5 shows that energy required for heating of single-family houses is 85.22% of the energy required for heating of the entire housing stock, while the their share accounts for 73.71% in the gross surface of the entire housing stock. Single-family houses, due to their shape factor, represent the category of energy inefficient residential buildings.

Data indicating the total annual energy need for heating in residential buildings in Bosnia and Herzegovina was obtained from the statistical survey on the total number of buildings, correlated with the calculated energy need for heating in actual typical buildings (Table 5).

Calculated values of energy consumption of the typical buildings were projected on the level of the entire category with the introduction of the coefficient of the ratio between the average gross surface of a relevant category and gross surface of the typical building, and the coefficient of the ratio between net and gross surface of a typical building (Table 5). Individual housing typologies built in the period of 1971-80 are characterized by the highest values of the energy need for heating, requiring cca 37.74% of the total energy needed for heating of the entire residential building stock, wherein the single-family houses particularly stand out, with a share of 32.14% of the total energy needs for all residential buildings.

Table 4. Energy need for heating of residential buildings in B&H (kWh/m2a)

	INDIVIDUAL HOUSING		**COLLECTIVE HOUSING**			
Construction period	**SFH**	**TH**	**MFH**	**AB1**	**AB2**	**H**
	Single-family house	Terraced house	Multi-family house	Attached apartment building	Apartment block	High-rise building
up to 1945	452.34	183.16	230.73	176.08		
1946-60	473.96	321.27	216.19	158.75	176.71	
1961 -70	464.90	196.42	188.44	153.06	170.10	193.37
1971-80	381.59	199.04	146.79		129.85	125.64
1981-91	135.93	219.20	189.20	93.62	110.87	
1992-2014	127.61		65.22	68.06	54.81	

Source: Analysis of data collected for the purpose of the research project Typology of Residential Buildings in B&H, 2016

Table 5. Total energy need for heating of residential buildings in B&H (MWh/a)

	INDIVIDUAL HOUSING		COLLECTIVE HOUSING					
Construction period	SFH	TH	MFH	AB1	AB2	H	TOTAL	TOTAL
	Single-family house	Terraced house	Multi-family house	Attached apartment building	Apartment block	High-rise building	All types	
up to 1945	244.439	8.433	17.488	20.151			290.512	1,48%
1946-60	719.865	28.327	195.151	58.365	24.688		1.026.397	5,24%
1961 -70	2.752.871	87.198	327.081	80.437	284.792	58.326	3.590.706	18,32%
1971-80	6.350.897	156.250	189.255		662.681	34.574	7.393.657	37,74%
1981-91	3.101.309	89.498	115.571	3.022	197.170		3.506.571	17,90%
1992-2014	3.528.879		116.191	35.918	105.050		3.786.038	19,32%
TOTAL	16.698.261	369.706	960.738	197.893	1.274.382	92.900	19.593.880	100,00%
TOTAL	85,22%	1,89%	4,90%	1,01%	6,50%	0,48%	100,00%	

Source: Typology of Residential Buildings in B&H, 2016

Energy Saving Potential of the Refurbishment Building Envelope and Energy Performance Indicator of the Existing Single-Family Houses

The third segment of this paper indicates the possibilities of reduction of energy need for heating after applying standard measures of refurbishment of building envelope, keeping the same compactness of the building.

Standard improvement measures correspond with the typical building refurbishment procedures in B&H (improvement of thermal characteristics of walls and ceilings), as well as possible replacement of the existing windows with new ones, with better characteristics (defined minimal values of heat transfer coefficient) - Table 6.

Comparison of Table 5 and Table 7 indicates that energy need of 19,593,880 MWh (70.53 PJ) could be reduced to 8,771,954 MWh (31.58 PJ) if the set of standard improvement measures is implemented. Research results demonstrate that the single-family houses, even after implementation of the standard improvement measures, remain to be the dominant category in the overall reduced consumption of en-

Table 6. Energy Efficiency Improvement Measures

CONSTRUCTION TYPE	IMPROVEMENT
Exterior wall	Thermal insulation thickness: 10 cm (λ=0,041W/mK)
Ceiling toward non-heated attic	Thermal insulation thickness: 10 cm (λ=0,041W/mK)
Ceiling toward non-heated basement	Thermal insulation thickness: 10 cm (λ=0,041W/mK)
Flat roof	Thermal insulation thickness: 20 cm (λ=0,041W/mK)
Sloped roof	Thermal insulation thickness: 20 cm (λ=0,041W/mK)
Windows	Windows with double insulation glass (U=1,6 W/m^{2}K)

Source: Typology of Residential Buildings in B&H, 2016

Table 7. Energy need for heating of residential buildings after implementation of standard measures in Bosnia and Herzegovina (MWh/a)

	INDIVIDUAL HOUSING		COLLECTIVE HOUSING					
CONSTRUCTION PERIOD	SFH	TH	MFH	AB1	AB2	H	TOTAL	TOTAL
	Single-family house	Terraced house	Multy-family house	Attached apartment building	Apartment block	High-rise building	All types	
up to 1945	80.221	4.142	6.396	6.433			97.191	**1,11%**
1946-60	261.376	12.203	60.137	15.687	7.130		356.533	**4,06%**
1961 -70	1.017.597	35.182	117.788	23.982	88.851	14.632	1.298.032	**14,80%**
1971-80	2.232.026	85.606	68.938		263.950	10.927	2.661.448	**30,34%**
1981-91	1.658.914	33.184	35.099	1.003	90.838		1.819.038	**20,74%**
1992-2014	2.379.318		76.753	20.680	62.961		2.539.712	**28,95%**
TOTAL	**7.629.451**	**170.318**	**365.111**	**67.785**	**513.730**	**25.560**	**8.771.954**	**100,00%**
TOTAL	**86,98%**	**1,94%**	**4,16%**	**0,77%**	**5,86%**	**0,29%**	**100,00%**	

Source: Typology of Residential Buildings in B&H, 2016

ergy, accounting for 86,98%, hence this category has the potential for highest absolute savings, while the relative savings amount for 54%. The study has shown that the individual single-family houses built in the period from 1981 to 2014 (490,874 buildings with a gross floor area of 80,936,618 m^2) have the energy need for heating of 6,630,188 MWh, while those built in the period 1971-1980 (statistically estimated number of 194,076 buildings, with a gross floor area of 26,704,411 m^2) are associated with energy need to heating of 6,350,897 MWh.

The smallest relative savings are achievable in case of single-family houses built in period 1992-2014 (32.43%), and in multi-family houses/buildings from period 1992-2014 (33.77%), due to better thermal performance of envelope elements of buildings built in the specified period. The presented comparative analyses indicate that the greatest potential for savings of energy needed for heating, after the application of the standard improvement measures of energy performance of envelope and heating systems, can be found in the single-family houses, and more specifically, those which built in the period 1971-1980 especially stand out as a priority for refurbishment/reconstruction.

Table 8 shows the shape factor (A/V ratio) of representative samples of the individual housing typologies in Bosnia and Herzegovina, which is highly expressed in the existing buildings from all periods, while for the buildings built up to 1981 the values are particularly high. The A/V ratio indicates the smaller net areas of the heated residential space or that the surface of the envelope enclosing the heated space is larger than the volume of the space.

A/V ratio analysis of the individual housing typologies in Bosnia and Herzegovina has yielded the following conclusions:

- The heated space is typically compact in the category of single-family houses built until 1980 (area of floor plan is below 72 m^2, which is characteristic for one-story buildings of dimensions less than 9 m x 8 m);

Table 8. Energy need for heating in kWh/m²a and shape factor of representative residential buildings (SFH and TH) in Bosnia and Herzegovina before and after implementation of standard measures

TYPES	PERIODS	Net area of the heated spaces	Heated space volume	Shape factor- f_0= A/V ratio	Trans-mission losses Ht (before)	Ventilation losses Hv (before)	Energy need for heating Qh,nd,interm (before)	Trans-mission losses Ht (after)	Venti-lation losses Hv (after)	Energy need for heating Qh,nd,interm (after)
Building category	Construction period	m²	m³	-	W/K	W/K	kWh/m²a	W/K	W/K	kWh/m²a
SFH Single-family house	up to 1945	68.4	213.4	0.91	541.42	78.65	452.34	187.93	35.75	148.45
	1946-60	43.2	98.2	1.49	318.08	35.85	473.96	120.64	16.30	172.09
	1961 -70	55.6	139.1	1.36	396.65	41.95	464.90	148.14	23.30	171.85
	1971-80	67.8	149.2	1.04	405.92	44.99	381.59	145.79	25.00	134.11
	1981-91	101.4	250.2	0.83	229.25	50.28	**135.93**	123.27	41.90	**72.71**
	1992-2014	121.1	298.7	0.92	238.13	50.04	127.61	155.04	50.04	86.04
TH Terraced house	up to 1945	135.9	407.6	0.71	427.11	122.90	183.16	232.23	68.28	89.97
	1946-60	36.4	93.2	1.36	209.74	27.85	321.27	94.29	15.47	138.40
	1961 -70	118.8	286.1	0.69	372.13	72.77	196.42	155.77	47.88	79.25
	1971-80	131.2	328.4	0.91	394.35	83.61	199.04	223.52	55.01	109.05
	1981-91	81.1	205.1	0.95	308.19	41.22	219.20	117.50	34.35	81.27

Source: Analysis of data collected for the purpose of the research project Typology of Residential Buildings in B&H, 2016

- Most of the single-family built after 1980 are two-story buildings with, partially heated on both levels (inside buildings there are unheated spaces - garages, storages);
- Terraced houses have a lower shape factor than the single-family houses, because they share two exterior/facade walls with other heated spaces of other terraced houses;
- Most of the terraced houses were built after 1960 and they do not have a compact heated space (as they include the non-heated spaces – garage, storage, attic).

Table 7 shows that the in the case of application of standard improvement of the energy characteristics of the envelope, wherein the compactness, or the shape factor of the heated space will remain unchanged, the transmission and ventilation losses are reduced, as well as the energy needed for heating by 1.5 to 3 times. By introducing the standard improvement measures for envelopes of the existing residential buildings, it is possible to reduce the total annual energy needed for heating by 55.23%.

According to the Rulebook on energy audit and energy certification of Republic of Srpska entity the energy class of a building is calculated relative to the limit value of the permitted energy class C. Currently, in the Rulebook permitted energy class C for residential building of individual housing is $Q_{H,nd} \leq$ 65 kWh/m²a, while permitted energy class C for residential building of collective housing is $Q_{H,nd} \leq 60$ kWh/m²a. According to EN 15217 Annex B, the energy performance (EP) indicator for the permitted energy class C is between the reference value $Q_{H,nd}$ obtained on the basis of the lowest energy requirements defined by energy performance regulations (R_r) and the average of the reference value R_r and the reference value of the considered building type (R_s), i.e. $R_r \leq EP < 0.5\ (R_r+R_s)$.

If the design value of the internal temperature in single-family house were 20°C and the reference value of the considered building type with average f_0=0.83 (R_s) 135.93 kWh•m^{-2}a^{-1}, the lowest energy requirement as defined by the energy performance regulations (R_r) would be 72.71 kWh•m^{-2}a^{-1}, resulting in an EP indicator for the permitted energy class C for existing single-family houses ranging 70.00 kWh•m^{-2}a^{-1} $\leq$ EP < 140.00 kWh•m^{-2}a^{-1}.

As the valid legislative in B&H did not affect certification of buildings up to 200 m^2, because involved Directive 2002/91/EC and Directive 2006/32/EC, the incorrect threshold of the allowed energy class C did not stand out. Following the initiative of investors who demanded the energy certificates, together with energy auditors, irregularities in the definition of energy classes for this type of building were identified. The introduction of the requirements from the Directive 2010/31/EU and Directive 2012/27/EU into the legislative framework moves the boundary of certification of buildings and certification covering buildings up to 50 m^2. In this case, the typologies of individual housing, single-family houses will fall under the energy certification.

CONCLUSION

Statistical surveys and quantitative studies indicate that the population in Bosnia and Herzegovina dominantly prefers single-family houses, which is evident in the structure of the existing residential building stock recorded in 2014, in which the buildings of individual housing account for 76% of the total gross floor area.

Survey of residential buildings in Bosnia and Herzegovina has shown that 23% of the buildings of individual housing are located in urban areas, while the prevailing share of 77% belongs to the rural areas. Furthermore, the energy need for heating of buildings of individual housing in the rural area of Bosnia and Herzegovina amounts to 12,480,471 MWh, accounting for 63.7% of the total energy need for heating of residential buildings.

The fact is that individual housing is the only housing type in B&H rural areas and future research should analyze questions related to energy production from renewable sources that could be found in the B&H rural settlements.

Shape factor (A/V ratio) of typical buildings of the individual housing from all of the periods of construction in Bosnia and Herzegovina is highly expressed, while in case of buildings built after 1981 the A/V ratio is particularly high. Further considerations lead to the conclusion and recommendation of the advantages of the design of terraced houses and semi-detached houses because of the lower A/V ratio (expected the lower energy need for heating).

The results show that, in terms of refurbishment of building stock, the priority buildings are those which originate from the period 1971-80, because the renewal of their envelope can perform highest impact on the greatest savings of energy need for heating.

The results of the research can be directly used to improve the Rulebook on energy audit and energy certification in subject area in Bosnia and Herzegovina. The research presented in this article aims to contribute to the methodological enhancements of the established policies and rulebooks guidelines in the domain of energy efficiency of the residential building sector.

REFERENCES

Agency for Statistics of Bosnia and Herzegovina. (2015). *Energy statistic*. Retrieved from http://www.bhas.ba/saopstenja/2015/APED_2015_001_01_BA.pdf

Arnautović-Aksić, D., Burazor, M., Delalić, N., Gajić, D., Gvero, P., Kadrić, D., … Zagora, N. (2016). Tipologija stambenih zgrada Bosne i Hercegovine/Typology of Residential buildings in Bosnia and Herzegovina. Sarajevo, B&H: Faculty of Architecture, University of Sarajevo.

CEN – European Committee for Standardization. (2007). *EN 15217, Energy performance of buildings — Methods for expressing energy performance and for energy certification of buildings*. Author.

EPISCOPE. (2012). *Typology Approach for Building Stock Energy Assessment*. Retrieved September 20, 2015, from http://episcope.eu/iee.project/tabula

EUR-lex. (2002). *Directive 2002/91/EC of the European Parliament and of the Council of 16 December 2002 on the energy performance of buildings*. Retrieved from https://eur-lex.europa.eu/legal-content/EN/TXT/?uri=CELEX%3A32002L0091

EUR-lex. (2006). *Directive 2006/32/EC of the European Parliament and of the Council of 5 April 2006 on energy end-use efficiency and energy services and repealing Council Directive 93/76/EEC*. Retrieved from https://eur-lex.europa.eu/legal-content/EN/TXT/?uri=celex%3A32006L0032

EUR-lex. (2010). *Directive 2010/31/EU of the European Parliament and of the Council of 19 May 2010 on the energy performance of buildings*. Retrieved from https://eur-lex.europa.eu/legal-content/EN/TXT/?uri=CELEX%3A32010L0031

EUR-lex. (2012). *Directive 2012/27/EU of the European Parliament and of the Council of 25 October 2012 on energy efficiency, amending Directives 2009/125/EC and 2010/30/EU and repealing Directives 2004/8/EC and 2006/32/EC*. Retrieved from https://eur-lex.europa.eu/eli/dir/2012/27/oj

Federal Ministry of Physical Planning. (2009). *Rulebook on Technical Requirements for Thermal Protection of Buildings and Rational Use of Energy*. Official Gazette Federation B&H, no. 49/09.

Gajić, D. (2014). *Energy optimisation of the building envelope on the representative samples of the old residential buildings in Banja Luk. (Doctoral dissertation). University of Belgrade, Faculty of Architecture*. Retrieved from http://nardus.mpn.gov.rs/handle/123456789/6493

Giebeler, G., Fisch, R., Krause, H., Musso, F., Petzinka, K.-H., & Rudolphi, A. (2009). *Refurbishment Manual*. Birkhäuser. doi:10.11129/detail.9783034614337

Hegger, M., Fuchs, M., Stark, T., & Zeumer, M. (2008). Energy Manual: sustainable architecture. Basel, CH: Birkhäuser.

International Energy Agency. (2014). *Energy balances of non-OECD Countries 2014 Edition*. Retrieved from https://www.iea.org/statistics/relateddatabases/energybalancesofnon-oecdcountries

ISO – International Organization for Standardization. (2008). Thermal performance of buildings-Calculation of energy use for heating. *EN ISO*, *13790*, 2008.

Jovanović-Popović, M., Ignjatović, D., Radivojević, A., Rajčić, A., Đukanović, L., Ćuković Ignjatović, N., & Nedić, M. (2013). Nacionalna tipologija stambenih zgrada Srbije. Belgrade, Serbia: Faculty of Architecture, University of Belgrade.

Ministry of Construction, Transport and Infrastructure. (2018). *Rulebook on Energy Efficiency of Buildings and Rulebook on the conditions, content and method of issuing certificates on the energy performances of buildings*. Belgrade, Serbia: Official Gazette no. 69/2012 and 48/2018. Retrieved June 10, 2013, from http://www.ingkomora.org.rs/strucniispiti

Ministry of Construction and Physical Planning. (2015). *Technical regulation on the rational use of energy and thermal protection in buildings*. Zagreb, Croatia: Official Gazette HR no. 128/15, Appendix B, Table 8.

Ministry of Spatial Planning, Construction and Ecology. (2015). *Rulebook on minimum requirements for energy performance of buildings*. Banja Luka, B&H: Official Gazette RS no. 30/15.

Ministry of Spatial Planning, Construction and Ecology. (2015). *Rulebook on the methodology for calculating energy performances of buildings*. Banja Luka, B&H: Official Gazette RS no. 30/15.

Ministry of Spatial Planning, Construction and Ecology. (2015). *Rulebook on energy audits and issuing energy certificate of buildings*. Banja Luka, B&H: Official Gazette RS no. 30/15.

MOFTER. (2012). *Bosnia and Herzegovina First National Energy Efficiency Action Plan 2010-2018 final draft*. Sarajevo, B&H: Ministry of Foreign Trade and Economic Relation (MOFTER).

Moran, P., Goggins, J., & Hajdukiewicz, M. (2017). Super-insulate or use renewable technology? Life cycle cost, energy and global warming potential analysis of nearly zero energy buildings (NZEB) in a temperate oceanic climate. *Energy and Building, 139*, 590–607. doi:10.1016/j.enbuild.2017.01.029

Olesen, B. W. (2012). Revision of EN 15251: indoor environmental criteria. *REHVA Journal, 49*(4), 6–12. Retrieved from https://www.rehva.eu/fileadmin/hvac-dictio/04-2012/revision-of-en-15251_rj1204.pdf

Pampuri, L., Cereghetti, N., Galbani Bianchi, P., & Caputo, P. (2017). Evaluation of the space heating need in residential buildings atterritorial scale: The case of Canton Ticino (CH). *Energy and Building, 148*, 218–227. doi:10.1016/j.enbuild.2017.04.061

KEY TERMS AND DEFINITIONS

A/V Ratio: The ratio of compactness of heating zone in the building. The ratio of thermal envelope area and the conditioned volume of building.

Building Envelope: The physical separator between the conditioned and unconditioned environment of a building including the resistance to air, water, heat, light, and noise transfer.

Compactness Ratio: The ratio of thermal envelope area and the conditioned volume of building.

Energy Characteristics: Parameters for determining the energy efficiency indicators of the building, which affect the thermal losses and thermal gain in the building.

Energy Efficiency Rulebooks: Set of rules on energy efficiency in buildings. Regulate the thermal protection of the building envelope or energy characteristic of building envelope, rational energy consumption, energy audits and energy certification of buildings.

Energy Need for Heating: The energy required for thermal loss of the building (transmission and ventilation losses through the building envelope), taking into account the heat gain from solar radiation, users and appliance.

Final Energy: The total energy consumed by end users and that energy consumption is paid.

Shape Factor: Building shape factor is the variable for assessing compactness of building.

Single-Family Houses: A stand-alone house (also called a single-detached dwelling, detached residence or detached house) is a free-standing residential building.

ENDNOTES

1 The named research project included all the quantitative aspects of recording and assessing residential sector from the EE perspective, except for the leading indicator of energy efficiency.

2 Electricity is mostly being produced using the non-renewable energy sources.

Chapter 13
(R)urban Synergy vs. Climate Change:
The Impact of ICT Networks on the Process of Adaptation and Mitigation

Aleksandra Stupar
Faculty of Architecture, University of Belgrade, Serbia

Vladimir Mihajlov
Faculty of Architecture, University of Belgrade, Serbia

Ivan Simić
Faculty of Architecture, University of Belgrade, Serbia

ABSTRACT

The chapter provides an insight into the relationship between synergic (r)urban systems, information networks and climate change, discussing the emerging ideas, and concepts related to the increasing use of information networks in the process of climate adaptation and mitigation. Emphasizing the strategic role of both digital and material information flows, supported by ICT tools, the chapter focuses on two main domains of data exchange and knowledge transfer: the public communication of climate change and the connectivity and interaction within (r)urban hybrid systems. Underlining the issues of effectiveness, accessibility, and low-carbon outcomes of synergic (r)urban reactions to climate shift, recent environmental and technological trends are considered in accordance with the preferred spatio-functional flexibility of emerging (r)urban hybrid settlements. The chapter also identifies and analyses three areas of ICT applicability, targeting the role of information networks in the anticipated climate-friendly development: human behavior, ecological awareness, and general efficiency.

DOI: 10.4018/978-1-5225-9932-6.ch013

INTRODUCTION

The beginning of the 21st century has increased the global awareness of climate change, emphasizing a necessity of reducing GHG emissions (IPCC, 2007c, 2013). Adaptation to climate change, as the ability of environment to support change without creating hazardous situations (Giddens, 2009), has become an important ideological issue between the neoliberal and green perception of future development. The consequences of global warming have triggered a hyper-production of studies, discussions and theories, focused on a growing number of problems and challenges related to nature, mechanisms, causes and effects of this process. Although the emphasis has mostly been on mitigation, the importance of adaptation has been highlighted after the adoption of the Cancun Adaptation Framework (2010) which supports the development of national adaptation plans and strategies. The improvements, although limited, have been mostly noticed on the level of urban settlements (Stupar and Mihajlov, 2016), stimulating changes toward carbon-neutral environment, energy efficiency, green modes of living, sustainable mobility and urban systems (European Commission 2013a, 2013b).

Considering the contemporary urbo-centric comprehension of global(ized) activities, both the role and the character of rural areas have been significantly redefined. The decreasing share of agriculture in their overall economy has instigated their transformation, altering their traditional characteristics. Stressing the role of green urbanism and cohesion (via social and technical networks), the efficient relationship between urban settlements and their rural 'backup' is identified as a base for higher overall sustainability (Beatley, 2000; Pflieger and Rozenblat, 2010; Simić i Bajcetic, 2016; Wood, 2007). Simultaneously, the characteristics of (r)urban hybrid systems, such as diversity, sustainable transportation, density, land use, passive solar design, greening and compact form (Jabareen, 2006), represent important factors in achieving the necessary dynamic of urban-rural relationship.

The former dichotomy between urban and rural was based on differences in functional and physical characteristics of these two types of settlements. Rural settlements were defined as predominantly agricultural, both on the level of activities and the level of land use. The relationship between urban and rural settlements was based on a simple reciprocal exchange in which the rural counterpart was highly dependent on industrial and commercial products from cities. However, the rural settlements in Western Europe have been gradually developed and transformed, creating a complex and dynamic network of interdependencies and providing equal economic opportunities for their population. Agriculture is not their main function, since nowadays they include other activities, once inherent only to cities. On the other hand, urban and peri-urban areas are hosting an increasing number of diverse agricultural activities. Urban horticulture is particularly important within the framework of sustainable development, taking into account all the benefits it can bring to climate change adaptation, ecological stewardship, the development of green infrastructure and benefits for local community (Krasny and Tidball, 2015).

Gradually, the urban-rural dichotomy has been replaced by urban-rural continuum, with dissolving boundaries enabling the visible and invisible flow of people, capital, goods, information and technology between the urban and the rural (Davoudi and Stead, 2002). Furthermore, the development of infrastructure networks and the expansion of urban activities within and beyond the limits of urban regions, gradually erase physical differences between these two entities. Their previous division is re-evaluated by the contemporary ecological paradigms, transcending the limits between built and natural environment and emphasizing the continuity of linkages which living world creates between them (Park, 2015; Schmitz et al, 2018). The movement of people additionally contributes to the urban-natural continuum. Therefore, contemporary cities could be defined as hybrid ecological-sociological systems with specific climate patterns, grouping of predominant species of animals and plants, related human structures and allocation of land cover (Picket et al., 2013). Rural and natural hinterland is viewed through a model of complex mosaic composed of different combinations of the above mentioned patterns. Fine resolution of this mosaic allows us to observe relation between a city and its rural landscape through the conceptual model of urban-rural gradient.

The evolving urban-rural continuum, influenced by the process of growing urbanization, has to be considered in the process of climate change adaptation and mitigation, both on the morphological level and on the level of urban-rural connectivity. Therefore, the climate sensitive rural development policies have been created within the EU framework, dealing with the wide range of economic, environmental and social challenges instigated by climate change. In Spain, which has a wide belt of rural areas that stretch from the snowy peaks of the Pyrenees to the sun soaked beaches of Andalucía, climate impacts endanger growing rural economy. Therefore all 17 of the regional Spanish Rural Development Programmes (RDPs) offer climate action opportunities to help rural areas mitigate and adapt to warmer and drier weather conditions (Gomez-Limon, 2010). A combination of different climate factors is forecast to influence future changes for agricultural production systems in Sweden, Finland and other Nordic countries. Scandinavian stereotypes of long winters and snowy climate may soon be changing as temperatures rise and seasons shift. Climate change will affect rural areas in Nordic countries in several ways linked to changing temperature patterns. Farmers from the north are already seeing the effects of climate change and have been using assistance from the RDPs to make positive climate action contributions (Parissaki, 2010). South-Eastern Europe will be affected by climate changes resulting from reduced precipitation and increased temperature. A range of measures in the RDPs of South East Europe countries cover operations that address climate change. Adaptation actions in the RDPs of EU Member States from South East Europe are dominated by water management objectives aiming to upgrade the efficiency of irrigation systems and improve the effectiveness of water storage capacities, as well as prevent the depletion and possible degradation of existing underground reservoirs. Climate change mitigation in these countries is addressed by investments in energy saving equipment, the conversion of agricultural land into forests, organic farming and soil conservation techniques (Witold, 2010). In Slovenia, for example, adaptation actions include using nets to protect fruit orchards against hail damage and mitigation projects contain actions such as a horticultural business using geothermal energy to heat orchid production systems. Sharing a number of objectives with other European Structural and Investment Funds (ESIF), the EU Framework for rural development programs promotes resource efficiency, supports shift toward a low-carbon and climate-resilient economy and emphasizes the importance of social inclusion, poverty reduction and economic development in rural areas (European Commission 2013a, 2013b). The higher level of connectivity between urban and rural areas, recognized as an important element of the Common Agricultural Policy (CAP), gradually becomes a necessity which might be achieved by stimulating a hybrid character of urban-rural systems supported by information networks.

Underlining the strategic role of material and digital information networks in the process of climate mitigation and adaptation, the chapter will focus on a relationship between (r)urban systems, information networks and climate change, manifested both in the domain of public communication of climate change and on the level of (r)urban synergy. Based on the review of key literature and Internet searches, it provides an insight into recent trends and practices, discussing both the possibilities and the effects of (un) conventional use of ICT flows in understanding and reducing the consequences of climate change. The selected literature covers three main topics - the causal relationship between climate change and developing (r)urban synergy; adaptation and mitigation strategies/policies; and the application of information networks in (r)urban areas. Internet searches were directed to three major areas of networks applicability in the process of climate adaptation and mitigation - behavioral changes of users, environmental awareness and general efficiency. The results of the analysis were classified according to web-site content, mode of use and anticipated impact/audience.

THE BACKGROUND: BETWEEN THEORY AND PRACTICE

In this chapter we will use the single term ICT network primarily for various IT software systems for networking groups and individuals through applications, web platforms and services that use internet and mobile network as the primary infrastructure for information and communication flow. The causal relationship between climate change (and the processes of adaptation and mitigation it triggers), ICT networks and (r)urban setting could be analyzed through several focal points.

Adaptation and mitigation represent two major, globally acknowledged, strategic responses to climate change. The first refers to adjustments of natural or human systems to actual or expected climate change, as well as to sever weather-related disasters and their increasing frequency. Adaptation also copes with the effects of climate change that can no longer be prevented, minimizing the impacts, exploring new opportunities and building adaptive capacity by increasing the ability of individuals, groups or organizations to adapt to changes. The process of adaptation includes implementation of related decisions, transforming the adaptive capacity into action (IPCC, 2014). This phase should be especially supported by ICT networks, as social and information facilitators.

Mitigation strategies relate primarily to long-term targeted activities focused on the reduction of global warming and related impacts, especially GHG emissions. Considering this, information and communication technologies can play a crucial role in decreasing the emissions generated by the sectors of production and economy, since their implementation could improve overall efficiency. The effect of ICT networks could also be visible in large-scale changes of social and economic behavior, including the changing patterns of trade, production and consumption, as well as both global and local engagement of citizens in decision-making (Yonazi et al., 2012).

Although adaptation and mitigation are always distinguished as independent strategies in official documents and policies on climate change, they are interconnected as the joint actions aiming at the growing climate challenges. Therefore, the application of ICT can serve both as an emergent platform referring to spontaneous actions taken by actors affected by climate-generated stimuli, or a planned and/or policy-driven adaptation process which includes strategies. These strategies provide a general plan of action, with policies and measures addressing climate change impacts within a given context (UNDP, 2004).

The role of information networks in adaptation and mitigation could be manifested as an important resource of data related to global warming, available to different groups of consumers. The simultaneous and interlinked existence of the material/physical environment and its electronic/digital counterpart has already been identified as the main feature of modern comprehension, influencing the nature of our interaction and needs, while simultaneously shaping different spatial patterns and typologies (Aurigi and De Cindio, 2008; Bucher and Finka, 2011; Castells, 2004: Drewe, 2000; Fusero, 2009; Graham and Marvin, 1996; Light, 1999; Mitchell, 2000 etc.). Consequently, all our efforts focused on climate-friendly actions have to be based on the logic of global networking, as a prerequisite for an efficient multi-sectored response to climate shifts. However, in order to increase the usability of ICT networks, their flow should be transparent (Booher and Innes, 2002) and easily interpreted, understood and incorporated into decision-making processes (Lemos et al., 2012). Consequently, the Intergovernmental Panel for Climate Change (IPCC), emphasizes that the process of adaptation and mitigation to climate change should have a significant impact on all the governmental levels of decision making process, i.e. how they manage data and make them accessible to citizens and communities (IPCC, 2007). Information networks and ICT sector, therefore, play the major role in education and the dissemination of multidisciplinary knowledge, engaging social capital for strengthening ecological and social responsiveness to climate change risks.

Describing the so-called E-topia, Mitchell (2000) envisioned the development path of a green human settlement with smart elements, which we gradually accept and follow. He distinguished several principles - dematerialization, demobilization, mass customization, intelligent operation and soft transformation - which would stimulate life with a minimized production of waste and reduced traffic and pollution (enabled by ICT networks). These changes, anticipated almost 20 years ago, were also considered as a platform enabling intelligent adaptation, automated personalization and the creation of efficient, responsive markets for available resources. The current development paths of urban and rural settlements demonstrate the application of numerous elements proposed by Mitchell, while their elaboration could be found in different documents, emphasizing the importance of low carbon economy and smart solutions. For example, 'SMART 2020 report' by The Climate Group (2008) promotes 'Clean revolution' enabled by ICT and focused on the SMART framework. The report also recognized basic steps/imperatives - standardize, monitor, account, rethink and transform - necessary for reducing the carbon footprint and GHG emissions. Similarly, the ARUP's report suggests smart solutions for low-carbon cities in which the reduced emission should be achieved simultaneously with increased living standard and economic success (Hill et al., 2009). Both documents suggest that information and communications technologies could provide new users' experience and smart behavioral change also influencing the development of low carbon urban economies. Consequently, the increased intensity and extent of connections are underlined as important factors in providing a continuous communication/feedback related to different activities and services, while smart interfaces enable efficient and personalized experience.

Within the EU context, "Smart Villages" is the current sub-theme of the broader European Network for Rural Development (ENRD) thematic work on 'Smart and Competitive Rural Areas'. It is a relatively new concept within the realm of EU policy making that refers to rural areas and communities which build on their existing strengths and assets as well as on developing new opportunities (ENRD, 2017). In Smart Villages traditional and new networks and services are enhanced by means of digital, telecommunication technologies, innovations and the better use of knowledge, for the benefit of inhabitants and businesses. Digital technologies and innovations may support quality of life, higher standard of living, public services for citizens, better use of resources, less impact on the environment, and new opportunities for rural value chains in terms of products and improved processes (European Commission, 2016).

The emerging (r)urban hybrid settlements should also follow these principles, especially focusing the problem of agricultural activities becoming increasingly present in urban area. They are not only endangered by climate change, but they also generate significant contributions to GHG emissions. At the EU level, agriculture participates with 10% of the total CO_2 emissions which equals to the total amount of CO_2 emitted by all forms of transport (Eurostat, 2016). A constant global increase of inhabitants exerts pressure on agriculture, extending its territories in order to meet growing food needs. This process has been done at the expense of forests and other ecologically valuable areas, making a vicious cycle in which agriculture represents both a victim (given the negative effects of global warming on food supply) and a perpetrator (one of the main causes of climate change) (Perrone, 2018). The impact of (r)urban systems on climate change is also noticeable on the level of daily migrations, due to highly dynamic fluctuations of population between urban and rural poles (Nawrotzki et al, 2015). Therefore, agriculture and transport represent the focus of ICT activities related to adaptation and mitigation in (r) urban systems. Considering this, the Food and Agriculture Organization (FAO) promotes sustainable agriculture through the concept of Climate-Smart Agriculture, proposing a series of social and environmental measures aimed at creating a sustainable agricultural system that optimizes and stabilizes crop yields. These practices also tackle the effects of climate change, such as desertification and sea-level rise. Among them, organic agriculture plays an essential role as it respects natural cycles, drastically reducing human impact (Lifegate, 2018). In order to optimize a system with so many local and regional actors and involved sectors, it is necessary to apply efficient ICT tools. Moreover, ICT tools are used for modeling and simulation of yield for each type of crop in specific climatic conditions, in order to optimize economic and human resources (FAO, 2009). This has to be coordinated both on the local as well on the regional and (inter)national level.

The synergy of information and technology, integrated in both urban and rural systems via ICT networks, could play a significant role in the mitigation of climate risks and adaptation to changing conditions. The main task is to connect environment and people by supporting an efficient exchange through learning and understanding of environmental processes and (r)urban systems.

NETWORKS OF/FOR 'GREEN' INFORMATION: UPGRADING THE (R)URBAN SYNERGY, FACING THE CLIMATE CHANGE

The modern technologies directly influence the ability of both urban and rural settlements to be resource-efficient and oriented toward the reduction of GHG emission, while open information have a significant role in minimizing carbon footprint. Information flows reinforce formal and informal social networks, especially on a local level which represents an optimal scale for creating and implementing adaptation strategies (Kern and Alber, 2008) and developing adaptive capacity (Pelling and High, 2005). Information networks also support communication and cooperation between all levels of governance and stakeholders, which is very important for raising awareness (Kousky and Schneider, 2003), identifying potential climate risks and adaptation priorities (Bulkeley, 2010), managing complex ecosystems (Bodin and Crona, 2009) and implementing collaborative planning (Healey, 2006). The digital form enables better (and instant) detection of changes, increases the efficiency of data transmission and analyses, and provides a better understanding of processes, their potentials and setbacks. The use of readily-available technologies and real-time systems for information transparency, monitoring and decision-making is increasing, and this process creates a promising setting for new trends of environmental responsibility, eco-oriented economic development, energy efficiency and community support. Connecting social, technological and natural systems, while targeting sectors of services, mobility and government, new technologies influence environment (both natural and built) and society through different media, systems, tools, gadgets and applications able to detect, collect, analyze and process different kinds of data. The impact of widely available information becomes visible in our everyday life - from a perception of environment and improvement of (r)urban performances, to communication and movement of individuals, groups and goods. Consequently, we can distinguish three groups of initiatives in which the link between (r)urban entities, ICT networks and adaptation and mitigation strategies could be perceived i further strengthened. The first one targets behavioral changes, the second one supports the process of environmental awareness, while the third one provides higher efficiency of activities.

Changing the Behavior

The multiplying problems generated by our carbon-intensive way of life certainly influence gradual modifications of behavior, changing the global comprehension of environmental problems. Playing an important role in dissemination of information and knowledge, numerous projects, initiatives and strategies have been conceived, discussed and implemented during the last few decades (Balaban, 2012; Castán Broto and Bulkeley, 2012, Hunt and Watkiss, 2011; Stupar and Mihajlov, 2016), while advanced technological performances of information systems support these processes. The variety of target groups, as well as local socio-cultural, economic and technologic circumstances, demand and/or generates different possibilities of access, data-processing, decision-making and acting.

Communication, transmission and exchange of information have to be conducted via both physical and electronic networks, in order to provide wider and deeper coverage of climate problems. Therefore, we can identify two basic approaches for raising eco- and climate- consciousness, depending on a transmission/reception method and anticipated initial scale and timing of impact (Stupar and Mihajlov, 2016). The first one (digital gateways) is based on new modes of communication, allowing global visibility of presented examples, fostering accumulation of world-wide knowledge and skills, as well as their mediation and transmission towards unknown recipients. This approach is usually applied in for establishing online resources of innovative practices. These 'gateways' usually act as passive inductors of behavioral changes and climate-friendly trends, although they include a certain level of interactivity between their creators, administrators and users. In general, they tackle issues of environmental changes, sustainability, resilience, transition and/or transformation of life styles. One of the most prominent examples of this approach is The Climate Change & Environment Network (CC&E Network), hosted by the Global Programme Climate Change and Environment (GPCCE) of the Swiss Agency for Development and Cooperation (SDC). It provides a knowledge exchange platform for practitioners, organizations and SDC staff and partners working on environmental issues and climate change problems. The network currently consists of professionals with different backgrounds (about 300), including governmental organizations, NGOs, multilateral organizations, companies, consulting firms, research and science institutions. The main objective of the CC&E Network is to promote the integration of both environmental themes and climate change mitigation/adaptation aims into policies, strategies, programs and projects of development cooperation. The network also promotes achievements and operational experiences, stimulating sharing of knowledge among CC&E professionals. Supporting cooperation on different levels, it targets climate challenges, environmental problems and smart development (WeAdapt, 2019).

The second approach to developing climate-awareness and changing carbon-intensive behavioral patterns is based on real-time diffusion (education programs and projects conducted *in-situ*). It provides interchange in real space, exclusively demanding an active response of already defined users. Usually conceived as a part of a specific project, with expected timeline and detectable outcome, the real-time diffusion actually 'produces' case-studies which could be overlapped, combined or just transmitted via digital 'gateways'. One of the most inspiring examples, highly applicable in the concept (r)urban synergy, represents the project The new climate generation, launched by the Municipality of Copenhagen in 2009, as a part of Copenhagen's climate plan 'Carbon neutral in 2025' (2012). Triggered by the important global event COP15, the project included 1500 school students, their teachers and nurseries, in order to raise general awareness about climate. It was also used as a platform for training so-called 'climate ambassadors', able to transmit and use new knowledge ('Copenhagen: The New Climate Generation', 2017). Similarly, the Causeway Rural & Urban Network (CRUN, 2016) enables the sustainability of local communities providing different services, trainings and collaboration. These activities are used for raising environmental consciousness, organizing climate-sensitive (r)urban projects and supporting the exchange of ideas and knowledge applicable both in urban and rural realm.

The interchange of knowledge, which supports the creation of smart technological solutions for sustainable agriculture, could also be conducted by public-private partnership, as in the case of the New Climate Center in Brazil. Bringing together entrepreneurs and farmers, the Center, supported by the team of World Bank Group, has enabled family farmers of Santa Catarina (a mountainous state in southern Brazil) to apply the principles of climate-smart agriculture. The team initiated launching of NITA (Nucleus for Technological Innovation for Family Agriculture), as a part of infoDev program which trains staff in designing thinking methods which engage stakeholders and co-create projects. NITA fosters small businesses in developing and commercializing climate-smart solutions (infoDev, 2018) and currently gathers 11 private, public and academic institutions, incubated by the initiative Green Competitiveness Launchpad (NITA, 2017). In this case, the State of Santa Catarina acts as a specific role model which contributes to the new position of Brazil (ex-food importer), as one of the largest global exporters of coffee, soybeans, beef, sugar cane and ethanol.

Some cases combine both approaches, using technological backup as a tool for a more intensive knowledge exchange and empowering of different users. For example, the initiative Digital Bangladesh underlines the role of technology in education, health, job placement and poverty reduction, enabling more efficient implementation of democracy, human rights and transparency, while aiming at the overall improvement of living conditions (The World Folio, 2015). Therefore, the government has equally emphasized four elements of Digital Bangladesh Vision - human resource development, people involvement, civil services and business, and introduced multimedia classrooms on different levels of education, providing a more efficient transfer of knowledge via online sources. Another example, based on similar principles but specifically focused on urban-rural relationship, is the project Information Network Village (INVIL, 2017), initiated in Korea during the 1990s. Fostering self-sufficiency of settlements, this project supports the sustainable growth of rural communities via high-speed internet access and education.

Although ICT provides access to climate change expertise to different stakeholders, the biggest challenge remains the knowledge gap between them. Changing the behavior requires further mediation between experts and government officials, citizens, educators, students / pupils, and other social groups. Procedures need to be further developed in order to utilize and implement expert knowledge available through ICT and to use it more effectively in the activities of different social spheres.

Sensing the Environment: Protecting the Resources

The initiatives labeled as climate-friendly, climate-sensitive or smart-climate have recognized the role of digital systems as key-elements of our raising awareness. They are able to detect environmental data (via sensors) and make them immediately visible and globally available (via networks). Relying on two basic types of interfaces - personal (smart phones, notebooks, tablets etc.) and public (wi-fi nodes, urban touch-screens, info-beamers) users are able to have an insight into urban and natural resources, processes and activities influencing their attitude toward environmental protection and growing climate challenges. Besides global leaders in this filed (IBM, Cisco, CH2M HILL), a number of smaller multi-professional firms and non-profit groups also deal with similar activities (e.g. MySociety, Code for America, Open

Knowledge Foundation etc.), creating software, web-services and applications focusing on environmental conditions, transportation, urban services and resources. EU and European Space Agency (ESA) are the establishers of the Copernicus Earth observation program, with the aim to establish large free and open-access environment and climate data platform for the ultimate benefit of all European citizens. It offers information services based on satellite Earth Observation and in situ (non-space) data. Vast amounts of global data from satellites and from ground-based, airborne and seaborne measurement systems are being used to provide information to help service providers, public authorities and other international organizations improve the quality of life for the citizens of Europe (Copernicus, 2017).

Monitoring both (r)urban and natural environments, their dynamic and variables, has become an important issue in the age of climate change. It provides better understanding of atmospheric processes, their influence on our lives, as well as a number of causal relationships influencing the future of our habitat. Although the focus of climatological observations is nowadays shifted to urban areas covered by denser networks of sensors, the continuously collected data are used on global level for further research of climate impacts - from spatial patterns and infrastructural systems, to social and health issues. Furthermore, the applicability of interactive web service, such as 'Urban EcoMap' (implemented in Amsterdam and San Francisco), could be extended to (r)urban settlements and areas since it displays environmental footprints for selected zones. The data includes average annual emissions per household, as well as a share of energy, waste and transportation.

Considering that transportation sector represents one of major emitters of GHG (IPCC, 2014), reducing its negative impact on environment has become an imperative. The relation between ICT networks and transportation has been frequently used for increasing the efficiency of movement, providing a higher visibility of real-time information (about choices, availability and surrounding activities), enabling multimodality and monitoring the condition. In the context of cities and surrounding/complementing rural areas, this aspect demands special attention. Currently, there is a number of web-services and applications combining sensors, public displays and smart mobile phones (e.g. 'CalTrans Quick Map' in San Francisco), which is also a win-win combination in the domain of services. Dealing with problems of maintenance, management, quality or accessibility of (r)urban infrastructure and activities, they enable real-time interaction with users affecting their environmental awareness and participation. Good examples of this practice are projects 'Trash Track' - related to waste management, or 'City Sourced' - a platform for identifying and reporting to urban problems, which could be used in the complex hybrid systems of (r)urban settlements. In line with these initiatives, it is interesting to mention an application - 'Personal action plan calculator'. Its role is to measure ecological footprint and carbon emissions of every user, providing a personalized action plan which includes practical tips for reducing negative impact on the environment.

Having in mind that the global share of energy sector in total GHG emissions is more than a quarter (IPCC 2007b, 2014; Eurostat, 2016), a growing number of policies and strategies is oriented toward energy transition, reduction of energy consumption and possible vulnerabilities and risks which should be identified and mapped (Burton et al., 2002; Satterthwaite et al., 2009). The IBM's weather modeling technology 'Hybrid Renewable Energy Forecasting' (HyRef) certainly fits well into this category, since it increases the reliability of renewable energy resources providing data related to wind and solar forecasting (one month in advance, or in 15-minute increments). Simultaneously, it enables higher level of data integration into the power grid and influences reduction of carbon footprint (IBM, 2013). Similar concept has been used by the Philippine Atmospheric, Geophysical and Astronomical Service Administration (PAGASA), which provides official Philippine weather forecasts, flood forecasting, research

& warning services on national level. The service is part of the Philippine Digital Strategy (PDS) 2011-2016, based on the CICT's ICT-Roadmap which stimulates the application of ICT by both government and the private sector in order to achieve the development agenda (GOVPH, 2018) and provide easy access, use, efficiency and quality of services, as well as a high level of privacy and security standards for every user (Alampay et al., 2008). Furthermore, the service also contributes to the national priority of introducing and boosting ICT4D in education, agriculture, livelihood and disaster prediction/prevention/reduction of effects.

Apart from these application and projects, there is a number of initiatives specifically related to rural development simultaneously targeting the issues of resources/environment, networking, innovation, collaboration, funding and program implementation. The main European digital platform represents The European Network for Rural Development (ENRD, 2017), while numerous national portals additionally elaborate this issue (UK Rural Networks, Rural Development Program of Croatia etc.). There are also many ICT tools, developed within climate resilient agriculture, which analyze different processes under changing climate conditions (local crop modeling, management of economic impacts on agriculture sector etc.). Some tools are site-specific and used in order to simulate the growth of specific crops, verifying their variations under different climate change scenarios, but they could be applied at national and/or regional level through a link to an appropriate Geographic Information System (GIS). Interesting examples, highlighted by FAO (2009), are WOFOST (developed by the Centre for World Food Studies, CFWS and the University of Wageningen), GOSSYM/COMAX (developed by the Universities of Clemson and Mississipi and the Agriculture Department of United States) and APSIM (Agricultural Production Systems SIMulator), developed by the Agricultural Production Systems Research Unit (APSRU).

Unlike in urban areas, rural areas mostly do not have adequate ICT infrastructure. In many regions of developing countries problem is poorly developed electric and mobile telephony network. Possible solutions to overcome this limitation are off-grid technical solutions of autonomous and sustainable power plants that would serve the needs of maintaining an ICT network in remote areas. It is also suggested to use satellite links that do not require complicated and expensive on-ground infrastructure. Another potential difficulty may be the lack of awareness and education on the use of ITC devices by certain social groups such as farmers, the elderly or the poor.

Managing the Efficient Life

The imperative of efficiency influences information services and applications, as well as the physical features of settlements and their interconnectivity. From transportation, energy and infrastructure, to food production/consumption and solid waste management, this kind of challenge has generated a number of creative solutions with different territorial and (cross) sectored impacts. Oriented toward individuals, specific groups or raised to a level of municipal, regional or transnational systems they facilitate or complement management, directly or indirectly contributing to the reduction of GHG emission.

The initiatives related to transportation, especially important in the context of complex (r)urban systems, include carpooling/rideshare networks and freight transport, potentially rationalizing transport costs, lowering fuel usage, decreasing traffic volume and congestion. A good example of this practice is a web-service 'RoadSharing.com', which emphasizes its cross-European, environmental, economic and social dimensions. It enables travelers to arrange share rides for free, while saving money, polluting less and socializing. Similar services could be found on regional and national level, mirroring the increased interest in this model of time/resource management (e.g. 'timskavoznja.com' - focused mostly on countries of ex-Yugoslavia). Simultaneously, there are networks aiming for the rationalization of freight/cargo transport (e.g. 'Cargo Shipping Networks' or 'Fixemer portal').

The comprehensive EU transportation project COMPASS, completed in 2015, aimed to provide an answer on future traveling demands and needs, conditioned by recent socio-economic, ICT and ITS trends, favoring the integration of multimodal transport solutions (Bielefeldt et al., 2013). The final outcome of the conducted assessment was supposed to contribute to new solutions leading to de-carbonization of transport activities, validating ICT solutions with stakeholders in the national contexts. The application of ICT was considered as a tool for gathering behavioral data, improving traveling survey and fostering harmonization. The case study of Scottish LATIS model (Land Use and Transport Integration in Scotland), a co-modal/intermodal public transport system and car sharing/pooling system, showed that measures decreasing overall car usage have more impact on the reduction of carbon emissions than measures focused on improving of public transportation travel-times (Stewart, 2015).

Apart from transportation challenges, which increase within growing (r)urban systems, the imperative of energy efficiency could be found in many projects focused on smart grid concept (for ex. the case of "Model City Mannheim" or "Smart Rural Grid"). Although mostly focused on cities, the concepts and solutions they promote regarding interconnectedness of different networks, energy companies, producers and households, are highly applicable within urban-rural synergies. They improve the efficiency of energy supply; reduce energy consumption and carbon-footprint (MOMA, 2014). Furthermore, the contemporary society is gradually shifting toward interlinked systems of production, transport, consumption and waste, comprising problems of carbon emission and energy efficiency while focusing on food. Initiatives such as 'Food miles', consider the environmental impact of food systems, while some (e.g. 'Philadelphia Fair Food Project', 'San Francisco's farmers' markets') stimulate more interaction between urban and rural settlements, local farmers and restaurants, developing solar-based and organic agriculture. European Commission is working on an EU Communication on Sustainability of the Food System. Slow Food, an international association that is advocating a holistic approach and a sustainable food system, has welcomed this initiative and is taking an active part in the debate, bringing its vision of food sustainability based on its philosophy of "good, clean and fair" (Slowfood, 2018).

The software eLum, created by Moroccan entrepreneurs, also represents an interesting tool leading toward carbon-free future, although currently targeting industrial and commercial customers. It employs artificial intelligence to help manage both on-grid and off-grid energy costs, while providing an energy storage system coupled with a photovoltaic production unit that runs on Energy OS (infoDev, 2016). The flexibility created by off-grid solutions, would be highly applicable in (r)urban constellation, enabling significant reductions of energy costs and higher efficiency of agricultural production and complementing services.

In spite of all these creating and innovative initiatives and tools, their scope is currently limited and still cannot cause a significant environmental impact. However, their possible global applicability and popularity might improve the challenging situation. Meanwhile, the globally accepted trend of urban farming represents another stream of (r)urban activation, which uses information networks in all three identified categories of application, both in the form of digital gateways or as a mean of real-time diffusion.

CONCLUSION

The latest technological trends of Open Network Environment, Internet of Things (IoT) or Cloud Computing, combined with the Open Data approach, Big Data challenge and environmentally friendly, low/no-carbon concepts, have created an innovative setting for climate sensitive development of both urban(ized) and natural environments. The complex nature of growing (r)urban synergy, as a result of numerous external and internal processes in the contemporary world, demands even more elaborated approach to climate change problems and challenges, demanding accurate and immediate sensing of actual condition and environmental changes. Connecting the different realms and users, while merging the knowledge and needs of global society and its local specificities, technological backup and information networks facilitate intensified flows of valuable data which are analyzed and processed in order to guide our behavior, actions and decisions toward climate mitigation and adaptation. The integral approach to urban and rural development, frequently emphasized as a preferred option for the climate-friendly future, represents a multi-disciplinary reaction to growing environmental concerns. Based on material and digital information networks, the concepts and tools created within this framework directly influence emerging visions of our future.

Table 1 summarizes previously analyzed examples of the application of ICT in various climate change adaptation and mitigation strategies. As we can notice, there is a great variety of ICT networks, applied technologies and diverse user focus groups. Also, there is a wide range of goals directed towards different adaptation or mitigation targets. However, despite all the above mentioned confirmations of great progress of the implementation of ICT for the adaptation and mitigation strategies, certain problems are still awaiting for adequate solutions.

First of all, there is a problem of social capacity to adopt new technologies for climate change circumstances. Therefore, education has the highest importance for building social capacity. It should be keep in mind that certain sections of population, such as elderly and poor, will be more challenging to tackle. Another type of issue is of a technological / infrastructural nature, which can make it difficult to expand ICT networksIt should be noted that these problems are even more pronounced in remote rural areas where underdeveloped energy and ICT infrastructure represents a serious problem, as well as generally less educated population regarding new technologies. Another technological challenge is sustainable energy power supply of ICT infrastructure, as well as the optimization of its consumption.

Adapting to changing climate, mitigating negative effects and breeding highly wired and eco-conscious generations might indeed have a positive effect on our environment. However, the global attention still has to reach its peak finally enabling humanity to erase a boundary between intellectualization and action. Multiplying the points of information access and upgrading their performances, while simultaneously extending their reach and real impact, is certainly a next challenge for developing built environment and its digital alter-ego. Therefore, the options analyzed in this article - due to their high accessibility, development flexibility and usability - should be considered as a starting point for a further integration of available approaches and tools into new strategies of climate-responsible (r)urban development.

Table 1. Overview of the analyzed ICT examples sorted by type and aims in mitigation and adaptation

ICT network strategies	Name	Type	Adaptation/ Mitigation aim
Changing the behaviour	• The Climate Change & Environment Network (CC&E Network) • Copenhagen's climate plan 'The new climate generation' • Causeway Rural & Urban Network (CRUN) • Nucleus for Technological Innovation for Family Agriculture (NITA) • Digital Bangladesh Vision/ Information Network Village (INVIL)	• Climate change knowledge exchange web platform • Elementary school education program • Web platform for local communities providing different services, trainings and collaboration • A part of infoDev program which trains small businesses staff • Technological backup tool for education of citizens in urban rural area	• Mitigation/adaptation aims into policies, strategies, programs. • 1500 school students, their teachers and nurseries, in order to raise general awareness about climate • Raising environmental consciousness, organizing climate-sensitive (r)urban projects and exchange of ideas and knowledge • Designing climate-smart solutions. • More intensive knowledge exchange and empowering of different users
Sensing the environment, protecting resources	• MySociety, Code for America/ Open Knowledge Foundation • European Space Agency Copernicus Earth observation program • Urban EcoMap/ CalTrans Quick Map • Trash Track • IBM's Hybrid Renewable Energy Forecasting' (HyRef), (PAGASA) • The European Network for Rural Development (ENRD) • WOFOST, GOSSYM/COMAX, APSIM.	• Software, web-services and applications focusing on environmental conditions, transportation, urban services and resources • Earth observation and remote sensing tech. • GIS interactive web service ... • Weather modeling technology • The main European digital platform for RD • ICT tools for climate resilient agriculture	• Enabling private and public users to have an insight into processes and activities influencing their attitude toward environ. protection and growing climate challenges • Establishing large free and open-access environment and climate data platform • Displays environmental footprints for selected zones • Waste management • Providing data related to wind and solar forecasting • Rural development funding and implementation • Simulating crop growth under changing climate conditions
Managing the efficient life	• RoadSharing.com • EU transportation project COMPASS • Model City Mannheim/ Smart Rural Grid • Food miles/ Philadelphia Fair Food Project/ Slow Food • eLum	• Carpooling and freight transport web platform • EU multimodal transport model for gathering data • Smart grid project • Interlinked network of production, transport, consumption and waste • Software tool with artificial intelligence for optimizing on-grid and off-grid energy costs	• Lowering transport costs, fuel usage, decrease traffic and congestion • Providing an answer on future traveling demands • Improve energy efficiency and reducing carbon footprint • Stimulating more interaction between urban and rural settlements, local farmers and restaurants • Reductions of energy costs and higher efficiency of and agriculture production and complementing services.

ACKNOWLEDGMENT

The article is realized as part of the project “Studying climate change and its influence on the environment: impacts, adaptation and mitigation” (43007) financed by the Ministry of Education and Science of the Republic of Serbia within the framework of integrated and interdisciplinary research for the period 2011-2019.

REFERENCES

Alampay, E., Heeks, R., & Soliva, P. (2003). *Bridging the Information Divide: A Philippine Guidebook on ICTs. The National College of Public Administration and Governance, University of the Philippines and the Institute of Development Policy and Management.* University of Manchester.

Aurigi, A., & De Cindio, F. (2008). *Augmented urban spaces: articulating the physical and electronic city*. Aldershot, UK: Ashgate.

Balaban, O. (2012). Climate Change and Cities: A Review on the Impacts and Policy Responses. *METU Journal of the Faculty of Architecture*, *29*(1), 21–44.

Beatley, T. (2000). *Green Urbanism – Learning from European Cities*. Washington, DC: Island Press.

Bielefeldt, C., Bak, M., Borkowski, P., Carreno, M., Matthews, B., Stewart, K., …Schnell, O. (2013). COMPASS: optimised co-modal passenger transport for reducing carbon emissions – Final results and Conclusions. Edinburgh, UK: European Commission.

Bodin, Ö., & Crona, B. (2009). The role of social networks in natural resource governance. *Global Environmental Change*, *19*(3), 366–374. doi:10.1016/j.gloenvcha.2009.05.002

Booher, D. E., & Judith, E. I. (2002). Network power in collaborative planning. *Journal of Planning Education and Research*, *21*(3), 221–236. doi:10.1177/0739456X0202100301

Broto, C., Bulkeley, H., & Bulkeley, V. (2013). A survey of urban climate change experiments in 100 cities. *Global Environmental Change*, *23*(1), 92–102. doi:10.1016/j.gloenvcha.2012.07.005 PMID:23805029

Bucher, U., & Finka, M. (2011). *The Electronic City*. Berlin: BWV, Berliner Wissenschafts-Verlag.

Burton, I., Huq, S., Lim, B., Pilifosova, O., & Schipper, E. (2002). From Impacts Assessment to Adaptation Priorities: The Shaping of Adaptation Policy. *Climate Policy*, *2*(3), 145–159. doi:10.3763/cpol.2002.0217

Castells, M. (2004). Space of flows, space of places: materials for a theory of urbanism in the information age. In S. Graham (Ed.), *The Cybercities Reader* (pp. 82–93). New York: Routledge, Taylor & Francis Group.

Causeway Rural & Urban Network. (2015). Retrieved March 11, 2017. http://crun.org/?doing_wp_cron=1491135028.5418028831481933593750

City of Copenhagen. (2012). *CPH 2025 Climate Plan: A Green, Smart and Carbon Neutral City - draft version 4 June 2012*. Retrieved March 11, 2017. http://subsite.kk.dk/sitecore/content/Subsites/CityOfCopenhagen/SubsiteFrontpage/Business/Growth_and_partnerships/~/media/F5A7EC91E7AC4B-0891F37331642555C4.ashx

City Sourced Mobilizing Civic Engagement. (2015). *City Sourced: Mobilizing Civic Engagement*. Retrieved November 11, 2015. https://www.citysourced.com/

Copernicus. (2017). *About Copernicus*. Retrieved November 11, 2015. https://www.copernicus.eu/en/about-copernicus

DAC & Cities. (2017). Copenhagen: *The New Climate Generation*. Retrieved March 11, 2017. http://www.dac.dk/en/dac-cities/sustainable-cities/all-cases/education/copenhagen-the-new-climate-generation/

Davoudi, S., & Stead, D. (2002). Urban-Rural Relationships: An introduction and a brief history. *Built Environment*, *28*(4), 269–277.

Drewe, P. (2000). *ICT and urban form. In Urban planning and design – Off the beaten track, Design Studio "The Network City"*. Delft University of Technology: Faculty of Architecture.

ENRD. (2017). *Smart Villages*. Retrieved February 11, 2019. https://enrd.ec.europa.eu/enrd-thematic-work/smart-and-competitive-rural-areas/smart-villages_en

European Commission. (2013a). An EU strategy on adaptation to climate change – Communication from the Commission to the European Parliament, the Council, the European Economic and Social Committee and the Committee of the Regions (COM (2013) 216 final). Brussels: European Commission.

European Commission. (2013b). Guidelines on developing adaptation strategies (SWD (2013) 134 final). Brussels: European Commission.

European Commission. (2016). *EU Action for Climate Villages*. Retrieved December 18, 2018. https://ec.europa.eu/agriculture/sites/agriculture/files/rural-development-2014-2020/looking-ahead/rur-dev-small-villages_en.pdf

European Network for Rural Development. (2017). Retrieved March 11, 2017. https://enrd.ec.europa.eu/

Eurostat. (2016). *Greenhouse gas emissions by source sector (source: EEA)*. Retrieved December 18, 2018. http://appsso.eurostat.ec.europa.eu/nui/show.do?dataset=env_air_gge&lang=en

Fixemer. (2017). *Fixemer e Logistic*. Accessed March 11, 2017. http://www.fixemer.com

Food and Agriculture Organization of the United Nations (FAO). (2009). *Information and Communication Technologies for climate change adaptation, with a focus on the agricultural sector*. Retrieved December 18, 2018. http://www.fao.org/docs/eims/upload/295345/Sala%20ICT-climate%20change%20Agriculture.pdf

Fusero, P. (2009). *E-City: Digital Networks and Cities of the Future*. Barcelona: LIST Laboratorio.

Giddens, A. (2009). *The Politics of Climate Change*. Cambridge, UK: Polity Press.

Gomez-Limon, H. A. (2010, May) Climate change effects and climate-related Rural Development Programme support in Spain. EU Rural Review - The Magazine from the European Network for Rural Development, 24-30.

Government of Phillipines (GOVPH). (2018). *The Philippine Atmospheric, Geophysical and Astronomical Services Administration*. Retrieved December 18, 2018. http://bagong.pagasa.dost.gov.ph/

Graham, S., & Simon, M. (2006). *Telecommunication and the City: Electronic Spaces, Urban Places*. London: Routledge.

Healey, P. (2006). *Collaborative planning: Shaping places in fragmented societies* (2nd ed.). London: Palgrave MacMillan. doi:10.1007/978-1-137-08600-6

Hill, D., Lean, D., Watts, M., & Buscher, V. (2011). *ARUP - The smart solution for cities*. Retrieved March 11, 2017. http://www.arup.com/~/media/Files/PDF/Publications/Research_and_whitepapers/UrbanLife_SmartSolutionForCities_July11.ashx

IBM. (2013). *Made in IBM Labs: IBM Drives the Future of Renewable Energy with New Wind and Solar Forecasting System*. Retrieved in March 11, 2017. http://www03.ibm.com/press/us/en/pressrelease/41310.wss

infoDev. (2016). *eLum's Innovation Illuminates Energy Savings for Moroccan Businesses*. Retrieved December 18, 2018. http://www.infodev.org/articles/elums-innovation-illuminates-energy-savings-moroccan-businesses

infoDev. (2018). *New Climate Center in Brazil Brings Together Entrepreneurs and Farmers*. Retrieved December 18, 2018. http://www.infodev.org/articles/new-climate-center-brazil-brings-together-entrepreneurs-and-farmers

INVIL project. (2001). *Information Network Village (INVIL) project*. Retrieved March 11, 2017. https://www.oecd.org/governance/observatory-public-sectorinnovation/innovations/page/informationnetworkvillageinvilproject.htm#tab_implementation

IPCC. (2007). Chapter 4. Adaptation and mitigation options. In Summary for Policymakers. In: Climate Change 2007: Synthesis Report. Contribution of Working Groups I, II and III to the Fourth Assessment Report of the Intergovernmental Panel on Climate Change. IPCC.

IPCC. (2014). Climate Change 2014: Impacts, Adaptation, and Vulnerability. *Contribution of Working Group II to the Fifth Assessment Report of the Intergovernmental Panel on Climate Change*. Original version: http://ipcc-wg2.gov/AR5/report

IPCC/Intergovernmental Panel on Climate Change. (2007a). *Climate Change 2007: synthesis report contribution of Working groups I, II and III to the Fourth Assessment Report of the Intergovernmental Panel on climate change*. Cambridge, UK: Cambridge University Press.

IPCC/Intergovernmental Panel on Climate Change. (2007b). *Summary for Policymakers. Climate Change 2007: The Physical Science Basis. Contribution of Working Group I to the Fourth Assessment Report of the Intergovernmental Panel on Climate Change*. Cambridge, UK: Cambridge University Press.

IPCC/Intergovernmental Panel on Climate Change. (2007c). *Summary for Policymakers. Mitigation. Contribution of Working Group III to the Fourth Assessment Report of the Intergovernmental Panel on Climate Change*. Cambridge, UK: Cambridge University Press.

IPCC/Intergovernmental Panel on Climate Change. (2013). *The Physical Science Basis. Contribution of Working Group I to the Fifth Assessment Report of the Intergovernmental Panel on Climate Change*. Cambridge, UK: Cambridge University Press.

Jabareen, Y. R. (2006). Sustainable Urban Forms - Their Typologies, Models, and Concepts. *Journal of Planning Education and Research*, *6*(1), 38–52. doi:10.1177/0739456X05285119

Kern, K., & Alber, G. (2008). Governing climate change in cities: Modes of urban governance in multi-level systems. *Proceedings of the OECD Conference on Competitive Cities and Climate Change*, 171-196.

Kousky, C., & Schneider, S. (2003). Global Climate Policy: Will Cities Lead the Way? *Climate Policy*, *3*(3), 359–372. doi:10.1016/j.clipol.2003.08.002

Krasny, E., & Tidball, G. (2015). *Civic Ecology – Adaptation and Transformation from the Ground Up*. London: The MIT Press. doi:10.7551/mitpress/9780262028653.001.0001

Lemos, C., Kirchhoff, C., & Vijay, R. (2012). Narrowing the climate information usability gap. *Nature Climate Change*, *2*(3), 789–794. doi:10.1038/nclimate1614

LifeGate. (2018). *How agriculture and climate change are related: causes and effects*. Retrieved December 18, 2018. https://www.lifegate.com/people/news/agriculture-and-climate-change-causes-effects-impacts

Light, J. (1999). From City Space to Cyberspace. In M. Crang, P. Crang, & J. May (Eds.), *Virtual Geography* (pp. 109–130). London: Routledge.

Map, Q. (2017). *California Transportation Quick Map*. Retrieved March 11, 2017. http://quickmap.dot.ca.gov/

Mitchel, W. J. (2000). *E-topia: "Urban life, Jim - But not as we know it*. Cambridge, MA: MIT Press.

Modellstadt - Mannheim. (2014). *MOMA – The power supply system gets intelligent*. Accessed November 11, 2015. http://www.modellstadt-mannheim.de/moma/web/de/home/index.html

One Planet Living. (2017). *Personal Action Plan Calculator*. Retrieved March 11, 2017. http://www.oneplanetliving.net/take-action/personal-action-plan-calculator

Parisaki, M. (2010, May) Climate change adaptation and mitigation measures for Nordic rural areas. EU Rural Review - The Magazine from the European Network for Rural Development, 24-30.

Park, S. (2015). Spatial assessment of landscape ecological connectivity in different urban gradient. *Environmental Monitoring and Assessment*, *187*(7), 425. doi:10.100710661-015-4645-9 PMID:26065890

Pelling, M. (2011). *Adaptation to climate change: From resilience to transformation*. Abingdon, UK: Routledge.

Pflieger, G., & Rozenblat, C. (2010). Urban Networks and Network Theory: The City as the Connector of Multiple Networks. *Urban Studies Journal*, *47*(13), 2723–2735. doi:10.1177/0042098010377368

Pickett, S. T. A., Cadenasso, M. L., & McGrath, B. (2013). Resilience in Ecology and Urban Design – Linking Theory and Practice for Sustainable Cities. London: Springer.

Satterthwaite, D., Huq, S., Reid, H., Pelling, M., & Romero Lankao, P. (2009). Adapting cities to climate change in urban areas: the possibilities and constraints in low and middle income nations. In Adapting Cities to Climate Change (pp. 3-34). London: Earthscan.

Schmitz, C., Díaz, P., Ruiz-Labourdette, D., Herrero-Jáuregui, C., Molina, M., Montes, C., ... Schmitz, M. (2018). Modelling of socio-ecological connectivity. The rural-urban network in the surroundings of Madrid (Central Spain). *Urban Ecosystems*, *33*(6), 1119–1212.

Sharing, R. (2017). *Autostop and Carpooling for commuters, free Car sharing and Hitchhiking service*. Retrieved March 11, 2017. http://www.roadsharing.com/

Simić, I., & Bajcetic, R. (2016). Small towns as mediators towards spatial resilience: The case study of network of settlements in Vojvodina region. *FACTA UNIVERSITATIS Series. Architecture and Civil Engineering*, *14*(3), 331–342.

Simić, I., & Mihajlov, V. (2016). Towards the 'green city' concept: spatio - morphological development resources of small towns in Vojvodina. *Arhitektura+Urbanizam, 2016*(43), 35-45.

Slowfood. (2018). *Towards an EU Common Sustainable Food Policy*. Retrieved February 11, 2019. https://www.slowfood.com/sloweurope/en/topics/food-sustainability/

Stewart, K. (2015). Assessing the Carbon Impact of ICT Measures: A Case Study Investigation Using Latis Model. *International Journal of Transportation Science and Technology*, *4*(3), 277–294. doi:10.1260/2046-0430.4.3.277

Stupar, A., & Mihajlov, V. (2016). Climate change adaptation in Serbia: The role of information networks. *METU Journal of the Faculty of Architecture*, *33*(1), 37–59. doi:10.4305/METU.JFA.2016.1.3

Sustainable Aid in Agricultural Production (NITA). (2017). *Primiero Portal De Inovacoes Tecnologicas Para a Agricultura Familiar*. Retrieved December 18, 2018. http://nita.org.br/

The Climate Group on behalf of the Global eSustainability Initiative GeSI. (2008). *SMART 2020: Enabling the low carbon economy in the information age.* Retrieved March 11, 2017. http://www.theclimategroup.org/_assets/files/Smart2020Report.pdf

The European Network for Rural Development (ENRD)ENRD. (n.d.). *Thematic Work 2017*. Accessed March 11, 2018 https://enrd.ec.europa.eu/enrd-thematic-work_en

The World Folio. (2015). *Digital Bangladesh: an ICT revolution*. Retrieved December 18, 2018. http://www.theworldfolio.com/news/digital-bangladesh-an-ict-revolution/3603/

Timska voznja. (2017). *Timska vožnja*. Retrieved March 11, 2017. http://www.timskavoznja.com

Trash Track – MIT. (2017). *Trash Track*. Retrieved March 11, 2017. http://senseable.mit.edu/trashtrack/

United Nations Development Program (UNDP). (2004). *ICT and Human Development: Towards Building a Composite Index for Asia: Realizing the Millennium Development Goals*. Retrieved December 18, 2018. http://www.undp.org/content/dam/rbap/docs/Research%20&%20Publications/human_development/RBAP-HDR-2004-Building-Composite-Index-for-Asia.pdf

United Nations Framework Convention on Climate Change (UNFCCC). (n.d.). *Cancun Adaptation Framework 2010*. Retrieved February 18th, 2019. https://unfccc.int/resource/docs/2010/cop16/eng/07a01.pdf

Urban Eco Map. (2017). *Working together to Improve Urban Environments*. Retrieved March 11, 2017. http://urbanecomap.org/

WeAdapt. (2018). *SDC Climate Change & Environment (CC&E) Network*. Retrieved December 18, 2018. https://www.weadapt.org/knowledge-base/sdc-climate-change-environment-network

Witold, L. (2010, May) Climate change and rural areas in South East Europe. EU Rural Review - The Magazine from the European Network for Rural Development, 36-41.

Wood, J. (2007). Synergy city; planning for a high density, super-symbiotic society. *Landscape and Urban Planning*, *83*(1), 77–83. doi:10.1016/j.landurbplan.2007.05.006

Yonazi, E., Kelly, T., Halewood, N., & Blackman, C. (2012). *The transformational use of information and communication technologies in Africa*. Washington, DC: The World Bank and African Development Bank. doi:10.1596/26791

KEY TERMS AND DEFINITIONS

Adaptation: Wide range of actions aimed to adapt natural or urban environment to inevitable effects of climate change.

ICT Networks: Various IT systems for networking groups and individuals through software and applications, web platforms, and services that use internet and mobile network as the primary infrastructure for information and communication flow.

Mitigation: Wide range of actions aimed to mitigate the effects of climate change.

(R)urban Systems: Hybrid synergetic relationship between urban and rural environment.

Social Capacity: Ability of one society or one of its section to learn, adapt, cope and progress under some specific changing conditions.

Chapter 14
Research on First Mini Solar Power Plants to Produce Electric Power on the South Serbia

Mihailo Mitković
Faculty of Civil Engineering and Architecture, University of Niš, Serbia

Jelena Đekić
Faculty of Civil Engineering and Architecture, University of Niš, Serbia

Petar Mitković
Faculty of Civil Engineering and Architecture, University of Niš, Serbia

Milica Igić
Faculty of Civil Engineering and Architecture, University of Niš, Serbia

ABSTRACT

The solar radiation and energy potential in Serbia is 30% higher than in Central Europe, and the intensity of solar radiation is among the highest in Europe. Specifically suitable are the condition for usage of solar energy in southeast Serbia where the annual average of global radiation on a horizontal surface amounts to more than 4.2 kWh/m2 a day. This chapter discusses four photovoltaic solar power plants, two having been installed in the territory of the city of Leskovac and two in the Bosilegrad territory. The research is based on geographical location of solar power plants, orientation, and inclination of the panels in respect to the horizontal surface. All the processed solar power plants have the capacity 30-40 kW. The chapter is projection of two years of researching and measuring data on the initial investments in order reach profitability and investment return period. The solar power plants that have fitted angle of 33° (Fortuna) give higher production of electric power than in those panels that have angle inclination of 10° (Domit).

DOI: 10.4018/978-1-5225-9932-6.ch014

INTRODUCTION

The use of renewable energy sources (RES) used as heating and electric energy has been progressively increasing globally. The European Parliament adopted in 2001 the Directive 2001/77/EC on the promotion of electricity produced from renewable energy sources in the internal electricity market (European Parliament and the Council of the European Union, 2001). The directive encourages usage of renewable energy sources, with the goal to satisfy 12% of total energy consumption, while the share of renewable energy sources in the total consumption of the EU should have reached 22.1% in 2010. The Directive also provides the referential values for individual countries which must be reached, in order to meet the goals of the EU. Because of the lack of progress regarding the goals of 2010, a more comprehensive legal framework was made. The new Directive on renewable energy, enacted in 2009, the mandatory goal was set, according to which 20% of energy consumption in the EU must be met from the renewable energy sources until 2020, and it is laid out through obligatory national sub-goals, which take into account different starting points of the member states (European Parliament and the Council of the European Union, 2009). As for the solar energy, the white book of the EU of 1997, provided that until 2010 the total capacity of installed photovoltaic (PV) systems would reach 3GW. This goal was reached as early as in 2006 and the total capacities by the end of 2008 amounted to 9.5GW, which is three times more than planned (Stamenić, 2009; European Commission, 2009). Of this number, 5.3GW were installed in Germany, and 3.4GW in Spain, which means that more than half of European capacity is located in Germany. Since 2005, solar PV electricity generation capacity has increased from 1.9 GW in 2005 to 80.7 GW at the end of 2013, and the 2020 National Renewable Energy Action Plan (NREAP) target of 84.4 GW was exceeded by the end of 2014, reaching about 88.4 GW.

Serbia possesses considerable potential for usage of renewable energy sources, but currently mostly hydro-potential is used, while other forms of RES (wind energy, solar energy, geothermal energy and biomass) are not adequately used. Republic of Serbia adopted the National Renewable Energy Action Plan as a framework for promotion of energy generated from renewable sources and set mandatory national goals concerning share of energy from RES in the gross total energy consumption (27%) and (10%) in transport until 2020. According to the National Action Plan, it is planned that in 2015, the installed capacity of solar power plants is 5MW, and gross electric energy production 7GWh, and in 2020, 10MW, that is, 13GWh (Ministry of energy, development and environmental protection, 2013a). In accordance with the National Action Plan, the Energy Development Strategy of the Republic of Serbia was made for the period until 2025 with projections until 2030. The projected capacity of solar power plants until 2025 is 100MW, that is, 200 MW until 2030 (Ministry of energy, development and environmental protection, 2013).

SOLAR POTENTIAL SERBIA

As for the solar energy, Serbia has the resources which are well above the European average, along with the extremely favourable seasonal distribution. The energy potential of solar radiation is for around 30% higher in Serbia than in Central Europe, and the intensity of solar radiation is among the highest in Europe. The average solar energy of global radiation on the flat surface during the winter season ranges between 1.1 kWh/m2 in the north of the country and 1.7 kWh/m2 in the south, and during the summer season between 5.4 kWh/m^{2}in the north and 6.9 kWh/m2 in the south (Figures 1a, b, c).

Figure 1. Average energy of global radiation on the horizontal surface (kWh/m2): a) in January; b) in July; c) annual average
(Institute for interdisciplinary research, 2004)

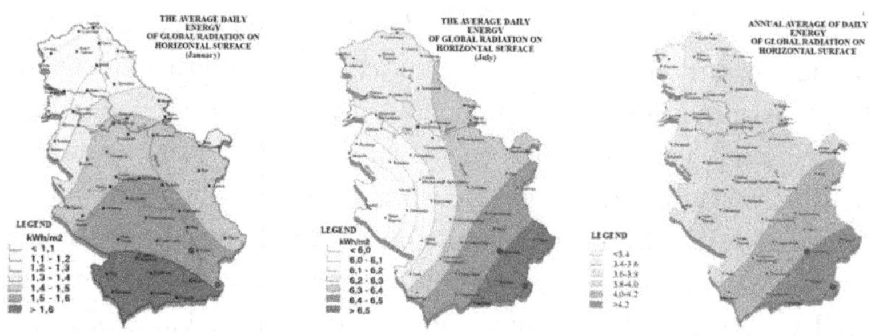

For comparison, average value of global radiation for the German territory is around 1,000 kWh/m2, while the average value of global solar radiation energy in Serbia ranges between 1,200 kWh/m2 a year in the north-west and 1,550 kWh/m2 in the southeast, while in the central area it is around 1,400 kWh/m2 a year. The highest value of annual sum of global solar radiation in Germany (1,130-1,240kWh/m2), coincides with the lowest vale in the north of Serbia (<1,240 kWh/m2) (Pavlović, et al. 2011; Stamenić 2009).

The total installed capacity do not correspond with the natural availability of solar radiation resources, so the most of PV capacity is installed in the German territory where the value of global radiation equals the European average, and it is considerably lower than the value for Serbia. According to these facts, Serbia has very favourable conditions for usage of solar energy, and its conversion to heating or electric power. In the White book of the Serbian electric power company (Marković, et al. 2011) the regulations of the EU relating to the renewable energy sources, as well as the legal framework and potential for use of RES in Serbia can be found. The decree on incentives for production of electric and heating energy ("Official Gazette of RS", no. 99/2009) the price of 23 c€/kWh was determined for buying off of electric energy produced in solar power plants of up to 5MW, which would be used for the following 12 years. In 2013, a new directive on incentives for privileged electric energy producers ("Off. Gazette of RS", no. 8/2013) was issued, which reduces the incentives, from 16.25 c€/kWh for ground power plants, to 20.66 c€/kWh for the power plants on structures, with installed power up to 0.03MW. For the power plants on structures with installed power (P) from 0.03-0.5MW the price is calculated per formula 20.941-9.383*P. In the Republic of Serbia several small electric power plants connected to the electric grid were installed: on the buildings of schools and faculties (Electric engineering school Rade Končarin Belgrade, secondary technical school Mihajlo Pupin in Kula, The Faculty of Technical Sciences in Novi Sad, The Faculty of Electronic Engineering and Faculty of Sciences and Mathematics of Niš) and privately owned companies and houses, which indicates that in Serbia, solar power plants are increasingly used for commercial purposes. According to the Study of solar and wind energy potential of Serbia (Institute for interdisciplinary research, 2004) the highest amounts of energy on the northern hemisphere are available in the period between April and September, and the energy influx on a certain surface depends on its inclination and orientation in respect to the sun rays, whereby the following facts apply:

- The highest annual energy gains are realized if the surface is oriented towards south and has a 30° angle, which is optimum of the periods March-April and August-September.
- The highest monthly energy gains are in May, June and July, if the surface is oriented towards south and has an inclination of 10 degrees.
- The highest monthly gains in the winter period from October to February are realized if the surface is oriented towards south and has an inclination of 60 degrees.

The paper presents the characteristics of four mini solar power plants in south-east Serbian and the results of the electric power output measured during two years. In three power plants, the panels are installed at the angle of around 30°, and on one of them the roof pitch of around 10° was used.

LOCATION AND CLIMATIC CONDITION

The paper treats solar power plants in Leskovac and Bosilegrad, in south-east Serbia, which has the most favourable conditions for solar energy harvesting. The city of Leskovac is located at 42°52' N and 21°57' E, in the center of the Leskovac valley. In topographical terms, the city of Leskovac is characterized by the flatland gradually declining towards north-east. The elevation above sea level of the valley varies between 210 and 240 m, and the city lies at around 230 meters above sea level (m.a.s.l.). The climate of Leskovac is moderate-continental, characterized by moderately warm summers and moderately cold winters, with two transitional periods, spring and autumn. Bosilegradis located at 42°30' N and 22°30' E. The municipality is characterized by hills and mountains with the elevation above sea level between 660 and 1,922 meters. The town enter is located at 696m.a.s.l. the Table 1 features a comparative presentation of data for Leskovac and Bosilegrad in the period 2005-2014.Bosilegradis located at 696m.a.s.l, and Leskovac at 230 m.a.s.l., which affects the climate, so from Table 1, it can be concluded that the temperature in Bosilegrad is always for 1-3°Clower than the temperature in Leskovac. The annual sunshine for the Leskovac area is 2,084h, whereby the maximum sunshine in July is around 313 hours, and the least sunshine in December is around 54h. There are no measurements for the Bosilegrad area, but sunshine increases towards south, and for the southernmost parts of Serbia is around 2,300h/year. The area of Leskovac is located in the zone with the average daily solar radiation energy of around 1.5-1.6 kWh/m2 in January, 6.4-6.5 kWh/m2 in July, i.e. with the annual average 4.0-4.2 kWh/m2, while in the Bosilegrad territory the daily solar radiation energy in summer is >6.5, i.e. >4.2 at the annual level (marks, Figures1a, b, c).

POWER PLANT DESCRIPTION AND DATA PROCESSING METHOD

The Table 2 provides the comparative display of characteristics of 4 solar power plants, two of which are located in Leskovac (E1-Domit andE2-Fortuna), and two in Bosilegrad (E3-Anitex andE4-Kodal), and the data for the designed model of a solar power plant for the area of Leskovac. All the power plants were built as a fixed system on a structure. The calculation of potential production was performed using PVGIS software, and the optimum inclination and panel orientation for the given area were used (inclination 32°, orientation -2°). The designed model was displayed with the goal to determine in what way the deviation from the proposed inclination and orientation affects the electric energy production.

Table 1. Average values of meteorological data for the area of Leskovac, for 2004-2013 period, obtained on the basis of Meteorological almanacs 2005-2014, Republic Hydrometeorological Service of Serbia

	Jan	Feb	Mar	Apr	May	Jun	Jul	Aug	Sep	Oct	Nov	Dec
	Leskovac											
AAT [°C]	0.9	2.0	7.0	12.27	16.7	20.5	22.5	22.3	17.2	11.3	6.7	2.2
P [mm]	47.3	49.4	54.7	63.2	79.6	64.2	48.4	45.5	41.3	63.8	56.1	68.5
SD[h]	76.1	72.5	144.5	182.8	232.1	268.2	311.9	305.8	195.3	144.4	96.6	54.3
AH[%]	80	76	69	68	70	66	62	62	69	76	79	82
AO[%]	66	70	61	57	55	46	34	29	48	51	58	73
	Bosilegrad											
AAT[°C]	-0.1	1.2	5.2	10.1	14.4	17.8	19.9	20.0	15.4	10.4	5.3	0.8
P[mm]	44.8	49.6	57.6	56.2	95.4	70.3	44.1	59.0	58.6	60.8	104.0	58.3
SD[h]	-	-	-	-	-	-	-	-	-	-	-	-
AH [%]	81	79	74	71	73	74	70	70	74	77	80	82
AO[%]	62	67	59	57	56	47	34	31	49	52	57	64
AAT –average air temperature, **P**-precipitation, **SD**- sunshine duration, **AH**-average humidity, **AO**-average overcast												

Table 2. Solar power plants characteristics and data for the designed models

	E1	E2	E3	E4	PVGISmodel
Location, elevation above sea level	Leskovac: 42°52'N; 21°57'E 228m.a.s.l.		Bosilegrad: 42°30'N; 22°30'E 696m.a.s.l.		43°0'0"N; 21°56'59"E 228m.a.s.l.
Power plant type	Fixed system on a structure				Fixed system
Panel type	monocrystal	polycrystal	polycrystal	polycrystal	crystal-silicon
Panel inclination (°)	10	30	30	30	32
Orientation	SW	S	S	S	S
Installed capacity(kW)	30	30	40.32	23	10

Table 3 presents measurements of two years (2014, 2015) of electric energy production by months, as well as the designed production for the power plant in the Leskovac area. The measurements were taken by reading the electricity meters by the Serbian electric power company. Since the power plants have different installed capacities (30kW; 30kW; 40.32kW; 23kW), in order to make the results comparable, calculated productions of electric energy per 1kW of installed capacity were used for all the diagrams. In order to make conclusion about the influence of inclination and orientation of the panels on electric energy production during the year, the measurement results for two solar power plants in Leskovac were used (one of which having the panel inclination of 10°, and the other 30°) with the designed production for that area (PVGIS model, panel inclination 32°). On the basis of the measurements accomplished so far (2014 and 2015) the average values per months and annual average values were calculated and displayed in comparison with the designed values (Figure 2a,b).

Table 3. The results of two years of measurements of energy production and designed energy production (PVGIS model) in kW

	E1 (30kW)		E2 (30kW)		E3 (40.32kW)		E4 (23kW)		PVGIS model (10kW)
	2014	**2015**	**2014**	**2015**	**2014**	**2015**	**2014**	**2015**	
January	1007	689	168	1517	2578	1770	1444	1019	582
February	1988	1789	2250	2576	3608	3569	2022	2052	746
March	3227	2333	2808	2744	4789	3227	2731	1930	1000
April	2929	4562	2870	4217	4004	5402	2298	3111	1200
May	4398	3705	4307	4228	5340	5779	3139	3353	1350
June	4848	4724	4490	4563	5715	6036	3338	3527	1370
July	5095	5721	4728	5390	6178	6805	3590	3984	1520
August	4772	4763	4847	4783	6369	6061	3645	3527	1490
September	2992	3023	3375	3340	4335	4648	2490	2631	1270
October	1886	1880	2655	2617	3757	3726	2232	2129	988
November	1006	1458	1608	2276	2366	3666	1349	2144	634
December	667	876	1108	1460	1808	3330	1101	2065	462
Total annual	34815	35523	36726	39711	50847	54019	29379	31472	12600

Figure 2. Production of electric energy (kW) per 1kW of installed capacity, for power plants in Leskovac (E1, E2, PVGIS model): a) distribution by months; b) total annual production.

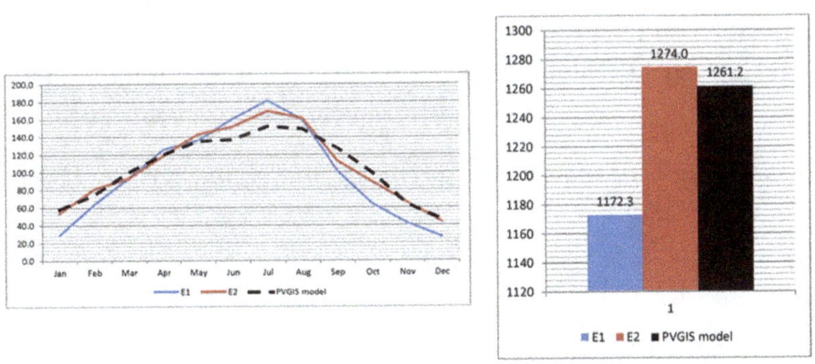

By comparative observation of the results of total annual production E1 and E2 in equal climate conditions and elevation above sea level, it becomes obvious that for the solar power plants in Leskovac, it depends on the town planning disposition: favourability of orientation and inclination of roof planes. Power plant E2 has a more favourable southerly orientation, while Power Plant E1 has a less favourable south-west orientation. It is confirmed by PVGIS model whose production curve significantly coincides with the curve for E2 (Figure 2a). The total annual production of E2 is for 11% higher than of E1 (according to the data from Table 3 and Figure 2b). If the data from the Table 2 are added, about the type of panels (for E1 – polycrystal, and for E2 – monocrystal panels), it would make the difference even greater, since the monocrystal panels produce more energy. E1 has a higher production only in summer months, when the incident angle of sunrays is more favourable for lower inclinations of roofs/panels. On the other hand, in the months with less sunlight, the difference in production is more favourable for E2. This causes the production line of E2 to have less rises and falls during the year, that is, lower differences between the maximum and minimum at the monthly level. In 2015, it is for E1 for 6.5 times greater in summer than in winter, and for E2 it is only 3.7 greater. This means that PVGIS is reliable, and it is the matter of town planning and architectonic conditions what inclinations and panel orientations can be realistically constructed in practice. If the panels are fitted above the roof cover, so as to coincide with the inclination and orientation of the roof, observing its formal characteristics, then the better production can be accomplished only by the choice of the panels with the highest solar energy conversion efficiency, which is the case of E1 power plant. On the other hand, if it is imperative to have the best energy, that is, financial performance, it is best to choose the recommended inclination and orientation of the panels, and to design the structure according to the best architectonic and town planning specifications.

For the purpose of confirming these statements and for evaluation of the impact of spatial-geographical distribution in terms of elevation above sea level, the research was expanded to solar power plants E3 and E4, which were constructed at a significantly higher elevation above sea level. The difference is 500m (the Bosilegrad power plants are at 730 m.a.s.l. and the Leskovac power plants at around 230m.a.s.l.). Since all four power plants are located in the settlement valleys in southwest Serbia, with negligible differences in terms of latitudes and longitudes, with a small approximation, we can assume that other natural conditions mostly depend on the difference in elevations above sea level of these locations.

Firstly, we will observe the graphic display of the lines of annual production for all four power plants (Figure 3) averaged by months for two years of production (2014 and 2015.). It is clearly visible that the power plants in Bosilegrad have almost the same production per unit of installed power, and that their lines almost coincide. On the other hand, the power plants in Leskovac have mutual deviations in production, which was proved earlier, but also the individual deviations in comparison with the Bosilegrad power plants. Power plant E1 still has the highest extremes, maximum production of in summer (in July 7% higher than E2, and around 11% higher than E3 and E4) and the minimum in winter (December as

Figure 3. Diagram of distribution of annual electric energy production (kW) by 1kW of installed capacity by months, average for 2014-2015

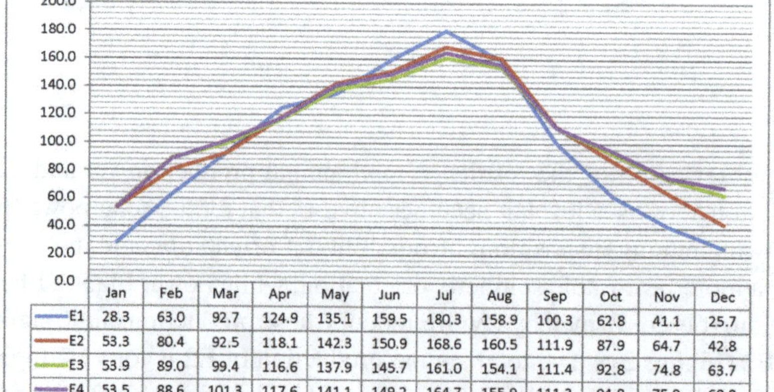

	Jan	Feb	Mar	Apr	May	Jun	Jul	Aug	Sep	Oct	Nov	Dec
E1	28.3	63.0	92.7	124.9	135.1	159.5	180.3	158.9	100.3	62.8	41.1	25.7
E2	53.3	80.4	92.5	118.1	142.3	150.9	168.6	160.5	111.9	87.9	64.7	42.8
E3	53.9	89.0	99.4	116.6	137.9	145.7	161.0	154.1	111.4	92.8	74.8	63.7
E4	53.5	88.6	101.3	117.6	141.1	149.2	164.7	155.9	111.3	94.8	75.9	68.8

Figure 4. Total annual production of electric energy (kW) per 1kW of installed capacity

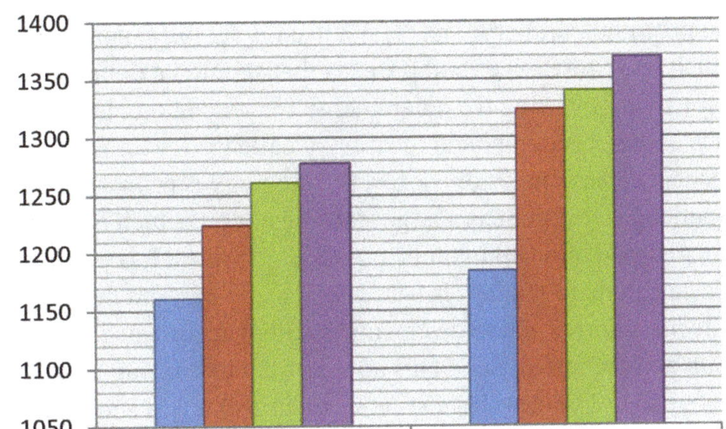

	2014	2015
E1	1160.5	1184.1
E2	1224.2	1323.7
E3	1261.1	1339.8
E4	1277.3	1368.3

much as 2.7 times less in comparison to E4), while E2 has values between E1 and the power plants at around 500m higher elevation above sea level, E3 and E4. In the non-extreme months, the production of the power plants having better orientation and inclination of panels demonstrate better results, and those are E1, E2 and E3. The data present us with the ability to consider the spatial influence, regarding the elevation above sea level. Firstly, we will observe it by the months. May and September are the months when energy production of the power plants at all elevations is almost equal. Three summer months between them are better for the power plants in the plains, while during the seven remaining months, the mountainous power plants feature higher energy production. The true picture can be observed in Figure 4, where total annual energy production (kW) is given per 1kW of installed capacity. In both observed years, the order of lowest to highest production is: E1, E2, E3 and E4. Since there was less sunlight in 2014 in comparison to 2015the total production is proportionately lower for all power plants. From the figure, one can clearly observe that the power plants at higher elevations have higher energy production both on the level of one and multiple years, which will have a considerable cumulative effect after 25 years of service life of solar panels. In the mentioned examples, this is due to the area of Leskovac in non-summer months, because of lower elevation above sea level and of the river basin has more fog in the morning hours, and more smog and dust due to higher pollution. Bosilegrad and other mountainous locations are more favourable in this respect.

The example of justifiability of financial investment are the data for the mini solar power plant Domit (E1) in Leskovac. The investor company constructed in 2012 the first power plant in the south of Serbia, having power 30kW, which is connected to the distribution system of the Serbian electric power company, for which the subvention tariff is guaranteed by the state for 12 years since commission, at a price of 20.66 of c€ per 1kWh. The total annual income is around 7,500 € (Mitković, et al. 2015). Regarding that the value of the investment is around 45,000 €, it is clear that the investment is returned after six years.

FUTURE RESEARCH DIRECTIONS

Generally speaking the potential that Serbia has is not used in its full potential and specially the South of Serbia. There are much more researches and projects that should be done. One of the key interests is making a better place for living by using the solar energy for producing electric energy. We should use all capacity that we have and that geographical location. Challenge for time that we are living at is making the world to fully accept clean sources of energy. Solar energy is primary sources and its potential is inexhaustible. In the South of Serbia there is a large stock of silicon, which is the basis for the production of photocells, which are grouped into photovoltaic panels or modules. It is possible to create a factory of cheap panels for mass usage in that region. It is necessary to research and develop a system of optimization of urban and architectural forms and to design and build in accordance with these principles, in the case of new construction and reconstruction. This applies both to housing and other facilities. Develop systems of mass and cheap use of photovoltaic systems on greenhouses and greenhouses in agricultural production, especially in the region of Southeastern Serbia and the environment, whose biggest economic potentials are in agricultural production. This would ensure the economic prosperity of the population of this underdeveloped area and halt the unfavourable demographic emptying. Further research and massive use of the effects of solar cells in the construction of the infrastructure: pathways - filling stations for electric vehicles, solar parking lots with chargers, solar public lighting, etc.

CONCLUSION

Solar radiation resources are significantly above European average and with extremely favourable seasonal distribution. The energy potential of solar radiation in Serbia is for 30% higher than in Central Europe and the intensity of solar radiation is among the highest in Europe. However, the total installed capacity does not correspond to the natural resources. The construction of solar power plants of any size in our country is in an initial phase. The best insolation for their construction can be found in southeast Serbia, where the annual average of global radiation on a horizontal surface amounts to more than 4.2 kWh/m2 a day.

The results of two years operation of four mini solar power plants on the structures have been researched. Two of those plants are installed in the territory of the city of Leskovac,

and two in the territory of Bosilegrad. The results of production of electric energy were analyzed from the town planning and architectonic aspects (inclination and orientation of roof planes and photovoltaic panels) and from the spatial aspects comprising geographic characteristics of the location and environment, i.e. elevation above sea level of the power plants. In the first case, it was established that the best harvest was made if the panels are fitted according to the proposed technical specifications, and those are the inclination of the panel of around 33° and strictly southern orientation. These conditions are always the most acceptable for the structures where architectonic and esthetic qualities must be achieved as well, so a deviation would cause a decreased production, which means that the power plants with a panel inclination less than 10° will produce less energy. In the second case, the production results indicate an increased production, with the increase of elevation above sea level. This means that when choosing the location for construction of power plants, in terms of space and planning, hilly and mountainous regions with better solar exposition should be chosen, since they have better ventilation and less fog. The analysis of investment return indicates the period of around six years, which is half of the time during which the subvention price is guaranteed, and the contract signed by the state.

REFERENCES

European Commission. (1997). *Energy for the future: Renewable sources of energy, White Paper for a Community Strategy and Action Plan*. European Commission.

European Commission. (2009). *PV Status Report 2009*. Luxembourg: Office for Official Publications of the European Communities.

European Commission. (2014). *PV Status Report 2014*. European Commission, Luxemburg: Publications Office of the European Union.

European Parliament and the Council of the European Union. (2009). Directive 2009/28/EC of the European Parliament and of the Council of 23 April 2009 on the promotion of the use of energy from renewable sources and amending and subsequently repealing Directives 2001/77/EC and 2003/30/EC. *Official Journal of the European Union, L, 140*(16).

European Parliament and the Council of the European Union. (2001). *Directive 2001/77/EC of the European Parliament and of the Council of 27 September 2001 on the promotion of electricity produced from renewable energy sources in the internal electricity market*. Official Journal of the European Communities L283.

Institut za interdisciplinarna istraživanja. (2004). *Studija energetskog potencijala Srbije za korišćenje sunčevog zračenja i energije vetra*. Univerzitet u Beogradu, Institut za interdisciplinarna istraživanja.

Marković, D., Čeperković, B., Vlajčić, A., & Resl, S. (2011). Bela knjiga Elektroprivrede Srbije. JP Elektroprivreda Srbije: Sektor za odnose s javnošću, Beograd.

Ministarstvo energetike, razvoja i zaštite životne sredine. (2013). *Nacionalni akcioni plan za korišćenje obnovljivih izvora energije Republike Srbije*. Ministarstvo energetike, razvoja i zaštite životne sredine, Beograd.

Ministarstvo energetike, razvoja i zaštite životne sredine. (2013). *Strategija razvoja energetike Republike Srbije za period do 2025.godine sa projekcijama do 2030.godine*. Ministarstvo energetike, razvoja i zaštite životne sredine.

Mitković, P., Đurić, J., Dinić, M., & Mitković, M. (2015). Mini solar power plants as a potential for development of the South of Serbia. Serbian Ceramic Society Conference- Advanced Ceramics and Application IV: Program and the book of abstracts. Belgrade: Serbian Ceramic Society, 102.

Pavlović, T., Milosavljević, D., Aleksandar, R., & Pavlović, M. (2011). Comparison and assessment of electricity generation capacity for different types of photovoltaic solar plants of 1 MW in Sokobanja, Serbia. *Thermal Science*, *15*(3), 605–618. doi:10.2298/TSCI110322065P

Stamenić, Lj. (2009). *Korišćenje solarne fotonaponske energije u Srbiji. Unecoop.* Retrieved from http://solarna.unecopn.org/studija.pdf

KEY TERMS AND DEFINITIONS

Electric Generators: Devices in which the occurrence of electromagnetic induction is used to obtain alternating current and voltage.

Energy Efficiency: Energy efficiency is the sum of measures and actions in all areas of life whose ultimate goal is minimal energy consumption, provided that the level of work and life remains the same or improves.

Environment: The environment or the human environment represents everything that surrounds us, that is, everything that directly or indirectly relates to human life and productive activity.

Global Radiation: Represents the sum of direct solar radiation and diffuse radiation resulting from reflected or dispersed sunlight.

Region of Southeast Serbia: Administrative region in Serbia, which are includes: ethnological, linguistic, geographic, and other characteristics.

Renewable Energy: Energy that is generated from renewable sources, which are wind, sunlight, tides, rain, geothermal heat.

Solar Power Plant: The system which converts sunlight into electricity.

Section 4

Urban–Rural Cooperation in Territorial Development

Chapter 15
New Concepts and Instruments for the Urban-Rural Continuum: Exploring Through Education

Ratka Colic
Faculty of Architecture, University of Belgrade, Serbia

Danijela Milovanović Rodić
Faculty of Architecture, University of Belgrade, Serbia

Marija Maruna
https://orcid.org/0000-0001-9015-0683
Faculty of Architecture, University of Belgrade, Serbia

ABSTRACT

The chapter presents and discusses the process and results of collaboration between an academic institution and local government in students' masters' theses and integrated urban development projects for urban-rural continuum of the territory of the City of Pančevo, Serbia. The chapter has four main parts: 1) a brief overview of the theoretical framework for multilevel governance; 2) the background, focusing on the national policy; 3) applied educational model; and 4) 11 students' projects dealing with the urban-rural continuum. Recommendations are made for continuing application of and research into potential teaching methods that promote meaningful interaction and practical and problem-oriented instruction that contribute to an understanding of contemporary problems of balanced urban-rural development. The conclusion discusses process and product outcomes, differentiating benefits for all those involved, as well as the obstacles and challenges encountered.

DOI: 10.4018/978-1-5225-9932-6.ch015

INTRODUCTION

In Europe's complex and rapidly evolving development context, characterised amongst other things by threats to the vitality of rural areas, as well as by urban sprawl and declining quality of life in urban areas due to congestion and pollution, the interdependencies between city and country have become a particularly important topic in key EU development policies.

The European Spatial Development Perspective (ESDP) (EC, 1999), as an indicator of the 'policy of concern' for the future of rural areas throughout Europe, was an early instrument of 'urban-rural partnership'. This is a concept applicable to both rural communities and areas blighted by uncontrolled urban sprawl, and its key tenets are the provision of basic public services and public transportation in rural areas; fostering partnership between cities and rural settlements and creation of 'functional regions'; integration of rural zones into urban areas' spatial development strategies; encouragement of and support for co-operation between rural areas, as well as between these and small and medium-sized towns (at both the national and international levels), to implement projects and exchange experiences; and establishment of networks between public institutions, small and medium-sized businesses, and local communities in both urban and rural areas. Urban-rural connections are referenced as strategic aspects in both the Urban Agenda for the EU (2016) and the New UN-Habitat Urban Agenda (2016). These international policies promote sustainable patterns of urbanization, including adequate living standards, economic growth, environmental protection, and a balanced system of urban and rural communities. They also emphasise decentralisation and a more active role of local governments in an inclusive society where the entire population ought to have access to basic services. Goal 11, as one of the 17 Sustainable Development Goals that together make the United Nations Agenda 2030, refers to cities that need to be sustainable, inclusive, safe and responsive. Within goal 11, sub-goal 11a is also defined, which refers to the support of linkages between urban, peripheral and rural areas.

The concept of a spatial and functional relationship between urban and rural areas is based on the movement of people, capital, goods, information, and technology, and the interdependencies of these factors can be examined by means of sociology, economics, geography, and spatial planning (Davoudi & Stead, 2002; Zonneveld & Stead, 2007). According to Davoudi & Stead (2002), urban-rural relationships must be viewed in the context of overall globalisation trends (production, finance, trade, and labour markets), whereby there has to be a shift away from the conventional view of rural areas as exclusively agricultural. Urban-rural linkages are a key segment of territorial development policies exactly because of their major potential to advance the quality of life for both urban and rural populations. These trends are all the more pronounced in post-socialist social and economic transition, and as such this approach should be viewed in relation to the complexity of the local context (Milovanović, Čolić & Maruna, 2017).

The sustainability of urban and rural areas is based on alignment and synergy of their individual development paths that are based on their distinct resources, potentials, and development constraints. In the context of spatial planning and programming, the urban-rural relationship can be made more sustainable through better co-ordination of transportation, land use, and open space planning; limiting growth and increasing the density of existing settlements based on the principles of a green and compact city; constructing blue and green infrastructure; safeguarding agricultural land; and promoting domestic production and the locally-based economy (Nilsson *et al.*, 2014).

Apart from an approach that entails delivering public services in rural communities, modern-day sustainability of urban and rural areas is based on harmonious development through promotion of economic growth, urban renewal, improvement of the quality of life, environmental protection, and better urban governance (Milovanović Rodić, *et al.* 2011, Maruna & Čolić, 2015). This entails integrating economic, environmental, social, cultural, and spatial development objectives and viewing the issue of governance as a cross-cutting issue. These aspects are all addressed in Serbia's national urban development policy of 2018 (Trkulja *et al.,* 2018).

THEORETICAL BACKGROUND

The governance model has seen alterations over the past twenty years, spurred by the acceleration and increasing unpredictability of changes to the development context. The key features of the new governance model are more flexibility in decision-making and greater involvement of various stakeholders in articulating and executing decisions, including a greater range of modalities for stakeholder participation. This has allowed those affected by decision-making to enter territory that was previously the sole domain of the authorities (public agencies at various levels of government and within multiple departments with divided and very often overlapping powers). This change in the value system has brought about an altered understanding of the role of public agencies, whose position has been shifting from 'holding power and making decisions' to initiating, motivating, informing, empowering, organising, and facilitating participation and dialogue by various spatial stakeholders: public agencies at all levels of government and various departments; private businesses; civil society organisations; and citizens.

The integration of different government levels (local, regional, and national, as well as European, in the context of EU integration) and other stakeholders into a consistent and integrated governance system constitutes the backbone of multilevel urban governance. This concept was defined in the Leipzig Charter on Sustainable European Cities (EurActiv 24/05/07), which for the first time highlighted the significance of multilevel governance for the execution of urban policies. The Toledo Declaration (2010) was adopted to address the context of the 2008 global financial and social crisis that has affected nations and cities throughout the globe; this document elaborates on these issues and makes recommendations for the development of instruments tailored to this model of governance.

To attain integrated urban development, multilevel urban governance can be put into practice by means of cross-cutting policy instruments that facilitate: (1) formulation of solutions based on the integration of environmental, social, economic, spatial, organisational, and cultural objectives of development; (2) constant communication and collaboration between various levels of government; (3) constant and constructive participation by stakeholders from the private and civil sectors; (4) support for the establishment of partnerships between the public, private, and civil sector; and (5) support for integrated local actions (EU, 2009). Different types of multilevel governance agreements occur in situations where the powers and responsibilities of different levels of government are divided. They are characterised by participation, focus on consensus, accountability, transparency, efficiency and effectiveness, equity and inclusiveness, and the rule of law (Tasan-Kok & Vranken, 2011).

INTRODUCING NEW CONCEPTS AND INSTRUMENTS THROUGH EDUCATION

National Policy

In parallel with the development in Serbia of the first local integrated urban development strategies (Čolić, 2015), the introduction and implementation of innovations in Serbia's urban governance, including the formulation of a national policy for sustainable and integrated urban development, was tracked.

The topic of urban-rural connections has been recognised in the Strategy for Sustainable and Integrated Urban Development of Serbia to 2030 (Trkulja *et al.,* 2018; Government of Serbia, 2019). The overarching objective of this Strategy is sustainable and integrated development of urban settlements ensured through improved economic, social, and cultural development, urban renewal, environmental protection and adaptation to climate change, affordable and high-quality housing, and preservation and appreciation of architectural heritage and urban identity.

The Strategy defines the key problems facing urban development in Serbia: uncontrolled urban sprawl; increasing differences in the quality of life and accessibility between urban and most rural settlements; declining regulation of urban settlements; deteriorating economies and social structures of populated places in post-communist transition; inefficiency of the local government financing system; increasing risk of poverty and/or social exclusion, especially for young people and vulnerable groups; inadequate transportation, infrastructure, utilities, and social protection arrangements and public spaces; unequal quality of the environment, healthcare, and safety, and lack of adaptation of urban settlements to climate change; inefficiency of urban planning and governance; democratic deficit in participation; and luck of transparency in decision-making in the governance of urban and rural settlements.

The Strategy advances the claim that the management of urban-rural connections has largely been neglected (p. 48). Serbia's rural population is on the decline: in the 2000s, it fell by more than 10% to about 40% of the total population. Lack of due care for rural-urban connections has increased disparities in the quality of life and accessibility between urban and rural settlements. This is reflected in a long-standing trend for depopulation of rural communities and unplanned sprawl of urban areas. The quality of life in liminal and peri-urban areas is lower than in city centres, but higher than in rural areas – so perpetuating negative trends with regard to regional, urban, and rural development.

There are five strategic directions of urban development proposed in the Strategy: sustainable economic development; urban renewal; social welfare (quality of life, health, and safety); quality of the environment; and urban governance. The goals in these directions are: 1) enhanced and integrated strategic urban framework for sustainable, innovative, and inclusive local economic growth, employment, competitiveness, and greater living standards; 2) enhanced and balanced quality of spatial regulation and accessibility; 3) enhanced quality and accessibility of social services, reduced risk of poverty, social inclusion, and demographic renewal of populated places; 4) enhanced quality of the environment, health, and security and large-scale adaptation to climate change; and 5) improved effectivity and efficiency of urban governance.

These goals are to be attained by means of batches of measures, sets of key inter-connected activities that contribute to sustainable and integrated development. Strengthened and enhanced governance of urban-rural development (batch 2.5) is planned to be achieved through: 1) development of public services, transportation and technical infrastructure, suburban public transport (bus and rail, subsidised over a medium-term period of between 5 and 7 years), and utilities in rural areas; and 2) identification and monitoring of ecosystem services provided by rural areas to urban settlements.

The spatial dimension for operationalising measures designed to implement the Strategy is considered through priority areas of intervention. These comprise different forms of intervention and transformation (primarily the physical development of parts of urban settlements, such as city centres, quarters, zones, blocks, liminal and rural areas, neighborhoods, architectural units, or streets) and a variety of processes and exercises to be undertaken in parallel with economic development, job creation, overall social progress and the improvement in quality, identity, and efficiency of the built environment, including adaptation to enhance environmental resilience. The priority areas of intervention include national programmes in support of sustainable and integrated urban development for: (1) industrial and commercial zones and brownfield sites; (2) illegally constructed and unregulated liminal urban zones (urban sprawl) and degraded rural areas; (3) threatened urban structures, and city centres; (4) parts of urban settlements with significant concentrations of social problems – social inclusion and poverty reduction; (5) settlements or parts of settlements faced with environmental issues and effects of climate change; (6) spatial units containing cultural and architectural heritage that constitutes major reference points of cultural and historical development of urban settlements or their clusters.

The Strategy follows an integrated and co-ordinated territorial approach to solving key and urgent issues of urban development so as to enhance quality of life by maximising the utilisation of territorial/urban capital and creative resources and promoting growth of the (innovative) urban economy and efficient urban governance. The national framework envisages two new planning instruments: Local Integrated Urban Development Strategy and Integrated Urban Development Project. These constitute innovative policy instruments for Serbia that are significant for the country's European integration; they will permit the preparation and implementation of urban development strategies and projects financed from various sources, including national, EU, and global funds.

The Strategy for Sustainable and Integrated Urban Development of Serbia to 2030 (2018) emphasises the significance of balanced urban-rural development through: (1) goals and batches of measures; (2) national programs in support of balanced and integrated urban development, especially for degraded rural areas; (3) application of various multi-level governance instruments (Strategic Direction #5), and (4) creation of integrated urban development projects.

Apart from recognising the importance of the urban-rural continuum, the national framework emphasises the application of key principles of the multilevel urban governance concept, namely co-ordination, co-operation, participation, and integration of sectoral policies, stakeholders, and sources of finance. The issue of urban-rural connections is set to increase in importance now that Serbia has gained access to the EU's 2014-2020 IPARD programme. In addition, this is the first time that national policy has designated integrated urban development projects as implementing instruments.

Multilevel governance is a governance concept that can allow the articulation and execution of appropriate responses to these challenges posed by urban-rural development.

Applied Educational Model

Below we will present the education model applied and the results achieved in the Integrated Urbanism study programme at Belgrade University's Faculty of Architecture, where one of the objectives was to familiarise students with the new EU urban governance instruments and promote their critical re-assessment. The goal of the programme is to build capacity of future planning professionals to allow them to better face the socio-economic context of Serbia's post-socialist transition. The key strategic precept of the education model followed in the Integrated Urbanism master's course is the creation of a flexible platform for introducing professional experiences from various disciplines (natural sciences, social sciences and humanities, and arts) and fields (academia, practitioners, public institutions, international bodies, activists, the market, etc.) into the teaching process in its entirety. The process of developing masters' projects, which will be presented in this paper, has been conceived in line with this commitment, which has allowed sound professional dialogue and practice-oriented learning through collaboration, not only for students but for all involved. This has made the teaching process a testing ground for the re-assessment of existing urban governance ideas, concepts, and instruments, and the introduction of new ones into practice (Maruna *et al.,* 2018; Milovanović Rodić *et al.*, 2018).

The master's thesis production process outlined below is primarily intended to develop solutions rooted in professional practice that benefit in terms of quality and complexity from the involvement of a multitude of different stakeholders. Knowledge creation is directly conditioned by the thematic framework, which is selected so as to represent the most up-to-date professional experiences and issues of importance to the planning profession globally, whilst at the same time corresponding to the current issues faced by urban planning practice in Serbia. At the same time, the overall thematic framework provides a platform for permanent communication between all actors during the development of students' final projects. Knowledge is constructed in the process of developing solutions through an exchange of professional experiences from various fields (Milovanović Rodić *et al.*, 2013). Different forms of organised discussion – such as meetings, group debates, and workshops – create a broad-based working environment and create space for open professional dialogue. Moreover, collaboration by students attending different study programmes and coming from a variety of professional contexts made a major contribution to the quality of their learning and masters' theses. On a number of occasions (in 2014, 2015, and 2017), support was provided by international organisations and foreign universities, which promoted and facilitated collaboration and guaranteed the relevance of the thematic framework (Maruna & Čolić, 2015).

The exercise was rooted in the local context, and local experts were thoroughly involved in the production of the masters' theses. This provided all participants with a precious opportunity for knowledge-sharing and learning: it allowed students to come up with valuable solutions for practice and develop professional experience, whilst also benefiting both practitioners and academics. Communication was part of the learning process, and the final exhibition served to showcase the results and allowed participants to summarise experiences and reflect on the process (see Figure 1).

More generally, the methodological approach to education used in the study programme contributes to the establishment of broad-based dialogue within the profession to re-assess the position of planning professionals in Serbia's altered socio-economic context and re-define their position in urban development practice. Ever since it was first founded in 2012, the Integrated Urbanism master's programme has been following, testing, and piloting innovations in urban development planning and governance. The teaching process has looked at a number of urban-development related issues: climate resilience and risk disaster management, multilevel governance, public property and public interest, localization of Sustainable Development Goals (SDGs), cultural identity, and urban governance instruments (Maruna&Čolić, 2014; Čolić *et al.*, 2015; Maruna *et al.*, 2015; Milovanović Rodić *et al.*, 2016, Milovanović Rodić& Stojić, 2018).

Students' Master Projects for Urban-Rural Continuum

This section presents masters' projects produced by students of the Integrated Urbanism master's programme at the University of Belgrade's Faculty of Architecture in a collaborative process undertaken by the City Pančevo and the Faculty of Architecture. The main topic for students' assignments was 'Integrated Urban Governance Instruments', with particular emphasis on 'Multi-Level Urban Governance Instruments'.

The students were asked to come up with an integrated project that constituted a sustainable solution to an identified problem in the given development context, as well as to specify instruments for its formulation and implementation. These integrated projects could be distinct in terms of both theme and content, and could deal with all or part of the territory of Pančevo. They could focus on spatial, institutional, or programme design, and could be both one-off or recurring, as well as both long- and short-term initiatives.

The starting point for understanding the territory was the City of Pančevo Development Strategy, 2014-2020 (2014) the city's current development framework. This document recognises a number of under-utilised potentials for the city: geographical position astride two rivers; well-developed network of rural communities; cultural and historical heritage; numerous brownfield sites; educated population; etc. According to the Strategy, the local context is also characterised by increasing unemployment, poorly-developed industrial areas; declining economic activity in the city centre; inadequate tourism offering; poorly-developed tourism infrastructure; degraded environment; low resilience to climate change; insufficient communication between businesses, the civil sector, and the local authority for development planning; and poor governance of urban-rural connections.

The quality of Pančevo's environment is a critical aspect for development. Due to the multiple threats facing the environment, development ought to focus on sustainable waste management systems, polluter mapping, use of alternative energy, and greater energy efficiency. Pančevo benefits from a climate favourable for agriculture and possesses significant holdings of arable land, which accounts for nearly 83% of the city's territory. This constitutes major potential for growth that can be enhanced by improving local agricultural infrastructure and employing new modes of farming (such as organic and biodynamic agriculture, etc.).

Apart from the city proper, the local authority's territory is also home to nine rural communities, which hold a great deal of potential not only for agriculture but also for rural tourism. In the future, emphasis is expected to be placed on balanced urban-rural development, improving the business environment and supporting small and medium-sized businesses, enhancing the quality of the environment, developing infrastructure in rural areas, industrial zones, and greenfield locations, revitalising brownfield sites and neglected locations and structures, and developing tourism.

To allow students to better understand the real environment for urban development applicable to both Pančevo and other Serbian communities, they were put in touch with city officials and staff of local utility companies. Communication with the public sector was complemented by collaboration with business associations, the Pančevo Business Council and the Regional Chamber of Commerce. These workshops were also attended, in a consulting capacity, by representatives of national planning and urban governance institutions: Ministry of Construction, Transportation and Infrastructure, Chamber of Engineers, Belgrade Urban Planning Institute, and Standing Conference of Towns and Municipalities (SCTM).

There was communication with outside stakeholders throughout the process (Figure 1). This took place in various ways, and with various purposes and outcomes:

Figure 1. Process of the production of masters' projects, Pančevo, 2016
(source: Authors)

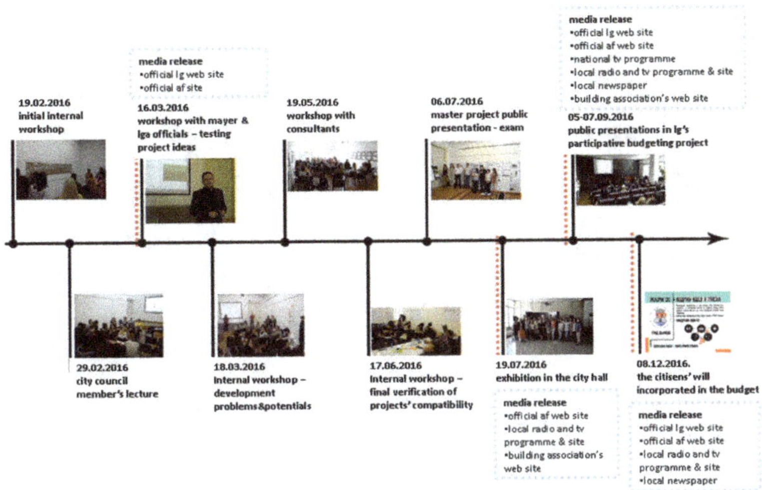

- A presentation of spatial and urban plans was held with the support of the city authorities, followed by an organised tour of city of Pančevo and the outlying villages guided by representatives of the Pančevo Construction and Development Directorate;
- The Pančevo Business Council met students at the headquarters of a large local private firm to present entrepreneurs' views of the business environment;
- On behalf of the team that produced the City of Pančevo Development Strategy, the City Councillor for Labour, Employment, and Social Policy held a lecture on its context, content, and preparation process;
- An SCTM expert delivered a lecture at the Faculty of Architecture on programme budgeting at local authorities as an instrument that allows better allocation of funds for development purposes;
- Two workshops were held at the Faculty of Architecture where students presented the findings of their research, ideas, and proposed solutions. The Mayor of Pančevo, city officials, and staff of utility firms, the Business Council, as well as national and Belgrade institutions took an active part in these sessions;
- A workshop was held at the SCTM where experts in various fields delivered thematic presentaitons and consultations about sources of finance, ways to apply for various funds, and best practices in areas covered by the students' papers;
- The Pančevo Business and Economic Development Secretary provided access to data, documents, and officers of city institutions crucial for understanding the territory. Each student was given tailored access with reference to their chosen topic. The city officers were given proper authority and informed about the context and purpose of communication and collaboration with the students, who contacted them personally with requests for assistance;
- A public presentation of students' projects was held at Pančevo City Hall. Apart from the representatives of the institutions involved, this event was also attended by interested citizens (Figure 2);

Figure 2. A public presentation of students' projects
(source: Authors)

- All of these events were covered by local media, with features broadcast by the local television station and published on the city's official web site and in *Pančevac*, the local newspaper.

This collaboration permitted significant progress in teaching integrated urban development and greatly contributed to the quality of the students' theses.

The section below presents students' projects and problems, solutions, and key tenets for balanced urban-rural development of the City of Pančevo. These are structured in line with the main objectives of the national urban development policy as following: (1) sustainable economic development; (2) urban renewal; (3) social welfare; (4) quality of the environment; and (5) urban governance.

Figure 3. IUDP 1: Farming co-operatives to promote organic production, student Rastko Krstić

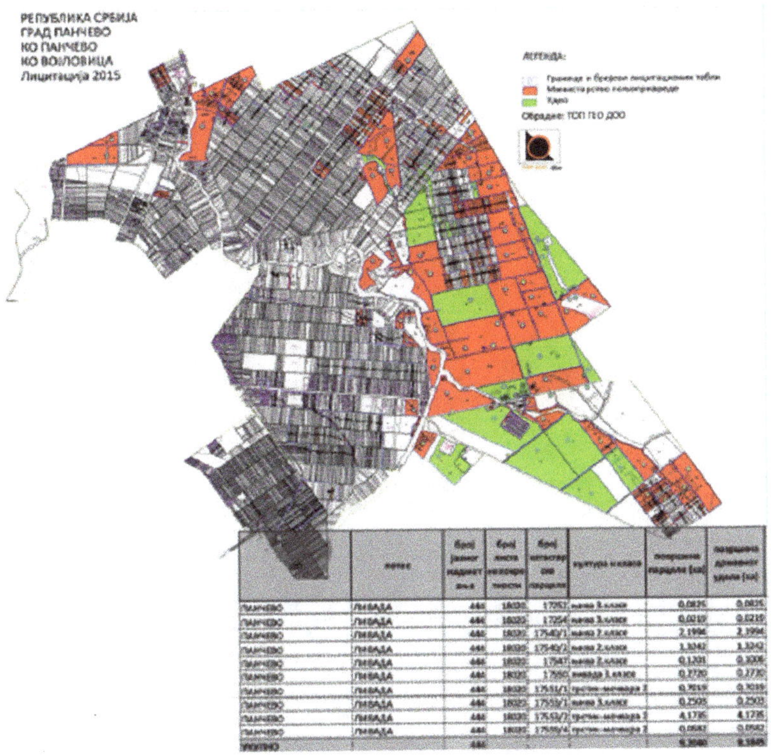

OBJECTIVE 1: SUSTAINABLE LOCAL ECONOMIC DEVELOPMENT

IUDP1: Farming Co-Operatives to Promote Organic Production

Problem: Small farms are not competitive in the market, and consequently their revenues are low. The issues of small farm size and obsolete equipment are compounded by the vulnerability of the soil to inappropriate disposal of municipal and farming waste.

Project Proposal: In this project the development of small farms is based on the greater revenues that farms can achieve by joining together into co-operatives. Apart from improving competitiveness and market access and facilitating the purchase of farming equipment, co-operatives would promote farmer awareness and encourage the acquisition of new knowledge and skills.

Urban-Rural Synergy: Integration of supply of and demand for farm production; support for the local circular economy

IUDP2: Rural Tourism as a Factor in the Development of Pančevo's Rural Areas

Problem: The village of Ivanovo faces low living standards; country-city migration and depopulation. The village fronts the Nadel and Danube Rivers, and its territory includes a river island that has been declared a nature protected area. There are many holiday homes in the area, which has a tradition as a

Figure 4. IUDP 2: Rural tourism as a factor in the development of Pančevo's rural areas, student Katarina Pantić

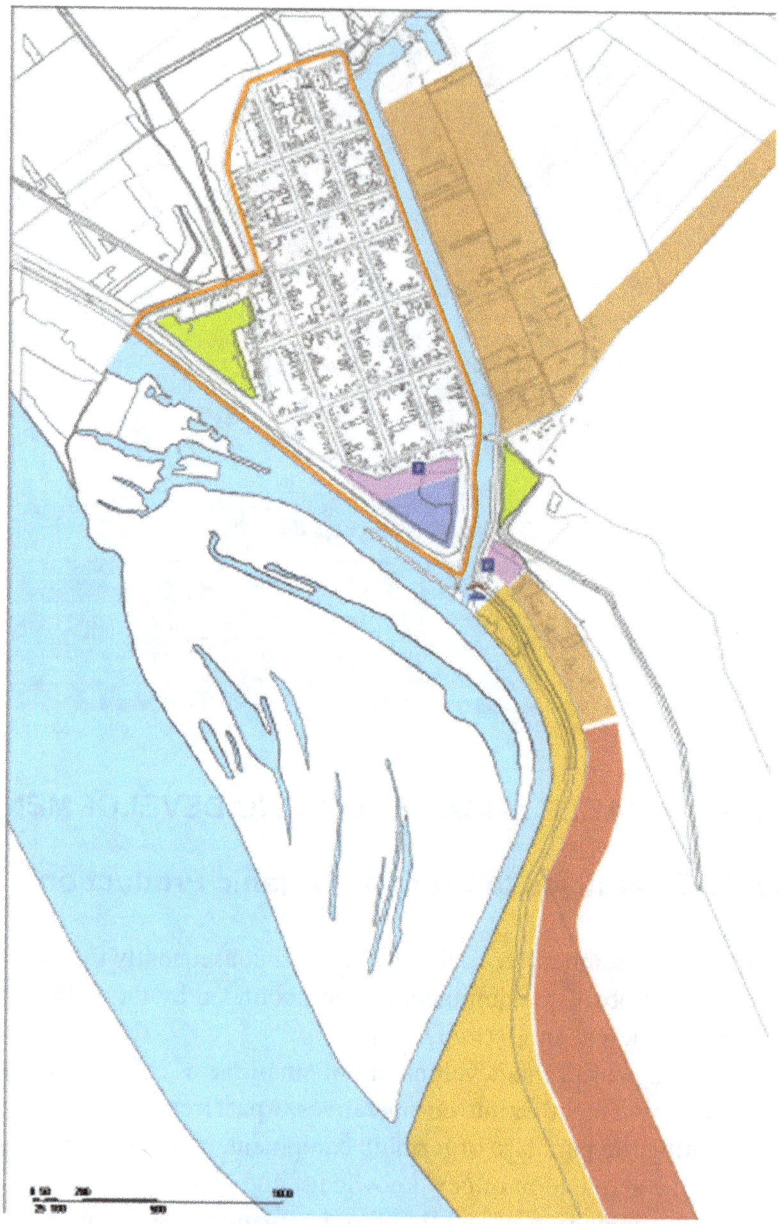

day-trip destination. The village's tourism offering and infrastructure are under-developed, and it lacks accommodation capacity.

Project Proposal: Under this project, rural development is to be based on greater revenues from tourism that the village can earn by developing recreational fishing, activating existing accommodation capacities and creating new ones, constructing a small boat marina, and developing an integrated offering that links leisure with entertainment. These efforts are expected to improve local products and services and enhance the quality of life.

Figure 5. IUDP 3: Supporting social entrepreneurship by allowing access to underused town-owned office space, student Božena Stojić

Figure 6. IUDP 4: Establishment of a civic association to support participation in activation and maintenance of public spaces, student Sofija Nedeljković

Urban-Rural Synergy: Integration of urban and rural tourism offering; improvement of quality of life in rural area; and reduction in the rate of daily migration.

IUDP3: Supporting Social Entrepreneurship by Allowing Access to Underused Town-Owned Office Space

Problem: Publicly-owned office space is under-utilised, which has had an adverse effect on economic activity and general dynamism of the town centre. On the other hand, non-governmental organisations (NGOs) are unable to rent publicly-owned premises for their social enterprises at concessionary prices unless they intend to engage in business.

Project Proposal: This project entails support for social entrepreneurship by (1) mobilising and incentivising NGOs to create social enterprises; (2) providing financial support and education for social enterprises, and allowing them access to unused town-owned office space; and (3) amending the institutional and regulatory environment.

Urban-Rural Synergy: Empowerment and social inclusion (association and revenue generation) of vulnerable groups, including rural women and elderly people.

OBJECTIVE 2: URBAN RENEWAL

IUDP4: Establishment of a Civic Association to Support Participation in Activation and Maintenance of Public Spaces

Problem: Dilapidation and poor state of repair of facades, children's playgrounds, parks, and other urban green spaces indicates a lack of effectiveness on the part of the authorities and the absence of appropriate mechanisms. These neglected spaces reduce the quality of living, denying local residents the use of public spaces and preventing them from engaging in activities there.

Project Proposal: Create socially and environmentally enhanced public spaces through community-initiated interventions. The focus here is on institutionalising neighbourhood committees, participatory bodies that set priorities and implement intervention programmes in public spaces.

Urban-Rural Synergy: Establishment of bodies at the lowest level of governance based on direct civic association designed to ensure efficient vertical co-operation with higher levels of authority.

OBJECTIVE 3: SOCIAL WELFARE

IUDP5: Ensuring Inclusion of Pančevo's Cultural Offering Into the Danube Route Cultural Network

Problem: Under-utilisation of resources to develop the cultural offering of the town, in particular in outlying rural areas; poor attendance and overlap between events; absence of prior announcements for seasonal programmes; mismatch between supply and demand; and poor audience participation in the development of the town's cultural content.

Figure 7. IUDP 5: Ensuring inclusion of Pančevo's cultural offering into the Danube Route cultural network, student Miloš Nikolić

Figure 8. IUDP 6: Sustainable system to manage hazardous household and farming waste, student Sara Bastoni

Project Proposal: This project aims at including Pančevo's cultural offering in the Danube Route cultural network. It envisages the creation of a network of cultural institutions and events in the town and nine outlying villages, and the deployment of measures to encourage activities at village cultural centres. Preserving the traditions, languages, and customs of these multi-ethnic-confessional areas may make them more attractive not only for the local population, but at the international level as well.

Urban-Rural Synergy: Positioning of Pančevo's cultural offering, networking between cultural institutions at the local, national and supra-national (Danube region) level, and integration and alignment of their activities.

OBJECTIVE 4: ENVIRONMENT PROTECTION AND CLIMATE RESILIENCE

IUDP6: Sustainable System to Manage Hazardous Household and Farming Waste

Problem: The issues addressed by this project are the lack of a system for selective collection of household and farming waste; low public engagement and awareness of this issue, its extent, consequences, and efforts to resolve it; degraded areas; and threats to public health.

Project Proposal: This project defines a sustainable process for managing waste throughout its lifecycle. It is based on principles of environmental design, which require products to not have adverse impacts on the environment at any stage of their lifecycles. The solution is tailored to the needs and capacities of Pančevo.

Urban-Rural Synergy: Establishment of an integrated waste management and environmental protection, particularly with regard to soil and water.

IUDP7: Construction of a Biogas Facility in Pančevo

Problem: Pančevo is threatened by high air pollution levels caused by the local chemical industry, as well as, in winter, district heating facilities and individual households that burn solid fuel for heating. On the other hand, farmland accounts for 83% of the town's territory and by-products of farming are routinely either burnt or discarded.

Project Proposal: The project proposes that organic farming waste should be used to make biogas, which would then be utilised as fuel for district heating across the territory of Pančevo, as well as for the manufacture of fertiliser, a by-product of biogas production.

Urban-Rural Synergy: Establishment of an integrated waste management; renewable energy; and environmental protection, particularly with regard to air.

IUDP 8: Mapping Environmental Polluters and Creating a GIS Database for Pančevo

Problem: Data on the city's polluters and their individual and cumulative impacts are not available to the local population.

Project Proposal: These data will be made public through the development of a GIS database, mapping, and on-going polluter monitoring. GIS database of pollution (industrial pollution, air and water

Figure 9. IUDP 7: Construction of a biogas facility in Pančevo, student Mihailo Žikić

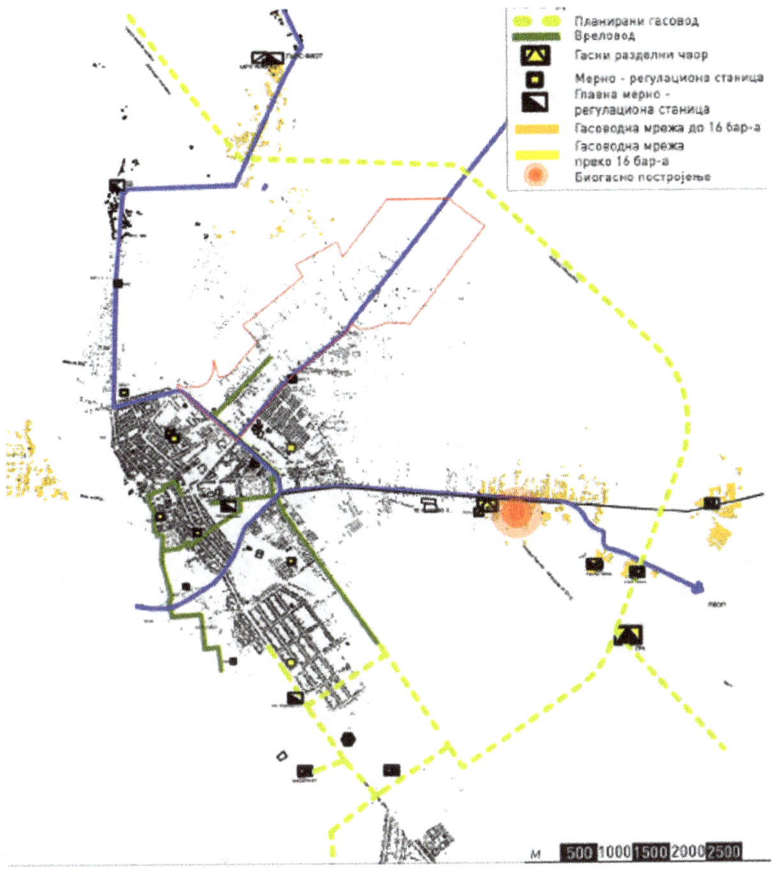

Figure 10. IUDP 8: Mapping environmental polluters and creating a GIS database for Pančevo, student Iva Mihaela Bapkova

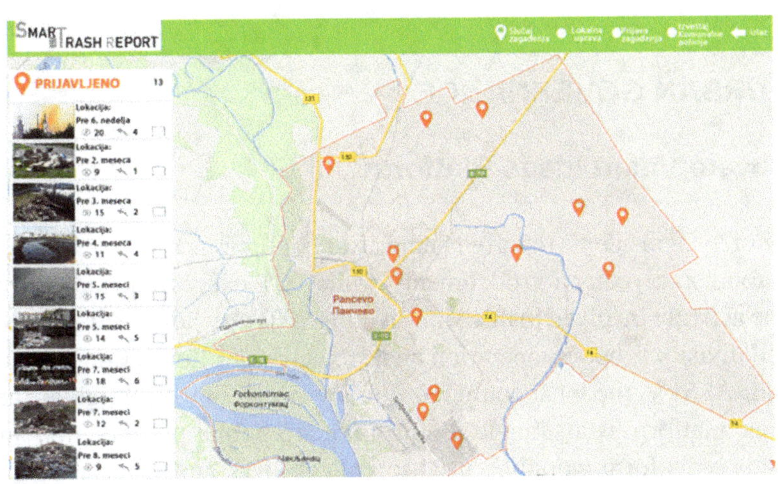

Figure 11. IUDP 9: E-PA development ideas platform, student Bogdan Nikolić

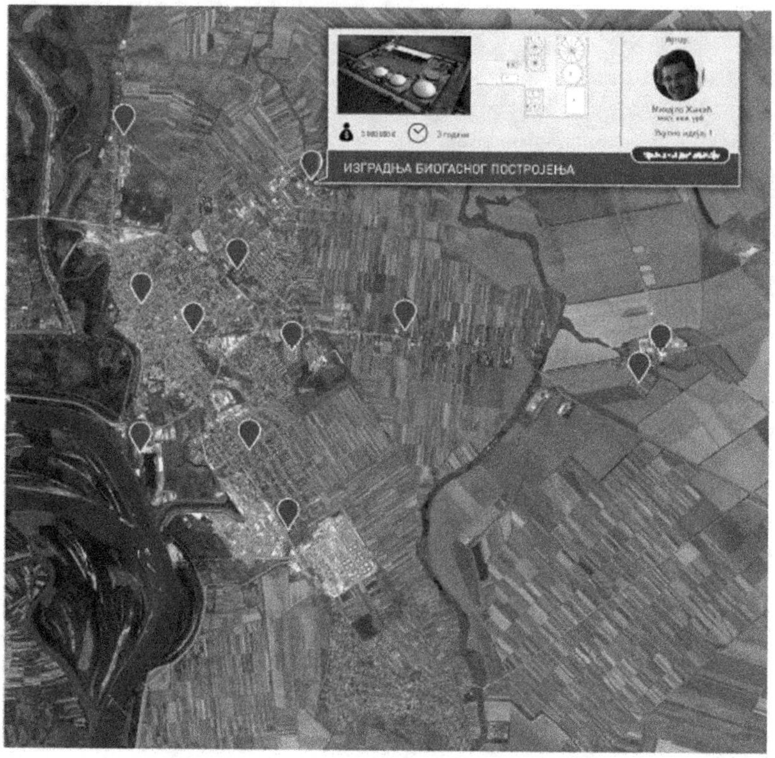

pollution, municipal waste, noise pollution, soil pollution, and the cumulative impact of all polluters in Pančevo) and a Smart Trash Report app will raise awareness and allow participation and transparency in environmental protection.

Urban-Rural Synergy: Establishment of a system to monitor cumulative impacts of polluters in urban and rural areas of Pančevo.

OBJECTIVE 5: URBAN GOVERNANCE

IUDP 9: E-PA Development Ideas Platform

Problem: Only 32% of local residents reported being aware of options for taking part in the development of planning and environmental protection documents, whilst 93% expressed their desire for learning more about how to become actively involved in decision-making. The city administration is clearly structured into departments and divisions, but co-operation and exchange of information are lacking.

Project Proposal: At 61%, the local computer literacy rate is twice as high as the Serbian average. As such, an electronic platform is a suitable tool to improve public participation and transparency in decision-making. This e-platform would be used to collect ideas and involve all interested parties in the planning process. These ideas will contribute to innovative solutions, new partnerships, and new sources of finance.

Figure 12. IUDP 11: Integrated branding as an instrument for improving the city's image, student Dragana Joksimović

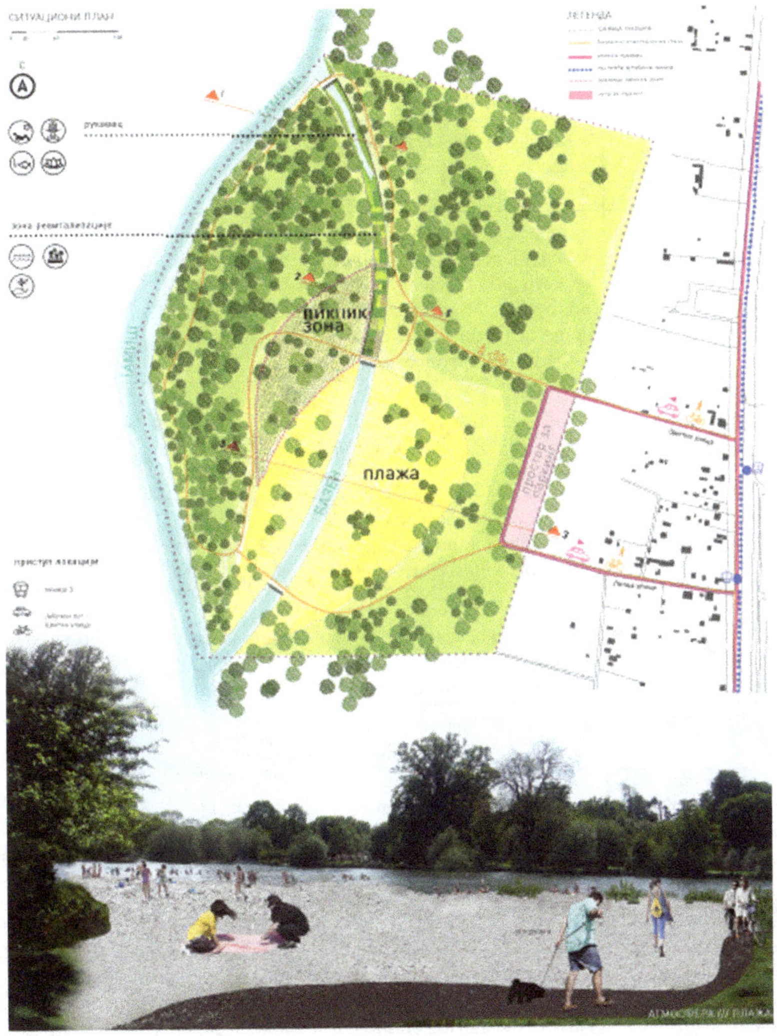

Urban-Rural Synergy: Establishment of a shared and accessible public information platform for all of Pančevo.

IUDP 10: Monitoring Implementation of the Strategy: Development of New Monitoring and Reporting Models

Problem: There is no entity specifically tasked with implementing the development strategy in Pančevo. The Implementation Monitoring Team is made up of staff whose primary responsibilities lie elsewhere in the city administration. In this context, the interim reporting model, project and activity monitoring, and final reports play a particularly significant role.

Table 1. Multi-level governance instruments in IUDPs for balanced urban-rural development

Integrated urban development projects	Urban-rural synergy themes	Multi-level governance instruments
IUDP 1: Farming co-operatives to promote organic production	• Protection of agricultural land and promotion of domestic production • Better land use	• Public-private partnership • Networks • Participation
IUDP 2: Rural tourism as a factor in the development of Pančevo's rural areas	• Busting of local economic development • Reduced unemployment • Better land use and planning of open spaces	• Networks • Participation
IUDP 3: Supporting social entrepreneurship by allowing access to underused town-owned office space	• Improvement of basic services • Better land use and planning • Green infrastructure • Protection of agricultural land and promotion of domestic production • Urban-rural partnership	• Public-private partnership
IUDP 4: Establishment of a civic association to support participation in activation and maintenance of public spaces	• Better management of public places	• Networks • Participation • Neighborhood Committee
IUDP 5: Ensuring inclusion of Pančevo's cultural offering into the Danube Route cultural network	• Improved social services • Improved cultural identity of the municipality • Improved tourist offering	• Networks • Participation
IUDP 6: Sustainable system to manage hazardous household and farming waste	• Green infrastructure • Protection of agricultural land and promotion of domestic production • Urban-rural partnership	• Public-private partnership
IIDP 7: Construction of a biogas facility in Pančevo	• Energy efficiency • Urban-rural partnership	• Public-private partnership
IUDP 8: Mapping environmental polluters and creating a GIS database for Pančevo	• Environment protection • Better risk disaster management	• Networks • Participation • GIS open platform
IUDP 9: E-PA development ideas platform	• Strengthening urban governance • Improving transparency in decision-making • Better land use and planning of open spaces	• Participation • E-platform
IUDP 10: Monitoring implementation of the Strategy: Development of new monitoring and reporting models	• Strengthening urban governance • Improving transparency in decision-making • Better monitoring of the Strategy implementation	• Participation • Participatory monitoring
IUDP 11: Integrated branding as instrument for improving the city's image	• Improved communication strategy	• Networks • Integrated branding office

Project Proposal: Support for the local authority in selecting, managing, and implementing projects; contribute to transparent, participatory, and effective governance and capacity-building for efficient implementation of the City of Pančevo Development Strategy; on-going and systemic project monitoring; results analysis; alignment of vision, budget, objectives, and activities; support for the local authority when applying for EU funds; and systematic, on-going, and appropriate public communication of completed, current, and future projects of the City of Pančevo.

Urban-Rural Synergy: Improved effectiveness and efficiency of local urban governance.

Figure 13. Participatory budgeting
(source: Authors) June 2016

Figure 14. Students' workshop: thematic grouping, complementarity of IUDPs, sources of funding
(source: Authors)

IUDP 11: Integrated Branding as an Instrument for Improving the City's Image

Problem: The city lacks a consistent image that could promote co-operation with outside partners and enhance economic power.

Project Proposal: Create a consistent City of Pančevo brand by mapping, presenting, and understanding the cultural values that make up its identity.

Urban-Rural Synergy: Creation of an integrated city branding office; creation of a stakeholder involvement platform; strengthening horizontal and vertical connections between authorities.

Students' projects address different challenges: devastation of agricultural land; migration of youth from rural to urban areas; poor social and technical public services, low income in rural areas; low quality of waste management; neglect of natural and cultural heritage. Also, their projects rely on underused potentials of endogenous values, urban identity and financing.

Public-private partnerships are planned to be complemented by other governance instruments that ought to enhance the effectiveness and efficiency of governance. Various types of participative instruments will be used to ensure integration, dialogue, and joint action by organisations and authorities at various levels and with private-sector and civil society stakeholders with the ultimate aim of attaining holistic and sustainable development. Participatory instruments are policies (such as regulations, plans, strategic documents, projects, etc.) that are designed to be produced in a process which fosters, promotes, and motivates the participation of a wide variety of public, civil society, and private-sector organisations, groups, communities, and members of the public in their formulation and implementation.

Finance is a key precondition for integrated projects. Particular attention was therefore devoted to budgeting each individual project, with reliance primarily on outside funding. The intention here is to relieve the local budget as much as possible, since existing projects under the Strategy are nearly completely funded from the city's coffers, or, indirectly, by the city's institutions and locally-owned enterprises (Mikavica & Čolić, 2017). Possible sources of finance that have been identified include the national budget, public-private partnerships, international financial instruments and donor programmes, and EU funds (IPA, Danube Transnational Programme, cross-border co-operation programmes, etc.). This approach results in the provision of three sources of finance, on average, for each programme, so significantly increasing their feasibility.

The students' projects envisage both pooling of resources and an integrated approach, which is achieved by aligning the proposed approaches with the integration of sectoral policies (as expressed in various types of documents) defined and implemented at various levels of governance that concern the territory of Pančevo (Milovanović Rodić, *et al.*, 2016). Moreover, to ensure the synergies of these proposed projects were fully understood, at the close of the semester a final student workshop was held where they were asked to map the projects (to identify their spatial coverage), group them thematically (relative to objectives of the local Strategy), assess their complementarity with other on-going initiatives, list all sources of funding, and specify the governance instruments involved (Figure 3). The students also proposed a slogan and logo for the entire batch of projects for Pančevo.

Masters' projects produced by students of the Integrated Urbanism master's programme are a set of mutually compatible projects intended for the territory of Pančevo (Figure 4). Some deal predominantly with spatial interventions and alterations to physical structure, whilst others focus on institutional design and mechanisms that facilitate change. Nevertheless, whatever their primary focus and initial impulse, all integrate various developmental aspects (spatial, environmental, social, and economic) to ensure sustainability of the solutions proposed.

SOLUTIONS AND RECOMMENDATIONS

Inspired by these results, the City of Pančevo and the Faculty of Architecture continued collaborating after the students graduated. The best-rated projects were incorporated into Pančevo's participatory budgeting process for 2017 with the involvement of the former students. The winning initiative was selected for implementation by the residents of Pančevo: a total of RSD 5mn (EUR 42,000) was set aside in the city's 2017 budget for its execution. Presentations of the projects, workshops, and exhibitions received media coverage, which gave the results visibility and incited public interest for taking part in the participatory budgeting project (Figure 5).

Partnership between the Faculty of Architecture and the City of Pančevo has allowed students to enhance their knowledge of balanced urban-rural development and how urban governance instruments are applied. It has enabled them to face modern-day challenges and complex development-related problems in an local context fraught with uncertainty. The Faculty will seek options for securing the continuity of this mode of teaching and maintaining such contacts with practitioners.

FUTURE RESEARCH DIRECTIONS

Better understanding and operationalisation of urban-rural relations requires on-going work and research. One of the options for this is the teaching model developed by the Integrated Urbanism master's programme at the Faculty of Architecture.

The students benefited from practice-oriented learning through collaboration when producing their integrated urban development projects. The objective was to enhance the quality of these contributions whilst focusing on contemporary topics, concepts, and methodology, in line with contemporary challenges and policies, and in co-operation with outside experts and local representatives. The results clearly show that such a programme can play a major role in continuing to advance the quality of the Integrated Urbanism master's course by contributing to academic knowledge whilst at the same time offering an opportunity to develop professional competences.

This generation of students has proven that the potential was there: apart from focusing on their development, additional work must be invested in creating opportunities for the students to practice and improve their knowledge and skills. These results bear out the need for continuing exploration and use of teaching methods that ensure effective interaction and emphasis on problems, which in turn contributes to mastery of the contemporary issues of balanced urban-rural development.

CONCLUSION

The teaching model applied proved to be justified in these conditions - both because of the operationalization of contemporary concepts in the local context, and also because of the additional legitimization of the development of a national policy that recognizes the IUDP as an instrument of implementation. The opportunity to work on a real, practical case within the framework of collaboration with different actors gave added value to the learning exercise and final project ideas.

Firstly, it promoted the understanding that the local development strategy is a set of complementary integrated projects for both urban and rural areas. These projects rely on the synergy of development paths that are based on the distinct potentials and constrains of rural and urban areas. Secondly, objectives of the projects were related to: (1) better management of both agricultural and construction land; (2) improvement of technical and social services; (3) climate resilience and environmental protection;

(4) support for employment of low-income groups (women and youth in rural areas); (5) improvement of natural and cultural heritage in rural areas, and (6) better urban governance. And, thirdly, besides the projects themselves, these ideas include different types of networks between public institutions, new types of partnership between private and public sector, and, as well, public participation and improved transparency in order to reach better governance in both urban and rural areas. Here, we relied on the principles of multilevel urban governance that are crucial for attaining integrated urban development – co-operation, participation, and integration of sectoral policies, stakeholders and financing.

The teaching model outlined above and students' projects promote the use of participatory multilevel governance instruments to achieve balanced urban-rural development. Their objective is to facilitate linkages, dialogue, and joint action between authorities at different levels of government, as well as between these bodies and stakeholders in the private and civil sector. The students' projects presented above are grounded in the understanding that balanced urban-rural development can only be attained if: (1) there is active and continuous collaboration between various stakeholders in designing and implementing solutions; (2) different sectoral policies are integrated; (3) co-ordination is allowed between different levels of government to implement integrated urban projects; (4) sources of financing are envisaged; (5) monitoring of implementation is ensured; and (6) institutions are able to change and adapt to an evolving development context and new challenges.

This practice constitutes innovation in linking academic education with practice and government, and holds significant potential for raising the capacities of all sides to face complex development-related problems, including balanced urban-rural continuum. Notwithstanding the uncertainties that lie ahead, Serbian cities have needed and will need professionals able to intelligently and creatively inspire and take part in planning, designing, and building sustainable cities and villages.

NOTE

The IUDPs were developed by the following students: Rastko Krstić, 'Farming co-operatives to promote organic production'; Katarina Panić, 'Rural tourism as a factor in the development of Pančevo's rural areas'; Božena Stojić, 'Supporting social entrepreneurship by allowing access to underused town-owned office space'; Sofija Nedeljković, 'Establishment of a civic association to support participation in activation and maintenance of public spaces'; Miloš Nikolić, 'Ensuring inclusion of Pančevo's cultural offering into the Danube Route cultural network', Sara Bastoni, 'Sustainable system to manage hazardous household and farming waste'; Mihailo Žikić, 'Construction of a biogas facility in Pančevo'; Iva Mihaela Bapkova, 'Mapping environmental polluters and creating a GIS database for Pančevo'; Bogdan Nikolić, 'E-PA development ideas platform'; Mihailo Mikavica, 'Monitoring implementation of the Strategy: Development of new monitoring and reporting models', and Dragana Joksimović, 'Integrated branding as an instrument for improving the city's image'.

This work was carried out within the framework of the projects "Environmental Climate Change Research: Monitoring of Impact, Adaptation and Mitigation" (43007) and "Spatial, ecological, energy and social aspects of settlement development and climate change - mutual influences" (TR 36035) funded by the Ministry of Education, Science and Technological Development of the Republic of Serbia.

REFERENCES

Čolić, R. (2015). Integrated Urban Development Strategy as an Instrument for Supporting Urban Governance. *Serbian Architectural Journal SAJ*, *7*(3), 317–342.

Čolić, R., Maruna, M., Milovanović Rodić, D., & Lalović, K. (2015). *Integralni urbani projekti za upravljanje rizikom od poplava na primeru Obrenovca: katalog izložbe završnih radova generacije studenata 2013/14* [Integrated Urban Projects for Disaster Risk Management: Case Study of Obrenovac: Final Works' Exhibition Catalogue: Generation of Students 2013/2014]. Beograd: Univerzitet u Beogradu, Arhitektonski fakultet.

Davoudi, S., & Stead, D. (2002). Urban-Rural Relationships: An introduction and a brief history. *Built Environment*, *28*(4), 269–277.

EC. (1999). *European Spatial Development Perspective: Towards Balanced and Sustainable Development of the Territory of the EU*. Retrieved May 15, 2019, from http://europa.eu.int/comm/regional_policy/sources/docoffic/official/reports/som_en.htm

EU. (2007). *Leipzig Charter on Sustainable Cities*. Retrieved May 15, 2019, from http://ec.europa.eu/regional_policy/archive/themes/urban/ leipzig_charter.pdf

EU. (2009). *The Committee of the Regions' White Paper on Multilevel Governance*. Retrieved May 15, 2019, from www.europarl.europa.eu/meetdocs/2009_2014/documents/regi/dv/cdr89-2009_/cdr89-2009_en.pdf

EU. (2010). *Toledo Declaration*. Retrieved May 15, 2019, from www.mdrap.ro/userfiles/declaratie_Toledo_enpdf

EU. (2016). *Urban Agenda for the EU*. Retrieved May 15, 2019, from http://ec.europa.eu/regional_policy/sources/policy/themes/ urban-development/agenda/pact-of-amsterdam.pdf

Government of Serbia. (2019). *Sustainable Urban Development Strategy of Serbia to 2030*. Official Gazette of the Republic of Serbia, No. 47/2019.

Maruna, M., & Čolić, R. (Eds.). (2014). Integralni urbani projekti za razvoj centra Kragujevca: Katalog izložbe završnih radova generacije studenata 2012/13 [Integrated Urban Projects for Kragujevac Inner City Development: Final Works' Exhibition Catalogue: Generation of Students 2012/2013]. Beograd: Univerzitet u Beogradu, Arhitektonski fakultet.

Maruna, M., & Čolić, R. (Eds.). (2015). Inovativni metodološki pristup izradi master rada: doprinos edukaciji profila urbaniste [The innovative methodological approach to the development of master work: A contribution to the education of urban planners profile]. Beograd: Arhitektonski fakultet & GIZ/ AMBERO Beograd.

Maruna, M., Čolić, R., Fokdal, J., Zehner, C., Milovanović-Rodić, D., & Lalović, K. (2015). Collaborative and Practice Oriented Learning on Disaster Risk Management in Post-Socialist Transition Countries. In *Proceeding from the XVI N-AERUS Conference Who wins and who loses? Exploring and learning from transformations and actors in the cities of the South* (pp. 408-421). Academic Press.

Maruna, M., Milovanović Rodić, D., & Čolić, R. (2018). Remodelling Urban Planning Education for Sustainable Development: The case of Serbia. *International Journal of Sustainability in Higher Education, 19*(4), 658–680. doi:10.1108/IJSHE-07-2017-0102

Mikavica, M., & Čolić, R. (2017). Monitoring implementacije strategije razvoja – primer Pančeva [Monitoring implementation of city development strategy – the case study of Pančevo], Arhitektura i Urbanizam, 44, 36-46.

Milovanović Rodić, D., Čolić, R., & Maruna, M. (2017). Multilevel Governance Instruments for Achieving Balanced Urban-Rural Development. In *Proceedings of 4th International Academic Conference Places and Technologies 2017. Keeping up with Technologies in the Context of Urban and Rural Synergy* (pp. 332-340). Academic Press.

Milovanović Rodić, D., Čolić, R., & Maruna, M. (2018). The Role of University in a Policy Making Process - Introducing Integrated Urban Projects for Effective Urban Governance in Serbia. In E. Anguillari & B. Dimitrijević (Eds.), *Integrated urban planning: directions, resources and territories* (pp. 63–83). TU Delft Open.

Milovanović Rodić, D., Lalović, K., & Nenadović, A. (2011). Virtual social networks as a booster for reaching livable villages: examining Serbian cases. In *Proceedings of 47 th ISOCARP Congress proceedings Liveable Cities: Urbanising World, Meeting the Challenge* (pp. 1-14). Academic Press.

Milovanović Rodić, D., Maruna, M., & Čolić, R. (2016). *Instrumenti upravljanja integralnim urbanim razvojem na primeru grada Pančeva* [Instruments for Integrated Urban Governance- Case of Pancevo]. Beograd: Arhitektonski fakultet Univerziteta u Beogradu.

Milovanović Rodić, D., & Stojić, B. (2018). Integralni pristup za održivi razvoj dunavskog regiona: prikaz pedagoškog modela i ostvarenih rezultata – master teza [Integral Approach for the Sustainable Development of the Danube Region: Pedagogical model and Achieved results - Master Thesis]. In *Proceedings of 14th Summer School of Urbanism* (pp. 72-82). Bjeljina: UUS, RGZ, GIZ.

Milovanović Rodić, D., Živković, J., & Lalović, K. (2013). Changing Architectural Education for Reaching Sustainable Future: A Contribution to the Discussion. *Spatium (Belgrade), 29*(29), 75–80. doi:10.2298/SPAT1329075M

Ministry of Agriculture and Environmental Protection Republic of Serbia. (2014). *Republic of Serbia IPARD Programme for 2014-2020.* Retrieved May 15, 2019, from https://ipard.co.rs/doc/IPARD%20 2014-2020.pdf

Nilsson, K., Sick Nielsen, T., Aalbers, C., Bell, S., Boitier, B., Chery, J. P., ... Loibl, W. (2014). *Strategies for Sustainable Urban Development and Urban-Rural Linkages. European Journal of Spatial Development.*

Tasan-Kok, T., & Vranken,, J. (2011). *Handbook for Multilevel Urban Governance in Europe. Analysing Participatory Instruments for an Integrated Urban Development.* The Hague: European Urban Knowledge Network.

Trkulja, S., Colic, R., & Maksin, M. (2018). *Sustainable and Integrated Urban Development Strategy of the Republic of Serbia until 2030.* Novi Sad: Artprint Media.

UN. (2015). *The 2030 Agenda for Sustainable Development.* Retrieved May 15, 2019, from https://sustainabledevelopment.un.org/?menu=1300

UN HABITAT. (2016). *New Urban Agenda.* Retrieved May 15, 2019, from https://www2.habitat3.org/

Zonneveld, W., & Stead, D. (2007, August). European Territorial Cooperation and the Concept of Urban - Rural Relationships. *Planning Practice and Research, 22*(3), 439–453. doi:10.1080/02697450701666787

KEY TERMS AND DEFINITIONS

Integrated Urban Development Projects: IUPSs can be targeted at spatial/physical intervention measures, but are usually a combination of spatial, environmental, cultural, social, economic aspects, and initiatives.

Multilevel Urban Governance: A concept highlighting both shifts in horizontal relations between state and society and changes in the vertical relations between actors located at different territorial levels.

Practice-Oriented Education Through Collaboration: Inspiring pedagogy and a learning environment that allows students to actively engage with real problems.

Chapter 16
Contemporary Rural Settlements:
New Paradigm for Planning and Constructing Rural Settlements Based on Complexity Theory

Dženana Bijedić
https://orcid.org/0000-0002-4134-0058
University of Sarajevo, Bosnia and Herzegovina

Rada Čahtarević
University of Sarajevo, Bosnia and Herzegovina

Denis Zvizdić
https://orcid.org/0000-0002-4134-0058
University of Sarajevo, Bosnia and Herzegovina

Adna Proho
University of Sarajevo, Bosnia and Herzegovina

ABSTRACT

Throughout history, built environment developed intuitively on empirical experience led by trial and error. Such approach provided resilience and evolvement of patterns that guided spatial organization. Newer interventions in the rural environment resulted in disconnected spatial fragments. By comparing vernacular and contemporary planning and construction practice in the natural park Blidinje, the authors tried to identify the reason for which contemporary interventions resulted in new patterns in spatial planning, ones completely unfamiliar for this geographical area. They identified the reason in the fact that men started to treat the natural space as a commodity, forcing stakeholders to be led primarily by economic principles. Such principles are rigid and linear, instead of contextual in this matter. At the same time, the environmentally sound approach should respect complexity of whole endeavor aiming to achieve diversity and variability. The models developed based on complexity theory and self-organization should preserve continuity and integrity of the place and man.

DOI: 10.4018/978-1-5225-9932-6.ch016

INTRODUCTION

Intuitive and empirical knowledge of architectural rural and vernacular heritage is related to slower development over many generations, providing strong resilience and maintenance of the inherited patterns of living. Individual needs are incorporated in tradition by variability and adaptability to the local changes and natural resources. Traditional vernacular patterns are a result of synergy and interconnection of all spatial elements, from material to functional and structural, as a product of self-organization and self-preservation in the complex environment.

The consequences of some contemporary interventions in traditional settings could produce conflict and degradation of natural environment with built structures. In the case study of the recent architectural planning and built objects in the area of Nature Park Blidinje, we examined the impact of contemporary planning to the natural and traditional vernacular environment.

BACKGROUND

Authors argue that newer interventions, which follow the urban planning rules and procedures, in the rural environment have not achieved the synergy between the built and the given environment. The elaboration aims to justify and to prove this by comparing two artificial interventions in the natural environment – the old, vernacular and the new, planned one. At the same time, we also tried to identify the causes that led to this, discussing the social context of the socio-techno-economic development of society.

Artificial elements - livestock houses, stalls, hooves, cultivated land, communications, were until recently an excellent example of anthropogenic type of interventions done by generations in a full harmony with the natural environment, both in functional and purpose sense. Acting in the communion with the mountain, respecting its laws and its rhythm, always with a sense of measure, the seasonal inhabitants of the Čvrsnica and Vran-planina preserved the simple and viable philosophy of living (Vego, 260).

PHILOSOPHY AND ETHICS REQUIREMENTS FOR UNDERSTANDING COMPLEXITY OF NATURAL ENVIRONMENT

Impressed by the possibilities that science has given to the control of nature and the overall human environment, changed the way of thinking that ancient thinkers had when intervening in the space (Aristotle's model of science). Namely, humans became more convinced in their superiority in creation, monitoring, and control of the events and stood out and stood above nature, ignoring any restrictions from that world. They accepted modern, Cartesian, theory of science based on rationalization, a priori separating the experience from the theory, and a claim that only pure reason separated from the material world can objectively observe it. In this sense, classical philosophy starts with clearly defined principles or "premises", which are considered absolute truths. Following the logic, all the conclusions of such reflection and their consequences, though often absurd, were accepted.

Contrary to this, evolved the axiomatic method in philosophy (formally similar, but different from the modern mathematical axiomatic method), which is very carefully examining and accepting the initial assumptions. It formulates the theory upon a logical re-examination. The main characteristic of this method is to confirm theory by the experience. If it turns out that practice cannot support the derived logical upgrade, other axioms are introduced, until the experiential confirmation of theory comes to fruition.

The axiomatic method, in a sense, is a return to Aristotle's model, which also claims that we must always move from practice to theory, and vice versa, with awareness of the real world. When the axiomatic method is combined with careful and objective empirical testing of the consequences deduced from the axioms, when these axioms concern empirical subjects, we have the outlines of the scientific method, which first fully became clear to Galileo.

In the natural environment system, interaction, interdependence and the consequences of entering into the interdependencies of its subsystems are so intertwined that abstraction, isolation, and neglect of these complex states become impossible. The natural system in which we deeply dissipate by interpolating the system of built environment puts the architect into a position of deep re-examining new alternatives to solving problems that man creates. Such a new way of thinking should begin in the light of relatively recent discoveries and theories in science that emphasize the much greater complexity of the world, the far greater complexity of the relationship, then it has been thought for many years.

Theories that are deeply rooted in the foundations of the classical, Cartesian model of world thinking and the Newtonian mechanic universe are primarily: Einstein's theory of relativity, Heinsberg principle of uncertainty, and the theory of chaos and fractals (Peitgen et al, 2004). The chaos theory is a deterministic universe that respects the basic laws of physics, but with the awareness of its predisposition to the mess, complexity, and unpredictability. This theory can help us understand the behavior of complex systems that seem unpredictable, with its support it enables us to understand the complex forms of organization of natural systems. The first possibility offered to us to think in the accordance with the chaos theory is the word fractal, described by Benoit Mandelbrot and derived from the Latin word '*fractus*' meaning fragmented.

The world's metaphor of the world-machine was a modern way of thinking that reflected the achievements of the 'Scientific Revolutions' established on the foundations of discoveries of Galileo and Newton and new methods of reasoning by Bacon and Descartes. Guided by this paradigm, many scientists have approached the interpretation and discovery of nature through separate entities that they treated as parts of the machine, where the links of particular entities were not considered as important. A similar relationship was also experienced in the field of art, and the re-examination of such attitude, initiated by studying of fractal and chaos, John Brigs (Lačević, 2004) formulates this:

The question is whether we are going to inhabit the world shaped (as we have long believed) by lifeless intertwining mechanical fragments that are trigger by mechanical laws and are awaiting our edifice and control? Or will we inhabit the world - the one that fosters fractals and chaos - who is alive, creative and diverse, because parts of it are united, inseparable and born of unpredictability, which in the ultimate sense is beyond our control? What scientists have the intrinsic attachment to new control shows that scientists have for a long time taken away the joy of free focus on the visual dimensions of their work with nature.

Today, scholars and scientists largely accept the paradigm shift from mechanical to organic, in terms of dramatic changes in world views, where architecture is no exception. It is evident that through the history of architecture many of the methodologies depict certain aspects of organic reflection and thus shaped the architectural vision of a new paradigm.

The sustainable, green, ecological approach to the emergence of architecturally defined space must be understood as a philosophy. If we want our facilities to be compatible with the principles of sustainability of the world, we have to accept the principle that our work is not simply a list of what we need to fulfill in a building process. This is why holistic ethics, that involves all participants in the process of dialogue by realizing the backlinks of the process, continuously upgrades and improves. The effort is to find out the approach that will reflect the efficiency and diversity of nature by creating self-regulating design solutions, full of spirit. By acknowledging the environmental issue and introducing the notion of ecological responsibility in the process of creating a built environment we come out of the scientific-technical domain, in which architectural creativity has moved from accepting the achievements of the scientific-technological revolution. In a way, we return to the time of the beginning of construction when a man was a true part of nature, that return means turning again to the artistic component of this form of human creativity (Bijedić, 2012). The architecture balancing the relationship between science and technology (engineering) and talent (art), put extra responsibility to our professional conduct, or as Fatima Lačević, concluding her 'Sketch of Aesthetic Response to Chaos and Fractals', claims:

Science and art put a demand for intimacy with the cosmos, not control of the human environment. That is why the theory of chaos and the new aesthetic are essentially hopeful for the salvation of the world. (Lačević, 2004).

The Mechanistic and Organic Paradigm of Nature

Natural space is a material reality in which different forms of life exist and develop, encompassing the living and nonliving world, as the natural environment. A mechanical deterministic image of the world based on scientific discoveries in the 16th century, including the works of Kepler, Galileo, and Newton, are giving the power to the mankind towards understanding and controlling the processes of natural space, intended for the purposeful use to humans. Nature in modern society becomes the subject of research and exploitation, and the domination of man. By separating human and the natural environment, man has disrupted many natural processes, disturbing the delicate balance in nature. Industrial civilization, by its interventions in a natural environment, based on mechanical principles, destroys the organic balance of the living and the inanimate world.

The development of modern civilization has troubled the balance of man and his environment, which is reflected by the degradation of natural ecosystems, in return having effects on the quality of life and man himself. The relation between man and his natural environment is the result of the vision of the world shaped through various religious, cultural and philosophical approaches and development of human civilization, as an essentially differentiated image of man and nature.

An anthropocentric mechanistic image that characterizes technological civilization is opposed by the organic picture of the world, which rests on the complexity and unity of natural processes involving human beings. Such organic picture of the world is portrayed by the majority of the Eastern philosophy, through Buddhist and Hindu principles, that all things flow infinitely into one another.

The organic picture of the world can also be found in the works of some Greek philosophers. The most famous are theses of the philosopher Heraclitus, describing the nature as the flow of energy, and not substances that are constantly transformed. In Plato's works, a human being is defined as a microcosm that reflects the macrocosms and is an integral part of the cosmic organism, and the universe is a world of itself, a world of change. In the work of some philosophers of the 18th and 19th centuries, an organic picture of nature is emphasized, for example, in Holbach's work "The System of Nature", where he presents complex natural dynamics as a chain of causes, the effects that flow from one another." (Clarke, 1993).

Turning to nature was characteristic of the Romanticism at the end of the 18th century, denying the simplification or control of nature, subjecting its processes to strict scientific laws.

The Romantic movement authors and philosophers as J.J. Russeau and Fichte, emphasizes the existence of continual changes in nature, and the connection of all phenomena of nature into a single whole. Whitehead in his organic, process philosophy, defines natural processes as creative rather than deterministic, nature as a process of continuous energy flow, giving a holistic picture of the world on which, the later principles of ecology are based.-(Clarke, 1993). Organic units according to Neigel (Neigel, 1974) are defined as such systems whose characteristic is the connectivity of all parts, so that any change in one, causes a change in all parts, so such entities cannot be analyzed only from the additive point of view. The natural space represented in organic paradigm as a whole, is an open, complex dynamic system.

The Complexity of the Living Space

Ecosystems, and indeed the global biosphere, are prototypical examples of complex adaptive systems (Levin, 1998).

New theories of complex dynamic systems have been applied in studying the natural processes and its dynamic, according to similarities between natural and other complex informational and physical systems based on abstracted network modeling and nonlinear interactions (Parrot, 2000).

In the living environment, living and non-living parts are interconnected by dynamic flows and cycles. Ecological, living space is not a static structure, but a dynamic interactive whole. This whole involves multifaceted systems that interact with each other as open structures.

Through the continuous circulation of matter and energy, the spatial structures of the natural environment are in the continuous process of formation, dynamic development, change and growth.

Natural forms as the results of continuous transformations represent a part of a complex structural system made up of very diverse units, in which each element represents what it is only in relation to other parts of the systemic whole. The subsystems are integrated into more complex units of a higher order. Natural systems are in a continuous state of activity, energy exchange and information with their environment, which is also part of the system at a certain level.

The complexity of natural structures cannot be reduced to simple properties or elemental parts, to simplified laws, and to separate categories. The boundaries of individual ecosystems are not clearly determined since all subsystems are integrated into complex ecological complexes of the higher order.

As a result of complex processes, natural forms are consistently variable in content, composed of different elements, making very complex visual structures, including great variety and innumerable variations, without strictly defined boundaries. Natural forms are non-linear, open structures, which are in a continuous development process of growth or decomposition, changing through spatial-temporal forms, in constant cyclical dynamics. Not only living forms but also forms of geological forms - mountains, rivers, and geological landscapes - are changing under the influence of various forces and actions. The entire, seemingly fixed "surface" of the earth, in the scales that transcend human understanding of time, oscillates, without fixed shape. The landscape we see today is only the current view in geological time. (Shimer, 1959.).

More recently, the problems of instability and the different spatial-temporal scales of the landscape is emphasized. "Non-equilibrium ecology" explores the processes that determine the landscape patterns on different scales, introducing an interest in chaos theory and fractal geometry. "Nature itself is dynamic and highly variable, its patterns at one particular place and time contingent on preceding events, its trajectory through time is open-ended and not tending at the equilibrium point." (Adams, 1996).

As the essential features of complex adaptive systems Levin (1998) mentions: diversity and individuality of components, localized interactions among those components, autonomous process of selection, continual adaptation, and the absence of global control and the emergence of hierarchical organization.

The integrity of the natural environment is reflected through the interconnection and interaction of all its parts. Complexity and self-organization of the natural systems supporting variations, adaptability, and developmental changes, exchanging energy and information with the environment, are the reflection of the dynamic processes and interaction on the multileveled spatial and temporal scales of the heterogeneous components in mutual interdependencies (Levin, 1998). Each individual natural system exists in a complex network of various conditions and environmental influences, with which it forms a dynamic whole. Organic natural space possesses dynamic stability that is not dependent on classical physical balance but is the result of more complex dynamic developing processes that include a time dimension. Spatial entities with their activity establish a complex communication network of interactions, creating a complex system of the individual and its surroundings. Each individual is a whole which itself represents an independent field of related activities.

Natural space is not homogeneous and infinite, and continuity is established at a complex hierarchical level. Its extension is not linear, but it can be viewed as gradual, at certain levels of complexity having natural boundaries, with differentiated areas or subfields. Time is not a complementary phenomenon of natural reality but is synthesized in a spatial-temporal dimension, in the form of a series of spatial levels of a certain continuous configuration, as a stratification of developmental dynamics.

The natural space is in a continuous process of formation, development, and degradation, continuous changes, whether it is his living or inanimate structures, the coast of the sea or the forest area. Self-regulation and self-organization are characterized by natural systems, which maintain balance through complex dynamic processes. These are very sensitive equilibrium systems, which can disrupt the smallest environmental impact, leaving lasting consequences and causing unpredictable changes.

As life is not a stable balance, it is important to emphasize the implication of how ecosystems are maintained. As the state of the environment constantly changes in time and space, ecosystems recognize the many development paths they can follow. This situation is even more pronounced due to the fact that ecosystems, as complex non-linear systems, are undefined and can be modified in unobtrusive ways, responding to minor disorders in the environment or system conditions, i.e. they have high self-regulation ability. It is therefore important to analyze the ecosystem, to understand them as dynamic, variable entities. As all ecosystems are exchanging matter and energy, whether they are the same or absorb, we consider them open systems.

Human Intervention in Natural Space

Human actions are part of the web of influences on ecological change, not external equilibrium - disturbing impacts. (Adams, 1996)

The overall development of the human organism is the result of adaptation to the environment, with limited possibilities of compensating for certain disorders. The human organism, as well as every natural system has its limitations, both spatial and temporal. The biological structure of the human body is a complex system that is sensitive to all external influences with a certain tolerance framework for the changes it is exposed to.

Each natural system has its own limitations, which are not linear in nature, so we cannot speak of strictly determined boundaries, but also not about the unlimited expansion of any natural system. This limitation allows the coexistence of an extremely large number of different natural systems, at different levels of complexity, time and spatial flows and various functional activities. The diversity and richness of forms and shapes in nature are made possible by the limited expansion of any particular system.

Forcing the unlimited linear expansion of technological spatial systems into the natural space, human intervention in space, neglected the organic principles of growth and self-regulation, leading to disturbance in the balance of his natural environment.

Within the framework of modern civilization, architecture represents the activity that contributed mostly to changes in the environment. Linear expansion of artificial space, at the expense of natural, treating natural space as a global homogeneous surface on which classical geometric principles are applied, this surface is fragmented without taking into account the organic unity of natural space. By applying linear modeling to natural space, treating it as a quantitative static homogeneous structure, its defragmentation becomes possible without taking into account the complexity or limits as characteristics of natural processes. This treatment of space results in the negation of geological, biological, climatic, ethnological, psychological aspects of the space.

The separation of man from nature is the result of the erroneous assumption of the separation or independence of natural phenomena from the very human. This led to the dehumanization of space at all levels, from biochemical contamination to psychological alienation. Fragmentation of urban space, mass production, standardization, and globalization made a man a helpless prisoner of space in which his perception and power of movement were restricted and debilitated.

In the human artificial space, principles that allow the development and renewal of life are ignored, the natural circulation of matter and energy is disabled, natural processes have been disturbed. The psychological and emotional character of the connection between man and his environment has been ignored, resulting in visual confusion, distorted orientation and the alienation of man from the source of his psychological stability.

By applying a classical rational, linear geometric representation to a natural space, treating it as a quantitative, homogeneous structure, its division becomes possible without taking into account the complexity or constraints characterizing natural processes. This treatment of space results in the negation of geological, biological, climatic, ethnological, psychological aspects of the space which leads to the congestion of all natural processes influencing in reverse human quality of life.

The development of spatial systems based on classical spatial modeling has reached its limits. A complex approach to the spatial conception requires the reintegration of the natural and urban environment, the interweaving of natural and artificial spaces, the opposition to the linear development of artificial spatial structures and the enabling of organic development and the growth of all spatial systems. All spatial systems must satisfy certain limits in order to provide a stable balance of intertwined relationships with other systems and the independence of functions on quantitative growth.

Classical geometric model is lacking the layering of space, the limitation of areas or zones, the correlation between certain levels and sub-levels of space, periodic and aperiodic dynamics, developmental and evolutionary dynamics, discontinuity or jumps between certain structural levels, and, finally, the dynamics of reproduction that are characteristic to the real, natural space. The inclusion of these spatial qualities into a more complex model would be more suitable for better integration within the environment. Natural spatial forms, formed as a result of complex processes, do not fit into deterministic linear geometric laws, treated as seemingly irregular. By the discovery of new geometry domains such as, for example, fractal geometry whose forms visually match many natural forms, and by the informational technology and modeling of complex dynamic patterns, the new horizons are emerging in many areas including architecture and spatial planning.

The Images of Nature as Commodities

Any framework that is going to engage with biodiversity conservation and the concept of nature will need to deal with the physical environment and especially with the place of people in the physical environment. However, it must also deal with ideas, images and representations… (Brockington, Duffy, Igoe, 2012)

The question of the true character of nature, whether it is just a social construction or something beyond human comprehension, have been raised in philosophy and natural sciences. In the period of Enlightenment, Nature as something wild, vitally in-deterministic, have been separated from human society and culture, forming nature/culture binary opposition (Adams, 1996.) But importance of socially conditioned representation of nature, are today emphasized, because socially and culturally constructed images and representations of nature have impact on human intervention in natural environment influencing and changing natural processes.

Brockington, Duffy, Igoe (2012) argued that landscape conservation and establishment of protected areas, in neoliberal capitalism serve to "market- based conservation programs and the faith in tourism as a driver for sustainability", producing mediated experience, as "ecotourism bubbles" or "green boxes of consumptive tourism". Nature becomes "the spectacle of Nature", as an image, a simplified model, the representation of the enjoyable, comfortable spectacle of domesticated wilderness. Nature is seen "from inside the box" as "Eco capitalists" and "Ecotourists" are intertwined, in the joint activity of enhancing growing "global economy of images" as commodities. Contemporary "Ecotourists" are "presented with nature in ways that render invisible the temporal and spatial scopes of ecological processes." (Brockington, Duffy, Igoe, 2012)

Imposing simplified, deterministic, domesticated and shallow image of nature onto the natural complexity as its vital wildness, human intervention in natural space isolate nature into the separated "green boxes", controlled and alienated from the network of environmental complexity and its true kernel.

COMPLEXITY PARADIGM IN MAN-MADE ENVIRONMENT

Complexity Theory in Spatial Planning and Design

Complexity theory has been developed in the middle of the 20th century, but it flourished towards the beginning of the new 21st century. Structure and dynamics of complex systems have been investigated in the work of H Simon, in the 60s, but the theory becomes widely spread in the different scientific fields in the late 90s. According to Simon, a complex system is "made up of a large number of parts that interact in a non-simple way. In such systems, the whole is more than the sum of the parts..." (Simon, 1962). Simon emphasized that hierarchical organization in complex systems is based on different types of connections on different levels and scales, forming the complex interactive structure. Nonlinearity, self-organization, emergence, and unpredictability are the main concepts connected to the complexity of science (Cilliers 2001, Heylighen, 1989). Complex systems are dynamic, multilayered, open, connected to its environment with unclear borders, highly interactive, transformable, and adaptable (Cilliers, 1998, Batty&Torrens, 2001.).

Self-organization is defined as spontaneous internal process, development of organised structure out of disordered parts, creating dynamical emergent whole. (Heylighen, 1989.) Self-organization is the active maintenance and formation of order in hierarchically organized levels without central control and management (Krohn, 1990). Self-organization is an organizational principle that permeates the living natural world. Natural systems, as open systems that exchange energy and matter with their environment, are complex designed self-organizing synergetic systems. The stability of natural self-organizing systems does not occur in a static or simple mechanic equilibrium state. Instability within such systems is a source of dynamic order, including hierarchical organizing levels. The very complex dynamics of such systems is the subject of study of the Chaos theory, which studies the existence of regularity in the state of matter far from equilibrium.

In states of instability, in which such systems lose their structure due to influence outside the system, there are unpredictable states in which the system itself "selects" a certain type of change by which the system returns to a state of arrangement that is different from the previous one, and thus creates its new form of organization. In order to achieve higher order, the system can behave in two ways: the system expands by introducing compatible elements into a new structure, or creating an order out of disorder where external disorders are incorporated into the structure of the system.

By discoveries of chaos theory and physics of dissipative systems that matter in some states is self-organized, choosing its future structure, creating its new form of organization, unpredictable and irreversible to the previous state, the physical matter can no longer be reduced to parts whose properties determine the character of the whole (Prigogine, 1982).

The Emergence Principle attributed to living organisms, where the characteristics of the entire organism cannot be derived from knowledge of the components, is extended to the behavior of the matter itself, linking both the living and the nonliving natural structures. This establishes a unique picture of natural phenomena, destroying the strictly withdrawn boundaries of the organic and inorganic world, opening up many areas of science to the new, more unified and holistic approaches that combine the dual structure of scientific theories.

Principles of the complex dynamical natural processes could be abstracted and simulated, enabling modeling understanding and exploration of different similar natural or artificial materials and systems (Batty, Torrens, 2001).

Biological forms have informational potential, revealing their unique structural complexity. Their development is taking place through several distinctive stages, and their shape are emerging from non-linear dynamic, through many transformations through different stages. One level of development leads to new levels, through new transformations. Small changes in environment can cause variability and diversity of future stages. One unique primary structure could flourish in multiple variations, producing rich and unpredictable environment.

New concepts and models should be applied in architectural planning design, integrating natural and artificial environment, on the basis on their complex nature, forming interconnected, evolutionary space as an emergent whole. The concept of space in the evolutionary model has a processed character, and it is in a constant flow of transformation, which is the result of exposure to a multitude of different actions. Global and local are not in a linear relationship, because the whole is not just the sum of its parts.

Holistic principles in design should integrate top-down and bottom-up processes, providing global development and dynamic stability on many spatial and temporal scales. New behavioral patterns on the higher level could emerge in such systems as a result of highly interactive parts on lower levels, producing unpredictable, and creative changes.

Complexity in Traditional Rural Settlements

Modern planning is based on top-down principles and rigid rational geometry aiming to achieve unity, predictability, and stability of future states. Unity is achieved by homogeneity and uniformity. Continuity is based on linearity and repetition, governed by one dominating macro-scale, without consideration of diversity and multiplicity of different natural scales.

Homogenous distribution on one spatial scale in modern planning is in conflict with natural principles of distribution and hierarchy of sizes. Macroscales in nature are distributed in proportional relations to smaller scales, forming complex unity (Salingaros, 1999). Interconnectivity is achieved on many scales simultaneously, providing robustness and antifragility of systems, by overlapping, mixing and forming connections by different amount of involvement on different levels. Evolving wholes. are formed by participation of all scales, on collective behavior producing emergent properties.

Empirical and traditional experience of the rural settlements and local traditions, providing flexibility in the context of natural processes produce complex spatial patterns without formal composition or order. Bottom-up growth of the traditional settlements were reflection of self-organized processes on different functional, economic and physical sub levels and scales, in mutual connective interdependence of natural and human life.

Individual builders of the rural settlements, using local materials, build traditional shapes and structures by the knowledge inherited through many generations. Traditional rural settlements, built without rational general planning and order, are founded in synchronicity with natural processes and environmental exchange of materials and energy. Individual actions were the main generator of planning, but individual interests as self-organization were in accordance with natural processes. The process of self-organization is characteristic of many physical and organic phenomena. Interconnectivity is spread through many levels including the systemic connections to the environment. The simplicity of basic order is transferred through many spatial-temporal levels and scales, resulting in a complex order, nonlinear and unstable. Smaller scales of individual spatial intervention produce bigger scales of collective settlements, loosely connected and highly interconnected to the natural environment.

Structure of used natural materials in traditional vernacular buildings is complex and heterogeneous, as a result of many multiplying and repetitive natural processes on many different levels. Simple laws and order are transferred through many levels, making a complex dynamic unity. Natural materials are part of the hierarchical network structure of the whole environment, with different types of connections on different levels, governing the adaptability and response to changes in the environment.

The short-term spatial pattern of rural settlement was variable and unfinished, allowing flexibility, adjustability and long-term persistence of living in accordance with a complex natural environment.

CASE STUDY - BLIDINJE NATURE PARK

Built Environment: Social and Spatial Context

Existing – Vernacular

In the past, the area of Blidinje Nature Park was not populated by a permanent population. Residents of lowland areas migrated to the mountais early in the summer to avoid the problem of adverse climatic conditions in the lowlands, large droughts and lack of water. The mountains gave them the opportunity to preserve livestock, produce meat, wool, and dairy products. However, some of the temporary inhabitants, by time, formed permanent settlements. When constructing the necessary buildings, they used local materials, respected orientation of site, as well as the terrain slopes, the winds, the outside temperatures, the neighbors, etc. They have lived for centuries in harmony with the nature, following a simple and sustainable life philosophy. As it is usually the case with successfully interpolated structures in to natural environment, unknown builder of that time created his living space carefully. For the purpose of making shelter for his family and livestock he used locally available resources, not dispersing material and energy in vain, while at the same time creating an expressive outdoor and indoor space.

As Salopek observed: "Regardless how, from the point of view of the civil historiography, the role of the peasantry seemed to be ahistorical. The fact that for centuries they lived in the natural surrounding that was barely modified, conveying their traditional life philosophy... The folk creator of that time has created his living space carefully, not wasting his strength. He was eloquent and expressive as life itself..." (Salopek, 1974).

In order to create adequate conditions for living, cattle breeding, dairy production, and agriculture, various types of buildings, such as shelters, houses and stables have been built.

The livestock house was first constructed as a ground floor rectangular, seasonal type building (mostly 4x6 m), positioned directly next to a cattle farm, on a landscaped ground and close to the springs of drinking water. In addition to the function of residence, milk and milk products were processed there. In this sense, the house is organized through two parts of the same surface. The first part, the fireplace – '*ognjište*', is intended for living, preparing for food and sleeping. The second part, 'mlikar', served to store dairy products. The apartment was entered through the low entrance door, which is also the only entrance on the building (Živković, 2015).

The construction of the cattle house was made of local wood and the structure was designed to be disassembled after the summer season. The disassembled material was preserved for reuse in the following year.

Due to the changes that followed, the cattle breeding was not so interesting for younger generations, the people needs changed, that was reflected in the organization of the inner space of the houses. First the fire place and the storage space for the greasy products were placed in separate buildings, and in the end of the evolution of the summer house it transformed into a residence building (4x7 m) for the whole year. The permanent residences unit included the space for storing dairy products. Parallel with this development the separated facilities for cattle, stables ware built. The stands were built on a location that was protected from wind blows and which was not threatened by the summer heat. The choice would normally be a downhill, south or southwest orientation. The characteristic stables ware built as two-story on a rectangular basis. The division into the floors supports the functional division: ground floor - cattle and the floor - storage of drowned and dried fodder. The entrance to the ground floor was buried on the one side of the stall, placed against the storm of the northeast wind, while the entrance to the upper floor is usually on the opposite side - directly from the ground. The roof pitch is approximately 60 degrees on the lateral surfaces, or more than 80 degrees on the front and back roof.

The flats and stalls lost their function in time, some of them were not used and the buildings are currently in ruinous condition, or only parts of the building remained. However, some buildings have undergone a transformation, and modifications of the physical structure according to the owners' needs and possibilities.

Existing – New

In relation to the construction period that left the mentioned architectural heritage, the later period produced different buildings and completely new spatial organization patters. Those houses were constructed in a manner that was more or less respecting, or less than respecting a local sites, i.e. given environment. Some structures were extension of the existing buildings, while some were built as separate structures.

The scientific and technological revolution increased man's power to manipulate nature. Development run by ever increasing human needs, resulted in selfishness, and arrogance towards tradition, heritage and natural resources that include land, water, materials. In that sense, the achievements of the science and technology are often in the service of narrow interests, not of the interest of the common good. As Al Gore emphasizes: "a sudden acceleration of the scientific and technological revolution, which has allowed an almost unimaginable magnification of our power to affect the word around by burning, cutting, digging, moving and transforming the physical matter that makes up the earth.", (Al Gore, 2008).

The consequence of contemporary economic policy is alienation of the human connection with the natural world. In the end man started abusing the Earth and its resources, although his life on the Planet directly depend on its capacity. The moral set of values is neglected, the main driver of economics is maximizing profits. This is mostly because, as Peter Russell in The Awakening Earth notes: "In most cases, science and technology are in the service of personal, collective or national interests, not the interests of humanity and the planet.", (Abadzic, 2001). In such a system, economic activities lose the necessary relationship with the ethic and neglected human's long-term needs.

The construction of apartment complex at the Blidinje Nature Park was done in accordance to the above described economic policy. From the aspect of commercialization, in the period from the beginning of the industrial revolution to the present day, the products of the process of creating a built environment, infrastructure and buildings of various forms, sizes and purposes have become commercial goods, producing more or less value, unified, separated from the place, and as a rule reduced to something that can be sold and purchased. Although architecture has always had (and will have) a strong artistic component,

Figure 1. Stone, traditional house vs. new apartment house

in parallel with the development of industrial technologies, based on scientific achievements, it is becoming more and more supportive of architectural creativity exclusively in the field of technology. This resulted in the replacement of the traditional, holistic approach to planning, designing and constructing built environment that accepted the knowledge gained through experience, with a new arrogant attitude towards requirements of space, human needs and environment in general (Bijedić, 2012).

At the Blidinje Nature Park area, the spatial parameters of individual houses were analyzed, which were: positioning in relation to the environment, access, position relative to neighbors, layouts, materials used, type of structure, and roof slopes. The houses of the ethnographic heritage are shown and compared with the newly built houses of the apartment complex.

Characteristics of the Ethnographic Heritage Houses

The folk builder of stone houses was considering the pitch of the terrain, the direction of wind when deciding of the property location. Buildings are incorporated into smaller groups, but their familiarity within the group is evident, each unit has its own outdoor open - green space. Access communications are also “spontaneous” where they are needed, and they also respect the terrain configuration. The landscape is still natural, complemented by stone structures that are integrated within it.

Characteristics of Newly Built Houses Within the Apartment Settlement

The apartment settlement is designed according to the principle of land division to the small parcels organized as an orthogonal grid, which forms rigid net of the access roads and paths. The area of the apartment complex is densely built, the space between buildings is narrow. As a consequence, the users of the apartment complex can hardly perceive the natural environment of the Nature Park, and also don’t have any intimacy. The settlement forms a new landscape dominated by buildings, while connection to natural surroundings is almost completely blocked by neighboring houses.

Figure 2. Spatial layout and view on the group of stone houses

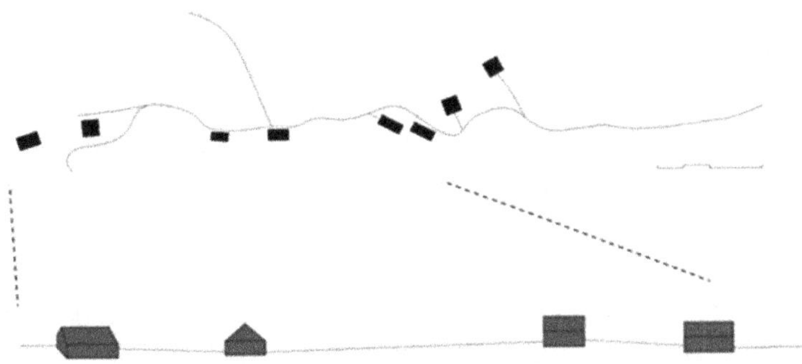

Figure 3. The apartment complex, disposition, and view of newly built houses

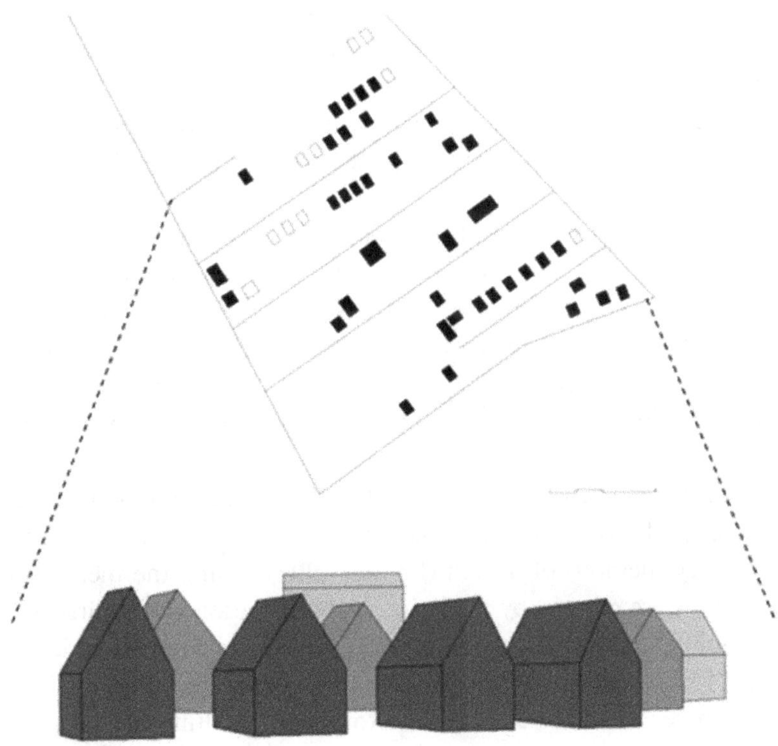

LESSONS LEARNED CASE STUDY

Based on this summarized comparison of the old and new spatial organization patterns and houses *per se*, it can be said that the house of the new age meets the needs for accommodation of the users. Houses are built with new building materials, including thermal insulation, modern energy efficient windows, HVAC systems, that makes them dependent on technology. The old house responded to the lifestyle of the inhabitants of the previous period.

When we come to the level of observation of spatial complexes, we notice more significant differences between actual and old models.

The construction of the buildings from ethnographic heritage promotes the construction in smaller groups in which each house had its own open space. This enabled their free orientation in space, as well as the possibility of integration into nature. Spatial organization of apartment houses settlement introduces new pattern that is strange to this specific location, thus created space that degraded and, in a way humiliated existing environment of the Blidinje Nature Park.

It is evident that there are two different ways of thinking about space. One puts men and nature in the center, while the other emphasizes economic gain. One, the old way, emerged spontaneously. The bottom-up principle emerging from the real needs of the user and creating a logical and methodic structure. Integrated into a natural environment and achieved synergy.

The second structure (apartment settlement) is located in the top-down area, it uses the orthogonal network to obtain densely constructed objects, recalling the way it is built in urban areas, where a small gap between the objects is caused by the available areas and a large population settling within. The landscape was altered, the ratio of nature changed, with the domination of unsuitable structures organized on unfamiliar manner.

Though it is a result of planned thinking, it can certainly be said that there was no awareness of the space in which it is built. The artificially imposed pattern is static and limited (it cannot be functional - substantiate theoretically), and nature to which man integrates - the future user is part of an open, dynamic and self-sustainable system where all elements of the whole depend on each other. Trying to copy self-organized patterns of rural settlements from the top-down, complex character of bottom-up spatial patterns are lost. Top-down planning is mostly based on linear homogeneous scales, distribution, and organization of individual spatial units. Homogeneous patterns of designed space are characterized by internal conflict. Only complex dynamics of nonlinear spatial heterogeneity could lead to synergy, forming emerging place supported by dynamic fluctuations based on the behavior of local individual residents (Allen, 2012).

Parallels to organic biological processes of generating forms through the genetic code and environmental influences, which result in a variety of variations in order to adapt to changes in the environment, can be used in evolutionary methods of design (Frazer, 1995). Using the methods of local traditional architecture, whose forms were created by evolution and adaptation to conditions in a given environment, architectural planning and design could takes place at the technological level in the form of the introduction of a certain formal genetic code, developed by a computer program, into a series of models representing simulated processes of the external environment, resulting in rich variations of the forms that arise as a result of different processes, rather than as compositions of previously defined forms. Spatial form exposed to various physical, climatic and biological processes, as an external environment that changes its form, becomes complex, variable and dynamic.

The traditional architecture shows a remarkable sense of meaning and relationship to nature, climate and topography. It confirms the wisdom, imagination, knowledge and skills of anonymous builders, which is encouraging and a source of inspiration for contemporary architecture. It is an evident dialogue with the surroundings in terms of architecture tailored harmonious integration of man and the environment in which we come to know about affordable synthesis of traditional and modern architecture, a harmonious relationship with the neighborhood and the nature and the human relationship between man, space, environment and nature (Zvizdić, 2012).

CONCLUSION

The concept of space that is evidenced in the traditional and vernacular heritage has a multivariate evolutionary character, which is a consequence of multiplied effects from the external environment. Space has a dynamic character, and it is in a constant flow of transformation, which is the result of exposure to a multitude of different processes and their effects. Successful integration of architecture and nature should not be based on optimal solutions and predictable performance, but on the open dynamical design that has evolutionary qualities, as holistic unity within processes and rhythms of the environment.

Architectural structures and spaces, having a complex character, do not have to present a perfect scheme to be complete. They represent an open unfinished whole. The degree of their integrity can be variable since the parts can be fragments of a larger whole at different levels. The simultaneous perception of many levels as a whole does not bring the simplicity of perceiving unity, but rather requires more than analytical observation of variable characteristics. Uniqueness and unity of spatial forms include different levels of hierarchical character, whether it is the unity of complementary principles, the unity of the individual part and the multitude, the unity of parts achieved by directing the continuity… Mutual connectivity and interdependence of many various scales create open and flexible patterns.

The traditional rural heritage and their underlying development of spatial patterns should become model for the generative principles of the more complex approach to planning in the natural environment, achieving a holistic integration of nature and human-built structures. Providing continuity and preserving identity and integrity of individual and collective spirit of the place and human life, new intervention in space should be based on complexity, diversity, and variability. The contextual and evolutionary approach should consider different levels of interconnections, on different spatial and temporal scales of planning and design research, toward unique directions and applications for every specific location, social and cultural niche.

REFERENCES

Abadžić, N. (2001). *Doba ekologije*. Sarajevo, Bosnia and Herzegovina: Vlastita naklada.

Adams, W. M. (1996). Future nature. Vision of conservation. London: Earthscan publications.

Allen, P. M. (2012). The Visible Expression of Co-evolving Complexity. In Complexity Theories of Cities Have Come of Age. Berlin: Springer. doi:10.1007/978-3-642-24544-2_5

Batty, M., & Torrens, P. M. (2001). *Modeling Complexity: The Limits to Prediction*. In 12th European Colloquium on Quantitative and Theoretical Geography, St-Valery-en-Caux, France.

Bijedić, D. (2012). *Arhitektura, Holizam umjesto optimalizacije, Integralni pristup u arhitektonskom stvaralaštvu*: Sarajevo, Bosnia and Herzegovina: Univerzitet u Sarajevu.

Brockington, D., Duffy, R., & Igoe, J. (2012). Nature Unbound: Conservation, Capitalism and the Future of Protected Areas. London: Routledge. doi:10.4324/9781849772075

Clarke, J. J. (1993). Nature in question. London: Earthscan publications.

Frazer, J. (1995). Evolutionary architecture. London: Architectural Association.

Gore, A. (2008). *Earth in the Balance Forging a New Common Purpose*. London: Earthscan.

Heylighen, F. (1989). Self-Organization, Emergence and the Architecture of Complexity. In *Proceedings of the 1st European Conference on System Science* (pp. 23-32). Paris: AFCET.

Krohn, W., & Kuppers, G. (1990). *Selforganization: Portrait of a Scientific Revolution. Sociology of sciences yearbook*. Dordrecht: Kluwer Academic Press. doi:10.1007/978-94-017-2975-8

Lačević, F., (2004). Umjetnost u dobu znanosti, Skeč o estetskom odgovoru na kaos i fraktale, smisao komparacije dva motrišta. In *Radovi – knjiga XIII – 2004*. Sarajevo, Bosnia and Herzegovina: Univerzitet u Sarajevu, Filozofski fakultet

Levin, S. (1998). Ecosystems and the Biosphere as Complex Adaptive Systems. *Ecosystems (New York, N.Y.)*, *1*(5), 431–436. doi:10.1007100219900037

Nejgel, E. (1974). *Struktura nauke*. Beograd: Nolit.

Prigogine, I., & Stengers, I. (1982). *Novi savez*. Zagreb: Globus.

Salingaros, N., & West, B. J. (1999). A universal rule for the distribution of sizes. *Environment and Planning. B, Planning & Design*, *26*(6), 909–923. doi:10.1068/b260909

Salopek, D. (1974). *Arhitektura bez arhitekta*. Zagreb, Croatia: Mala arhitektonska biblibiblioteka SAH-e.

Shimer, J. A. (1959). *This sculptured earth*. New York: Columbia University Press.

Simon, H. (1962). The Architecture of Complexity. *Proceedings of the American Philosophical Society*, *106*(6), 467–482.

Vego, J. (2000). Buildings of the Ethnographic Heritage in Blidinje Nature Park. *Prostor*, *8*(2), 269–272.

Živković, Z. (2015). *Tradicijska kamena kuća dalmatinskog zaleđa, priručnik za obnovu i turističku valorizaciju*. Zagreb: Ministarstvo turizma Republike Hrvatske i Ministarstvo vanjskih i europskih poslova.

Zvizdić, D. (2012). Lessons Learned from the Bosnia and Herzegovina Urban and Architectural Heritage, As a road map for sustainable urban development. In Proceedings of Islamic Urban Heritage Research Preservation and Management, Summer School. Istambul, Turky: IRCICA.

Chapter 17
Urban-Rural Synergy in Shrinking Regions Through Application of e-co Model

Tijana M. Vujičić
Faculty of Architecture, Civil Engineering, and Geodesy, University of Banja Luka, Bosnia and Herzegovina

Dijana Simonović
https://orcid.org/0000-0002-0692-7515
Faculty of Architecture, Civil Engineering, and Geodesy, University of Banja Luka, Bosnia and Herzegovina

ABSTRACT

Many regions worldwide are faced with the problem of shrinkage, manifested through demographic decline, economic loss, and perforation of the urban tissue. Starting from the assumption that the shrinking phenomenon is present in Bosnia and Herzegovina, the research deals with the diagnosis of the shrinking condition in urban and rural areas of the northwest region of the country. In the light of the shrinkage problems, the chapter defines a new alternative model—e-co—which should enable the recovery from the crisis and restore the vitality of the region. The chance for recovery is recognized in the processes of rural urbanization and urban ruralization, the integration of rural values and urban advantages, the modern globality, and traditional locality.

INTRODUCTION

Since the industrialization period until today, there have been a number of negative demographic, economic and spatial changes in the Balkans that produced a disparity among towns and regions. In such a context, prominent are the areas of intensive growth and regions affected by complex shrinking processes. Faced with the challenges as a result of negative impacts, some of them benefit from the post-socialist transitional processes, whereas a large number suffer losses. The causes, consequences and shrinking

DOI: 10.4018/978-1-5225-9932-6.ch017

rates in the Balkan countries have not been sufficiently examined, and even less are being examined and offered possible solutions to the problems these areas are facing. Key actors that are supposed to manage, alleviate and direct these negative changes, such as representatives of national and local authorities, spatial and urban planners, scientists, and researchers have failed here. There is no integrated recovery policy or strategy either at the national or the local level, and society is far from thoughtful planning and managing spatial dynamics of demographic, social and economic processes (Vujičić, 2018a). Spatial and urban politics continue to incite growth and development although the situation and trends indicate reverse processes. The shrinkage understood as a process of complex negative demographic, economic and spatial changes is the reality of the towns in the Balkans and extinct villages are the most visible legacy (Vujičić & Đukić, 2015; Djukić, Antonić, & Vujičić, 2017). With respect to such context of problems, it is necessary to review the inherited paradigms and approaches to planning within the professional, academic and scientific milieu, as well as to create innovative management models at all institutional levels. Along with this, the scientific community should propose and suggest possible solutions to identified problems and offer a more general picture to both the general public and the representatives of the public administration sector.

Therefore, the main aim of this research is to offer a diagnosis of the shrinking condition in Bosnia and Herzegovina (B&H) and to create instruments for recovery from the crisis, i.e. to define an adequate model for managing shrinking processes. Such an alternative management model adapted to the needs of shrinking regions should allow them simultaneous management (*steering*) of the processes of *rural urbanization* and *urban ruralization*, which means to offer a chance for recovery and strengthening of shrinking regions resilience in the face of uncertainty. Furthermore, the recovery, primarily understood as the population growth of not only shrinking cities but also villages, is realized through "sublimation of rural values and urban advantages" (Ralević & Simonović, 2017, p. 215). As such, the proposed model should enable the experience of rural life in urban areas and vice versa, i.e. *urban rurality* and *rural urbanity*. The research starts from two assumptions. Initially, the focus of the research is to prove the assumption that the region is in stagnation, with the shrinkage problem being considered to affect rural areas far more significantly than urban ones. The aim of the research is also to prove that the current trend of flows of people will lead to the regional shrinkage. Secondly, it is assumed that it is possible to create alternative management model, which offers an adequate framework for improving the vitality of the shrinking regions, on the basis of analyses of the traditional forms of organization of life in cities, villages and peri-urban areas, as well as thorough analyses of contemporary practices of the organization of work driven by a growing e-lance economy.

METHODOLOGY

The paper is divided into three sections. The first part presents a theoretical and historical background, which elaborates the global and local context of problems and processes of transformation of urban and rural landscapes, and the key theoretical terms and concepts that lead to the creation of models for the regeneration of shrinking areas. The second part investigates the shrinking conditions in Bosnia and Herzegovina on the basis of the proposed analytical model, which follows in the sequel. The final part of the paper presents the *e-co* model, explains its principles and characteristics, defines its possibilities for application, as well as spatial patterns advocated by the model - urban rurality and rural urbanity.

In terms of methodology, the theoretical part is primarily based on the analysis of the theoretical sources that explain the global characteristics and local specificities of the shrinking and scientific sources that provide insights into possible solutions for mitigation or problem-solving. In this regard, the paper sheds light on morphogenesis of the settlements from the pre-industrial period to the present; reads the urban, rural, and peri-urban spatial patterns and its transformations; and presents the typical features of shrinking process in the northwest region of Bosnia and Herzegovina. Peri-urban areas and pseudo-urban settlements (Milić, 2002) of city Banja Luka are being considered here as 'test areas' for seeking patterns of urban-rural synergy. It is these specific areas that have the elasticity for integrating traditional values with the advantages of modern lifestyle and the ways of using space (Ralević & Simonović, 2017, p. 215). On the other hand, the paper explores modern forms of economy, exchange of goods and services embedded in an *e-lance economy* (Hooker, Lewis, Smith, Worrell, & Yoon, 2008). The principles of this growing branch of the economy are the basis for the development of the management model for the recovery of not only urban but also rural areas.

In order to prove the hypothesis of the stagnation of cities, villages and regions in general, proposed below is a model for identification and estimation of the level of shrinkage. Shrinking indicators are classified into three key groups: population, economic and spatial. Population criteria include: a) general population statistics, b) household statistics, c) vital statistics, and d) statistics of emigrations from the observed area in a period of time. The economic criteria include data on the number of the employees/ unemployed and the share of the unemployed in the total population. The spatial criteria include data on abandoned and underused areas of the city - brownfield land and negative changes in population density. The most parameters are monitored directly or indirectly through official statistics; however, a large number of them are covered by statistics only at the level of municipalities. When it comes to the level of settlements, the statistical information has been collected only for the population, its age structure, and the number of households. Therefore, stagnation thresholds are defined according to criteria for which statistics are available at the settlement level, while criteria covered by statistics for municipal level are used to deepen the analysis, identify causally related relationships and influencing factors that lead to a shrinking process.

A general approach to estimating stagnation by any criterion is based on a simple ratio of values from two reference censuses (a recent one and one taken earlier e.g. 2013/1991). This ratio is presented as a coefficient which, if more than 1.05, indicates growth by a certain criterion; the coefficient of less than 0.95 denotes a drop, while its value ranging 0.95 - 1.05 indicates stability. To monitor the overall demographic trends, the 60-year period between 1953 and 2013 is taken as a wide analytical period, with three sections in the years of official censuses taken in 1953, 1991 and 2013. If developmental stagnation trends of towns, villages, municipalities, and regions are regarded as dynamic non-linear processes, it is very important to define the tipping points. It is indisputable that the civil war of 1992-1995 is an important milestone in Bosnia and Herzegovina, and hence the census year of 1991 is identified as a tipping point in relation to which the parallels will be underlined in the research and changes made in terms of pre- and post-war period in all aspects important for the subject of research. These two periods are identified with what is referred to above as the industrial period (before the war) and the post-industrial period (after the war). The northwest of Bosnia and Herzegovina is a test area for exploring the presence, range, causes, and effects of the urban shrinkage, which includes 17 municipalities.

HISTORICAL AND THEORETICAL BACKGROUND

Shrinking Phenomenon

The shrinking phenomenon is recognized predominantly as a symptom of a structural crisis within urban areas expressed through a loss of population and an economic downturn (Jürgen, 1993; Martinez-Fernandez, Kubo, Noya, & Weyman, 2012; Pallagst, Wiechmann, & Martinez-Fernandez, 2014; Wiechmann & Wolff, 2013). The term *shrinking cities* have become common in an academic vocabulary as a phrase implying a process of complex negative changes affecting towns and producing multidimensional negative effects in the economic, demographic, social and physical domains (Pallagst, Wiechmann, & Martinez-Fernandez, 2014). Understanding of the shrinking is place-related and time-related (Haase, Rink, Grossmann, Bernt, & Mykhnenko, 2014), and it is usually difficult to make a clear difference between the causes and consequences leading to the shrinking process, yet the first and key symptom of the urban stagnation is the decline in population (Pallagst, Wiechmann, & Martinez-Fernandez, 2014).

Investigations show that many European cities entered a shrinking process back in the 1960s and that the phenomenon of stagnation in the 1990s was more pronounced than the phenomenon of growth and urban development (Turok & Mykhnenko, 2007). Additionally, Oswalt (2005) argues that shrinking cities are not an exception or deviation, but he rather characterizes the phenomenon as a global one. At the beginning of the 21st century, the phenomenon of urban stagnation is increasingly prevalent in Central and Eastern Europe, the United States, developed countries of Asia, North America, and in Australia (Rieniets, 2006). The research conducted in Europe indicates that 3563 out of 7035 of European cities shrank in the period 1990-2010 (Wiechmann, 2013).

De-population and economic downturn have a huge impact on the quality and vitality of the so-called *perforated cities*, which arise from these negative changes (Lütke-Daldrup, 2001). The term *perforated cities* was first used by the Head of Leipzig City Planning Department, Engelbert Lütke-Daldrup in 2001 to describe Leipzig as a city facing the problems of failed and abandoned spaces and a large number of vacancies. The coin *perforated city* explained nothing more than the physical manifestation of the stagnation of the city and the emergence of a new urban form spreading through the eastern part of Leipzig. The urban landscape of the perforated city was hollowed out by holes and bits of deserted land. These spaces, the so-called brownfield, occupy significant areas within the perforated cities and are most often found within old industrial zones. On the other hand, due to the outflow and reduction of the population, the extinction of households takes place, which is manifested in the space through the abundance of dispersedly located, abandoned housing units. This results in densely built urban areas unaccompanied by an appropriate density of population and concentration of jobs. Such a manifestation of the spatial distribution of the population and economy is completely new and represents a deviation from the concept of a traditional European compact city, which remains in plans as the primary goal (Florentin, 2010, p. 87).

Although urban stagnation is not a new phenomenon, it has not been sufficiently explored and only recently has it become interesting. In the Balkans, the understanding and perception of the concept is unclear and does not attract much attention in the political circles and of the general public. Politicians and representatives of local governments avoid this topic for the following two reasons: a) in many Slavic countries, the term shrinking is difficult to translate, and b) viewed in the context of public relations, it often carries negative connotations and meanings, which do not go hand in hand with policies based on economic prosperity and urban development (Stryjakiewicz, et al., 2013). Furthermore, since the period of active industrialization and accelerated urbanization of cities, urban planning has a quasi-causal connection with urban growth, and its methods and visions have been justified by the assumption of linear growth and development (Vujičić, 2018a). However, economic and demographic indicators clearly show that the era of stable continuous development was only a historical episode (Rieniets, 2006). Contemporary urban planning is completely unprepared for the new challenges caused by the urban stagnation process. It does not have adequate visions, experiences, planning methods and tools adapted to address this issue. In the context of the development and increasing dispersion of the shrinking phenomenon, this research starts from a reversed perspective. The city's evolutionary route is not seen here as a linear process of a constant rise and prosperity, but as a process of facing complex multifaceted challenges of today that often involve the stagnation and decay of cities. Furthermore, planning is viewed as an open device that allows adaptation and offers adequate instruments to respond to these challenges.

Bearing in mind that processes of stagnation always take place in urban and rural areas, settlements, territories and landscapes, and that previous research has considered this phenomenon as one affecting discretely and separately either the urban or rural context, we should call attention to present-day theoretical research that brings into focus the links and trends connecting the two (Lynch, 2005).

Transformations of Urban and Rural Landscapes From Pre-Industrial to Post-Industrial Phase

Considering a wider temporal, thematic and spatial context of the stagnation process in the former Socialist Federal Republic of Yugoslavia (SFRY,1963-1992), the course, causes and characteristics of the process of transformation of the urban, rural and peri-urban landscapes, settlements and characteristic spatial patterns, from the pre-industrial to the post-industrial phase will be explained here. During this period, which was dynamic in terms of demographic, economic, political and territorial changes, villages and cities were constituted and developed through their mutual relations and relations with the environment. In late phases of the process of developmental transformation, new specific pseudo-urban settlements which represent 'the village and the city' are emerged and developed in peri-urban areas, at the same time being 'neither the village nor the city' (Ralević & Simonović, 2017, p.215). Observed is their cause-and-effect type of connection with the existing towns and villages in the immediate vicinity. It was also noted that they possess the desirable vitality and adaptability of the settlement to the dynamics of change, applicable in shrinking areas. For this reason, the specific urban, rural and pseudo-urban spatial patterns are the first to be considered here, being the reference for later interpretation of the transformation of space, settlements, and areas from pre-industrial to post-industrial period.

In the pre-industrial development period, rural and urban spatial patterns were the characteristic ones, with villages and towns emerging, developing and existing as well-defined, distinct settlement types. The basics of coding villages and towns are founded on a review of different types of codes: source codes, initial codes, planimetric and structuring codes (Ralević & Simonović, 2017). Furthermore, the authors accepted to equalize urban with a city or town, and rural with a village. According to the source codes, the village is an organic creation rising from the very nature and being based on the genetic code of its inhabitants - the villagers. The city is an artificial creation, which is linked to production and technology. From the perspective of initial codes, "the village itself is a process of constant adaptation to new changes. The city strives to be an act that occurs in a short period given to cross - and achieved perpetuity" (Ralević & Simonović, 2017, p. 218). Planimetric codes of villages are characterized by the archaicity of composition, groups, and ensembles; urban planimetry is associated with geometric patterns and aspiration towards the organization of the form. The structural code of the village is clear: the center of the village, longitudinal and transverse axis. In the form of a skeleton, the structure of the village integrates residential and production spatial units. The structure of the town is a distribution matrix incorporating variously organized urban functions and activities. The organizational coding of the village is based on the inseparability of business and residential activities, while the town is characterized by quite the opposite – spatial separation of activities (Ralević & Simonović, 2017) (Figure 1).

As previously expounded, during the pre-industrial era, the rural and urban spatial patterns developed in parallel and, at the same time, in a mutual functional balance keeping the original individual character, structure and position in the macro-regional settlement area (Figure 2). The traditional spatial order, in addition to generating the described characteristic codes of the village and the city, implied clear boundaries between urban, rural and natural areas. In addition, the notion of a place was clearly recognized in a territorial sense. However, with the industrial revolution, the urbanization process gained intensity and dynamics, establishing a new spatial order – pseudo-urban, which relativize the border between urban, rural and natural and blurs the territory and the concept of the place. A pseudo-urban order generates new pseudo-urban patterns in peri-urban areas in which the characteristics of rural and urban are interwoven, but where new and specific characteristics are also recognized. The spatial and social effects and pseudo-urban patterns produced by the new order prevail in suburban areas, evolving as conurbation forms, 'urban polyps' and other spatial consequences of the modern urbanization (Figure 2) (Milić, 1998). However, new settlements were developed in peri-urban areas being neither city nor village i.e. both the city and the village at the same time. The area possesses features characteristic to the village in terms of changeability and adaptability. At the same time, it is characterized by a discontinuity of the process of changes, a feature of the town. In peri-urban areas and pseudo-urban settlements, located between the village and the town or between towns, the archaic planimetric region is dominant in the rural planimetric code, while the features and patterns inherent in the urban planimetric code are reduced to a minimum. In the peri-urban area, new settlements are based on the structural code of a village formed at the crossroads of the roads or along the existing watercourses.

Figure 1. Model of village and town
Source: Tijana Vujičić and Dijana Simonović

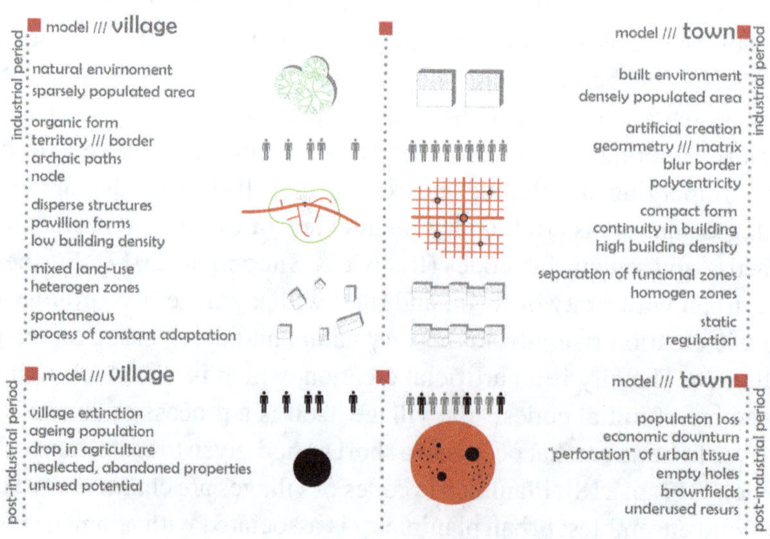

Figure 2. Diagram of flows of people: municipal and regional level
Source: Tijana Vujičić and Dijana Simonović

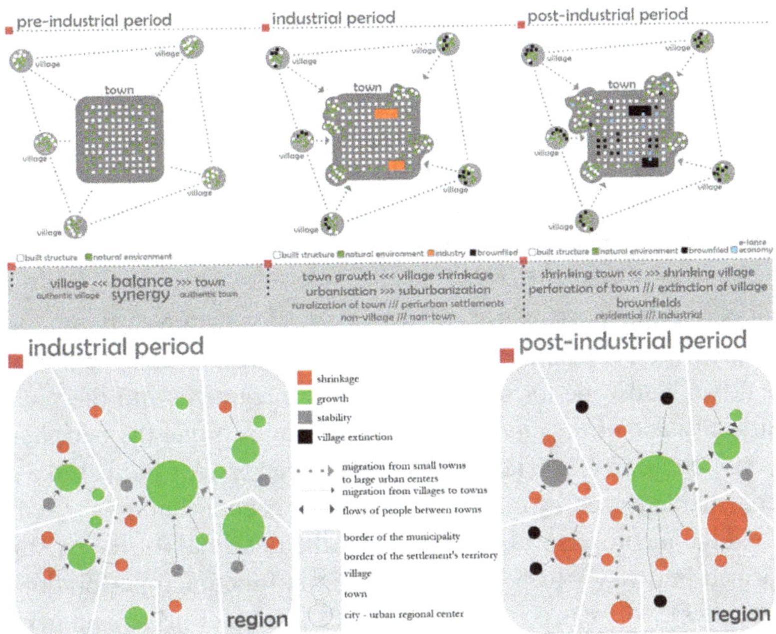

Morphogenetic Characteristics of Banja Luka's Peri-Urban Landscape

In the context of the theme *Urban-Rural Cooperation in Constructed Landscape*, and of the efforts to consider the previously examined characteristic patterns of space – urban, rural and pseudo-urban – at the most general level of spatiality, next we interpret the process of transformation of Banja Luka's urban and rural landscape. Our goal is to identify the characteristics and determine the potential of the three types for interconnection, sublimation and correlation. In the pre-industrial era, the urban, peri-urban and rural landscape of Banja Luka was characterised by harmony, amalgamation, fusion and complementarity. The man-made and natural landscape existed in unison, constituting a harmonious and congruous whole, in which the constructed, man-made gradually changed to the natural, without clear demarcation. The urban landscape mainly came into being following the principles of linearity and policentricity: the urban cores moved downstream the River Vrbas and the major longitudinal thoroughfare. In terms of morphology, the urban landscape was characteristically funnel-shaped, fully integrated in and arising from the natural and geographical features of the area (Simonović, 2010).

In the industrial era, the concept of Banja Luka's urban landscape was readjusted to incorporate the modernist idea of the park city or garden city, also typical of the majority of Bosnian-Herzegovinian cities. This concept differs from Le Corbusier's modernist ideas of urban landscape and functional subordination of space and buildings to geometric spatial order, which is located beyond or above the actual space, i.e. landscape. They are more like *Stadtlandschaft*, the concept of Central European cities, whose foundation is the open, free landscape, with groups of unobtrusively arranged habitation units. The process of urbanisation of Banja Luka after WWII saw the arrival of a new population from the city's rural surroundings, who gradually adapted to the way of life of industrial workers, military and civil servants, but who nonetheless succeeded in keeping ties with the rural upcountry they had left to move to the city. Between the 1960s and 1990s, urban Banja Luka expanded, with more natural landscape transformed, while making use of what topographically assisted the expansion: new industrial plants were erected in the industrial zones along the River Vrbas, compounded with the construction of unplanned residential neighborhoods on the outskirts and country houses in the rural zones around the city (Simonović, 2014). An unreasonable increase in urban area over a long period of time was caused by the following: the character of the settlement, the method of construction in use and habits of the population (with the prevailing urban-morphological type of a free-standing detached house on the plot and the great need of the inhabitants for a garden, vegetable garden, etc.) i.e. the identity of Banja Luka as a town of urban gardening (Finci, 1967).

In the most recent period, the post-industrial era, the transformation of Banja Luka's landscape has been characterised by two tendencies being part of the same process, of Lefebvre's 'implosion-explosion': new and changeable urban landscape forms and structures, integrated with the social and natural context, inside the urban landscape of the city and beyond its limits, along the periphery and its rural and natural surroundings (Lefebvre, 2003). The growth of the urban area of Banja Luka since the end of the 20th century has been characterized by a low degree of the general building density, a small average population density, and the territory several times larger than optimal for the existing population. The city area has transformed in a way which has resulted in the functional perforation of its urban tissue (present as brownfield sites), whereas the urban tissue of the greater urban and peri-urban Banja Luka is organically expansive, but perforated in the spatial and physical sense (present are greenfield sites).

The territorial increase in the urban area that is directed by plans rarely coincides with the directions of the urban tissue spreading, which is not regulated by an urban plan. The following three distinctive spatial-morphological types of the new urban-rural form have been developed: a) the type of predominantly residential area with single-family stand-alone houses on compact farming plots; b) the mixed type of settlement with concentration of production, storage and service facilities by the main road and the residence in the background, with single-family stand-alone houses on compact building plots resulting from the fragmentation of former agricultural plots; and c) a specific type of settlement with a combined concentration of residential, agricultural, production, storage and service facilities by the main road, which was formed by dividing former agricultural plots into a compact plot module intended for residing, and a large module of building plots intended for commercial purposes. The conurbation form was developing gradually, being continually connected with the city proper into which it was integrating. The landscape concept is being phased out from Banja Luka building plans, regulations and legislation (Simonović, 2014). We believe this concept can be revitalized with the application of an *e-CO* model of spatial, social, economic and environmental community, one resistant to change.

Introducing the principle of the urban-rural synergy into creating the response to the settlement shrinkage issue is based on an interpretation of not only spatial-morphological transformations of the urban landscape but also specific social and cultural, organizational, and functional alterations in the settlements as a result of the urban-rural blending. Observed were the following specificities that served as the basis of the urban-rural synergy (Simonović, 2015):

- The changes are most pronounced in areas between villages and towns i.e. in the areas of less expensive building land (transformed farming land), beyond the 'reach' of the spatial regulations or within the 'blind', uncovered area of the inter-municipal regulatory instruments. These are the areas with new hybrid settlements that are 'both the town and the village', with forms and patterns brought by the inhabitants and workers from their former places of living. The urban forms are predominant in the beginning, with a gradual increase of introducing the rural forms after the settling of the inhabitants.

- The new hybrid urban-rural settlements were mostly inhabited during and after the conflict in B&H, not only by rural but also urban population from towns of different levels of development. These settlements were formed by grouping new households of the former locals, neighbors, relatives from the same 'neighborhood' or the region, and those of the same origin or affiliation. In the beginning, they accepted the existing rules of physical regulation of an autochthonous settlement (organized ordering of the plots and houses next to the main road, following the principles of the urban geometrical order) to gradually introduce the 'old' rules and forms of the use and division of the land and spatial organization applied in their former places of living.
- the proportion of the unemployed in the overall structure of the inhabited population is large, with a small percentage of farmers and craftsmen or other service workers, and a small percentage of those working in the city or in the area of conurbation forms, so that the connection of the inhabitants of these areas with the city of Banjaluka – its population, cultural patterns, and participation in the development of identity – is very weak and they mainly remain 'invisible', both to the population of the downtown of Banja Luka, as well as to the city and entity administration, with the exception of the secondary school and university student population, in relation to whom there is a greater measure of social openness
- The features of these specific urban satellites are dependence and incompleteness, as well as the lack of centrality and public and community spaces, dependence on the city and isolation at the same time.

The described forms are differentiated as satellite settlement forms of a specific nature and identity, which were developed through the transformation of the settlements themselves. As these settlements do not have developed qualities of spatial integrity and functional independence in relation to the city of Banja Luka, their identity cannot be viewed separately from the identity of Banja Luka despite certain authentic identities they undoubtedly possess. Moreover, the results of earlier research on the development of urban identity of Banja Luka as a landscape city (Simonović, 2010) show that this is a common inherent process of the urban identity development, which is accompanied by 'spillages', 'migration' and 'transmission' of the crucial identity distinctions from the city core to satellite settlements, resembling the system of communicating vessels. This attitude includes the beginning of the idea of the reasonableness of the aspiration to achieve vital and resilient models for the recovery of shrinking settlements by applying the urban-rural synergy.

Urban-Rural Synergy for E-Society

Considering the specificity of spatial patterns in urban, rural and peripheral areas, it was noted that peri-urban areas are some kind of 'test areas' for seeking a model of sublimation of development forces inherent in both urban and rural areas. For the creation of an urban-rural synergy, it is necessary to point out the traditional values cherished by the village and the modern urban achievements. The individual qualities of the urban and rural forms represent the basic development potentials of the model for the recovery of the settlement from the shrinking process. Thus, synergy is realized through the unity of urbanity and rurality, i.e. rural urbanity and urban rurality (Konjar, Kosanović, Popov, & Fikfak, 2018).

However, as a legacy of traditional spatial patterns, their functional and spatial equilibrium, which existed until the industrial period of the urbanization process, is also emphasized here. In addition to the values of traditional spatial patterns, the specificities of pseudo-urban forms reflected in their resilience and adaptability to change are distinguished as the basic developmental energy potentials. They are probably the key to integrating traditional values with the advantages of modern lifestyle and way of using space. The future development or recovery of stagnating towns and villages should take place simultaneously and in a functional balance. A necessary condition is to maintain an individual but no longer the original character and structure of the village and the town, but also the process of their recovery. However, their positions in the macro-regional settlement area become an open field for alternatives.

Here, this free space for integration, networking, and synergy acquires a virtual dimension, and the former traditional concepts of networks, communications, mobility, and spatial-physical connectivity are replaced by ICT-based virtual networking (Hislop, 2008). In this way, the key mobility problems associated with the modern heritage of separating work and living places, which did not exist in the traditional order, especially in the village, are diminishing or gradually disappearing. The most important thing is that it is inherent in the genetic and biological human code to connect the place of work and the residence, which is attainable today using ICT and *e-lance economy* (Hooker, Lewis, Smith, Worrell, & Yoon, 2008).

In industrial period jobs moved from the farm to the factory, in postindustrial period it has moved "from the cube to the cloud" (Houlne & Maxwell, 2013). Globalization and information economy suggest that (work) place is no longer relevant (Sassen, 2003). Criticizing the achievements of globalization that produce geography of centrality and marginality, Sassen (2003) points out that, in an era where it is often thought that the notion of the local falls behind the global, digital technology can serve as a support for local initiatives, and digital networks can produce counter-geographies of globalization. Drawing a parallel to this, one can argue that such virtual networking can also contribute to the processes of counter-urbanization and/or counter-metropolization, i.e. the regeneration of smaller towns and villages. The digitization of work and relativization of work-place, inherent to the global economy, open the field for a conceptualization of framework for networking between cities and villages, modeled on the experiences of global inter-city networking (Sassen, 2003; Sassen, 2010; Sassen, 2016). These alternative networks can be seen here as a trigger for urban-rural synergy development.

RANGE OF SHRINKING IN BOSNIA-HERZEGOVINA

Growth and Shrinking in the NW Region of B&H

Based on the proposed model, a multispectral analysis was carried out to identify shrinking municipalities, cities and villages, understand the processes and impact factors leading to the decline, and assess the level of shrinkage within the region. The proposed model was applied and tested in the area of northwest Bosnia and Herzegovina, which is territorially and administratively a part of the Republic of Srpska, one of the two independent entities of Bosnia and Herzegovina. According to the current administrative and territorial division of the state, entities are further divided into municipalities, while settlements or habitations are units of the lowest order. The growth and shrinking trends are monitored in a sixty-year period through two key stages. The first phase, which lasted until 1991, was marked by population growth, industrialization, and intensive urbanization (Figure 3). The second phase, from 1991 to the present, is characterized by a substantial decline in population, de-industrialization, and stagnation of most cities and villages of the region.

In the subject area, the process of intensive urbanization took place during the aftermath of the Second World War, in particular in the 1960s and 1970s of the 20th century (Institut for Statistics of Federation of Bosnia and Herzegovina [ISFBH], 1999b). According to the official census data, most cities and municipalities in the northwest of the Bosnia and Herzegovina reached its maximum number of inhabitants in 1991, which was followed by a subsequent decline in the population (Federal Statistical Office of Socialist Federal Republic of Yugoslavia [FSOSFRY], 1975; ISFBH, 1998; Republika Srpska Institute of Statistics [RSIS], 2017a). The population of the majority of the analyzed towns at least doubled in the period 1953 - 1991 (ISFBH, 1999b; RSIS, 2017a; FSOSFRY, 1975). The constant growth in urban population is mostly the result of pronounced migrations from villages to towns, and the relocation of

Figure 3. Population loss/increase in municipalities, towns and villages in the periods 1953-1991, 1991-2013
Source: Tijana Vujičić

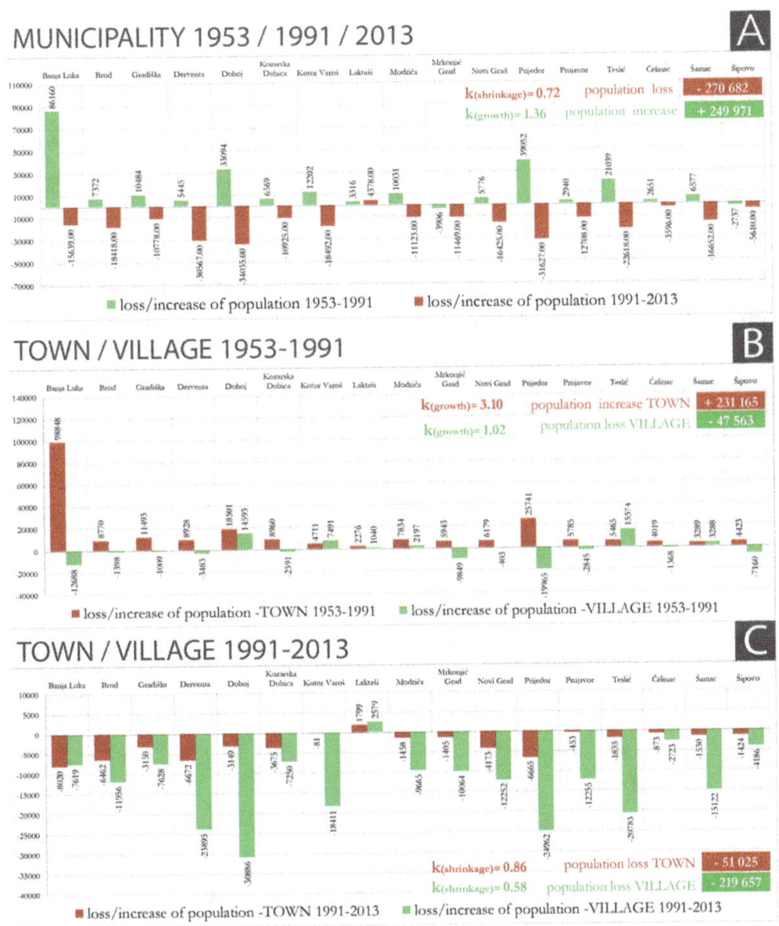

residents from other municipalities (Figure 2) (ISFBH, 1999a). Although the urban population grew during this period, it is evident that the village population was generally stable. The fact that the population at the municipal level grew until 1991 points to the fact that the population growth was positive at both urban and rural level. On the other hand, there is a noticeable increase in the population in Banja Luka, Doboj and Prijedor, the major urban centers, which is the result of emigration of the population from smaller towns.

In the period 1991 - 2013 there was a significant decline in the population in towns and villages, i.e. whole municipalities (Figure 3). There was a decline in the population in 16 out of 17 municipalities, while only Laktaši municipality had a positive trend (Figure 4). The municipalities and towns from the northwest region lost respective 270 682 and 51 025 inhabitants (Figure 3) (ISFBH, 1999b; RSIS, 2017a). At the same time, a drastic decline in the rural population took place. In the period 1991-2013, all the villages of the respective municipalities in this region lost 219 657 inhabitants or almost nine small towns with a population of 25 000 (ISFBH, 1999b; RSIS, 2017a). These data are alarming and indicate an extremely high level of stagnation at all spatial levels.

Observed for the entire 60-year period from 1953 to 2013, the population of towns was growing, especially in Banja Luka, Doboj and Prijedor. At the same time, the rural population declined much more than the urban population grew. Such a trend is the result of several chronologically traceable factors. Namely, from 1953 to 1991, demographic trends were positive at both the city and village level; nevertheless, a large part of the rural population moved to towns. During and after the civil war in B&H, a large part of the population immigrated to other countries, while a significant number of people were killed during the war. Generally speaking, this led to a decline in the overall population in all municipalities of the NW region (except for Laktaši). At the same time or immediately after the war, there was a second wave of urbanization and movement of the population from villages to towns. However, regardless of the fact that there was a decrease in the number of inhabitants both at the level of towns and municipalities after 1991, the share of the urban population continued to grow, while the share of rural population in the total population of the municipality decreased. This points to the fact that the process of urbanization is still continuing so that the outflow of the rural population is still present, but their inflow into towns is insufficient to produce the urban population growth in relation to the pre-war period. In other words,

Figure 4. Northwest region of Bosnia and Herzegovina: population loss/increase
Source: Tijana Vujičić

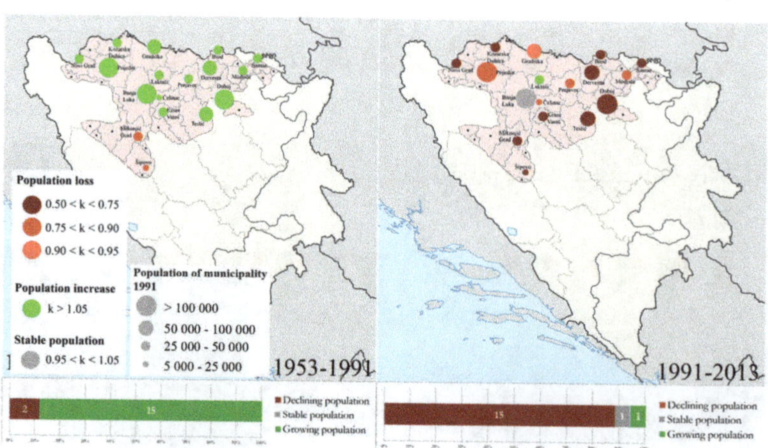

there was a reallocation of the remaining population in municipalities, which mitigated the decline of the urban population, but there was a major decline in the population of rural areas or a decline of the population of municipalities in general. This resulted in the extinction of the village and concentration of population and activities in towns.

It is concluded that in the 60-year period (1953-2013) the population of towns grew to be higher in all the cities of the NW region in 2013 in relation to the 1953 population (Figure 3, Figure 4). However, the population of municipalities in 2013 significantly decreased compared to the 1953 population. Even 10 out of 17 municipalities have fewer inhabitants than in 1953. The rural population in 2013 is lower in 16 out of 17 municipalities than in 1953.

The conclusion is that, at the level of towns and villages, a significant decline in the population in most peripheral municipalities of the region took place in the post-industrial phase (after 1991). On the other hand, the municipalities located centrally, which are urban centers, saw the growth in urban population, unlike rural ones. In the period 1991-2013, the region lost 28% (270 682) of the total 1991 population; the number of urban and rural population decreased by respective 14% (51 025) and 42% (219 657) (Figure 3) (Vujičić, 2018a). All these indicators confirm the assumption of the general regional shrinkage. In order to define adequate models for solving or alleviating this issue, it is necessary to examine the causes of such negative demographic changes i.e. the causes of shrinking. Therefore, presented below are the results of investigating other negative factors that have brought cities, villages, municipalities, and regions into such a bad situation.

Causes and Effects of Stagnation

Reduction of the population in the NW region of B&H is caused by a number of factors that are divided into two groups: sudden reversals that occurred in the early 1990s and the slow socio-economic changes that began in the 21st century. The first group represents dynamic, sudden negative changes, one of the biggest being the civil war which in the period 1992-1995 has led to great suffering, displacement of the population, dissolution of companies, destruction of structures, natural and cultural goods and great environmental damage. Certain studies show that in all investigated municipalities of the NW region of B&H, a total of 17 072 inhabitants or 1.82% of the 1991 population died (Tokača, 2012). There are no official data on population from the immediate post-war period, and those collected by the RSIS (2013b) are based on estimates. Observed collectively and according to these estimates, in the five-year period (1991-1996), the region lost 179 866 inhabitants (RSIS, 2013b), which is a huge decline equivalent to the extinction of slightly more than 7 cities of 25 000 inhabitants. However, since these data are based on estimates, they, nevertheless, cannot be considered sufficiently reliable to conclude and interpret the consequences of the war. The absence of a census in the post-war period made it even more impossible to assess the state of towns and municipalities in B&H. Only 18 years after the end of the war, a census was conducted to provide a clear picture of the situation in the post-war B&H (RSIS, 2017a).

The war was a dynamic period facing a drastic decline in all segments of the society as a whole. However, the B&H conflict was a turning point and a trigger for a series of slow and gradual changes that took place after the end of the conflict, which caused further degradation of towns (Vujičić & Đukić, 2015). The changes are sometimes the cause and effect at the same time, which shows their complexity and interconnectedness. These changes led to almost equal, if not even more serious loss of population of the region in the postwar period (S.Pašalić & D.Pašalić, 2016; Pašalić, Vuković, Jovičić, & Stević, 2017). Through this research, a comprehensive analysis was carried out of a multitude of stagnation indicators from the post-war period which served to identify the causes of stagnation, the key ones being the following: natural population decline and negative migrations.

The decline in natural population increase is the result of a number of factors manifested through the following:

- An increase in the age when getting married for the first time - in the ten-year period 1996-2016, the groom's average age increased by 1.5 years, and the bride's by 2 years (RSIS, 2017c)
- Deviation from the traditional concept of being married - the proportion of unmarried men and women in the total population of the region older than 15 in 2013 was on average 25-30% of the population, while the share of the population living in a common-law marriage was 7% (RSIS, 2017b)
- An increase in primiparas' age - from 2001 to 2015, the average age of primiparas increased by 3 years, from 24.20 to 27.20 years, and trends indicate a tendency to further increase (RSIS, 2002-2016)
- Fertility rate reduction - total fertility rate, which takes into account all women over 15 years of age, is worryingly low and, in most municipalities of the region in 2013, it was far below the rate that ensures population replacement rate (2.1), amounting to 1.3 -1.4 (RSIS, 2017b; RSIS, 2017a)
- The decline in the participation of a fertile group of women in giving birth - the results of the 2013 census show that the proportion of women who have not given birth in the population of women over 15 years of age is 25% on average (RSIS, 2017b). On the other hand, it is indicative that in the total number of births (female population over the age of 15), the largest proportion (60%) are the women who are currently over 50, which indicates that the women born in 1963 and earlier gave twice more births than the women whose current age is 50 and less. The highest percentage of women who have not given birth (~50%) are women with university education.

Such trends produce a drastic decline in natural population increase and the overall aging of the population, that is pointed out also by other researchers: S.Pašalić and D.Pašalić (2016), S. Pašalić, Vuković, Jovičić, and Stević (2017), and Tanović Lukić, S. Pasalić, and Golijanin (2014). The decline in natural population increase is the key symptom of stagnating towns, and it is addressed in the analysis as the difference between the live-born and the deceased. The results of the study show alarming data on a drastic decline in population growth. Worrying is the data on the population growth balance in the period 1996-2015, which show that the analyzed northwest B&H municipalities lost the population of 17 861 in the period 1996-2015 due to the negative natural population increase (RSIS, 2002-2016). A comparative analysis of the age structure of the population according to the results of the census of 1971 and 2013 within the municipalities in the region shows that the proportion of people older than 60 has doubled, while the proportion of people younger than 19 is twice lower on average. Such a situation is a great burden for the whole society, which affects the life and activities of the community in general. It points to the need to review management policies, reform public services, develop new activities - the so-called social entrepreneurship, as well as building a public suprastructure for the needs of the elderly.

In addition to the pronounced negative natural migrations of the population within the municipalities of the area in consideration, negative external migration trends are highlighted as an important cause of depopulation (S. Pašalić & D. Pašalić, 2016). In the post-war period, the extent of migrations has reached such a high level that many small towns and municipalities are in danger of extinction. Although the trend of going to work abroad was also present in Bosnia and Herzegovina in earlier periods of development, the issue of migrations has now deteriorated significantly. Namely, it was mainly the male population who would go abroad in the 20th century, while the rest of the family would stay in the homeland. Today, a trend is visible in Bosnia and Herzegovina that whole families leave the country in search of better jobs, better conditions to raise children and generally better living conditions for the whole family. The exact number of emigrants to other countries is unknown, and the data on the number of emigrants from B&H are obtained from the migration statistics of the countries of admission. According to the data of the Ministry of Human Rights and Refugees of Bosnia and Herzegovina (2014), from 1998 to mid-November 2013, 61 752 inhabitants of Bosnia and Herzegovina renounced citizenship in order to obtain the citizenship of another state. The data of the Agency for Identification Documents, Registers and Data Exchange of Bosnia and Herzegovina (2017) show that 11 525 inhabitants from the area of 17 investigated municipalities in the NW region of B&H canceled their (permanent) residence from B&H to register abroad in the period 2005-2017. However, it is assumed that the number of people who have emigrated is significantly higher, especially given the fact that BH citizens staying abroad for more than one year but not intending to permanently reside in the country they are residing in at present, and who have strong ties with their homeland (family, property, business etc.), are not required to cancel their residence in Bosnia-Herzegovina

The results of previous analyses show that significant migrations of the population took place in the post-war period as well as deviation from the traditional forms of marriage and family, which eventually led to the decline in population growth and aging of the overall population. Many towns in the region have become towns of the old and lost their vitality, while villages are in danger of total extinction. The cause of such transition is an issue to be addressed by sociologists and demographers, while at the same time the administrative institutions should develop models to mitigate negative social and demographic trends or implement such policies to stop the trends and turn them into the population growth. The following are recognized as key causes of such negative trends: a) a low standard of living (unemployment, unstable jobs, low salaries, poverty), b) failure to provide a minimum of life conditions (the place to live - an apartment or house), c) even if the place to live exists, it is often of a low standard and level of comfort (small area, poor structure). Such negative social and demographic trends on the developed area i.e. a town are reflected in the following: a) decline in natural population increase, b) extinction of households, c) negligence of housing units d) extinction of industry and occurrence of new industrial brownfields, e) negligence of infrastructural systems - costly, irrational and infeasible investments in their maintenance, f) reduction or termination of social functions intended to the young - closing schools and kindergartens, g) vanishing of urban public spaces etc.

Such condition requires an urgent reaction of the whole society, primarily representatives of the national and local authorities. Creating various strategic documents focused on mitigating and stopping negative trends (demographic renewal program, population policy, economic recovery and restructuring strategy, integral spatial recovery strategy) is an initial step on the way to solving the problem of shrinking. The key approach to solving this multi-dimensional issue is a proactive action on the causes of such a situation. Proactive here means an approach integrating inherited traditional values and patterns of life in the countryside and modern forms of economy and exchange of goods characteristic to cities - a synergy of the past and the future, urban and rural.

RE-URBANIZATION AND RE-RURALIZATION AS A SOLUTION TO CHALLENGES OF SHRINKING

In relation to the detailed examination of the issue of shrinkage in general, the diagnosis of the conditions in the specific northwest area of B&H and the forecast of future trends, this research should contribute by proposing an adequate model to manage the shrinking processes. Proposed below are solutions for recovery from the crisis, general and specific recommendations are given, and instruments are formulated for the operationalization of the created *e-co* model. An alternative model being advocated should ensure not only a constructive confrontation with the challenges of stagnation and degradation of settlements, but also preventive and visionary action, and continuous adaptation to the urban development process. Additionally, this research offers re-urbanization and re-ruralization concepts as a response to declining trends. In other words, it remodels the process of developing villages and towns endangered by shrinking or extinction.

E-CO Model

The *e-co* model is defined by three interwoven dimensions *e*, *CO* and *eco* (Figure 5), which should synchronously operate to facilitate the recovery of towns and villages in shrinking regions.

Through the *e* dimension, the model addresses the issues recognized in the following: a) spatial concentration and centralization of jobs in urban areas supported in the traditional industrial economy; b) loss of jobs resulting from the shutdown of industries in the post-industrial period; and c) growth in mobility in relation to a home-workplace. To address the identified issues, *e-co* model addresses strengthening and developing the *e-lance economy* (Hooker, Lewis, Smith, Worrell, & Yoon, 2008). Internet-based industries should provide the dispersion of workplaces and population within and outside towns and reduce the level of commuting (Hislop, 2008), or provide access to and retention of population in shrinking areas. Such type of economy is based on the exchange of information and networking within the virtual space (Sassen, 2003). It generates the categories of working cloud space and virtual networks as the counterpart to the workplace and traffic connections, aiming to reduce physical mobility and retaining the population in shrinking areas. Virtualization of the workplace relativizes the physical dimension of the traditional workspace (Sassen, 2003), that is, creates a *working non-place*. According to the model, it can be built into an apartment or an industrial / business facility as well as into urban or rural context/space. The holder of virtual networks in this model is the Internet, understood as the *information highways* - communication infrastructure providing information exchange (Hislop, 2008). In the human dimension, the model introduces new categories of workers: e-lancer, web-worker, home-worker, and freelancer.

Figure 5. E-co model
Source: Tijana Vujičić and Dijana Simonović

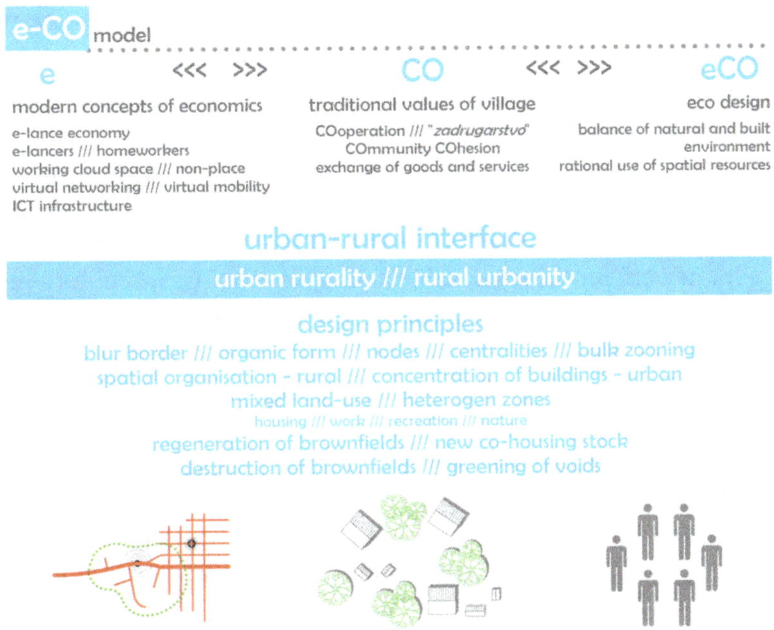

Through the CO dimension, the model addresses the problem of the loss of ethos, inherent in Dinaric world prevailing in the Balkans, the loss of identity and mentality characteristics of the individual and the community as a result of adjusting to the Western model of life (Mihaljević, 1995). Unlike the Western model, characterized by the individuality of a person that methodically and entrepreneurially provides for themselves and their family, where private property and family is considered a basic social and biological cell (Mihaljević, 1995), the *e-co* model renews the traditional lifestyle of Dinaric type, whose main determinants are collectivism, communion and association. In this context, *e-co* model emphasizes the following: a) individuation, understood as a process of fulfilling each individual, b) strengthening the spirit of collectivism inherent in the people of the Balkan cultural context, and c) establishing disturbed synchrony between the individual and the collective. Cooperating, exchanging, assisting - those values mark the operating dimension of the *e-co* model that is realized and operationalized through the concept of a cooperative (in Serbian *zadruga*) founded in the physical and spatial framework and operationalized in a virtual dimension through various e- platforms. In the context of housing, *e-co* supports the *cohousing* concept, based on the original principles of organization of life in the countryside in the Balkans. This concept facilitates interaction among neighbors - among generations and ensures social, practical, economic and environmental benefits, and involves the creation and sharing of common spaces (external and internal). The pastime dimension of life by the *e-co* model implies the strengthening of social cohesion and the spirit of community within the neighborhood and is supported through eco-friendly interventions in space. Observed on an individual level, the *e-co* model advocates a free daily routine, balanced relationship between free and working time, and a natural biological rhythm.

Through the *eco* dimension, the model responds to the problem of broken links between the human and nature, inherent in the urban style of life in densely built areas. In the context of shrinking, or more precisely its spatial manifestation in the form of urban tissue perforation, the *e-co* model proposes the activation of these voids according to the principle of acupuncture of green spaces, that is, it advocates balancing the natural and developed environment. The model introduces different forms of soft landscaping: urban agriculture, urban gardens, urban forestry and urban wilderness (Rößler, 2008; Djukić, Antonić, & Vujičić, 2017), but emphasizes those that strengthen the CO dimension of the model and the pattern of life i.e. those that stimulate social interaction. On the other hand, the model advocates the regeneration of brownfield space rather than building new buildings. Recycling old, abandoned buildings should provide a framework for the sustainable and rational use of space. As a resource, brownfield structures can serve various alternative functions that badly need the accommodation space both in urban and rural environments (e.g. co-working spaces, care centers, pop-up shops).

The *e-co* model primarily aims at alleviating and stopping the main causes of stagnation, primarily negative natural population increase and migrations. In this regard, target group is the population of young, (IT) educated people who are dissatisfied with life in developed urban centers and who are looking for alternative lifestyles. The hectic lifestyle of urban people has disrupted the natural balance and pace of life. The lifestyle in such urban centers is marked by an unbalanced relationship between working and leisure, a lost connection with nature and the community, as well as an extensive number of short or long journeys (to work, kindergartens, schools, hospitals, recreational facilities, etc.). Living by the Western model significantly affects the mental and physical health of urban people (Mihaljević, 1995). Seen spatially, urban places are an unhealthy environment to live and work in. Nowadays, the towns where the target population live are characterized by constant traffic jams, lack of parking spaces, polluted air, life in small housing units below the required level of comfort, insufficiently developed public social infrastructure (public spaces, playgrounds, parks, etc.) and degraded neighborhood i.e. lack of community within the settlements.

The *e-co* model promotes an alternative *e-co* lifestyle that strives for a balanced daily routine without programmed work-home ratio and a strong bond with nature and the community. The principles of sharing, association, and exchange promote the social interaction within these neighbors; spatially, they manifest themselves through new forms of common spaces. The work and exchange of goods are operationalized in the virtual *e* dimension. Commuting is reduced to a minimum, while spatial interventions support different forms of organizing free time activities of individuals and the community. Furthermore, the model targets those groups of users who are eager to connect with nature and who, being aware of the low level of food safety, want to produce healthy food their family themselves. The *e-co* human is modern on the outside, and the traditional inside.

Urban Rurality and Rural Urbanity

The proposed *e-co* model is the result of a proactive approach to the processes of developing towns and villages that are in the process of shrinking or extinction, where proactivity means their development in the desired direction. The desired direction in the 'steering' between different model options means the selection of those offering an optimal response to specific, current problems of shrinking or extinction in the considered area, as well as possible new, changeable contextual conditions and challenges. Regarding the problems of shrinking of towns, the extinction of villages and the development of the urban-rural continuum in the considered area of the northwestern Bosnia and Herzegovina, the contribution of this research is the simulation of the process of remodeling their development and operationalization i.e. creating the most appropriate response to the causes and forms of the recorded problems.

Among the causes of the problems identified by this research are those who are in the domain of changing the identity, the meaning of places, spatial patterns, and lifestyles in the observed area. It is the very spectrum of problems that is the source of the idea of sublimation of urban (modern) advantages and rural (traditional) values, which is here applied through the *e-co* lifestyle. This component of the proposed *e-co* model for the recovery of shrinking towns and villages in extinction is constructed based on the results of the research of pseudo-urban patterns of peri-urban areas i.e. of revealed features and legality of the development of the urban-rural continuum in the observed northwestern area of Bosnia and Herzegovina. Urban studies into the development of the spatial identity of towns belonging to this cultural and spatial context (Milić, 2002; Savić, 2013; Simonović, 2010, Simonović, 2014; Simonović, Novaković & Vujičić, 2011) are unique in terms of emphasizing the dichotomy of traditional and modern, in terms of ethos and on the spatial, identity, mentality, and other levels.

Significant results of this research relate to the specificity of new, pseudo-urban spatial patterns i.e. the facts on the fulfillment of the criteria of vitality, toughness, and resistance to the dynamics of the development process. It is characterized by dynamism and spontaneity of the process of creation and expansion, the absence of planned control and planners' actions. It is formed by overlapping two traditionally different types of building density, characterized by different densities, land use methods, space purposes, utility levels, and lifestyles (Milić, 2002). The vitality of these patterns has provided the basis for conceptual responses to the shrinking and negative development problems of towns and villages in the northwest of Bosnia and Herzegovina. Combining mutually opposing processes of *urban ruralization* and *rural urbanization*, the development of these areas often produces negative spatial and social effects. However, in an effort to produce the positive effects of these processes and not to cause the loss of inherent features and values/advantages of either the village or the town, the concept of an optimal urban-rural community was constituted. It is essential that the directions for the development of towns and villages facing shrinking or extinction are directed towards the sublimation of rural and traditional values (through empirical and customary standards) with modern forms of business and the exchange of goods inherent to towns (using normative-regulatory standards). For the purpose of recovery, observation, planning, and design of these settlements should carry on through both focuses - rural and urban (Ralević & Simonović, 2017). It is targeted to produce an urban-rural synergy with a source in the past and the outcome in the future, through the integration of "rural values and urban advantages" (Ralević & Simonović, 2017, p. 215).

The view that the alternative *e-co* model provides conditions for the revival of shrinking communities or those nearing extinction within the observed area was formed through the clarification of the vitality of the new pseudo-urban forms investigated in the peri-urban area of the city of Banja Luka. Key landmark of the spatial identity – a permanent presence of rural elements in the urban form of Banja Luka (Simonović, 2010) – was recognized as urban rurality and, through sublimation with rural urbanity, incorporated into the *e-co* model. The concept of urban rurality has been developed based on the pseudo-urban form of the peri-urban region of Banja Luka (the model is explained in greater detail in the chapter on the transformation of the urban and rural landscape between the pre-industrial and post-industrial eras). The new urban-rural structure, basically a conurbation, continuous across to the city core which it has integrated itself into, also forms a spatial basis for building a synergy between the urban and rural. It is important to point out the significance of connections and trends between the urban, rural and peri-urban (Lynch, 2005), and of creating synergetic relations between them as based on their interconnections and common development trends. Beside connections and trends, elements relevant for the understanding and application of the proposed *e-co* concept can be found in the *rural-urban interface*, as explained by Satterthwaite (2000) in his concept of the *rural-urban continuum.*

CONCLUSION

Shrinkage is a ubiquitous phenomenon that leads to significant negative consequences in all domains of society as a whole. Taking the shrinkage of the region as a precondition, the paper presents a model for identification and evaluation of shrinking processes applicable not only to the Bosnia and Herzegovina context but also in general. Using the proposed model in the territory of 17 municipalities in the north-west of Bosnia and Herzegovina, it has been proven that the problem of shrinkage also affects villages and towns i.e. the whole region. The worst affected villages are those threatened by extinction. The key causes of their shrinkage are the village-to-town migrations, while a decline in natural population increase and emigrating abroad are identified as the main causes of shrinkage in towns. Reduction and aging of the population are the key consequences of these negative changes that pose a serious challenge to the whole society. Their spatial manifestation is a large number of abandoned housing units - extinct households, but also a large number of dissolved companies. Brownfield spaces are an omnipresent pattern that follows the processes of shrinkage and produces a functional perforation of the developed environment (Đukić, Simonović, & Vujičić, 2014; Simonović, Novaković, & Vujičić, 2011). This paper searched for the answer on how to solve the existing problems.

Basing the research on the analysis of traditional forms of life organization specific to towns, villages and peri-urban areas in the Balkans (Mihaljević, 1995; Milić, 1998; Milić, 2002; Simonović, 2010; Simonović, 2015), as well as contemporary global work organization practices (Hislop, 2008; Hooker, Lewis, Smith, Worrell, & Yoon, 2008; Sassen, 2001; Sassen, 2003; Sassen, 2016), this paper offers answers to the problems of shrinkage through a new, alternative concept. Understood as a systemic approach and strategic commitment, the proposed *e-co* model should help towns and villages recover from the crisis and restore their vitality. In uncertain circumstances, this model should offer an adequate framework for adapting and learning in the process of managing negative changes. The mission of this *e-co* approach is to reduce the vulnerability of this area and build the adaptive capacity of the community (Vujičić, 2018b). Sustainable and resilient towns and/or villages are achieved here through the synergy of urban and rural, sublimation of rural values and urban advantages. The authors of this paper see the opportunity for the recovery of towns and villages in the development of such an offer of amenities to enable the experience of rural life in urban areas and vice versa i.e. urban rurality and rural urbanity. Three dimensions of the model – *e*, *CO* and *eco* – provide integrity and focus of the approach and offer a framework for operational action within problem areas.

The *e-lance economy* is the driving force, the nucleus for development of the *e-co* community and regeneration of the shrinking areas. It enables the dispersion of jobs at different spatial scales and levels, from global, regional, national, to local. The digitization of work has produced new job categories and job organization systems (Sassen, 2003). The key driver of all activities in the information economy is the Internet. Strengthening and developing of the digital communication infrastructure are the key prerequisites for the implementation of the *e-co* model, which requires a strong commitment of the government and local communities to invest in its development. The spatial dispersion of jobs inherent in the *e-lance economy* tends to reduce migration from village to city, i.e. smaller cities to larger urban centers. Simultaneously, investment in the development of communication infrastructure aims to provide conditions for the arrival and retention of population in shrinking areas. Along whit this, the *e-co* model advocates an alternative lifestyle, rooted in primordial Balkan mentality, which establishes the disturbed synchrony between individual and collective, the human and nature, and work-home balance. Coexistence, co-operation, exchange, mutual assisting (among neighbors, among generations) and community cohesion are values advocated by the model and operationalized through various forms of association in the form of a cooperative (in Serbian *zadruga*). The concept of the cooperative (*zadruga*) presents a platform for local initiative, cooperation, exchanges and action, initiated and led by the local community itself, which can be operationalized in a virtual dimension through various forms of e-platforms.Life in an *e-co* community implies a spatial concept of neighborhood organization similar to what is known in the West as cohousing. In the Balkan context, this concept of *e-co* community receives local specificities that integrate the original values of the Balkan village and the modern advantages of the global city. The regeneration of urban and rural landscapes affected by the shrinkage implies the development of such an offer of amenities to enable the experience of rural life in urban areas - urban rurality - and vice versa, the experience of urban life in rural areas - rural urbanity. The balance between the natural and built environment and rational use of spatial resources are fundamental principles of *e-co* design. The spatial potential, necessary for the implementation of the concepts proposed by the *e-co* model, is found in the abandoned spaces of residential, industrial, military and other brownfields, as well as in the voids within the perforated city fabric.

The model presented in this paper is a conceptual systemic approach that can contribute to the recovery of areas affected by shrinking processes. However, it does not seek to solve all the issues and problems of towns and villages in decline and even though it possesses universal values its adjustments to specific context are necessary. Proactively acting, it aims to eliminate or mitigate the key causes of shrinkage and to offer reactive responses to the consequences of stagnation. Since it represents a general framework for action, future research should offer its further elaboration through the following: a) various program content options embedded in *e*, *CO* and *eco* dimensions, as well as b) different model options - which involve combinations of different programs at the spatial and temporal levels. In this respect, the future research may focus on the following: developing the co-housing model, the model of the cooperative, the model for developing the e-lance economy, the soft landscaping model, the model for the categories of the population that depend on the assistance and care of others (elderly persons and children), and the like. In addition to this type of focused and purposeful studies, it is necessary to develop general control models to enable the implementation of the *e-co* concept. In order for this vision to be feasible, cooperation between the administration, civil society and scholars is necessary. Continuous learning by doing is the basic characteristic of the development process of this concept, and monitoring and evaluation are key tools for measuring efficiency. Developing indicators and monitoring parameters is also one of the fields for future research. Adaptations based on the learned are inevitable in the concept implementation process. As the general and system framework, the *e-co* concept creates a platform for interdisciplinary action of various professions, thereby encouraging co-operation and dialogue. In such a context, *e-co* is seen as an open-lab system that considers the implementation of the idea as a continuous process of research and learning by all the stakeholders involved. To what extent the *e-co* vision will be achievable depends on the capacity of the community to learn and adapt in the process of implementation.

REFERENCES

Agency for Identification Documents, Registers and Data Exchange of Bosnia and Herzegovina. (2017). *Data on the canceled nationalities from B&H due to migration abroad*. Banja Luka: Author.

Đukić, A., Antonić, B., & Vujičić, T. M. (2017). Urban shrinkage in a 'shrinking' Serbia – the approach to a global phenomenon in a local context. *Geodetski Vestnik, 61*(4), 614–629. doi:10.15292/geodetski-vestnik.2017.04.614-629

Đukić, A., Simonović, D., & Vujičić, T. (Eds.). (2014). *International Scientific Monograph BROWNINFO. Toward a Methodological Framework for Brownfield Database Development*. Banja Luka: University of Banja Luka, Faculty of Architecture Civil Engineering and Geodesy.

Đukić, A., Vujičić, T., & Antonić, B. (2018). Greening as an Approach for Urban Renewal of Shrinking Cities. In P. Mitković, M. Dinić Branković, M. Tanić, A. Mirić, & V. Milosević (Ed.), *Proceedings of 2nd International Conference on Urban Planning - ICUP2018* (pp. 291-297). Niš: Faculty of Civil Engineering and Architecture University of Niš.

Federal Statistical Office of Socialist Federal Republic of Yugoslavia. (1975). *Census of Population and Housing 1971. Population and Households in 1948, 1953, 1961 and 1971 and dwellings in 1971. Results by municipalities and settlements*. Belgrade: Author.

Finci, J. (1967). *Stanovanje u Urbanističkom planu Banjaluke* [Housing in the Urban Plan of Banja Luka]. Banjaluka: Urbanistički zavod Banjaluka.

Florentin, D. (2010). The "Perforated City": Leipzig's Model of Urban Shrinkage Management. *Berkeley Planning Journal, 23*(1), 83–101.

Haase, A., Rink, D., Grossmann, K., Bernt, M., & Mykhnenko, V. (2014). Conceptualizing urban shrinkage. *Environment & Planning A, 46*(7), 1519–1534. doi:10.1068/a46269

Hislop, D. (Ed.). (2008). *Mobility and Technology in the Workplace*. New York: Routledge. doi:10.4324/9780203894354

Hooker, R., Lewis, C., Smith, H., Worrell, J. M., & Yoon, T. (2008). Governing E-Collaboration in E-Lance Networks. In N. Kock (Ed.), *Encyclopedia of E-Collaboration* (pp. 324–329). Hershey, PA: IGI Global. doi:10.4018/978-1-59904-000-4.ch050

Houlne, T., & Maxwell, T. (2013). *The New World of Work: From the Cube to the Cloud*. Irving, TX: Inspire on Purpose Books.

Institut for Statistics of Federation of Bosnia and Herzegovina. (1998). *Results of the Census of Population, Households, Flats and Agricultural Holdings 1991. Population comparative data 1971, 1981 and 1991*. Sarajevo: Author.

Institut for Statistics of Federation of Bosnia and Herzegovina. (1999a). *Results of the Census of Population, Households, Flats and Agricultural Holdings 1991. Population by municipalities*. Sarajevo: Author.

Institut for Statistics of Federation of Bosnia and Herzegovina. (1999b). *Results of the Census of Population, Households, Flats and Agricultural Holdings 1991. Population by settlements*. Sarajevo: Author.

Jürgen, F. (1993). A Theory of Urban Decline: Economy, Demography and Political Elites. *Urban Studies (Edinburgh, Scotland)*, *30*(6), 907–917. doi:10.1080/00420989320080851

Konjar, M., Kosanović, S., Popov, S. G., & Fikfak, A. (2018). Urban/Rural Dichotomy and the Forms-In-Between. In A. Fikfak, S. Kosanović, M. Konjar, & E. Anguillari (Eds.), *Sustainability and Resilience Socio-Spatial Perspective* (pp. 149–169). Delft: TU Delft Open.

Lefebvre, H. (2003). *The Urban Revolution* (R. Bononno, Trans.). Minneapolis, MN: University of Minnesota Press.

Lütke-Daldrup, E. (2001). Die perforierte Stadt. Eine Versuchsanordnung [The perforated city. A test arrangement]. *Stadtbauwelt*, *150*, 40–45.

Lynch, K. (2005). *Rural-Urban Interaction in the Developing World*. New York: Routledge.

Martinez-Fernandez, C., Kubo, N., Noya, A., & Weyman, T. (2012). *Demographic Change and Local Development: Shrinkage, Regeneration and Social Dynamics*. OECD Local Economic and Employment Development (LEED). Retrieved March 22, 2017, from http://www.carltd.com/sites/carwebsite/files/Demographic%20Change%20and%20Local%20Development.pdf

Mihaljević, G. (1995). *Imati i biti: ekonomija i kuća* [Having and being: the economy and the house]. Beograd: Arhitektonski fakultet Univerziteta u Beogradu.

Milić, V. (1998). Fizička struktura predela [Physical structure of the landscape]. In I Letnja skola arhitekture "Petnjica 97" Arhitektura predela ... do razmere 1:1" (pp. 92-98). Beograd: Arhitektonski fakultet Univerziteta u Beogradu i Istraživačka stanica Petnjica.

Milić, V. (2002). Pseudourbani prostor između sela i grada: trendovi u Republici Srpskoj [Pseudo-urban spaces between villages and towns: Trends in the Republic of Srpska]. In Zbornik ljetnje škole urbanizma Šipovo 2002 (pp. 65-73). Banja Luka: Urbanistički zavod Republike Srpske.

Ministry of Human Rights and Refugees of Bosnia and Herzegovina. (2014). *Overview of Emigrations from Bosnia and Herzegovina*. Sarajevo: Author.

Oswalt, P. (Ed.). (2005). Shrinking cities Volume 1 – International Research. Ostfildern-Ruit: Hatje Cantz Verlag.

Pallagst, K., Wiechmann, T., & Martinez-Fernandez, C. (2014). *Shrinking Cities International Perspectives and Policy Implications*. New York: Routledge.

Pašalić, S., & Pašalić, D. (2016). Migration losses of Bosnia and Herzegovina and demographic aging. *International E-Journal of Advances in Social Sciences*, *2*(6), 650–658. doi:10.18769/ijasos.280343

Pašalić, S., Vuković, D., Jovičić, M., & Stević, D. (2017). Dynamics and directions of demographic changes in Bosnia and Herzegovina. *International E-Journal of Advances in Social Sciences*, *3*(7), 40–48. doi:10.18769/ijasos.309482

Ralević, M., & Simonović, D. (2017). Innovative approaches in the process of re-integration of city and village. In Dž. Bijedić, A. Krstić-Furundžić, & M. Zečević (Ed.), *Book of Conference Proceedings 4th International Academic Conference Places&Technologies 2017: Keeping up with technologies in the context of urban and rural synergy* (pp. 215-226). Sarajevo: Faculty of Arcitecture University of Sarajevo.

Republika Srpska Institute of Statistics. (2013b). *Estimation of the number of inhabitants by the municipalities of Republika Srpska 1996-2012*. Banja Luka: Author.

Republika Srpska Institute of Statistics. (2017a). *Results of the Census of population, households and dwellings in the Republic of Srpska 2013*. Banja Luka: Author. Retrieved from http://www2.rzs.rs.ba/static/uploads/bilteni/popis/gradovi_opstine_naseljena_mjesta/Rezultati_Popisa_2013_Gradovi_Opstine_Naseljena_Mjesta_WEB.pdf

Republika Srpska Institute of Statistics. (2017b). *Results of the Census 2013, Marital status and fertility*. Banja Luka: Author. Retrieved August 15, 2017, from http://www.rzs.rs.ba/front/article/2365/?left_mi=None&add=None

Republika Srpska Institute of Statistics. (2017c). *Population Statistics Annual Release-Marriages and divorces in Republika Srpska, No. 109/17*. Banja Luka: Author. Retrieved August 15, 2017, from http://www.rzs.rs.ba/front/article/2463/

Republika Srpska Institute of Statistics. (2002-2016). *Bulletin Demographic Statistics 5-19*. Banja Luka: Author. Retrieved August 13, 2017, from http://www.rzs.rs.ba/front/category/2/132/?&add=None

Rieniets, T. (2006). *Shrinking cities—growing domain for urban planning*. Retrieved August 28, 2013, from http://aarch.dk/fileadmin/grupper/institut_ii/PDF/paper_presentation_EURA2005.pdf

Rößler, S. (2008). Green space development in shrinking cities: opportunities and constraints. *Urbani Izziv, 19*(2), 147-152. Retrieved September 15, 2018, from https://www.jstor.org/stable/24906040?seq=1&cid=pdf-reference#references_tab_contents

Sassen, S. (2003). *Protugeografije globalizacije* [Counter-geographies of Globalization] (T. Medak, Trans.). Zagreb: Multimedijalni institut.

Sassen, S. (2010). Global inter-city networks and commodity chains: Any intersections? *Global Networks, 10*(1), 150–163. doi:10.1111/j.1471-0374.2010.00279.x

Sassen, S. (2016). Digitization And Work: Potentials and Challenges in Low-Wage Labor Markets. In J. Stollmann, K. Wolf, A. Brück, S. Frank, A. Million, P. Misselwitz, & C. Schröder (Eds.), *Beware of smart people! Redefining the smart city paradigm towards inclusive urbanism* (pp. 33–44). Berlin: Universitätsverlag der TU Berlin. doi:10.14279/depositonce-5074

Savić, J. (2013). *Moderno i vernakularno u procesu urbane transformacije Banjaluke* [The modern and the vernacular in urban transformation of Banjaluka] (Unpublished doctoral dissertation). Beograd: Arhitektonski fakultet Univerzitet u Beogradu.

Simonović, D. (2010). *Pejzažni gradovi: poređenje razvoja urbanog identiteta gradova Banjaluke i Graca* [Landscape cities: A comparison between the development of urban identities of Banja Luka and Graz]. Banjaluka: Arhitektonsko-građevinski fakultet Univerziteta u Banjoj Luci.

Simonović, D. (2014). *Istraživanje mogućnosti obnove i unapređenja identiteta pejzažnog grada Banjaluke putem urbanističke regulacije* [Research into the possibility of rehabilitation and upgrade of the urban identity of Banja Luka as a landscape city through urban regulation] (Unpublished doctoral dissertation). Banjaluka: Arhitektonsko-građevinsko-geodetski fakultet Univerziteta u Banjoj Luci.

Simonović, D. (2015). Interweaving of Banjaluka's urban and rural landscapes. In A. Fikfak, E. Vaništa Lazarević, N. Fikfak, M. Vukmirović & P. Gabrijelčič (Eds.), *Book of Conference Proceedings of Places&Technologies 2015 - Keeping up with technologies to make healthy places book of conference proceedings* (pp. 194-200). Nova Gorica: Faculty of Architecture University of Ljubljana.

Simonović, D., Novaković, N., & Vujičić, T. (2011). Toward a strategy of regeneration of urban landscape: brownfiled as a strategic resource. In *Proceedings of I International Conference ECOLOGY OF URBAN AREAS* (pp. 439-449). Zrenjanin: Faculty of Technical Sciences Mihajlo Pupin.

Stryjakiewicz, T., Schmeidler, K., Vajdovich Visy, E., Jaroszewska, E., Marcinczak, S., … Đukić, A. (2013). Shrinking Cities in Post-Communist Europe. In K. Pallagst (Ed.), Cities Regrowing Smaller - Fostering Knowledge on Regeneration Strategies in Shrinking Cities across Europe. CIRES Synopsis Report.

Tanović Lukić, M., Pasalić, S., & Golijanin, J. (2014). Demographic Development of Bosnia and Herzegovina from the Ottoman Period Till 1991 and the Modern Demographic Problems. *Procedia: Social and Behavioral Sciences, 120*, 238–247. doi:10.1016/j.sbspro.2014.02.101

Tokača, M. (2012). *Bosanska knjiga mrtvih. Ljudski gubici u Bosni i Hercegovini 1991-1995* [Bosnian Book of the Dead. Human Losses in Bosnia and Herzegovina 1991-1995]. Sarajevo: Istraživačko dokumentacioni centar Sarajevo.

Turok, I., & Mykhnenko, V. (2007). The Trajectories of European Cities, 1960-2005. *Cities (London, England), 24*(3), 165–182. doi:10.1016/j.cities.2007.01.007

Vujičić, T. (2018a). *Methodological framework of integral adaptive management for shrinking cities - case study Prijedor* (Unpublished doctoral dissertation). Faculty of Architecture University of Belgrade, Serbia.

Vujičić, T. (2018b). Shifting Forward: Resilience Thinking in Out-of-Order Urban System. In A. Fikfak, S. Kosanović, M. Konjar, & E. Anguillari (Eds.), *Sustainability and resilience socio-spatial perspective* (pp. 43–72). Delft: Delft University of Technology, Netherland.

Vujičić, T. M., & Đukić, A. (2015). Methodological framework for shrinking cities case study research: Northwest region of Bosnia and Herzegovina. *Geodetski Vestnik, 59*(3), 520–536. doi:10.15292/geodetski-vestnik.2015.03.520-536

Wiechmann, T. (2013, September 12). *Shrinking Cities in Europe - Evidence from COST Action "Cities Regrowing Smaller" (CIRES).* Retrieved December 9, 2014, from http://www.shrinkingcities.eu/fileadmin/Conference/Presentations/01_Wiechmann.pdf

Wiechmann, T., & Wolff, M. (2013). Urban Shrinkage in a Spatial Perspective – Operationalization of Shrinking Cities in Europe 1990 - 2010. In *AESOP / ACSP 5th Joint Congress 2013 Planning for Resilient Cities and Regions.* Dublin: Association of European Schools of Planning (AESOP).

KEY TERMS AND DEFINITIONS

Brownfield: Previously developed or built-up spaces in a state of visible abandonment and inefficient use, with the capacity for renewal and sustainable development.

E-CO Model: A focused strategic approach based on the e-lance economy, community cohesion, and eco-friendly design.

E-Lance Economy: New type of economy based on fluid, virtual networks, where independent e-lancers use the Internet to sell their services or work to network and exploit specific business opportunities.

Perforated Area: A developed area affected by the shrinking processes with a number of abandoned, underused, and neglected buildings, the so-called brownfields.

Peri-Urban Settlement: Settlements located at the edge of towns, areas between the town and the village, and between towns. They are characterized by dispersed, conurbation form of perforated built-up areas, and the presence of pseudo-urban forms.

Rural Urbanity: The presence in villages and peri-urban areas of modern forms of the organization of work and living inherent in towns that offer the experience of the urban life in rural areas.

Shrinkage: A multidimensional issue affecting towns, villages and whole regions, manifested by demographic decline, economic loss and urban tissue perforation.

Urban Rurality: The presence in towns and peri-urban areas of traditional forms of the organization of work and living inherent in villages that offer the experience of the rural life in urban areas.

Urban-Rural Synergy: The integration of rural values and urban advantages, of the modern global and the traditional local, which traces its roots in the processes of rural urbanization and urban ruralization.

Chapter 18
Restructuring Centers of Former Rural Settlements and Weekend-Residence Zones:
Shaping Post-Socialist Suburbia in Niš, Serbia

Milena Dinić-Branković
Faculty of Civil Engineering and Architecture, University of Niš, Serbia

Jelena Đekić
Faculty of Civil Engineering and Architecture, University of Niš, Serbia

Petar Mitković
Faculty of Civil Engineering and Architecture, University of Niš, Serbia

Milica Igić
Faculty of Civil Engineering and Architecture, University of Niš, Serbia

ABSTRACT

The zone of post-socialist suburbia was predefined even during socialism, when rural settlements surrounding large cities were joined with city territory. After the fall of socialism, new intensive transformations unfolded in suburban areas. This chapter analyzes the centers of suburban settlements by using a segment of the administrative area of the City of Niš, Serbia as a case study. Selected settlements have high values of demographic growth in the post-socialist period: four settlements of rural origin, one sprawling settlement, and one settlement of former weekend-residences. This chapter is set to investigate current development patterns in suburbia and to formulate guidelines for designing suburban centers. Urban matrix for mixed-use centers is created, which shapes the center spatially and functionally. Vital centers of suburban neighborhoods that are integrated into the city network represent the only sustainable form of development of suburban areas and city as a whole.

DOI: 10.4018/978-1-5225-9932-6.ch018

INTRODUCTION

Niš is a former socialist city of medium size located in southeast Serbia. According to the results of the Census of 2011, the administrative area of the City of Niš has a population of 257.867 people, which reside in 5 urban and 35 so-called "other settlements" (Serb. *Ostala naselja*). The current administrative and legal regulation for the differentiation of settlements in Serbia fails to provide an accurate insight into the types of settlements in Serbia, which is evident in the example of Niš. According to the Nomenclature of Statistical Territorial Units of the Republic of Serbia (Statistical Office of the Republic of Serbia, 2010), four subdivisions of the City of Niš (settlements Niš-Medijana, Niš-Palilula, Niš-Pantelej, Niš-Crveni Krst) and adjacent settlement Niška Banja are declared "urban settlements". All other settlements that are located within the territory of the Master Plan (MP) of Niš 2010-2025, but outside of city bounds are defined as "other settlements", therefore non-urban (Statistical Office of the Republic of Serbia, 2011)[1]. Out of them, 34 have rural origins and 1 is workers' settlement created after World War II (settlement Nikola Tesla). "Other settlements" in the territory of MP of Niš are in various transformation stages – from traditional villages, through mixed-suburban settlements, to the settlements with certain urban characteristics.

The period after World War II was characterized by high demographic growth in the village settlements in the vicinity of Niš and certain parts of today's suburban zone (Dinić, Đurić, & Mitković, 2014). City of Niš experienced significant expansion of its territory in the post-socialist period and is still expanding. Master Plan of Niš 1995-2010 marked the process of annexation of surrounding rural settlements to the city. This MP covered an area of 150.36 km2, encompassing Niš urban area and 18 settlements nearest to the city. Current MP of Niš 2010-2025 has significantly increased its coverage - as much as 73% compared to the previous plan from 1995 (from 150.36 to 266.77 km2). Agricultural land at the city perimeter is being urbanized and suburban way of life is becoming increasingly popular. The latest MP of Niš includes 18 new villages, while one settlement was converted into an urban settlement by administrative decision (Niška Banja). The standing MP of Niš 2010-2025 states that the development of rural and suburban areas is an important element to consider, as well as the overall quality of life in rural areas on the periphery of the MP, which should provide a quality basis for all activities related to construction of buildings, use of space, traffic connections and communal infrastructure.

Previous research on this topic (Dinić, Đurić, & Mitković, 2014) defined the suburban belt of the City of Niš which includes 11 existing "other settlements": Deveti maj, Popovac, Medoševac, Gornji Komren, Donji Komren, Donja Vrežina, Brzi Brod, Nikola Tesla, Suvi Do, Gabrovac and Pasi Poljana. These settlements are in immediate contact with the city territory and the share of agricultural population in each of them is less than 2%. They have already undergone the transformation from rural into suburban in spatial, functional and demographic terms. The remaining "other settlements"- remote villages are not considered as suburbia in this research.

Former rural settlements are the dominant type in the suburban belt of the City of Niš. Around and in-between the settlements of rural origin, other suburban forms also appear, creating a myriad of urban patterns and peculiar spatial and social compounds. In the last couple of decades, sprawling occurred to a smaller extent, as well as the transformation of former weekend-residence zones into permanent housing. After the collapse of socialism, a "patchwork" development was initiated in the suburban belt, since all of these types of settlements have merged over time, both with each other and with the urban area of Niš. Consequently, boundaries between various settlement types are now very difficult to establish. These suburbanization processes are most visible and most intense in one particular area of the suburban belt of Niš, in the Cadastral Municipality Kamenica, called Kamenica 2. This populated area involves settlements Podvinik, Vinik and a part of Branko Bjegović, which are not administratively declared as independent settlements.

In such development circumstances, when suburban fabric entails multiple levels of different character, planning and designing post-socialist suburbia is particularly challenging. Given the significance of polycentric city development, which is often stated as a long-term strategic goal in planning of Eastern European cities (Tsenkova, 2007; Stanilov, 2007a), as well as Serbian cities (Vujović & Petrović, 2007), shaping central places in suburbia becomes increasingly important.

When it comes to Niš, during socialist period most of the rural settlements in today's suburban belt had a formed village center. A three-level functional typology of rural settlements was established, according to which villages are classified into "primary villages" (without a center), "villages with rural centers" and "centers of community of villages". However, due to the immediate proximity of the city and its impact, rural centers and centers of community of villages were quite passive in the network of centers, and all urban needs were being met in the city. This setting did not change much in the post-socialist planning documents, since the MP of Niš 2010-2025 retained similar terms as well as the overall concept in the "Network of centers of Niš". However, as these settlements no longer have rural but suburban character, the question arises as to whether the inherited rural model of "settlement with a center" and "center of community of settlements" is sustainable today. The planning and design of central places in suburban belt become even more complex when other suburban types are taken into account - sprawling settlements and converted weekend-residence zones. These settlements often lack any kind of central uses, and therefore require some restructuring of urban fabric. Another crucial point in planning and designing of suburban centers concerns assigning them an appropriate rank in a hierarchically structured network of centers at the entire city territory.

Having these issues in mind, this study analyzes the centers of various types of suburban settlements that are nowadays present in the suburban belt of the City of Niš: (1) former village settlements of Brzi Brod, Donja Vrežina, Donji Komren and Gornji Komren, (2) sprawling settlement Podvinik and (3) former weekend-residence zone Vinik (Figure 1). A brief overview of their development characteristics is given in this chapter, followed by an analysis of the current state of their centers and planning context. All of these settlements obtained urban characteristics in the transition period, but this transformation was not accompanied by the transformation of urban structure of their central areas. The following sections explore the design characteristics of suburban centers and discuss future research perspectives. Since it is of crucial importance for sustainable development of suburban areas to improve their centers and make them vital nexuses of social life, the aims of this research are to examine the spatial-functional characteristics of post-socialist suburban development patterns and to propose guidelines for designing the centers of suburban settlements.

METHODOLOGY

The choice of Niš's suburban landscape as the subject of study is premised on the fact that post-socialist transformations of small and medium-sized cities, particularly in southeastern Europe, have not been studied as thoroughly as those of large cities and metropolitan areas. This research is empirical and uses a review of relevant literature and planning documents, as well as practical experience involving field observation. The standing planning documents for the analyzed suburban area, which are considered in

Figure 1. Position of the analyzed village settlements, weekend-residence zone and sprawl settlement in comparison to the urban area of Niš, and their centers: 1. Donji Komren; 2. Gornji Komren; 3. Donja Vrežina; 4. Brzi Brod; 5. Podvinik; 6. Vinik

Source: Authors' drawings, based on: Master Plan of Niš 2010-2025 (2011), Plan of General Regulation of the Municipality of Medijana (2012), Plan of General Regulation of the Municipality of Crveni Krst–II phase (2013) and Plan of General Regulation of the Municipality of Pantelej–II phase (2015)

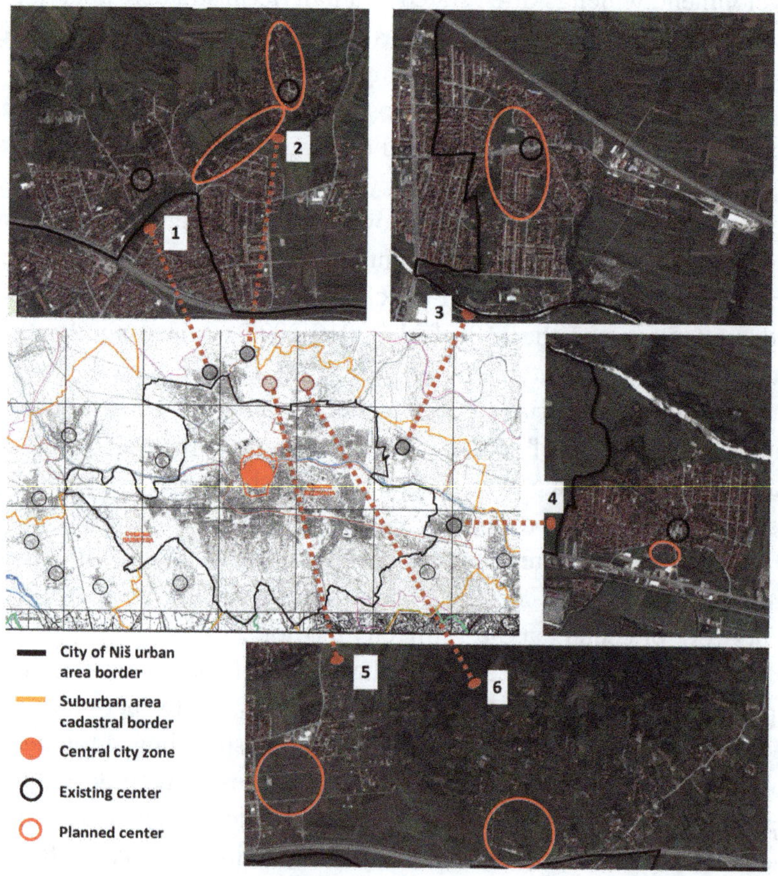

this research, are: Master Plan of Niš 1995-2010 (1995), Master Plan of Niš 2010-2025 (2011), Plan of General Regulation (PGR) of the Municipality of Medijana (2012), Plan of General Regulation (PGR) of the Municipality of Crveni Krst–II phase (2013) and Plan of General Regulation (PGR) of the Municipality of Pantelej–II phase (2015).

The suburban settlements were selected based on the criteria: (1) all of them had high values of demographic growth in the post-socialist period, between two Censuses 1991-2011, (2) they are located in adjacent segments of the suburban belt, thus forming a continuous landscape, and (3) they show most prominent features of post-socialist transformations and "patchwork" development. Since former rural settlements are the dominant type at the urban fringe, their share in the sample is the highest. The choice of settlements reflects on other areas where urban-rural interactions occur. Therefore, this portion of the suburban belt of Niš makes a suitable basis for analyzing the post-socialist suburban centers and establishing their design principles. For planning and designing of suburban centers the authors developed the theoretical framework "Urban matrix for mixed-use centers", based on literature review. Similar technique for mixed-use projects entitled "Block standards" was originally developed by Calthorpe & Fulton (2001).

BACKGROUND

During socialist period, the urban area border of socialist cities was clearly delineated, and it was separated by an agricultural belt from the surrounding villages (Hirt & Stanilov, 2007). During the 1960s and 1970s, numerous villages in the surroundings of large socialist cities were joined with urban territories (Stanilov, 2007b), in order to increase the share of urban population in the country and provide the reserve of land for future urban growth. Despite the proximity of the city, due to the limited finances, there were no investments in the rural settlements' infrastructure, and they retained their rural character.

In the transition period, an intensive spatial-functional restructuring of the socialist city unfolded. The cities were surrounded by a large number of small village settlements with a high share of non-agricultural populations. The existing network of rural settlements surrounding cities is stated as the most important historical and cultural reason of suburbanization and urban sprawl of post-socialist cities (Couch, Petschel-Held, & Leontidou, 2007). Since 1990, a formation of new suburban belt began in the agricultural area between the urban border and surrounding village settlements. Haphazard low-density urban growth occurred, and borders between the urban and rural become blurred. Even though the phenomenon of dispersed suburbia happened in Eastern Europe much later than in the Western Europe, it mostly retraced the capitalistic model of sprawl (Tammaru, Leetmaa, Silm, & Ahas, 2009; Stanilov 2007b; Sykora & Ourednicek, 2007, Sykora & Stanilov, 2014). Stanilov (2007a) concludes that "the Central and Eastern European cities are being transformed from high-density, monocentric settlements, dominated by high-rise public housing and communal modes of transportation, into sprawling, multi-nodal metropolitan areas reaching extreme levels of privatization of housing, services, transportation, and public space" (p. 7).

Residential suburbanization has become the most visible symbol of post-socialist urban transformation (Sykora & Ourednicek, 2007), even though it proceeded quite slowly. The boom of residential construction at the urban periphery of the post-socialist city is a result of various factors: the rise of commercial facilities at the expense of residential use in the city center, the restitution of land, the relaxation of land development controls, the establishment of an open land market and the diversification of residential choices (Stanilov, 2007b). As a result of a growing public preference for detached single-family houses with private yards, a substantial amount of new residential development occurred beyond the urban boundary (Hirt & Stanilov, 2007). Weekend residences and "second homes" in the rural-urban fringe also influenced the patterns of suburbanization significantly (Paul & Tonts, 2005; Serebrennikov, 2018), as people began to convert them to permanent homes in the 1990s. The former weekend residence zones and villages settlements gradually became proper suburbs, which are city-dependent in terms of jobs and services (Stanilov, 2007b). These suburbs are mainly occupied by thinly spread individual housing, and they often still lack basic public facilities and infrastructure. Although many of the former peripheral villages surrounding socialist cities were connected with public transit, the level and quality of this service is significantly below the aspirations of new suburban residents (Stanilov, 2007b). Therefore, the denizens are dependent on the use of cars to meet basic needs (Dinić-Branković, Bogdanović-Protić, Đekić, & Mitković, 2016).

In contemporary planning theory, a polycentric spatial structure, or de-concentration of urban functions from the main city center to secondary centers within the administrative territory of a city, represents the only sustainable form of development when it comes to positioning central places (Dinić & Mitković, 2016). So far, in the ongoing transformation of metropolitan areas in the post-socialist Central and Eastern European countries, the formation of sub-centers outside of the metropolitan core in a hierarchical network is not fully articulated (Stanilov & Sykora, 2014). Even though the suburban landscape of post-socialist metropolis was shaped by intense changes in demographic, cultural and socio-economic development, the restructuring of central uses in adjacent former rural communities did not occur, and these centers functionally correspond more to centers of rural agglomerations. It can be stated that nowadays large rural settlements usually attract residents from city territory due to their environmental quality, privacy and affordability, but still continue to exploit the infrastructure and services of the neighbor cities (Lazauskaite, Šarkiene & Skripkiene, 2015). However, the restructuring of suburban sub-centers in the post-socialist city is an urban planning issue of crucial importance for the process of shaping sustainable suburban neighborhoods.

Cities of Western Europe have initiated various planning strategies to combat the phenomenon of sprawling suburbia and guide spatial development, in order to prevent urban sprawl degrading the cultural and natural landscapes in the rural–urban fringe (Paul & Tonts, 2005). In the United States, where sprawling developments are most visible, new approaches and techniques in planning are also advised to control urban growth. For new Greenfield development the same principles are suggested as for infill development, such as walkable neighborhoods with diversity of everyday uses and housing opportunities, even though they are more challenging to achieve in more remote areas (Calthorpe & Fulton, 2001). In the case of the City of Niš, in regard to its firm tradition of compact development, the post-socialist planning documents also made an attempt to limit sprawl developments and to envision more viable suburban neighborhoods with local centers.

One of the most successful approaches to the efficient use of land and infrastructure in metropolitan areas, that is of particular interest for shaping the centers of suburban settlements, are block standards (Schmitz et al., 2003). The concept of standardized blocks that originated in the United States is a flexible planning technique aimed at creating a sense of place in centers of towns, villages and neighborhoods. Block standards define four typical suburban blocks: residential, civic, commercial and mixed-use, with specific land uses and densities prescribed for each type (Calthorpe & Fulton, 2001). Mixed- use blocks incorporate most of the significant retail in each center along with housing and office; commercial blocks primarily accommodate office and other employment uses but also allow some ground-floor retail; residential blocks focus primarily on a range of residential densities but also allow some other uses; while civic blocks provide for parks, public uses, and civic institutions (Calthorpe & Fulton, 2001). The share of block types varies to best suite different types of centers. The goal is to enable flexibility in design, while providing a mixed-use and walkable center.

CENTERS OF SUBURBAN SETTLEMENTS IN THE SUBURBAN BELT OF THE CITY OF NIŠ

Former Rural Settlements

In socialist times, the analyzed settlements of Brzi Brod, Donja Vrežina, Gornji Komren and Donji Komren had a formed village center, and the population densities of these villages were sufficient to enable the existence of settlement centers. In the post-socialist period, the settlements experienced considerable urban growth, while their centers lost a part of their uses and acquired some new uses in time (Table 1). It is evident that the cultural function is disappearing from the center – Cultural centers lost their use and alternative amenities were not formed.

The settlement of Donja Vrežina fused with continuous urban fabric in the 1970s and the urban boundary is now difficult to establish. This border is more administrative than it is spatial-morphological. Due to direct contact with the city and good traffic connections, Donja Vrežina experienced considerable urban growth, which unfolded with relatively high residential densities for suburbia. Brzi Brod also leans directly on Niš urban fabric, but it is separated from urban residential zone by a large complex of Water Supply Company. A favorable development circumstance is that this settlement adheres to the dominant traffic route on the east-west development axis, providing quick and easy access to the city. Even though Brzi Brod has not demonstrated that explosive population increase as Donja Vrežina, it had a steady and positive demographic growth. In both of these former villages the number of inhabitants corresponds to the traditional neighborhood unit, and medium population densities are also evident. Due to high demographic growth, both settlements feature emergence of new central functions throughout their territory (business, administration, services, catering, hotels), even though to a small extent. These uses are market driven and spontaneously generated, and are mostly located along major traffic routes, regardless of settlements' traditional centers.

Table 1. Functional structure and demographic growth of suburban settlements in the suburban belt of the City of Niš

Settlement	Central uses	Census 2011	Demographic growth (1991-2011)	Population density (2011)
Brzi Brod	1,2,3,5,6,7,9,10,12,13	4555	27%	61 inh./ha
Donja Vrežina	1,2,3,4,5,7,8,9,10,11,13,14	6629	149%	85 inh./ha
Gornji Komren	1,2,3,5,14	910	27%	29 inh./ha
Donji Komren	1,2,9,14	1831	~32%*	33 inh./ha
Vinik	4,10	~2000*	~100%*	13 inh./ha
Podvinik	/			

Uses: 1. Bus stop; 2. Store; 3. Inn; 4. Small industrial plants, warehouse; 5. Cultural center (not operational); 6. Post office; 7. Outpatient clinic; 8. Pharmacy; 9. Local community office; 10. Craft workshop, services; 11. Club; 12. Police station; 13. Business, administration; 14. Park
* Estimated value – due to the change of municipal cadaster borders between two censuses (1991-2011), the accurate number of inhabitants is not available.

The settlement Donji Komren merged with the city in the 1960s. The case of the settlements of Donji Komren and Gornji Komren is particularly interesting for this research, since other types of development occurred in the area around them. Namely, several settlements of different origin and status in this area are merged nowadays (Gornji Komren, Donji Komren, Branko Bjegović, Vinik and Podvinik), both mutually and with urban area of Niš, creating a heterogeneous territorial patchwork. It is therefore quite difficult to establish where each settlement begins and where it ends. Urban boundary to the City of Niš is also of administrative character, since it cuts through the illicitly built settlement Branko Bjegović. This settlement is thus partly located in the urban area and partly outside of urban boundary.

Even though they feature high values of demographic growth, Donji Komren and Gornji Komren have the number of inhabitants at the level of a primary residential unit, and lower population density. The causes for this are: peripheral position of the zone in respect to the network of city centers, very humble structure of existing settlement centers and explosive development of the neighboring settlements, in which urban housing is defined (Kamenica 2 zone – Branko Bjegović, Vinik and Podvinik). The center of Donji Komren lost some commercial amenities (retail), while the center of Gornji Komren lost its cultural use. In both settlements, the remains of former village centers are visible, but the diversity and concentration of central uses are very low. Some non-residential uses are present in Donji Komren, such as spontaneous retail and services to a small extent, but outside of the traditional center. A couple of new hotels were also built along the urban border because of easy access from the highway. On the other hand, Gornji Komren has experienced some minor population loss between the last two censuses 2002-2011 and has therefore even fewer central uses. Since this settlement is farthest from the city, it is probably one the reasons why it is less preferred than other settlements in this patchwork entity.

The functional structure of all of these former village settlements is somewhat changed in comparison to the rural one, but insufficiently for the formation of quality centers. The main reason is that the accelerated development of motorization made it possible to easily meet urban needs in the city. Even though these settlements in the post-socialist period experienced considerable demographic growth and obtained structural characteristics of urban settlements, their centers did not undergo necessary spatial-functional transformation. In terms of urban functions, accessibility, attractiveness and centrality they are much more modest than the centers in urban areas. The visual identity of these centers is characterized by a poor offer of services and low diversity of uses in the centers themselves.

Former Weekend Zone and Sprawl Settlements

Post-socialist sprawl within the administrative territory of the City of Niš is very difficult to delineate from former villages and weekend-residence zones, because these areas merged in time (Dinić, Đurić, & Mitković, 2014). At the foot of Vinik Hill, the first spontaneous construction of weekend cottages began in the eighth decade of the 20th century (Simonović, 1995), forming the original Vinik settlement. New construction of individual housing with low densities after the 1990s, associated with the period of transition, occurred mostly in the area between: (1) the city, (2) settlement Branko Bjegović with medium residential densities and good concentration of central uses, and (3) village settlements Gornji Komren, Donji Komren and Kamenica. The entire area Kamenica 2 belongs to the Cadastral Municipality of Kamenica village (MP of Niš, 2011). In the 1990s, the transformation of former weekend cottages into residential homes began at the south hillside (Vinik settlement), along with the construction of new individual houses on agricultural land at the west hillside (Podvinik settlement). Since the development of these settlements unfolds in the area between the established urban boundary of Niš and former villages, and since the urban territory has a lower value of demographic growth than the suburban territory, it can be stated that the development in this area has the characteristics of post-socialist urban sprawl.

Kamenica 2 zone is characterized by low densities and random growth. Most of the housing in both settlements, Vinik and Podvinik, was constructed without building permits, so there is no adequate infrastructure in the area. The area of Kamenica 2 that involves Vinik and Podvnik represents an entirely residential zone (prevalently individual housing on large plots), without any other uses, not even basic stores for daily supply. Therefore no settlement centers were created when settling was initiated. Construction of commercial services has only just begun along the peripheral traffic routes (Table 1). Denizens of both settlements partially satisfy their basic urban needs in the neighboring settlement Branko Bjegović (elementary school and kindergarten, health care station, daily supply). Pedestrian access to central uses and facilities is quite limited for the residents of Podvinik, and particularly of Vinik. All other urban needs are satisfied in the city, and the denizens are automobile-dependent. De-concentration of commercial functions in this area has not occurred yet, because the construction of new business and retail facilities took place at those peripheral locations in the city where the necessary infrastructure already existed.

Planning Context

The issue of planning and designing suburban centers in Niš was often resolved in a laissez-faire approach to urban development. While the suburban settlements were rural, their centers were not covered by planning regulations. As they started to gain urban character during 1970s, some former villages singled out as desirable residential environments due to favorable natural conditions, affordable prices, good public transport connections or proximity to the city. This development pattern did not imply hierarchically organized, properly structured and well-designed centers. Along with the subsequent transformation of weekend residence zones and the emergence of sprawl, these were the crucial incentives for planning institutions to get involved. MP of Niš 1995-2010 started the annexation of surrounding rural settlements to the city, including Brzi Brod, Donja Vrežina, Donji Komren and Gornji Komren. By establishing the limits of building area, development was defined and unplanned city expansion along the farmland was prevented. However, in the years to follow, post-socialist reality and the weakening of planning control mechanisms regarding urban growth policies resulted in unplanned developments and relocation of previously planned centers. Comparing the two MPs of Niš from 1995 and 2011, it can be observed that the planned location of local centers is altered in the cases of Donja Vrežina and Donji Komen, the planned location and scope of centers of Brzi Brod and Gornji Komren are slightly modified, while the sites of planned centers of Vinik and Podvinik remain unchanged.

In line with the standing planning regulations for the City of Niš, which were designed according to growth scenario, peripheral suburban settlements should absorb the planned demographic growth. Therefore, the existing settlement centers are improved, or new ones are formed (Figure 1), along with the increase of diversity and capacity of central services. In the settlement Brzi Brod a new settlement center is formed on undeveloped land south of the existing center (PGR of the Municipality of Medijana, 2012), thus shifting the central area closer to the major traffic route. In Donja Vrežina it is planned to considerably increase the scope of the existing village center at the expense of the adjacent undeveloped land (PGR of the Municipality of Pantelej–II phase, 2015). This development also concentrates along an important traffic artery. The existing center of the settlement in Donji Komren is extinguished due to explosive new developments in the vicinity. A new common suburban center Donji-Gornji Komren for the area is formed, which will include the existing center of Gornji Komren and a part of the undeveloped land (PGR of the Municipality of Crveni Krst–II phase, 2013).

The development of Kamenica 2 zone has been planned in order to control sprawl. Planning documents for this area (MP of Niš, 2011) envision that density will increase (housing of moderate and medium densities). It is also planned that public services (school, kindergarten) and new local centers in Vinik settlement and the part of Podvinik settlement will be organized on undeveloped land (Figure 1). These settlement centers are planned in proximity of the highway and should have good accessibility, but their realization has not yet begun.

HOW TO DESIGN THE CENTERS OF SUBURBAN SETTLEMENTS: SOLUTIONS AND RECOMMENDATIONS

Suburban settlements with low density single family housing, which rely on the city in all aspects, need to evolve into more sustainable forms of suburban life, with higher densities and a low-rank community center which serves the local population. The proposed development model of suburban settlements is actually a return to the traditional concept of a compact, pedestrian-oriented neighborhood with higher densities and a defined center, which accommodates services important for the social life of the community. Such examples of development are similar to urban villages, which are already well-known in global planning practice (Poundbury, England; Reston, Virginia, USA; Addison, Texas, USA; etc.). These contemporary development trends are also proposed for the settlements in the suburban belt of the City of Niš.

Suburban centers in this study, regarding their position and the size of gravitating area, should be modeled as the centers of a neighborhood unit. At this moment, the proposed model is applicable only to those former villages that are located within the suburban belt of the City of Niš, as well as to weekend zones and sprawling areas, and not to those remote satellites which still have rural character. When it comes to the center of a former village, in order to transform a rural settlement into an urban one, it is necessary to physically and functionally improve its traditional rural center, or to develop a brand new settlement center, following the mixed-use model designs. Implementation of mixed-use development (Schwanke, 2003) is also proposed for the creation of new centers in sprawling areas and in areas of former weekend residence.

The suburban settlement is defined as an urban extension of the city with clearly defined urban-rural limits. Therefore the suburban center needs to be integrated into the network of city centers, which is a feasible scenario in the case of Niš suburban belt. Suburban centers, according to their significance and position within a defined hierarchy of centers, should not have uses/services which could endanger the functioning of higher-rank urban centers, and they should gravitate directly to secondary urban centers. Since there is no interim level in this relation, or a district center (also local in character), the available uses/services in a suburban center may have slightly higher diversity compared to the center of urban neighborhood unit. Apart from the main settlement center, smaller mixed-use centers of primary residential units should also be created. These local centers serve approximately 1.000 inhabitants and enable satisfying daily needs.

A very important planning issue in development of suburban settlements concerns transportation options. Individual traffic between suburban settlements and the city should be reduced to a minimum. Due to its compact urban structure, the City of Niš has managed to retain the efficient traffic network system on the entire city territory, including the suburban belt. However, public transport to former rural satellites in the suburban belt is of low quality. It is also an important fact that a majority of native residents cannot afford a car, thus making them dependent on public transportation. Long time spent in commuting significantly reduces their quality of life. A regular bus/light railway line must be therefore provided for connection to the city in order to reduce the usage of individual vehicles and limit sprawl developments. A sustainable public transportation line is directly dependent upon higher population density values, but it is also closely related to the concentration of business and commercial activities. Both of these features also implicate the need to create a settlement center.

For the design of the suburban neighborhood center this research proposes to implement the "Urban matrix for mixed-use centers" as a flexible framework that defines the center in spatial and functional terms (Table 2). The matrix is designed to best suit the traditional way of using space in former rural settlements and the needs of new suburban residents in other post-socialist development patterns. In such a matrix, the basic spatial unit is smaller in size and can be conceived either as: (1) *urban block* (defined by roadways) or as (2) *urban assembly* (partly defined by roadways/cul-de-sacs and partly by pedestrian communications). By fragmenting central urban matrix into smaller units, the dimensions of the built structure are reduced and spatial flows are improved. Implementation of smaller volumes of buildings also creates a space of anthropometric scale that favors pedestrian traffic. In the immediate center, the urban blocks/assemblies should be easily walkable spatial units and should therefore have a perimeter of no more than 600m. The blocks in the outer-peripheral central zone can be slightly larger in size.

The blocks/assemblies of the defined urban matrix for the suburban center of the post-socialist city can be divided into four basic types:

1. **Mixed-use blocks/assemblies** contain the most important commercial services (retail, small shops), along with public services (public spaces and facilities), business (offices) and housing;
2. **Business blocks/assemblies** primarily contain bulky business premises, and only to some extent commerce, with potential big-box commercial structures and production (small plants);
3. **Public use blocks/assemblies** include educational institutions (schools, kindergartens), social institutions (accommodation of old and uncared for persons), health institutions (clinical complexes), open recreational areas and sport complexes;
4. **Residential blocks/assemblies** also permit certain commercial uses in the ground floor level of structures.

The immediate center contains mixed-use blocks and residential blocks with commercial uses, while the peripheral area of the center (transition zone between the center and housing) includes public use blocks, business blocks and residential blocks with services. According to their rank in the city centers hierarchy, the suburban settlement center should have the greatest share of residential blocks but also a significant share of mixed-use blocks. Each block type is assigned certain dimensions, and accompanying standards are defined: maximum blocks size, number of floors, required number of parking spaces, range of building coverage ratio and floor area ratio values, and minimum percentage of the block perimeter in which the buildings must be continuously lined on the alignment (continuous façade front). For residential blocks containing certain services, minimum population density is defined.

Table 2. Urban matrix for mixed-use centers in suburbia

Standardized elements			Share (% of center)	Max size of urban block (ha)	Building coverage ratio	Floor area ratio	Min street façade front (%)	Building height (no. of floors)	Popula-tion density (den/ ha)
IMMEDIATE CENTER		Mixed-use blocks/ assemblies	20-50	1	0,5-0,8	1,5-3,2	70	3-5	/
			Permitted uses[1]: small shop[2]/green market; services; restaurant, café; library, book store, gallery; local administrative office, post office; clubs, associations; outpatient clinic, pharmacy; business (small shops, bank branch office); open public areas (small park/ square, sports amenities); housing.						
	CENTER PERIPHERY	Residential blocks/ assemblies	30-60	1	0,5-0,7	1,5-2,8	60	3-4	120-300
			Permitted uses: owned or rented apartments[3]; individual houses on a small lot, semi-detached houses, terraced houses; commercial uses in the ground level.						
		Business blocks/ assemblies	max 10	1,2	0,4-0,7	/	60	1-2	/
			Permitted uses: business (large formats); commerce (big-box); production (small plants).						
		Public use blocks/ assemblies	5-10	1,2	/	/	/	/	/
			Permitted uses: park, sports-recreation functions, school, kindergarten.						
Parking			1 parking place per 1 apartment/70 m2, for mixed-use blocks it is advisable to construct an underground garage.						

[1] max 20% of business within the structure and min 30% of other non-residential uses
[2] shops for everyday and occasional need, max surface area of individual large commercial formats 1.000m2
[3] low rise high density housing

From the aspect of physical shaping, the main settlement center should be the nexus of urban structure with the highest concentration of services and highest intensity of development. It is equally important to clearly identify the settlement center both in the urban matrix and the cityscape of the settlement. This is possible through the implementation of adequate urban landmarks, such as recognizable buildings, height references, traditional motifs – plazas, squares (Llewelyn-Davies, Baxter, & associates, 2009). It is desirable to conceive the main settlement center around an open public space with the passive recreation use (Gupta, 2008), as it represents the focal point of the design. Meanwhile, the centers of primary residential units groups can be linear and positioned along frequent streets. A square with recognizable urban landmark in the main settlement center should be a hub for important public services (administration, post office), which is an association to the traditional way of using space. The square should be large enough to provide enough space for gathering and traditional manifestations, possibly a market, which gives small settlements a specific charm. Good visual connections should be provided towards the urban landmark in the square.

As mixed-use designs require higher densities, buildings should be designed Gnd. level+3 and Gnd. level+4 floors high in the center, facing open public areas. In this way, sufficient level of urbanity is achieved and mixed-use development is supported by the shape of physical structure. Higher floors create the necessary scenery for public areas, and also enable mixing uses along the vertical – commercial uses in the ground floor level and possibly on the first floor, with residential units on the upper floors. For this reason, the minimum number of floors is defined as Gnd. level+2. Mixing of uses should also be performed horizontally in the suburban center.

Housing is designed in the center of the suburban settlement in order to provide 24-hour vitality. Housing in the vicinity of services, public amenities and jobs contributes to healthy environment and reduces transportation needs. Since the neighborhood center is the lowest rank in the network of city centers, the share of housing can be high, up to 60%. *Business use* in the functional structure of the center provides jobs for local population and reduces pressure on the city. Business use must not be dominant within the mixed-use concept, so as not to endanger the business focus of the city and create featureless business environment in suburbia. The share of business in a suburban center should not exceed 10%, so that the remaining 30% would be available of more "vital" activities, such as retail, services and catering. Across the territory of a suburban settlement, *commercial use* should be positioned only in the established centers in order to create nexuses of activity. It is particularly important to control the size and spatial distribution of retail shops and services, in order to preserve the uses in existing urban centers and to maintain the established hierarchical structure of city centers. Otherwise, commercial facilities in suburbia may in time cause devastation of certain city centers. The surface area of individual large commercial formats should be limited to 1.000 m2 (West Dorset District Council, 2006) in the main settlement center and they should not be planned at other locations in a suburban settlement. For all the planned facilities having surface area larger than recommended, an impact study on the existing urban centers must be made (West Dorset District Council, 2006). The more significant services/amenities, such as commercial facility of a large format, shops for occasional supply or important cultural-entertainment amenities (theaters, cinemas) should first be considered in the city centers, and only afterwards in suburban ones. These facilities need to be properly integrated into the context of a suburban center and must be accessible by various modes of transportation. Large complexes having *educational u*se are not suitable for mixed-use designs, so they need to be planned outside the immediate center of the suburban settlement. These facilities, as well as kindergarten complexes, can partly appear as the center of a primary residential unit, with the accompanying uses such as retail, catering and services. *Hotel*, as a use which is preferred in mixed-use designs in highly urbanized centers, should not be planned for in the structure of uses of a suburban center. These settlements mostly do not have the potential for a hotel as one of three main uses of a mixed-used design, except in rural settlements having tradition in tourism. Suburban settlements are less attractive to tourists, so a road-side motel intended for cheaper and shorter stay is a more appropriate tourist facility.

The criterion of pedestrian accessibility is of crucial importance for suburban centers. A suburban settlement center, which is based on the neighborhood unit model, allows for the center to be accessible to all the people living in the settlement, in line with the service capacity, number of inhabitants (around 5.000) and service radius (max 600-800 m). Everyday walks to the central locations are an association to the traditional way of using space and an opportunity to establish social contact. Pedestrian and bicycle traffic in the settlement should be encouraged. Individual vehicle traffic as a dominant transportation mode has a very negative effect within the suburban settlement, since it often results in large distances and pedestrian unfriendly environments. The main road should have a tangential position in relation to the settlement central zone, so that center wouldn't be burdened with heavy traffic. Central streets should have a narrow profile in order to calm traffic. The width of the sidewalks should be large enough to accommodate certain ground floor uses if necessary (i.e. restaurants and cafes).

An often impediment to implementing walkable medium-density, mixed-use centers in suburbia is the lack of critical mass to support both: (1) central uses that are needed for the existence of the settlement center and (2) multi-storey building types for multi-family housing that are framing the central area. Foreign practice shows that Greenfield developments have difficulties in creating a critical market for retail, town homes and apartments, and often end up with a void in which central uses (retail, jobs) and multifamily housing wait for market demand to catch up (Calthorpe & Fulton, 2001). In the case of Niš, this "wicked problem" is present in a similar way - without an attractive and vital central place, variety of transportation options and job opportunities people are not often motivated to live in suburbia, and without the denizens there is no need for a local center.

Former rural settlements analyzed in this research make a good potential baseline for the proposed development matrix. This is due to valuable resources such as significant population value, affordable land prices and existing central places, but also due to some intangible values like inherited pedestrian-oriented mobility, traditional way of using space along with specific customs and events, and friendly neighboring social relations. On the other hand, existing compact rural structure which entails prevalently single family plots makes it difficult to implement multi-family housing in the traditional center. However, this issue is inextricably bound to other factors contributing to mixed-use development - proximity to job centers and services, as well as existing *genius loci* (quality ambient, urban landmarks, architectural heritage, traditional trades and activities, good town planning layout etc.). In a mixed-use neighborhood with a representative center where denizens are employed, market demand for higher-density housing will definitely exist, as well as a high concentration of central uses. In such circumstances the market acts as initiator of urban densification, so conversions of individual plots into multi-family housing are facilitated. It is therefore expected that development of former village centers would in time achieve the standards that are established in the urban matrix.

When it comes to implementing urban matrix for mixed-use centers in former weekend zones and sprawling areas, different issues are present. Given the low intensity of development, planning and creating for a center is easy, and so is the application of urban matrix. The crucial issue here is providing for a critical mass of users for the settlement center, which is a concern dependent upon many things besides the proposed matrix. In Niš, as in many post-socialist cities with the tradition in compact development, a lot of urban growth occurs "inwards", resulting in overcrowded urban areas. This is also partly due to the fact that suburban living requires higher incomes to achieve the standards of the quality of life, which not many people in Niš can afford yet. However, since suburban way of life is becoming popular, planning and development strategies should re-direct some of the growth towards suburbia, in order to improve living conditions in overburdened city areas. In such circumstances, retail and services would quickly follow and the implementation of a center based on this matrix would improve the settlement and the quality of life of suburban denizens.

Only by acting together, individual success factors in a mixed-use project achieve the overall synergy, which results in vital suburban places. It is therefore concluded that all issues of mixed-use development in a suburban center must be addressed simultaneously. This does not mean that the entire central project needs to be developed at the same time. In complex projects of regeneration or new development of city centers, phasing can be a great tool to construct just the right amount of central fabric that the market can absorb. This of course implies a pre-developed coherent vision of the central entity, which is realized in phases. The use of the discussed elements of urban design that support the proposed matrix (public open space, urban landmarks, traffic calming) is feasible in the analyzed suburban centers in Niš.

Urban matrix for mixed-use centers in suburbia is flexible enough to allow the creation of central places in both built environments and undeveloped land, by adapting the share of various types of blocks or assemblies to a particular site. It also leaves enough room within the central fabric to adapt to current market demands. The proposed matrix for suburban centers would be a useful instrument in the process of planning the network of city centers, by helping planners to envision the center in suburbia (in terms of size, structure, uses and flows) and thus decide upon the best possible site given the existing context. This urban matrix should also help in overcoming some disproportions between MP visions for suburban centers and actual urban development patterns.

FUTURE RESEARCH DIRECTIONS

A precondition for the implementation of mixed-use development in suburban centers in Serbia is to make a classification of "other" settlements into suburban settlements and those that do not have urban character. Creating a new statistical typology of settlements is not easy, since it should take into account numerous factors, such as the number of inhabitants, physical, functional, socio-economic and design characteristics of the settlements. However, of a new typology of settlements is necessary because the dichotomous classification that is in use nowadays does not correspond well to the actual state and is not best suited to the planning process. In the case of Niš, a variety of settlements that exists beyond the urban boundary confirms the need for a new classification. Given the fact that the settlements in the suburban belt of Niš have already gained urban character, they require special planning treatment and a different terminology.

Urban planning practice in Niš could explore the idea of the proposed urban matrix for mixed-use centers in suburbia as a pilot project in those suburban settlements with highest value of urban growth, such as the ones analyzed in this study. The results would provide a better insight into future patterns of shaping centers in suburbia, and would be useful to other post-socialist cities in the region as well.

For the implementation of the proposed urban matrix in suburban centers in Serbia it is also necessary to improve the standing legislation. Current planning practice divides the central zone into specific subareas at the level of the Plan of General Regulation, defining strictly the allowed and prohibited uses. This does not allow for a detailed overview of the center area and limits the possibilities of a vertical mixture of uses. Therefore, it is proposed to use the planning document of a lower level for the realization of the formulated guidelines, such as the Plan of Detailed Regulation. This Plan should be mandatory for centers of suburban settlements. The Plan of Detailed Regulation would include the proposed urban matrix as a tool for designing suburban centers, thus promoting mixed-use development in the entire central zone.

CONCLUSION

In Serbia, the planned mixed-use centers of suburban settlements are a relatively new development format, which will certainly gain importance in the future. A very intensive development period in the suburban belts of Serbian cities in the last two decades has created conditions for spatial-functional transformation of suburban settlement centers, which was illustrated in the case of Niš. For this reason, the standing planning documents provide for remodeling of suburban settlement centers through reconstruction of the existing central locations or development of new centers.

The framework established in this study identifies and shapes spatial-functional units in a suburban center, with the overall aim to understand and guide suburban development in the post-socialist period and help foster urban-rural synergy. Implementation of the proposed urban matrix for mixed-use centers in suburbia is relatively simple in realization on undeveloped land. When it comes to reconstruction of the already developed suburban fabric, there is a tendency that future development of reconstructed centers will approach these established standards. Remodeling of traditional rural centers is recommended as a rule of thumb, if they exist and if they possess a recognizable character. In those settlements which do not have that potential, it is proposed to construct a new settlement center which will create the much needed *genius loci* and form the focus of social life. Implementation of the urban matrix for mixed-use centers is possible in both cases, regarding the low intensity of development of former rural settlements and weekend-residence zones, as well as urban sprawl. Honoring of the specifics of genesis and character of suburban settlements is particularly important, in order to avoid the excessive development and "over-urbanized" centers.

Post-socialist cities such as Niš are experiencing urban densification and overcrowding of city area, while former rural settlements and weekend-residence zones make a significant share of today's suburbia. Limiting urban sprawl at the urban-rural fringe is also urgent issue in order to preserve rural landscapes. Since it is expected that these suburban settlements should absorb the anticipated demographic growth, the future development should be directed towards them for the rational usage of resources, especially of infrastructure. For these reasons, urban design of suburban centers is of great importance, and the proposed matrix makes a useful tool in planning and design. Such urban matrix could be adapted to the specific context of other suburban settlements on the regional level, particularly in post-socialist cities that are experiencing similar development patterns. The vital centers of suburban settlements, which satisfy everyday needs for retail, services, public amenities, recreation and establishment of the social contact, reduce the pressure on the main city center, reduce commuting costs and represent the only form of sustainable development of both suburban areas and city as a whole.

REFERENCES

Calthorpe, P., & Fulton, W. B. (2001). *The Regional City: Planning for the End of Sprawl*. Washington, DC: Island Press.

Couch, C., Petschel-Held, G., & Leontidou, L. (2007). *Urban Sprawl in Europe: Landscape, Land-Use Change and Policy*. Oxford, UK: Wiley-Blackwell. doi:10.1002/9780470692066

Dinić, M., Đurić, J., & Mitković, P. (2014). Rural settlements in suburban landscape of the post-socialist city – Niš case study. In *Proceedings of Landscape Transformations - an International interdisciplinary student conference* (pp. 136-143). Prague: Czech Technical University.

Dinić, M., & Mitković, P. (2016). Suburban design: From "bedroom communities" to sustainable neighborhoods. *Geodetski Vestnik*, *60*(1), 98–113.

Dinić-Branković, M., Bogdanović-Protić, I., Đekić, J., & Mitković, P. (2016). Post-socialist suburbanization and sprawl development patterns – Niš case study. *Facta universitatis, series Architecture and civil engineering, 14*(3), 355-366.

Gupta, P. K. (2008). *Creating Great Town Centers and Urban Villages*. Washington, DC: Urban Land Institute.

Hirt, S., & Stanilov, K. (2007). The perils of post-socialist transformation: Residential development in Sofia. In K. Stanilov (Ed.), *The Post-Socialist City: Urban Form and Space Transformations in Central and Eastern Europe after Socialism* (pp. 215–244). Dordrecht: Springer. doi:10.1007/978-1-4020-6053-3_11

Lazauskaite, D., Šarkiene, E., & Skripkiene, R. (2015). The assessment of development scenarios for suburban metropolitan districts under the multiple criteria methods. *Procedia Engineering*, *122*, 19–28. doi:10.1016/j.proeng.2015.10.003

Llewelyn-Davies, B., & ... (2009). *Priručnik za urbani dizajn*. Beograd: Prograf i Orion Art.

Paul, V., & Tonts, M. (2005). Containing Urban Sprawl: Trends in Land Use and Spatial Planning in the Metropolitan Region of Barcelona. *Journal of Environmental Planning and Management*, *48*(1), 7–35. doi:10.1080/0964056042000308139

Schmitz, A. (2003). *The New Shape of Suburbia: Trends in Residential Development*. Washington, DC: ULI-Urban Land Institute.

Schwanke, D. (2003). *Mixed-Use Development Handbook*. Washington, DC: ULI-Urban Land Institute.

Serebrennikov, E. N. (2018). Expansion features of Irkutsk suburban areas' development. *IOP Conference Series: Earth and Environmental Science. Sci.,* 190. 10.1088/1755-1315/190/1/012063

Simonović, D. (1995). *Enciklopedija Niša – priroda, prostor, stanovništvo*. Niš: Gradina.

Stanilov, K. (2007a). Taking stock of post-socialist urban development: A recapitulation. In K. Stanilov (Ed.), *The Post-Socialist City: Urban Form and Space Transformations in Central and Eastern Europe after Socialism* (pp. 3–17). Dordrecht: Springer. doi:10.1007/978-1-4020-6053-3_1

Stanilov, K. (2007b). Housing trends in Central and Eastern European cities during and after the period of transition. In K. Stanilov (Ed.), *The Post-Socialist City: Urban Form and Space Transformations in Central and Eastern Europe after Socialism* (pp. 173–190). Dordrecht: Springer. doi:10.1007/978-1-4020-6053-3_9

Stanilov, K., & Sykora, L. (2014). Postsocialist Suburbanization Patterns and Dynamics: A Comparative Perspective. In K. Stanilov & L. Sykora (Eds.), *Confronting Suburbanization: Urban Decentralization in Postsocialist Central and Eastern Europe* (pp. 256–295). Chichester, UK: Wiley-Blackwell. doi:10.1002/9781118295861.ch9

Statistical Office of the Republic of Serbia. (2010). *Administrative territorial division and NSTJ levels 1, 2, 3*. Retrieved August 19, 2019, from http://www.stat.gov.rs/sr-Latn/oblasti/registar-prostornih-jedinica-i-gis/administrativno-teritorijalna-podela-i-nstj-nivoi-1-2-3

Statistical Office of the Republic of Serbia. (2011). *Systematic list of settlements in the Republic of Serbia*. Retrieved August 19, 2019, from http://publikacije.stat.gov.rs/G2011/pdf/G201121009.pdf

Sykora, L., & Ourednicek, M. (2007). Sprawling post-communist metropolis: Commercial and residential suburbanisation in Prague and Brno, the Czech Republic. In M. Dijst, E. Razin, & C. Vazquez (Eds.), *Employment Deconcentration in European Metropolitan Areas: Market Forces versus Planning Regulations* (pp. 209–234). Dordrecht: Springer. doi:10.1007/978-1-4020-5762-5_8

Sykora, L., & Stanilov, K. (2014). The Challenge of Postsocialist Suburbanization. In K. Stanilov & L. Sykora (Eds.), *Confronting Suburbanization: Urban Decentralization in Postsocialist Central and Eastern Europe* (pp. 1–32). Chichester, UK: Wiley-Blackwell.

Tammaru, T., Leetmaa, K., Silm, S., & Ahas, R. (2009). Temporal and Spatial Dynamics of the New Residential Areas around Tallinn. *European Planning Studies*, *17*(3), 423–439. doi:10.1080/09654310802618077

Tsenkova, S. (2007). *Urban futures: Strategic planning in post-socialist Europe. In The Post-Socialist City: Urban Form and Space Transformations in Central and Eastern Europe after Socialism* (pp. 447–470). Dordrecht: Springer. doi:10.1007/978-1-4020-6053-3_22

Urban Planning Institute of Niš. (1995). Master plan of Niš 1995-2010. *Official Gazette of the City of Niš No. 13/95.*

Urban Planning Institute of Niš. (2011). Master plan of Niš 2010-2025. *Official Gazette of the City of Niš No. 43/11.*

Urban Planning Institute of Niš. (2012). Plan of General Regulation of the Municipality of Medijana. *Official Gazette of the City of Niš No. 72/12.*

Urban Planning Institute of Niš. (2013). Plan of General Regulation of the Municipality of Crveni Krst–II phase. *Official Gazette of the City of Niš No. 96/13.*

Urban Planning Institute of Niš. (2015). Plan of General Regulation of the Municipality of Pantelej–II phase. *Official Gazette of the City of Niš No. 44/15.*

Vujović, S., & Petrović, M. (2007). *Belgrade's post-socialist urban evolution: Reflections by the actors in the development process. In The Post-Socialist City: Urban Form and Space Transformations in Central and Eastern Europe after Socialism* (pp. 361–382). Dordrecht: Springer. doi:10.1007/978-1-4020-6053-3_18

West Dorset District Council. (2006). *Poundbury Development Brief.* Retrieved December 02, 2013, from http://www.dorsetforyou.com/media.jsp?mediaid=116539&filetype=pdf

KEY TERMS AND DEFINITIONS

Genius Loci: Specific character or atmosphere of a place.

Mixed-Use Development: Planned development that is characterized by physical and functional integration of complementary uses, and which favors pedestrian traffic.

Other Settlements: Settlements that do not have urban character, usually of rural origin. Specific term defined by Serbian legislation.

Polycentric Spatial Structure: A network of hierarchically structured centers within the administrative territory of the city, which implies de-concentration of urban functions from the main city center to secondary centers and local centers.

Post-Socialist: A development period of former socialist cities after the fall of communism in 1989, which is characterized by transition to market economies and considerable transformation of urban structure.

Sprawl: Development on Greenfield land outside of city bounds that far exceeds the needs of demographic growth. It is usually characterized by low-density, mono-functionality, automobile dependence and lack of organized centers.

Suburban Belt: Area adjacent to urban border of the post-socialist city, consisting of: former villages which have undergone the transformation from rural into suburban, weekend-residence zones and new suburban developments.

Weekend-Residence Zones: Weekend cottages in the vicinity of the city, intended for part-time stay and leisure.

ENDNOTE

1 The reality is that the structure of settlements in the administrative area of the City of Niš is far more complex than this relatively simplified division into "urban" and "other". There are significant differences between various "other settlements" in the MP area in terms of demographic and economic characteristics, accessibility and quality of services. Even though it is not defined by legislation, the term "suburban settlement" appears in planning documents (former MP of Niš 1995-2010 and the current MP of Niš 2010-2025).

Compilation of References

Abadžić, N. (2001). *Doba ekologije*. Sarajevo, Bosnia and Herzegovina: Vlastita naklada.

Adams, W. M. (1996). Future nature. Vision of conservation. London: Earthscan publications.

Agency for Identification Documents, Registers and Data Exchange of Bosnia and Herzegovina. (2017). *Data on the canceled nationalities from B&H due to migration abroad*. Banja Luka: Author.

Agency for Statistics of Bosnia and Herzegovina. (2015). *Energy statistic*. Retrieved from http://www.bhas.ba/saopstenja/2015/APED_2015_001_01_BA.pdf

Alampay, E., Heeks, R., & Soliva, P. (2003). *Bridging the Information Divide: A Philippine Guidebook on ICTs. The National College of Public Administration and Governance, University of the Philippines and the Institute of Development Policy and Management*. University of Manchester.

Ali, M. M., & Al-Kodmany, K. (2012). Tall Buildings and Urban Habitat of the 21st Century: A Global Perspective. *Buildings, 2*, 384-423. Retrieved February 16, 2019, from https://www.researchgate.net/publication/307762382_Tall_Buildings_and_Urban_Habitat_of_the_21st_Century_A_Global_Perspective

Allem, J. P., & Ferrara, E. (2016). The importance of debiasing social media data to better understand e-cigarette-related attitudes and behaviors. *Journal of Medical Internet Research, 18*(8), e219. doi:10.2196/jmir.6185 PMID:27507563

Allen, P. M. (2012). The Visible Expression of Co-evolving Complexity. In Complexity Theories of Cities Have Come of Age. Berlin: Springer. doi:10.1007/978-3-642-24544-2_5

Allender, S., Lacey, B., Webster, P., Rayner, M., Deepa, M., Scarborough, P., ... Mohan, V. (2010). Level of urbanization and noncommunicable disease risk factors in Tamil Nadu, India. *Bulletin of the World Health Organization, 88*(4), 297–304. https://www.who.int/bulletin/volumes/88/4/09-065847/en/ RetrievedFebruary162019. doi:10.2471/BLT.09.065847 PMID:20431794

Andeweg, M. T., Brunoro, S., & Verhoef, L. G. W. C. (2007). *COST C16 Improving the Quality of Existing Urban Building Envelopes, state of the art* (Vol. 2). Research in Architectural Engineering Series.

Angelova, I. (2016). *New economies in transition society and their impact on the urban physical environment-Research on the link between small businesses and the urban development* (Unpublished master thesis). Ss. Cyril and Methodius University, Skopje, Republic of North Macedonia.

Antić, M., Šantić, D., Kašanin-Grubin, M., & Malić, A. (2017). Sustainable rural development in Serbia – relationship between population dynamics and environment. *Journal of Environmental Protection and Ecology, 18*(1), 323–331.

Antonić, B. (2015). Место становања у савременој урбаној политици: како искористити европска искуства у Србији [The Place of Housing in Contemporary Urban Policy: How to Use European Experince in Serbia]. In V. Trifunović, D. Minić i Z. Popović (Eds.), *Zbornik radova: I Planski osnovi / II Urbane politike / III Izmene i dopune Zakona o planiranju i izgradnji i Zakona o državnom premeru i katastru* (pp. 125-134). Beograd: Udruženje urbanista Srbije.

Antonić, B. (2016). How to understand the history of housing planning in modern Serbia to achieve new quality in housing? In C. Hein (Ed.), *Proceedings of 17th IPHS Conference* – Volume 02 *The Urban Fabric* (pp. 165-174). Delft, the Netherlands: TU Delft - Faculty of Architecture. DOI: 10.7480/iphs.2016.2.1234

Antonić, B. (2017). The Role of housing Agencies in the Future Development of Serbia. *plaNext – Next Generation Planning, 5*, 30-46. Doi:10.24306/plnxt.2017.05.003

Antonić, B., & Vaništa Lazarević, E. (2019). Decentralised Mass Housing Policy in Socialist Yugoslavia. In M. Benkő & K. Kissfazekas (Eds.), Understanding Post-Socialist European Cities – case studies in urban planning and design (pp 172-187). Budapest: L'Harmattan.

Antonić, B., & Đukić, A. (2018). Phenomenon of the Shrinking Illegal Suburbs in Serbia: Can the Concept of Shrinking Cities be Useful for their Upgrading. *Habitat International, 75*, 161–170. doi:10.1016/j.habitatint.2018.03.010

Aranđelović, M., Videnović, A. (2016). Društveni značaj zadružnih domova u postsocijalističkom kontekstu. *Zbornik Matice srpske za društvene nauke, 4*(159-160), 947-960.

Arnautović-Aksić, D., Burazor, M., Delalić, N., Gajić, D., Gvero, P., Kadrić, D., … Zagora, N. (2016). Tipologija stambenih zgrada Bosne i Hercegovine/Typology of Residential buildings in Bosnia and Herzegovina. Sarajevo, B&H: Faculty of Architecture, University of Sarajevo.

Aronsson, L. (1994). Sustainable tourism systems: The example of sustainable rural tourism in Sweden. *Journal of Sustainable Tourism, 2*(1-2), 77–92. doi:10.1080/09669589409510685

Arrabaça, P. (2003). *Formas Urbanas na Cidade Construída: Agualva-Cacém, 1953-2001*. Universidade de Lisboa.

Asian Development Bank. (n.d.). *Competitive Cities: City Cluster Economic Development* (Brochure). Retrieved February 16, 2019, from https://www.adb.org/sites/default/files/competitive-cities-brochure.pdf

Atkinson, R. (2001). The emerging 'urban agenda' and the European spatial development perspective: Towards an EU urban policy? *European Planning Studies, 9*(3), 385–406. doi:10.1080/09654310120037630

Aurigi, A., & De Cindio, F. (2008). *Augmented urban spaces: articulating the physical and electronic city*. Aldershot, UK: Ashgate.

Bajlon, M. (1975). Stan u Beogradu. *Arhitektura urbanizam, 74-77*, 23-42.

Baker, N., Fanchiotti, A., & Steemers, K. (1993). *Daylighting in Architecture: a European reference book*. Brussels: James and James.

Balaban, O. (2012). Climate Change and Cities: A Review on the Impacts and Policy Responses. *METU Journal of the Faculty of Architecture, 29*(1), 21–44.

Balvanera, P., Caldero-Contreras, R., Castro, A. J., Felipe-Lucia, M. R., Geijzendorffer, I. R., Jacobs, S., ... Gillson, L. (2017). Interconnected place-based social–ecological research can inform global sustainability. *Current Opinion in Environmental Sustainability, 29*, 1–7. doi:10.1016/j.cosust.2017.09.005

Barata Salgueiro, T. (2001). *Lisboa, Periferia e Centralidades*. Oeiras: Celta. Retrieved from http://www.bulhosa.pt/livro/lisboa-periferia-e-centralidades-teresa-barata-salgueiro/%5Cnhttp://www.bulhosa.pt/livro/do-comercio-a-distribuicao-roteiro-de-uma-mudanca-teresa-barata-salgueiro/livro/lisboa-periferia-e-centralidades-teresa-barata-salgueiro/

Barata Salgueiro, T. (1992). *A cidade em Portugal: uma geografia urbana*. Porto: Afrontamento; doi:10.18055/Finis1874

Bardow, A., Nyvad, B., & Nauntofte, B. (2001). Relationships between Medication Intake, Complaints of Dry Mouth, Salivary Flow Rate and Composition, and the Rate of Tooth Demineralization in situ. *Archives of Oral Biology*, *46*(5), 413–423. doi:10.1016/S0003-9969(01)00003-6 PMID:11286806

Barta, G., Beluszky, P., Czirfusz, M., Gyiri, R., & Kukely, G. (2006). *Rehabilitating the brownfield zones of Budapest* (Discussion Paper No. 51, pp. 57-58). Pécs, Hungary: Centre for Regional Studies of the Hungarian Academy of Sciences Press.

Bataković, D. (Ed.). (2005). Histoire du peuple serbe [History of the Serbian People]. Lausanne: L'Age d'Homme.

Batty, M., & Torrens, P. M. (2001). *Modeling Complexity: The Limits to Prediction*. In 12th European Colloquium on Quantitative and Theoretical Geography, St-Valery-en-Caux, France.

Beatley, T. (2000). *Green Urbanism – Learning from European Cities*. Washington, DC: Island Press.

Bedrosian, T. A., & Nelson, R. J. (2013). Influence of the modern light environment on mood. *Molecular Psychiatry*, *18*(7), 751–757. doi:10.1038/mp.2013.70 PMID:23711982

Bell, D., & Leitinger, S. (2016*). Results of the applicability of the technology Mobile Eye Tracking in naturalistic cycling studies*. Bikealyze, Deliverable 2, unpublished, Salzburg Research.

Ben-Nakhi, A. E. (2002). Minimizing thermal bridging through window systems in buildings of hot regions. *Applied Thermal Engineering*, *22*(9), 989–998. doi:10.1016/S1359-4311(01)00121-1

Berson, A., Smith, S., & Thearling, K. (2000). *Building data mining applications for CRM*. New York: McGraw-Hill.

Berto, R. (2014). The Role of Nature in Coping with Psycho-Physiological Stress: A Literature Review on Restorativeness. *Behavioral Science*, *4*(4), 394–409. doi:10.3390/bs4040394 PMID:25431444

Bielefeldt, C., Bak, M., Borkowski, P., Carreno, M., Matthews, B., Stewart, K., ...Schnell, O. (2013). COMPASS: optimised co-modal passenger transport for reducing carbon emissions – Final results and Conclusions. Edinburgh, UK: European Commission.

Bijedić, D. (2012). *Arhitektura, Holizam umjesto optimalizacije, Integralni pristup u arhitektonskom stvaralaštvu*: Sarajevo, Bosnia and Herzegovina: Univerzitet u Sarajevu.

Bikas, D., Tsikaloudaki, K., Kontoleon, K. J., Giarma, C., Tsoka, S., & Tsirigoti, D. (2017). Ventilated Facades: Requirements and Specifications Across Europe. *Procedia Environmental Sciences*, *38*, 148–154. doi:10.1016/j.proenv.2017.03.096

Bjeličić, S. (1962). Stambena zajednica – Osnovna planska urbanistička jedinica grada. *Arhitektura urbanizam, 14*, 43-44.

Blagojević, Lj. (2007). *Нови Београд: оспорени модернизам* [New Belgrade: Contested Modernism]. Belgrade: Zavod za udžbenike.

Bodin, Ö., & Crona, B. (2009). The role of social networks in natural resource governance. *Global Environmental Change*, *19*(3), 366–374. doi:10.1016/j.gloenvcha.2009.05.002

Boeri, S., Lanzani, A., & Marini, E. (1993). *Il territorio che cambia : ambienti, paesaggi e immagini della regione milanese*. Milano: Abitare Segesta. Retrieved from http://www.worldcat.org/title/territorio-che-cambia-ambienti-paesaggi-e-immagini-della-regione-milanese/oclc/717038600&referer=brief_results

Bogdanov, N. (2007). *Mala ruralna domaćinstva u Srbiji i ruralna nepoljoprivredna ekonomija*. United Nation Development Program.

Booher, D. E., & Judith, E. I. (2002). Network power in collaborative planning. *Journal of Planning Education and Research, 21*(3), 221–236. doi:10.1177/0739456X0202100301

Borsay, P., Brown, V. C., & Palairet, M. R. (1993). Pedja J. Marković, Beograd i Evropa 1918–1941 [Belgrade and Europe 1918–1941]. *Urban History, 20*(2), 261–262. doi:10.1017/S0963926800010580

Boutayeb, A., & Boutayeb, S. (2005). The burden of non communicable diseases in developing countries. *International Journal for Equity in Health, 4*(1), 2. doi:10.1186/1475-9276-4-2 PMID:15651987

Bragança, L., Wetzel, C., Buhagiar, V., & Verhoef, L. G. W. (2007). *COST C16 Improving the Quality of Existing Urban Building Envelopes, facades and roofs* (Vol. 5). Research in Architectural Engineering Series.

Brauer, R., & Dymitrow, M. (2014). Quality of life in rural areas: A topic for the Rural Development policy? *Bulletin of Geography. Socio-Economic Series, 25*(25), 25–54. doi:10.2478/bog-2014-0028

BREEAM. (2016). *UK New Construction technical manual*. BRE.

Brenner, N. (2014). *Implosions/Explosions. Towards a study of planetary urbanization* (N. Brenner, Ed.). Berlin: Jovis.

Brès, A. (2015). *Figures Discretes de l'Urbain – À la reencontre des réseaux et des territoires*. Genéve: MètisPresses.

Brezzi, M., Dijkstra, L., & Ruiz, V. (2011). *OECD Extended Regional Typology: The Economic Performance of Remote Rural Regions. OECD Regional Development Working Papers, 2011/06*. Paris, France: OECD Publishing. Retrieved February 24, 2017, from doi:10.1787/5kg6z83tw7f4-en

Briedenhann, J., & Wickens, E. (2004). Tourism routes as a tool for the economic development of rural areas—Vibrant hope or impossible dream? *Tourism Management, 25*(1), 71–79. doi:10.1016/S0261-5177(03)00063-3

Brockington, D., Duffy, R., & Igoe, J. (2012). Nature Unbound: Conservation, Capitalism and the Future of Protected Areas. London: Routledge. doi:10.4324/9781849772075

Broman, M. (2017). Introduction. In K. Gunnartz (Ed.), Drivers and consequences of urban development, IVA project Good cities of the future. Stockholm, Sweden: The Royal Swedish Academy of Engineering Sciences (IVA).

Broto, C., Bulkeley, H., & Bulkeley, V. (2013). A survey of urban climate change experiments in 100 cities. *Global Environmental Change, 23*(1), 92–102. doi:10.1016/j.gloenvcha.2012.07.005 PMID:23805029

Browning, W. D., Ryan, C. O., & Clancy, J. O. (2014). *14 Patterns of Biophilic Design. Improving Health & Well-Being in the Built Environment*. New York: Terrapin Bright Green, LLC.

Brychtová, A., & Coltekin, A. (2014). An Empirical User Study for Measuring the Influence of Colour Distance and Font Size in Map Reading Using Eye Tracking. *The Cartographic Journal Maney Publishing*.

BS. (2008). *BS 8206-2:2008 Lighting for buildings. Code of practice for daylighting*. BSI.

Bucher, U., & Finka, M. (2011). *The Electronic City*. Berlin: BWV, Berliner Wissenschafts-Verlag.

Buhalis, D., & Amaranggana, A. (2015). Smart tourism destinations enhancing tourism experience through personalisation of services. In *Information and Communication Technologies in Tourism 2015* (pp. 377–389). Springer International Publishing. doi:10.1007/978-3-319-14343-9_28

Burton, I., Huq, S., Lim, B., Pilifosova, O., & Schipper, E. (2002). From Impacts Assessment to Adaptation Priorities: The Shaping of Adaptation Policy. *Climate Policy*, *2*(3), 145–159. doi:10.3763/cpol.2002.0217

Butler, R. W. (1980). The concept of a tourist area cycle of evolution: Implications for management of resources. *The Canadian Geographer. Geographe Canadien*, *24*(1), 5–12. doi:10.1111/j.1541-0064.1980.tb00970.x

Calthorpe, P., & Fulton, W. B. (2001). *The Regional City: Planning for the End of Sprawl*. Washington, DC: Island Press.

Canter, D. (1977). *The psychology of place*. London: The Architectural Press.

Castells, M. (2004). Space of flows, space of places: materials for a theory of urbanism in the information age. In S. Graham (Ed.), *The Cybercities Reader* (pp. 82–93). New York: Routledge, Taylor & Francis Group.

Causeway Rural & Urban Network. (2015). Retrieved March 11, 2017. http://crun.org/?doing_wp_cron=1491135028.5418028831481933593750

Cavaco, C. (1999). O turismo rural nas políticas de desenvolvimento do turismo em Portugal. In C. Cavaco (Ed.), *Desenvolvimento Rural: Desafio e Utopia* (pp. 281–292). Lisboa: CEG.

Čavrak, V. (2016). *Razvoj metropolskih policentričnih regija i koncept pametnog grada*. Retrieved from: https://www.researchgate.net/publication/303413058_RAZVOJ_METROPOLSKIH_POLICENTRICNIH_REGIJA_I_KONCEPT_PAMETNOG_GRADA

Cawley, M., & Gillmor, D. A. (2008). Integrated rural tourism. *Annals of Tourism Research*, *35*(2), 316–337. doi:10.1016/j.annals.2007.07.011

CEN – European Committee for Standardization. (2007). *EN 15217, Energy performance of buildings — Methods for expressing energy performance and for energy certification of buildings*. Author.

Chaloupka-Risser, Ch., & Füssl, E. (2017). The importance of communication between cyclists and other traffic participants and its potential in reducing traffic safety-critical events. Transactions on Transport Sciences 2017, 8(1), 24-30. doi:10.5507/tots.2017.004

Chaloupka-Risser, C., Risser, R., & Zuzan, W. (2011). *Verkehrspsychologie: Grundlagen und Anwendungen*. Wien: Facultas.

Chapman, P., King, S., & Underwood, G. (2001). *The Role of Experience in Searching Scenes, Accident research Unit*. School of Psychology, University of Notthingham.

Chen, M., Zhang, H., Liu, W., & Zhang, W. (2014). The Global Pattern of Urbanization and Economic Growth: Evidence from the Last Three Decades. *PLoS One*, *9*(8), e103799. doi:10.1371/journal.pone.0103799 PMID:25099392

Choay, F. (1994). La mort de la ville et le règne de l'urbain. In *La Ville: art et architecture en Europe, 1873-1993* (pp. 26–35). Paris: Éditions du Centre Pompidou; doi:10.1093/ndt/12.4.839

Christaller, W. (1933). Die Zentralen Orte in Suddeutschland. Jena, Darmstadt: Wissenschaftliche Buchgesellschaft.

CIHEAM. (2008). Development strategies for rural areas. In *Mediterra. The Future of Agriculture in Mediterranean Countries* (pp. 275–301). Montpelier: SciencesPo. Les Presse.

Cinzano, P., Falchi, F., & Elvidge, C. D. (2001). The first World Atlas of the artificial night sky brightness. *Monthly Notices of the Royal Astronomical Society*, *328*(3), 689–707. doi:10.1046/j.1365-8711.2001.04882.x

City of Copenhagen. (2012). *CPH 2025 Climate Plan: A Green, Smart and Carbon Neutral City - draft version 4 June 2012*. Retrieved March 11, 2017. http://subsite.kk.dk/sitecore/content/Subsites/CityOfCopenhagen/SubsiteFrontpage/Business/Growth_and_partnerships/~/media/F5A7EC91E7AC4B0891F37331642555C4.ashx

City Sourced Mobilizing Civic Engagement. (2015). *City Sourced: Mobilizing Civic Engagement*. Retrieved November 11, 2015. https://www.citysourced.com/

Clarke, J. J. (1993). Nature in question. London: Earthscan publications.

Čolić, R., Maruna, M., Milovanović Rodić, D., & Lalović, K. (2015). *Integralni urbani projekti za upravljanje rizikom od poplava na primeru Obrenovca: katalog izložbe završnih radova generacije studenata 2013/14* [Integrated Urban Projects for Disaster Risk Management: Case Study of Obrenovac: Final Works' Exhibition Catalogue: Generation of Students 2013/2014]. Beograd: Univerzitet u Beogradu, Arhitektonski fakultet.

Čolić, R. (2015). Integrated Urban Development Strategy as an Instrument for Supporting Urban Governance. *Serbian Architectural Journal SAJ*, *7*(3), 317–342.

Copernicus. (2017). *About Copernicus*. Retrieved November 11, 2015. https://www.copernicus.eu/en/about-copernicus

Corboz, A. (1983). Le Territoire comme palimpseste. *Diogene, 121*, 14-35.

Corner, J. (2006). Terra Fluxus. In C. Waldheim (Ed.), *The Landscape Urbanism Reader* (pp. 21–33). New York: Princeton Architectural Press.

Corrucini, R. S. (1991). Anthropological aspects of orofacial and occlusal variations and anomalies. In M. A. Kelly & C. S. Larsen (Eds.), *Advances in dental anthropology*. New York: Wiley-Liss.

Couch, C., Petschel-Held, G., & Leontidou, L. (2007). *Urban Sprawl in Europe: Landscape, Land-Use Change and Policy*. Oxford, UK: Wiley-Blackwell. doi:10.1002/9780470692066

Council of Europe. (2000). European Landscape Convention. *European Treaty Series, 176*.

Cox, E., & Longlands, S. (2016). *City systems: The role of small and medium-sized towns and cities in growing the northern powerhouse*. IPPR.

Crouch, D. (1994). Home, escape and identity: Rural cultures and sustainable touris*m. Journal of Sustainable Tourism*, *2*(1-2), 93–101. doi:10.1080/09669589409510686

Cruz, C. M. (2008). A importância do cadastro no desenvolvimento urbano português. In XI Colóquio Ibérico de Geografia. Alcalá de Henares.

Cullen, G. (1990). *Gradski pejzaz*. Beograd, Srbija: Gradjevinska knjiga.

Cvejić, J., Vasiljević, N., & Tutundzić, A. (2008). *Tipologija predela Beograda za potrebe izrade Evropske konvencije o predelima* [Typology of Belgrade's landscapes for the needs of application of European Landscape Convention]. Beograd: Univerzitet u Beogradu - Šumarski fakultet i Sekretarijat za zaštitu životne sredine Grada Beograda.

Cvejić, S., Babović, M., Petrović, M., Bogdanov, N. & Vuković, O. (2010). *Socijalna isključenost u ruralnim oblastima Srbije*. UNDP- Sector for inclusive development.

DAC & Cities. (2017). Copenhagen: *The New Climate Generation*. Retrieved March 11, 2017. http://www.dac.dk/en/dac-cities/sustainable-cities/all-cases/education/copenhagen-the-new-climate-generation/

Dave, K., Lawrence, S., & Pennock, D. M. (2003, May). Mining the peanut gallery: Opinion extraction and semantic classification of product reviews. In *Proceedings of the 12th international conference on World Wide Web* (pp. 519-528). ACM.

Davoudi, S. (2003). European briefing: polycentricity in European spatial planning: from an analytical tool to a normative agenda. *European Planning Studies, 11*(8), 979–999. doi:10.1080/0965431032000146169

Davoudi, S., & Stead, D. (2002). Urban-Rural Relationships: An introduction and a brief history. *Built Environment, 28*(4), 269–277.

Davoudi, S., & Stead, D. (2002). Urban-rural relationships: An introduction and brief history. *Built Environment, 28*(4), 269–277.

de la Solà-Morales, M. (1997). Territoris sense model. *Papers: Regió Metropolitana de Barcelona*, (26), 21–27. Retrieved from http://www.raco.cat/index.php/PapersIERMB/article/viewArticle/101847/0

de la Solà-Morales, M. (2004). Contra el modelo de metrópolis universal. In Á. M. Ramos (Ed.), *Lo urbano en 20 autores contemporáneos* (pp. 99–104). Barcelona: Edicions UPC.

de la Solà-Morales, M. (n.d.). Las formas de crecimiento urbano. Las formas de crecimiento urbano. *Universitat Politècnica de Catalunya.*

Delcourt, H. R., & Delcourt, P. A. (1988). Quaternary landscape ecology: Relevant scales in space and time. *Landscape Ecology, 2*(1), 23–45. doi:10.1007/BF00138906

Dematteis, G. (1995). *Progetto implicito : il contributo della geografia umana alle scienza del territorio.* Milano: Franco Angeli.

Đenadić, M., Muhi, B., & Jovanović, D. (2016). Rural tourism – Serbia's missed chance. *Economics of Agriculture, 2/2016*, 515–529.

Đerčan, B., Radaković, M., Ostojić, M., Mirković, M., Obrenov, S. & Vozar, J. (2017). Urban hierarchy in Serbia. *Researches Reviews of the Department of Geography, Tourism and Hotel Management, 46*(1), 12-24.

Dijkstra, L., & Poelman, H. (2014). *A harmonised definition of cities and rural areas: the new degree of urbanization. Regional and Urban Policy,* Working Papers 01/2014. Brussels: European Commission. Retrieved February 25, 2017, from-https://ec.europa.eu/regional_policy/sources/docgener/work/2014_01_new_urban.pdf

Dimitrova, A. (2018). *Contextual reading of industrial heritage* (Unpublished master thesis). Ss. Cyril and Methodius University, Skopje, Republic of North Macedonia.

Dinić, M., Đurić, J., & Mitković, P. (2014). Rural settlements in suburban landscape of the post-socialist city – Niš case study. In *Proceedings of Landscape Transformations - an International interdisciplinary student conference* (pp. 136-143). Prague: Czech Technical University.

Dinić-Branković, M., Bogdanović-Protić, I., Đekić, J., & Mitković, P. (2016). Post-socialist suburbanization and sprawl development patterns – Niš case study. *Facta universitatis, series Architecture and civil engineering, 14*(3), 355-366.

Dinić, M., & Mitković, P. (2016). Suburban design: From "bedroom communities" to sustainable neighborhoods. *Geodetski Vestnik, 60*(1), 98–113.

Directorate-General for Energy –European Commission. (2012). *EU energy in figures – statistical pocketbook 2012.* Brussels: European Union.

Djukic, A., Stupar, A., & Antonic, B. (2018). The Orthogonal Urban Matrix of the Towns in Vojvodina, Northern Serbia: Genesis and Transformation. In G. Carlone, N. Martinelli, & F. Rotondo (Eds.), *Designing Grid Cities for Optimized Urban Development and Planning* (pp. 128–156). Hershey, PA: IGI Global; doi:10.4018/978-1-5225-3613-0.ch008

Djukic, A., Stupar, A., & Antonic, B. (2018b). The consequences of urban policies in socialist Yugoslavia on the transformation of historic centres: The case study of cities in Northern Serbia. *Journal of Housing and the Built Environment*, *33*(3), 555–573. doi:10.100710901-018-9612-7

Domingues, Á. (2006). *Cidade e Democracia. 30 Anos de Transformação Urbana em Portugal*. Lisboa: Argumentum/ Ordem dos Arquitectos.

Đorđević, A. (1972). Ekspoze povodom usvajanja Generalnog urbanističkog plana 1972. *Urbanizam Beograda*, *19*, 13–19.

Đorđević, J., & Tošić, B. (2013). *Regionalno planiranje i razvoj* [Regional Planning and Development]. Novi Sad: University of Novi Sad.

Dozza, M., & Fernandez, A. (2014). Understanding Bicycle Dynamics and Cyclist Behavior From Naturalistic Field Data. *IEEE Transactions on Intelligent Transportation Systems*, *15*(1), 376–384. doi:10.1109/TITS.2013.2279687

Drewe, P. (2000). *ICT and urban form. In Urban planning and design – Off the beaten track, Design Studio "The Network City"*. Delft University of Technology: Faculty of Architecture.

Drucker, P. F. (1993). *Post-capitalist society*. New York, NY: Harper Business.

Duany, A., & Talen, E. (2002). Transect Planning. *Journal of the American Planning Association*, (68): 345–266.

Đukić, A. (2015). New Belgrade: visions, plans and realizations 1950-2014. In G. Doytchinov, A. Đukić & C. Ioniță (Eds.), Planning Capital Cities: Belgrade, Bucharest, Sofia (pp, 160-172). Graz: TU Graz Verlag.

Đukić, A., Vujičić, T., & Antonić, B. (2018). Greening as an Approach for Urban Renewal of Shrinking Cities. In P. Mitković, M. Dinić Branković, M. Tanić, A. Mirić, & V. Milosević (Ed.), *Proceedings of 2nd International Conference on Urban Planning - ICUP2018* (pp. 291-297). Niš: Faculty of Civil Engineering and Architecture University of Niš.

Đukić, A., Antonić, B., & Vujičić, T. M. (2017). Urban shrinkage in a 'shrinking' Serbia – the approach to a global phenomenon in a local context. *Geodetski Vestnik*, *61*(4), 614–629. doi:10.15292/geodetski-vestnik.2017.04.614-629

Đukić, A., Simonović, D., & Vujičić, T. (Eds.). (2014). *International Scientific Monograph BROWNINFO. Toward a Methodological Framework for Brownfield Database Development*. Banja Luka: University of Banja Luka, Faculty of Architecture Civil Engineering and Geodesy.

Duncan, J. (1990). *The City as Text: the Politics of Landscape Interpretation in the Kandyan Kingdom*. Cambridge, UK: Cambridge University Press.

Đurović, Đ. (1969). Sociološki pristup planiranju budućeg razvoja Beograda. *Urbanizam Beograda*, *4*, 19.

EC. (1999). European Spatial Development Perspective (ESDP). Towards Balanced and Sustainable Development of the Territory of the European Union. EC.

EC. (1999). *European Spatial Development Perspective: Towards Balanced and Sustainable Development of the Territory of the EU*. Retrieved May 15, 2019, from http://europa.eu.int/comm/regional_policy/sources/docoffic/official/reports/som_en.htm

Efstratoglou, S., Bogdanov, N., & Meredith, D. (2007). Defining Rural Areas in Serbia and Their Typology. *Proceedings of the 100th Seminar Development of Agriculture and rural areas in Central and Eastern Europe*, 553-562.

ENRD. (2017). *Smart Villages*. Retrieved February 11, 2019. https://enrd.ec.europa.eu/enrd-thematic-work/smart-and-competitive-rural-areas/smart-villages_en

EPISCOPE. (2012). *Typology Approach for Building Stock Energy Assessment*. Retrieved September 20, 2015, from http://episcope.eu/iee.project/tabula

ESPON and UHI Millennium Institute. (2011). *EDORA: European Development Opportunities for Rural Areas*. Luxembourg: ESPON & UHI Millennium Institute.

ESPON. (2017). *Policy Brief // Shrinking Rural Regions in Europe*. Retrieved from https://www.espon.eu/rural-shrinking

EU. (2007). *Leipzig Charter on Sustainable Cities*. Retrieved May 15, 2019, from http://ec.europa.eu/regional_policy/archive/themes/urban/ leipzig_charter.pdf

EU. (2009). *The Committee of the Regions' White Paper on Multilevel Governance*. Retrieved May 15, 2019, from www.europarl.europa.eu/meetdocs/2009_2014/documents/regi/dv/cdr89-2009_/cdr89-2009_en.pdf

EU. (2010). *Toledo Declaration*. Retrieved May 15, 2019, from www.mdrap.ro/userfiles/declaratie_Toledo_enpdf

EU. (2016). *Urban Agenda for the EU*. Retrieved May 15, 2019, from http://ec.europa.eu/regional_policy/sources/policy/themes/ urban-development/agenda/pact-of-amsterdam.pdf

EUR-lex. (2002). *Directive 2002/91/EC of the European Parliament and of the Council of 16 December 2002 on the energy performance of buildings*. Retrieved from https://eur-lex.europa.eu/legal-content/EN/TXT/?uri=CELEX%3A32002L0091

EUR-lex. (2006). *Directive 2006/32/EC of the European Parliament and of the Council of 5 April 2006 on energy end-use efficiency and energy services and repealing Council Directive 93/76/EEC*. Retrieved from https://eur-lex.europa.eu/legal-content/EN/TXT/?uri=celex%3A32006L0032

EUR-lex. (2010). *Directive 2010/31/EU of the European Parliament and of the Council of 19 May 2010 on the energy performance of buildings*. Retrieved from https://eur-lex.europa.eu/legal-content/EN/TXT/?uri=CELEX%3A32010L0031

EUR-lex. (2012). *Directive 2012/27/EU of the European Parliament and of the Council of 25 October 2012 on energy efficiency, amending Directives 2009/125/EC and 2010/30/EU and repealing Directives 2004/8/EC and 2006/32/EC*. Retrieved from https://eur-lex.europa.eu/eli/dir/2012/27/oj

European Commission. (1997). *Energy for the future: Renewable sources of energy, White Paper for a Community Strategy and Action Plan*. European Commission.

European Commission. (2008). *Poverty and Social Exclusion in Rural Areas. Directorate General for Employment, Social Affairs and Equal Opportunities, Unit E2*. Rome: European Commision.

European Commission. (2009). *PV Status Report 2009*. Luxembourg: Office for Official Publications of the European Communities.

European Commission. (2013a). An EU strategy on adaptation to climate change – Communication from the Commission to the European Parliament, the Council, the European Economic and Social Committee and the Committee of the Regions (COM (2013) 216 final). Brussels: European Commission.

European Commission. (2013b). Guidelines on developing adaptation strategies (SWD (2013) 134 final). Brussels: European Commission.

European Commission. (2014). *PV Status Report 2014*. European Commission, Luxemburg: Publications Office of the European Union.

European Commission. (2014). *Rural Development Programme 2014-2020*. European Commission.

European Commission. (2016). Cork 2.0 Declaration: A Better Life in Rural Areas. *Conference on Rural Development*.

European Commission. (2016). *EU Action for Climate Villages*. Retrieved December 18, 2018. https://ec.europa.eu/agriculture/sites/agriculture/files/rural-development-2014-2020/looking-ahead/rur-dev-small-villages_en.pdf

European network for Rural development. (2010). *Climate Action. EU Rural Review No 4*. European Commission.

European Network for Rural Development. (2017). Retrieved March 11, 2017. https://enrd.ec.europa.eu/

European Parliament and the Council of the European Union. (2001). *Directive 2001/77/EC of the European Parliament and of the Council of 27 September 2001 on the promotion of electricity produced from renewable energy sources in the internal electricity market*. Official Journal of the European Communities L283.

European Parliament and the Council of the European Union. (2009). Directive 2009/28/EC of the European Parliament and of the Council of 23 April 2009 on the promotion of the use of energy from renewable sources and amending and subsequently repealing Directives 2001/77/EC and 2003/30/EC. *Official Journal of the European Union, L, 140*(16).

European Parliament. (2013). *Regulation 1305/2013 of 17 December 2013 on the European Agricultural Fund for Rural Development, Article 18*. Author.

European Transport Safety Council 2015: Europe must boost safety for cyclists and walkers. (n.d.). Assessed 27.6.2016. http://etsc.eu/europe-must-boost-safety-for-cyclists-and-walkers/

European Transport Safety Council 2015: Flash, P. I. N. 29 Making walking and cycling on Europe's roads safer. (n.d.). Assessed 27.6.2016. http://etsc.eu/wp-content/uploads/ etsc_pin_flash_29_walking_cycling_safer.pdf

Eurostat - GISCO. (2013). *Eurostat. Statistical Atlas Eurostat regional yearbook 2013*. Retrieved March 10, 2017, from http://ec.europa.eu/eurostat/statistical-atlas/gis/viewer/?year=2013&chapter=01&mids=2,116&o=1,1&center=50.0958,20.00071,3&ch=113,114&

Eurostat. (2016). *Greenhouse gas emissions by source sector (source: EEA)*. Retrieved December 18, 2018. http://appsso.eurostat.ec.europa.eu/nui/show.do?dataset=env_air_gge&lang=en

EUROSTAT. (2016). *Rural development statistics by urban-rural typology*. EUROSTAT.

EUROSTAT. (2018). *Statistics on rural areas in the EU*. Retrieved from https://ec.europa.eu/eurostat/statistics-explained/index.php/Statistics_on_rural_areas_in_the_EU

Eusébio, C., Kastenholz, E., & Breda, Z. (2014). Tourism and susteinable development of rural destinations: a stakeholders` view. *Revista Portuguesa de Estudos Regionais, 2. Quatrimestre,* (36), 12-21.

Fang, Z., Li, N., Li, B., Luo, G., & Huang, Y. (2014). The effect of building envelope insulation on cooling energy consumption in summer. *Energy and Building*, *77*, 197–205. doi:10.1016/j.enbuild.2014.03.030

Federal Ministry of Physical Planning. (2009). *Rulebook on Technical Requirements for Thermal Protection of Buildings and Rational Use of Energy*. Official Gazette Federation B&H, no. 49/09.

Federal Statistical Office of Socialist Federal Republic of Yugoslavia. (1975). *Census of Population and Housing 1971. Population and Households in 1948, 1953, 1961 and 1971 and dwellings in 1971. Results by municipalities and settlements*. Belgrade: Author.

Fei-Fei, L., VanRullen, R., Koch, C., & Perona, P. (2005). Why does natural scene categorization require little attention? Exploring attentional requirements for natural and synthetic stimuli. *Visual Cognition, 12*(6), 893–924. doi:10.1080/13506280444000571

Felson, M., & Spaeth, J. L. (1978). Community structure and collaborative consumption: A routine activity approach. *The American Behavioral Scientist, 21*(4), 614–624. doi:10.1177/000276427802100411

Finci, J. (1967). *Stanovanje u Urbanističkom planu Banjaluke* [Housing in the Urban Plan of Banja Luka]. Banjaluka: Urbanistički zavod Banjaluka.

Fixemer. (2017). *Fixemer e Logistic*. Accessed March 11, 2017. http://www.fixemer.com

Florentin, D. (2010). The "Perforated City": Leipzig's Model of Urban Shrinkage Management. *Berkeley Planning Journal, 23*(1), 83–101.

Florida, R. (2005). *Cities and the creative class*. New York, NY: Routledge. doi:10.4324/9780203997673

Font, A. (1997). Anatomia de una metrópoli discontinua: la Barcelona Metropolitana. *Papers*, (26), 9–19.

Font, A. (Ed.). (2004). L'explosió de la ciutat. Morfologies, mirades i mocions sobre les transformacion territorials recents en les regions urbanes de l'Europa Meridional. Barcelona: Universidad Politècnica de Catalunya, Institut d'Estudis Territorials, Generalitat de Catalunya.

Font, A. (Ed.). (2012). Patrons urbanístics de les activitats econòmiques. Regió Metropolitana de Barcelona / Urban patterns of economic activities. Barcelona Metropolitan Region. Barcelona: Universidad Politècnica de Catalunya, Institut d'Estudis Territorials, Generalitat de Catalunya.

Font, A. (1999). *La construcció del territori metropolità. Morfogènesi de la Regió urbana de Barcelona*. Barcelona: Mancomunitat de Municipis de l'Àrea Metropolitana de Barcelona.

Food and Agriculture Organization of the United Nations (FAO). (2009). *Information and Communication Technologies for climate change adaptation, with a focus on the agricultural sector*. Retrieved December 18, 2018. http://www.fao.org/docs/eims/upload/295345/Sala%20ICT-climate%20change%20Agriculture.pdf

Formaneck, S. (2013). Skyscrapers and Economic Strategy and Development: The Case of Bahrain"s Proposed Murjan Tower. *Journal of Economics. Business and Management, 1*(4), 350–354.

Forman, R. (1995). *Land Mosaics: The Ecology of Landscapes and Regions*. Cambridge, UK: Cambridge University Press. doi:10.1017/9781107050327

Frazer, J. (1995). Evolutionary architecture. London: Architectural Association.

Frempton, K. (2004). *Moderna arhitektura, kritička istorija*. Beograd, Srbija: Orion art.

Friedmann, J. (2006). Four theses in the study of China's urbanization. *International Journal of Urban and Regional Research, 30*(2), 440–451. doi:10.1111/j.1468-2427.2006.00671.x

Furundžić, Z. N., Furundžić, P. D., & Krstić-Furundžić, A. (2019). Chapter Three – New Technology for the Promotion of Health in the Cognitive City. In E. Vaništa Lazarević, M. Vukmirović, A. Krstić-Furundžić, & A. Đukić, Keeping (Eds.), Keeping Up with Technologies to Create the Cognitive City (pp. 31-43). Newcastle upon Tyne, UK: Cambridge Scholars Publishing.

Fusero, P. (2009). *E-City: Digital Networks and Cities of the Future*. Barcelona: LIST Laboratorio.

Gajić, D. (2014). *Energy optimisation of the building envelope on the representative samples of the old residential buildings in Banja Luk. (Doctoral dissertation). University of Belgrade, Faculty of Architecture*. Retrieved from http://nardus.mpn.gov.rs/handle/123456789/6493

Garrod, B., Wornell, R., & Youell, R. (2006, January). Re-conceptualising rural resources as countryside capital: The case of rural tourism. *Journal of Rural Studies*, *22*(1), 117–128. doi:10.1016/j.jrurstud.2005.08.001

Garrod, B., Youell, R., & Wornell, R. (2004). *Links Between Rural Tourism and Countryside capital*. Cheltenham, UK: Countryside Agency.

Gavrilović, N. (1967). Komunalna opremljenost Beograda. *Arhitektura urbanizam, 41-42*, 27-34.

George, P. (2001). Paisagem – do urbano ao rural notas a partir de Lisboa…. *Finisterra*, *36*(72), 115–121. doi:10.18055/Finis1626

George, W., Heather, M., & Reid, D. (2009). *Rural tourism Development Localism and Cultural Change*. Channel View Publications.

Giddens, A. (2009). *The Politics of Climate Change*. Cambridge, UK: Polity Press.

Giebeler, G., Fisch, R., Krause, H., Musso, F., Petzinka, K.-H., & Rudolphi, A. (2009). *Refurbishment Manual*. Birkhäuser. doi:10.11129/detail.9783034614337

Godfrey, R., & Julien, M. (2005). Urbanisation and health. *Clinical Medicine*, *5*(2), 137–141. doi:10.7861/clinmedicine.5-2-137 PMID:15847005

Gomez-Limon, H. A. (2010, May) Climate change effects and climate-related Rural Development Programme support in Spain. EU Rural Review - The Magazine from the European Network for Rural Development, 24-30.

GoodWorkLabs. (2015). Retrieved from http://www.goodworklabs.com/how-mobile-app-benefits-travel-and-tourism-industry/

Gore, A. (2008). *Earth in the Balance Forging a New Common Purpose*. London: Earthscan.

Government of Phillipines (GOVPH). (2018). *The Philippine Atmospheric, Geophysical and Astronomical Services Administration*. Retrieved December 18, 2018. http://bagong.pagasa.dost.gov.ph/

Government of Serbia. (2019). *Sustainable Urban Development Strategy of Serbia to 2030*. Official Gazette of the Republic of Serbia, No. 47/2019.

Government of the Republic of Serbia – GRS. (1995). Закон о планирању и уређењу простора и насеља [Law on Planning and arranging Settlements and Spaces of Republic of Serbia]. Belgrade: Official Gazette of Republic of Serbia No. 44/1995.

Government of the Republic of Serbia – GRS. (2003). Закон о планирању и изградњи [Law on Planning and Construction of Republic of Serbia]. Belgrade: Official Gazette of Republic of Serbia No. 47/2003.

Government of the Republic of Serbia – GRS. (2006). Уредба о управним окрузима [Regulation on Administrative Districts of Republic of Serbia]. Belgrade: Official Gazette of Republic of Serbia No. 15/2006.

Government of the Republic of Serbia – GRS. (2010). Просторни план Републике Србије [Spatial Plan of the Republic of Serbia]. Belgrade: Official Gazette of Republic of Serbia No. 88/2010.

Government of the Republic of Serbia – GRS. (2014). Закон о планирању и изградњи [Law on Planning and Construction of Republic of Serbia]. Belgrade: Official Gazette of Republic of Serbia No. 72/2009, 81/2009, 64/2010, 24/2011, 121/2012, 42/2013, 50/2013, 98/2013, 132/2014 i 145/2014.

Government of the Republic of Serbia – GRS. (2014). Уредба о утврђивању јединствене листе развијености региона и јединица локалне самоуправе за 2014. годину [Regulation on establishing a single list of development of regions and local self-government units for 2014 of Republic of Serbia]. Belgrade: Official Gazette of Republic of Serbia No. 104/2014.

Graham, S., & Marvin, S. (2001). *Splintering Urbanism. Networked infrastructures, technological mobilities and the urban condition. Technology*. London: Routledge; doi:10.4324/9780203452202

Graham, S., & Simon, M. (2006). *Telecommunication and the City: Electronic Spaces, Urban Places*. London: Routledge.

Grčić, M. (1999). Funkcionalna klasifikacija naselja Mačve, Šabačke posavine i Pocerine [Functional classification of the settlements of Mačva, Šabačka Posavina and Pocerina]. *Glasnik: Serbian Geographical Society*, *79*(1), 3–20.

Grčić, M., & Grčić, Lj. (2002). *Mačva, Šabavčka Posavina i Pocerina*. Belgrade: University of Belgrade - Faculty of Geography.

Greco, I., & Bencardino, M. (2014). The paradigm of the modern city: SMART and SENSEable Cities for smart, inclusive and sustainable growth. In *Proceedings of the International Conference on Computational Science and Its Applications* (pp. 579-597). Cham, Switzerland: Springer International Publishing. 10.1007/978-3-319-09129-7_42

Green, J. (2016). *Biophilic Cities Lead the Way to Urban Sustainability*. The Dirt, American Society of Landscape Architects. Retrieved January 20, 2017, from https://dirt.asla.org/2016/06/06/biophilic-cities-forge-new-paths

Greffe, X. (1994). Is rural tourism a lever for economic and social development? *Journal of Sustainable Tourism*, *2*(1-2), 22–40. doi:10.1080/09669589409510681

Gregotti, V. (1981). La forme du territoire. *AA; l'Architecture d'Aujourd'Hui*, 218.

Gretzel, U., Zhong, L., & Koo, C. (2016). Application of smart tourism to cities. *International Journal of Tourism Cities*, *2*(2). doi:10.1108/IJTC-04-2016-0007

Griffiths, J., Jones, V., Leeman, I., Lewis, D., Patel, K. K., Wilson, K., & Blankenstein, R. (2000). *Oral Health Care for People with Mental Health Problems Guidelines and Recommendatio*ns. British Society for Disability and Oral Health. Retrieved February 20, 2017, from http://www.bsdh.org/documents/mental.pdf

Guimarães, E. T., Bragança, L., Almeida, M., & Mateus, R. (2015). Analysis of Portuguese Residential Buildings' Needs and Proposed Solutions, Connecting People and Ideas. *Proceedings of EURO ELECS 2015*.

Gullone, E. (2000). The Biophilia Hypothesis and life in the 21st Century: Increasing Mental health or Increasing Pathology? *Journal of Happiness Studies*, *1*(3), 293–321. doi:10.1023/A:1010043827986

Gupta, P. K. (2008). *Creating Great Town Centers and Urban Villages*. Washington, DC: Urban Land Institute.

Gustafsson, L., & Archer, J. (2013). A naturalistic study of commuter cyclists in the greater Stockholm area. *Accident; Analysis and Prevention, 58*, 286–98. Retrieved from http://www.sciencedirect.com/science/ article/pii/S0001457512002266

Haase, A., Rink, D., Grossmann, K., Bernt, M., & Mykhnenko, V. (2014). Conceptualizing urban shrinkage. *Environment & Planning A*, *46*(7), 1519–1534. doi:10.1068/a46269

Habitat, U. N. (1976). *The Vancouver Declaration on Human Settlements. United Nations Conference on Human Settlements*, Vancouver, Canada.

Häkkinen, T. (2012). Systematic method for the sustainability analysis of refurbishment concepts of exterior walls. *Construction & Building Materials, 37*, 783–790. doi:10.1016/j.conbuildmat.2012.07.084

Hall, D., & Richards, G. (2002). *Tourism and sustainable community development*. London: Routledge.

Haller, M., Pröll, B., Retschitzegger, W., Tjoa, A. M., & Wagner, R. R. (2000). Integrating heterogeneous tourism information in TIScover: the MIRO-Web approach. *ENTER 2000: 7th. International Congress on Tourism and Communications Technologies in Tourism*. 10.1007/978-3-7091-6291-0_7

Hall, M. C., & Page, S. J. (2006). *The Geography of Tourism and Recreation: Environment, place and space*. London: Routledge. doi:10.4324/9780203420249

Hamilton, F. E. I., & Carter, W. F. (2005a). Foreign direct investment and city restructuring. In I. Hamilton F.E., K. Dimitrovska Andrews, & N. Pichler-Milanovic (Eds.), Transformation of cities in Central and Eastern Europe: Towards globalization (pp. 116-152). Tokyo, Japan: United Nations University Press.

Hamilton, F. E. I., Pichler-Milanovic, N., & Dimitrovska Andrews, K. (2005b). Introduction. In I. Hamilton F.E., K. Dimitrovska Andrews, & N. Pichler-Milanovic (Eds.), Transformation of cities in Central and Eastern Europe: Towards globalization (pp.3-21). Tokyo, Japan: United Nations University Press.

Hamilton, I., Dimitrovska Andrews, K., & Pichler-Milanović, N. (Eds.). (2005). *Transformation of cities in central and Eastern Europe: Towards globalization*. Tokyo: UN University Press.

Hartig, T., & Kahn, P. H. (2016). Living in cities, naturally. *Science, 352*(6288), 938–940. doi:10.1126cience.aaf3759 PMID:27199417

Harvey, D. (1990). *The condition of postmodernity*. Cambridge, UK: Blackwell Publishers.

Healey, P. (2006). *Collaborative planning: Shaping places in fragmented societies* (2nd ed.). London: Palgrave MacMillan. doi:10.1007/978-1-137-08600-6

Hegedüs, J., & Tosics, I. (1996). *Transition of the housing sector in the East Central European countries*. Washington, DC: The Urban Institute. doi:10.1111/j.1467-940X.1996.tb00113.x

Hegger, M., Fuchs, M., Stark, T., & Zeumer, M. (2008). Energy Manual: sustainable architecture. Basel, CH: Birkhäuser.

Henderson, J. M., McClure, K., Pierce, S., & Schrock, G. (1997). Object identification without foveal vision: Evidence from an artificial scotoma paradigm. *Perception & Psychophysics, 59*(3), 323–346. doi:10.3758/BF03211901 PMID:9136264

Herce Vallejo, M., Magrinyà, F., & Miró, J. (2004). *Construcció de ciutat i xarxas d'infraestructures*. Barcelona: Edicions UPC.

Heylighen, F. (1989). Self-Organization, Emergence and the Architecture of Complexity. In *Proceedings of the 1st European Conference on System Science* (pp. 23-32). Paris: AFCET.

Hill, D., Lean, D., Watts, M., & Buscher, V. (2011). *ARUP - The smart solution for cities*. Retrieved March 11, 2017. http://www.arup.com/~/media/Files/PDF/Publications/Research_and_whitepapers/UrbanLife_SmartSolutionForCities_July11.ashx

Hirt, S., & Stanilov, K. (2007). The perils of post-socialist transformation: Residential development in Sofia. In K. Stanilov (Ed.), *The Post-Socialist City: Urban Form and Space Transformations in Central and Eastern Europe after Socialism* (pp. 215–244). Dordrecht: Springer. doi:10.1007/978-1-4020-6053-3_11

Hirt, S., & Stanilov, K. (2009). *Revisiting Urban Planning in the Transitional Countries*. Nairobi: UN Habitat.

Hirt, S., & Stanilov, K. (2014). *Twenty Years of Transition: The Evolution of Urban Planning in Eastern Europe and the Former Soviet Union, 1989-2009*. Nairobi: UN Habitat.

Hislop, D. (Ed.). (2008). *Mobility and Technology in the Workplace*. New York: Routledge. doi:10.4324/9780203894354

Hooker, R., Lewis, C., Smith, H., Worrell, J. M., & Yoon, T. (2008). Governing E-Collaboration in E-Lance Networks. In N. Kock (Ed.), *Encyclopedia of E-Collaboration* (pp. 324–329). Hershey, PA: IGI Global. doi:10.4018/978-1-59904-000-4.ch050

Hope, K. R. (1998). Urbanization and urban growth in Africa. *Journal of Asian and African Studies, 33*(4), 345–358. doi:10.1177/002190969803300403

Hough, M. (2007). Nature as Infrastructure: Strategies for Sustainable Regional Landscapes. *Places*, (19).

Houlne, T., & Maxwell, T. (2013). *The New World of Work: From the Cube to the Cloud*. Irving, TX: Inspire on Purpose Books.

Howard, G., Bogh, C., Goldstein, G., Morgan, J., Prüss, A., Shaw, R., & Teuton, J. (2002). *Healthy Villages: A guide for communities and community health workers*. Geneva: World Health Organization.

Hunter, W. C., Chung, N., Gretzel, U., & Koo, C. (2015). Constructivist research in smart tourism. *Asia Pacific Journal of Information Systems, 25*(1), 105–120. doi:10.14329/apjis.2015.25.1.105

Hydén, Ch. (1987). *The development of a method for traffic safety evaluation: The Swedish Traffic Conflicts Technique*. Department of Traffic Engineering, Lund University.

IBM. (2013). *Made in IBM Labs: IBM Drives the Future of Renewable Energy with New Wind and Solar Forecasting System.* Retrieved in March 11, 2017. http://www03.ibm.com/press/us/en/pressrelease/41310.wss

infoDev. (2016). *eLum's Innovation Illuminates Energy Savings for Moroccan Businesses*. Retrieved December 18, 2018. http://www.infodev.org/articles/elums-innovation-illuminates-energy-savings-moroccan-businesses

infoDev. (2018). *New Climate Center in Brazil Brings Together Entrepreneurs and Farmers*. Retrieved December 18, 2018. http://www.infodev.org/articles/new-climate-center-brazil-brings-together-entrepreneurs-and-farmers

Ingersoll, R. (2006). *Sprawltown – Looking for the City on its Edges*. New York: Princeton Architectural Press.

Institut for Statistics of Federation of Bosnia and Herzegovina. (1998). *Results of the Census of Population, Households, Flats and Agricultural Holdings 1991. Population comparative data 1971, 1981 and 1991*. Sarajevo: Author.

Institut for Statistics of Federation of Bosnia and Herzegovina. (1999a). *Results of the Census of Population, Households, Flats and Agricultural Holdings 1991. Population by municipalities*. Sarajevo: Author.

Institut for Statistics of Federation of Bosnia and Herzegovina. (1999b). *Results of the Census of Population, Households, Flats and Agricultural Holdings 1991. Population by settlements*. Sarajevo: Author.

Institut za interdisciplinarna istraživanja. (2004). *Studija energetskog potencijala Srbije za korišćenje sunčevog zračenja i energije vetra*. Univerzitet u Beogradu, Institut za interdisciplinarna istraživanja.

Institute of Urban Development Krakow (IUDK). (2013). *The State of European Cities in Transition 2013: Taking Stock after 20 Years of Reform*. Nairobi: UN Habitat.

International Energy Agency. (2014). *Energy balances of non-OECD Countries 2014 Edition*. Retrieved from https://www.iea.org/statistics/relateddatabases/energybalancesofnon-oecdcountries

INVIL project. (2001). *Information Network Village (INVIL) project*. Retrieved March 11, 2017. https://www.oecd.org/governance/observatory-public-sectorinnovation/innovations/page/informationnetworkvillageinvilproject.htm#tab_implementation

IOM - Institute of Medicine. (2010). *Promoting cardiovascular health in the developing world: A critical challenge to achieve global health*. Washington, DC: The National Academies Press. Retrieved January 18, 2012, from http://www.iom.edu/Reports/2010/Promoting-Cardiovascular-Health-in-the-Developing-World-A-Critical-Challenge-to-Achieve-Global-Health.aspx

IPCC. (2007). Chapter 4. Adaptation and mitigation options. In Summary for Policymakers. In: Climate Change 2007: Synthesis Report. Contribution of Working Groups I, II and III to the Fourth Assessment Report of the Intergovernmental Panel on Climate Change. IPCC.

IPCC. (2014). Climate Change 2014: Impacts, Adaptation, and Vulnerability. *Contribution of Working Group II to the Fifth Assessment Report of the Intergovernmental Panel on Climate Change*. Original version: http://ipcc-wg2.gov/AR5/report

IPCC/Intergovernmental Panel on Climate Change. (2007a). *Climate Change 2007: synthesis report contribution of Working groups I, II and III to the Fourth Assessment Report of the Intergovernmental Panel on climate change*. Cambridge, UK: Cambridge University Press.

IPCC/Intergovernmental Panel on Climate Change. (2007b). *Summary for Policymakers. Climate Change 2007: The Physical Science Basis. Contribution of Working Group I to the Fourth Assessment Report of the Intergovernmental Panel on Climate Change*. Cambridge, UK: Cambridge University Press.

IPCC/Intergovernmental Panel on Climate Change. (2007c). *Summary for Policymakers. Mitigation. Contribution of Working Group III to the Fourth Assessment Report of the Intergovernmental Panel on Climate Change*. Cambridge, UK: Cambridge University Press.

IPCC/Intergovernmental Panel on Climate Change. (2013). *The Physical Science Basis. Contribution of Working Group I to the Fifth Assessment Report of the Intergovernmental Panel on Climate Change*. Cambridge, UK: Cambridge University Press.

ISO – International Organization for Standardization. (2008). Thermal performance of buildings-Calculation of energy use for heating. *EN ISO, 13790*, 2008.

ISO. (2007). *EN ISO10211 Thermal bridges in building construction - Heat flows and surface temperatures - Detailed calculations*. Geneva: ISO.

ISO. (2016). *EN ISO 18523-2:2016, Energy performance of buildings - Schedule and condition of building, zone and space usage for energy calculation - Part 2: Residential buildings*. Geneva: ISO.

ISO. (2017). *Building components and building elements - Thermal resistance and thermal transmittance - Calculation method*. International Organization for Standardization.

Jabareen, Y. R. (2006). Sustainable Urban Forms - Their Typologies, Models, and Concepts. *Journal of Planning Education and Research*, *6*(1), 38–52. doi:10.1177/0739456X05285119

Jaffe, E. (2015). *Why City Noise Is a Serious Health Hazard: An audiologist explains why it's so much more than a mere annoyance*. City Lab. Retrieved December 20, 2016, from http://www.citylab.com/housing/2015/04/why-city-noise-is-a-serious-health-hazard/391194

Janković, N. (2017). *Transformation of Landscape Configuration in the Process of City Territory Development: Belgrade in The Second Half of XX Century* (Doctoral dissertation). Belgrade: University of Belgrade - Faculty of Architecture.

Jefferson, M. (1939). The Law of the Primate City. *Geographical Review, 29*(2), 226–232. doi:10.2307/209944

Jernejec, M. (1962). Stambena naselja danas i sutra. *Arhitektura urbanizam, 14*, 45.

Jing, X. (2006). Rural Tourism and Sustainable Community Development. *Proceedings of International forum on rural tourism, final report*, 73-80

Johnson, M., Charlton, J., Oxley, J., & Newstead, S. (2010). Naturalistic cycling study: identifying risk factors for on-road commuter cyclists. *Annals of Advances in Automotive Medicine/Annual Scientific Conference, 54*, 275–83. Retrieved from http://www.pubmedcentral.nih.gov/articlerender.fcgi? artid=3242541&tool=pmcentrez&rendertype=abstract

Johnson, M., Charlton, J., Oxley, J., & Newstead, S. (2010). Naturalistic cycling study: identifying risk factors for on-road commuter cyclists. *Annals of Advances in Automotive Medicine/Annual Scientific Conference, 54,* 275–83. Retrieved from http://www.pubmedcentral.nih.gov/articlerender.fcgi?artid=3242541&tool=pmcentrez&rendertype=abstract

John, W., & Mitchell, T. (2000). *Landscape and Power*. Chicago: University of Chicago Press.

Jovanović Popović, M., Ignjatović, D., Radivojević, A., Rajčić, A., Đukanović, Lj., Ćuković Ignjatović, N., & Nedić, M. (2012). *Atlas of Family Housing in Serbia*. Belgrade: Faculty of Architecture & GIZ.

Jovanović Popović, M., Ignjatović, D., Radivojević, A., Rajčić, A., Đukanović, Lj., Ćuković Ignjatović, N., & Nedić, M. (2013). *National Typology of Residential Buildings in Serbia*. Belgrade: Faculty of Architecture & GIZ.

Jovanović-Popović, M., Ignjatović, D., Radivojević, A., Rajčić, A., Đukanović, L., Ćuković Ignjatović, N., & Nedić, M. (2013). Nacionalna tipologija stambenih zgrada Srbije. Belgrade, Serbia: Faculty of Architecture, University of Belgrade.

Jürgen, F. (1993). A Theory of Urban Decline: Economy, Demography and Political Elites. *Urban Studies (Edinburgh, Scotland), 30*(6), 907–917. doi:10.1080/00420989320080851

Kaplan, R., & Kaplan, S. (1989). *The Experience of Nature: a Psychological Perspective*. Cambridge, UK: Cambridge University Press.

Kasismis, C. (2010). Demographic trends in rural Europe and international migration to rural areas. *Agriregionieuropa (Ancona), 6*(21). Retrieved from https://agriregionieuropa.univpm.it/en/content/article/31/21/demographic-trends-rural-europe-and-international-migration-rural-areas

Kaufman, C. A. (2015, August 10). *Stephen Hawking says we should really be scared of capitalism, not robots*. Retrieved January 14, 2019, from http://www.huffingtonpost.com/entry/stephen-hawking-capitalism-robots

Keivani, R., Parsa, A., & McGreal, S. (2002). Institutions and Urban Change in a Globalising World the Case of Warsaw. *Cities (London, England), 19*(3), 183–193. doi:10.1016/S0264-2751(02)00015-X

Kellert, R. S. (2008). Dimensions, Elements and Attributes of Biophilic Design. In S. Kellert, J. Heerwage, & M. Mador (Eds.), *Biophilic Design: The Theory, Science and Practice of Bringing Buildings to Life* (pp. 3–4). Hoboken, NJ: John Wiley & Sons. Inc.

Kellert, R. S., Heerwagen, H. J., & Mador, L. M. (2008). *Biophilic Design: The Theory, Science and Practice of Bringing Buildings to Life*. Hoboken, NJ: John Wiley & Sons. Inc.

Kern, K., & Alber, G. (2008). Governing climate change in cities: Modes of urban governance in multi-level systems. *Proceedings of the OECD Conference on Competitive Cities and Climate Change*, 171-196.

Kessler, R. C., McGonagle, K. A., Swartz, M., Blazer, D. G., & Nelson, C. B. (1993). Sex and depression in the National Comorbidity Survey. I: Lifetime prevalence, chronicity and recurrence. *Journal of Affective Disorders, 29*(2-3), 85–96. doi:10.1016/0165-0327(93)90026-G PMID:8300981

Khanna, P. (2016). How much economic growth comes from our cities? *World Economic Forum*. Retrieved February 20, 2019, from https://www.weforum.org/agenda/2016/04/how-much-economic-growth-comes-from-our-cities/

Kisilewicz, T. (2019). On the role of external walls in the reduction of energy demand and the mitigation of human thermal discomfort. *Sustainability (Switzerland)*, *11*(4), 1061. doi:10.3390u11041061

Kiss, E. (2002). Restructuring in the industrial areas of Budapest in the period of transition. *Urban Studies (Edinburgh, Scotland)*, *39*(1), 69–84. doi:10.1080/00420980220099078

Kiss, E. (2007). The evolution of industrial areas in Budapest after 1989. In K. Stanilov (Ed.), *The Post-Socialist City: Urban Form and Space Transformations in Central and Eastern Europe after Socialism* (pp. 147–172). Dordrecht, The Netherlands: Springer. doi:10.1007/978-1-4020-6053-3_8

Kjellstrom, T., Friel, S., Havemann, K., McGranahan, G., Mercado, S., & Sattherthwaite, D. (2008). *Our cities, our health, our future: Acting on social determinants for health equity in urban settings. Report to the WHO Commission on Social Determinants of Health from the Knowledge Network on Urban Settings Acting on social determinants for health equity in urban settings*. World Health Organization, Centre for Health Development.

Knowles, J., Adams, R., Cuerden, R., Savill, T., Reid, S. & Tight M. (2008). *Collisions Involving Cyclists on Britain's Roads: Establishing the Causes*. TRL (PPR 445).

Kojic, B. (1973). *Seoska arhitektura i rurizam*. Beograd, Srbija: Gradjevinska knjiga.

Konjar, M., Kosanović, S., Popov, S. G., & Fikfak, A. (2018). Urban/Rural Dichotomy and the Forms-In-Between. In A. Fikfak, S. Kosanović, M. Konjar, & E. Anguillari (Eds.), *Sustainability and Resilience Socio-Spatial Perspective* (pp. 149–169). Delft: TU Delft Open.

Kousky, C., & Schneider, S. (2003). Global Climate Policy: Will Cities Lead the Way? *Climate Policy*, *3*(3), 359–372. doi:10.1016/j.clipol.2003.08.002

Kovács, Z. (1998). Ghettoization or gentrification? Post-socialist scenarios for Budapest. *Netherlands Journal of Housing and the Built Environment, 13*(1), 63-81. Doi:10.1007/BF02496934

Kraay, J. H., & van der Horst, R. (1985). *The Trautenfels study. A diagnosis of road safety using the Dutch conflict observation technique DOCTOR*. Report 85-53, The Netherlands: SWOV.

Krasny, E., & Tidball, G. (2015). *Civic Ecology – Adaptation and Transformation from the Ground Up*. London: The MIT Press. doi:10.7551/mitpress/9780262028653.001.0001

Krier, L. (2008). Settlements of the future. In T. Haas (Ed.), *Beyond Urbanism and the Future of Cities*. Milan, Italia: Rizzoli.

Krohn, W., & Kuppers, G. (1990). *Selforganization: Portrait of a Scientific Revolution. Sociology of sciences yearbook*. Dordrecht: Kluwer Academic Press. doi:10.1007/978-94-017-2975-8

Kumar, K. (2005). *From post-industrial to post-modern society: New theories of the contemporary world, sec*. Cambridge, UK: Blackwell Publishing.

Kurekci, N. A. (2016). Determination of optimum insulation thickness for building walls by using heating and cooling degree-day values of all turkey's provincial centers. *Energy and Building*, *118*, 197–213. doi:10.1016/j.enbuild.2016.03.004

Kwon, S., Kim, J., & Oh, D. S. (2012). Measurement of Urban Competitiveness Based on Innovation Indicators in Six Metropolitan Cities in Korea. *WTR- World Technopolis Review, 1,* 177-185.

Lačević, F., (2004). Umjetnost u dobu znanosti, Skeč o estetskom odgovoru na kaos i fraktale, smisao komparacije dva motrišta. In *Radovi – knjiga XIII – 2004*. Sarajevo, Bosnia and Herzegovina: Univerzitet u Sarajevu, Filozofski fakultet

Lachenmayr, B. J. (1995). *Sehen und gesehen werden: Sicher unterwegs im Straßenverkehr*. Aachen: Shaker.

Lane, B. (1994). What is rural tourism? *Journal of SustainableTourism, 2*(1-2), 7-21

Law on territorial organization of the Republic of Serbia. *Official Gazette RS, No. 129/2007, 18/2016 and 47/2018*. pp. 237-325.

Lazauskaite, D., Šarkiene, E., & Skripkiene, R. (2015). The assessment of development scenarios for suburban metropolitan districts under the multiple criteria methods. *Procedia Engineering, 122*, 19–28. doi:10.1016/j.proeng.2015.10.003

Lefebvre, H. (1970). *La Révolution Urbaine*. Paris: Éditions Gallimard.

Lefebvre, H. (1991). *The production of space (D. Nicholson-Smith, Trans.)*. Oxford, UK: Blackwell Publishing. (Original work published 1974)

Lefebvre, H. (2003). *The Urban Revolution* (R. Bononno, Trans.). Minneapolis, MN: University of Minnesota Press.

Leite, J. S. (2016). *Ruas emergentes: Interpretação morfológica do contexto urbano português*. Universidade de Lisboa. Retrieved from https://www.repository.utl.pt/handle/10400.5/12443

Lélé, S. M. (1991). Sustainable development: A critical review. *World Development, 19*(6), 607–621. doi:10.1016/0305-750X(91)90197-P

Lemos, C., Kirchhoff, C., & Vijay, R. (2012). Narrowing the climate information usability gap. *Nature Climate Change, 2*(3), 789–794. doi:10.1038/nclimate1614

Levin, S. (1998). Ecosystems and the Biosphere as Complex Adaptive Systems. *Ecosystems (New York, N.Y.), 1*(5), 431–436. doi:10.1007100219900037

Li, B. L. (2000). Why is the holistic approach becoming so important in landscape ecology? *Landscape and Urban Planning, 50*(1-3), 27–41. doi:10.1016/S0169-2046(00)00078-5

LifeGate. (2018). *How agriculture and climate change are related: causes and effects*. Retrieved December 18, 2018. https://www.lifegate.com/people/news/agriculture-and-climate-change-causes-effects-impacts

Light, J. (1999). From City Space to Cyberspace. In M. Crang, P. Crang, & J. May (Eds.), *Virtual Geography* (pp. 109–130). London: Routledge.

Litman, T. A. (2017). *Urban Sanity: Understanding Urban Mental Health Impacts and How to Create Saner, Happier Cities*. Victoria Transport Policy Institute.

Llewelyn-Davies, B., & ... (2009). *Priručnik za urbani dizajn*. Beograd: Prograf i Orion Art.

Llop, C. (1997). Lògiques i espais projectuals d'una metròpoli, Barcelona 1976-1992. *Papers*, (26), 37–52.

Llop, C. (2016). *Ciudades, territorios metropolitanos y regiones urbanas eficientes: estrategias y propuestas de proyecto para la regeneración de la Ciudad_Mosaico_Territorial después de la Explosión de la Ciudad: la Región Metropolitana de Barcelona como laboratório*. Barcelona: Universitat Politècnica de Catalunya. Retrieved from https://upcommons.upc.edu/handle/2117/97155

Llop, C., & Bosc, S. (2012). *Working with the territory: strategies for the new territorialities / Travailler avec le territoire : stratégies pour les nouvelles territorialités*. Barcelona: CREPUD-MED / ACTAR.

Lukić, V. (2011). Commuting in the Settlement System of Serbia. *Stanovnistvo*, *49*(2), 25–50. doi:10.2298/STNV1102025L

Lukić, V. (2013). Veze između migracije i dnevne migracije [Links between migration and daily migration]. *Stanovnistvo*, *LI*(1), 69–90. doi:10.2298/STNV1301069L

Lütke-Daldrup, E. (2001). Die perforierte Stadt. Eine Versuchsanordnung [The perforated city. A test arrangement]. *Stadtbauwelt*, *150*, 40–45.

Lux, G. (2008). *Industrial development, public policy and spatial differentiation in Central Europe: Continuities and change* (Discussion Paper No. 62, pp. 26-28). Pécs, Hungary: Centre for Regional Studies of the Hungarian Academy of Sciences Press.

Lynch, K. (2005). *Rural-Urban Interaction in the Developing World*. New York: Routledge.

Macura, M. (1960). Sociologija stanovanja i industrijalizacija stambene arhitekture. *Arhitektura urbanizam, 2*, 31-33.

Maksimović, B. (1962). Stanovanje i civilizacija. *Arhitektura urbanizam, 18*, 43-46.

Maksimović, B. (1969). Funkcionalne vrednosti slobodnih prostora u blizini stana. *Arhitektura urbanizam, 56-57*, 96-104.

Maksimović, B. (1973). Kritika urbanističke perspketive Beograda iz 1948. Godine: Povodom izlaganja o planu rekonstrukcije Beograda. *Godišnjak Muzeja grada Beograda, 20*, 429-443.

Maksin Mićić, M., Milijić, S., & Nenković Riznić, M. (2009). Spatial and environmental planning of sustainable regional development in Serbia. SPATIUM International Review, 21, 39-52.

Malecki, E. J. (2003). Digital development in rural areas: Potentials and pitfalls. *Journal of Rural Studies*, *19*(2), 201–2014. doi:10.1016/S0743-0167(02)00068-2

Mangin, D. (2004). *La ville franchisé : Formes et structures de la ville contemporaine*. Paris: Éditions de la Villette.

Mangin, D. (2013). *Paris/Babele – Une mégapole européenne*. Paris: Éditions de la Villette.

Mangin, D., & Panerai, P. (1999). *Projet urbain*. Marseille: Parenthèses.

Mäntylä, M. V., Graziotin, D., & Kuutila, M. (2018). The evolution of sentiment analysis—A review of research topics, venues, and top cited papers. *Computer Science Review*, *27*, 16–32. doi:10.1016/j.cosrev.2017.10.002

Map, Q. (2017). *California Transportation Quick Map*. Retrieved March 11, 2017. http://quickmap.dot.ca.gov/

Marcuse, P. (1994). Not Chaos, But Walls: postmodernism and the partitioned city. In S. Watson & K. Gibson (Eds.), *Postmodern Cities and Spaces* (pp. 243–254). Cambridge, MA: Blackwell. doi:10.1111/j.1365-2184.1980.tb00458.x

Marković, D., Čeperković, B., Vlajčić, A., & Resl, S. (2011). Bela knjiga Elektroprivrede Srbije. JP Elektroprivreda Srbije: Sektor za odnose s javnošću, Beograd.

Marković, P. (1992). *Београд и Европа: 1918-1941* [Belgrade and Europe 1918-1941]. Belgrade: Savremena Administracija.

Markusen, A. (1996). Sticky places in slippery space: A Typology of industrial districts. *Economic Geography*, *72*(3), 293–313. doi:10.2307/144402

Marot, S. (1995). L'alternative du paysage. *Visiteur: Ville, Territoire, Paysage, Architecture. Revue Trimestrielle de Critique Des Situations Construites*, *1*, 54–80.

Martinez-Fernandez, C., Kubo, N., Noya, A., & Weyman, T. (2012). *Demographic Change and Local Development: Shrinkage, Regeneration and Social Dynamics.* OECD Local Economic and Employment Development (LEED). Retrieved March 22, 2017, from http://www.carltd.com/sites/carwebsite/files/Demographic%20Change%20and%20Local%20Development.pdf

Maruna, M., & Čolić, R. (Eds.). (2014). Integralni urbani projekti za razvoj centra Kragujevca: Katalog izložbe završnih radova generacije studenata 2012/13 [Integrated Urban Projects for Kragujevac Inner City Development: Final Works' Exhibition Catalogue: Generation of Students 2012/2013]. Beograd: Univerzitet u Beogradu, Arhitektonski fakultet.

Maruna, M., & Čolić, R. (Eds.). (2015). Inovativni metodološki pristup izradi master rada: doprinos edukaciji profila urbaniste [The innovative methodological approach to the development of master work: A contribution to the education of urban planners profile]. Beograd: Arhitektonski fakultet & GIZ/ AMBERO Beograd.

Maruna, M., Čolić, R., Fokdal, J., Zehner, C., Milovanović-Rodić, D., & Lalović, K. (2015). Collaborative and Practice Oriented Learning on Disaster Risk Management in Post-Socialist Transition Countries. In *Proceeding from the XVI N-AERUS Conference Who wins and who loses? Exploring and learning from transformations and actors in the cities of the South* (pp. 408-421). Academic Press.

Maruna, M., Milovanović Rodić, D., & Čolić, R. (2018). Remodelling Urban Planning Education for Sustainable Development: The case of Serbia. *International Journal of Sustainability in Higher Education, 19*(4), 658–680. doi:10.1108/IJSHE-07-2017-0102

Mason, P. (2016). *Postcapitalism: A guide to our future* (1st American ed.). New York, NY: Farrar, Straus and Giroux.

McHarg, I. (1969). *Design with nature.* American Museum of Natural History.

Mickovski, G., Radevski, A., & Karanakov, B. (2016). Transformation of industrial space in the context of post-socialist transformation of the city of Skopje. In *Proceedings of Conference Inclusive/Exclusive Cities* (pp. 264-281). Skopje: City of Skopje

Mihaljević, G. (1995). *Imati i biti: ekonomija i kuća* [Having and being: the economy and the house]. Beograd: Arhitektonski fakultet Univerziteta u Beogradu.

Mikavica, M., & Čolić, R. (2017). Monitoring implementacije strategije razvoja – primer Pančeva [Monitoring implementation of city development strategy – the case study of Pančevo], Arhitektura i Urbanizam, 44, 36-46.

Milić, B. (2011). *Ruralni razvoj - praktikum za lokalne aktere.* Stalna konferencija gradova i opština SKGO.

Milić, V. (1998). Fizička struktura predela [Physical structure of the landscape]. In I Letnja skola arhitekture "Petnjica 97" Arhitektura predela ... do razmere 1:1" (pp. 92-98). Beograd: Arhitektonski fakultet Univerziteta u Beogradu i Istraživačka stanica Petnjica.

Milić, V. (2002). Pseudourbani prostor između sela i grada: trendovi u Republici Srpskoj [Pseudo-urban spaces between villages and towns: Trends in the Republic of Srpska]. In Zbornik ljetnje škole urbanizma Šipovo 2002 (pp. 65-73). Banja Luka: Urbanistički zavod Republike Srpske.

Milić, V. (2006). *Urbanistički aspekti socijalnog stanovanja* [Urban Aspects of Social Housing]. Belgrade: Arhitektonski fakultet.

Milinković, S. (1969). Osnovne koncepcije planiranja slobodnih površina. *Urbanizam Beograda, 4,* 27–29.

Millennium Ecosystem Assessment. (2005). *Ecosystems and Human Well-being: Synthesis.* Washington, DC: Island Press.

Mills, E. (1966). *Social identity in high-density housing* (Fifth Year Architectural Thesis). School of Design North Carolina State University.

Milovanović Rodić, D., & Stojić, B. (2018). Integralni pristup za održivi razvoj dunavskog regiona: prikaz pedagoškog modela i ostvarenih rezultata – master teza [Integral Approach for the Sustainable Development of the Danube Region: Pedagogical model and Achieved results - Master Thesis]. In *Proceedings of 14th Summer School of Urbanism* (pp. 72-82). Bjeljina: UUS, RGZ, GIZ.

Milovanović Rodić, D., Čolić, R., & Maruna, M. (2017). Multilevel Governance Instruments for Achieving Balanced Urban-Rural Development. In *Proceedings of 4th International Academic Conference Places and Technologies 2017. Keeping up with Technologies in the Context of Urban and Rural Synergy* (pp. 332-340). Academic Press.

Milovanović Rodić, D., Lalović, K., & Nenadović, A. (2011). Virtual social networks as a booster for reaching livable villages: examining Serbian cases. In *Proceedings of 47 th ISOCARP Congress proceedings Liveable Cities: Urbanising World, Meeting the Challenge* (pp. 1-14). Academic Press.

Milovanović Rodić, D., Maruna, M., & Čolić, R. (2016). *Instrumenti upravljanja integralnim urbanim razvojem na primeru grada Pančeva* [Instruments for Integrated Urban Governance- Case of Pancevo]. Beograd: Arhitektonski fakultet Univerziteta u Beogradu.

Milovanović Rodić, D., Čolić, R., & Maruna, M. (2018). The Role of University in a Policy Making Process - Introducing Integrated Urban Projects for Effective Urban Governance in Serbia. In E. Anguillari & B. Dimitrijević (Eds.), *Integrated urban planning: directions, resources and territories* (pp. 63–83). TU Delft Open.

Milovanović Rodić, D., Živković, J., & Lalović, K. (2013). Changing Architectural Education for Reaching Sustainable Future: A Contribution to the Discussion. *Spatium (Belgrade)*, *29*(29), 75–80. doi:10.2298/SPAT1329075M

Milovanović, A. (2018). Treći Beograd: pregled razvoja urbanističke misli i delovanja u periodu od 1921. godine do danas. *Arhitektura i urbanizam, 45,* 16-25.

Minić, O. (1954). Razvoj Beograda i njegova arhitektura između dva rata. *Godišnjak Muzeja grada Beograda, 1*, 177-188.

Minić, O. (1955). U potrazi za likom Beograda. *Godišnjak Muzeja grada Beograda, 2*, 449-460.

Ministarstvo energetike, razvoja i zaštite životne sredine. (2013). *Nacionalni akcioni plan za korišćenje obnovljivih izvora energije Republike Srbije.* Ministarstvo energetike, razvoja i zaštite životne sredine, Beograd.

Ministarstvo energetike, razvoja i zaštite životne sredine. (2013). *Strategija razvoja energetike Republike Srbije za period do 2025.godine sa projekcijama do 2030.godine*. Ministarstvo energetike, razvoja i zaštite životne sredine.

Ministry of Agriculture and Environmental Protection Republic of Serbia. (2014). *Republic of Serbia IPARD Programme for 2014-2020.* Retrieved May 15, 2019, from https://ipard.co.rs/doc/IPARD%202014-2020.pdf

Ministry of Agriculture, Forestry and Water Management. (2009). *Plan for the rural development strategy for the period 2009-2013.* Retrieved January 17, 2019. http://www.agrif.bg.ac.rs/files/events/38/Plan%20strategije%20ruralnog%20razvoja%202009-20013.pdf

Ministry of Agriculture, Forestry and Water Management. (2014). Strategy for Agriculture and Rural Development of the Republic of Serbia 2014-2024. Official Gazette RS, No. 85/2014. pp. 36.

Ministry of Construction and Physical Planning. (2015). *Technical regulation on the rational use of energy and thermal protection in buildings*. Zagreb, Croatia: Official Gazette HR no. 128/15, Appendix B, Table 8.

Ministry of Construction, Transport and Infrastructure. (2018). *Rulebook on Energy Efficiency of Buildings and Rulebook on the conditions, content and method of issuing certificates on the energy performances of buildings*. Belgrade, Serbia: Official Gazette no. 69/2012 and 48/2018. Retrieved June 10, 2013, from http://www.ingkomora.org.rs/strucniispiti

Ministry of Finance and Economy. (2011). *Master plan for sustainable development of rural tourism in Serbia*. Retrieved January 15, 2019. https://futurehospitalityleaders.files.wordpress.com/2012/11/master-plan-odrzivog-razvoja-ruralnog-turizma-u-srbiji.pdf

Ministry of Human Rights and Refugees of Bosnia and Herzegovina. (2014). *Overview of Emigrations from Bosnia and Herzegovina*. Sarajevo: Author.

Ministry of Spatial Planning, Construction and Ecology. (2015). *Rulebook on energy audits and issuing energy certificate of buildings*. Banja Luka, B&H: Official Gazette RS no. 30/15.

Ministry of Spatial Planning, Construction and Ecology. (2015). *Rulebook on minimum requirements for energy performance of buildings*. Banja Luka, B&H: Official Gazette RS no. 30/15.

Ministry of Spatial Planning, Construction and Ecology. (2015). *Rulebook on the methodology for calculating energy performances of buildings*. Banja Luka, B&H: Official Gazette RS no. 30/15.

Ministry of Trade, Tourism and Telecommunications. (2006). Strategy for tourism development of the Republic of Serbia. *Official Gazette of Republic Serbia No. 91/2006.*

Ministry of Trade, Tourism and Telecommunications. (2016). Strategy for tourism development of the Republic of Serbia for the period 2016-2025. *Official Gazette of Republic Serbia No. 98/2016.*

Ministry of Trade, Tourism and Telecommunications. (2018). *Law on tourism – draft version*. Retrieved January 15, 2019. http://www.parlament.gov.rs/upload/archive/files/lat/pdf/predlozi_zakona/2018/3153-18%20-%20Lat.pdf

Mitchell, M., & Hall, D. (2005). Rural Tourism as Sustainable Business: Key Themes and Issues. In D. Hall, I. Kirkpatrick, & M. Mitchell (Eds.), *Rural Tourism and Sustainable Business* (pp. 3–14). Channel View Publications.

Mitchel, W. J. (2000). *E-topia: "Urban life, Jim - But not as we know it*. Cambridge, MA: MIT Press.

Mitković, P., Đurić, J., Dinić, M., & Mitković, M. (2015). Mini solar power plants as a potential for development of the South of Serbia. Serbian Ceramic Society Conference- Advanced Ceramics and Application IV: Program and the book of abstracts. Belgrade: Serbian Ceramic Society, 102.

Modellstadt - Mannheim. (2014). *MOMA – The power supply system gets intelligent*. Accessed November 11, 2015. http://www.modellstadt-mannheim.de/moma/web/de/home/index.html

MOFTER. (2012). *Bosnia and Herzegovina First National Energy Efficiency Action Plan 2010-2018 final draft*. Sarajevo, B&H: Ministry of Foreign Trade and Economic Relation (MOFTER).

Mojović, Đ., & Žerjav, B. (2011). *Stanovanje pod zakup* [Rented Housing]. Belgrade: Urban Development Program.

Moldovan, I., Moldovan, V. S., & Ilies, M. N. (2014). Iconic Architecture: Skyscrapers. In *Proceedings of 5th International Conference "Civil engineering – science and practice"* (pp. 1461-1468). Montenegro: Žabljak. Retrieved January 17, 2019, from https://www.researchgate.net/publication/308099136_ICONIC_ARCHITECTURE_SKYSCRAPERS

Möller, P. (2015). Young adults' perceptions of and affective bonds to a rural tourism community. *Fennia-International Journal of Geography*, *194*(1), 32–45.

Moran, P., Goggins, J., & Hajdukiewicz, M. (2017). Super-insulate or use renewable technology? Life cycle cost, energy and global warming potential analysis of nearly zero energy buildings (NZEB) in a temperate oceanic climate. *Energy and Building*, *139*, 590–607. doi:10.1016/j.enbuild.2017.01.029

Moudon, A. V. (1994). Getting to Know the Built Landscape: Typomorphology. In L. S. Karen Frank (Ed.), *Ordering Spaces: Types in Architectural Design* (pp. 289–311). New York: Van Nostrand Reinhold.

Moudon, A. V. (1998). The Changing Morphology of Suburban Neighborhoods. In A. Petruccioli (Ed.), *Typological Process and Design Theory* (pp. 141–157). Cambridge, MA: Aga Khan Program for Islamic Architecture.

Municipality of Blace. (2011). *Spatial plan of Blace Municipality* (Official Gazette of Blace Municipality No. 4/11). Blace, Serbia: Municipality of Blace.

Muñoz, F. (2004). *urBANALización. La producció residencial de baixa densitat a la província de Barcelona, 1985-2001.* Universitat Autònoma de Barcelona.

Muñoz, F. (2008). Paisajes de la movilidad: de los espacios multiplex a los aeropuertos low cost. *Papers*, (47), 68–73.

Muresan, I., Harun, R., Arion, F., Oroian, C., Dumitras, D., Mihai, V., ... Chiciudean, G. (2019). Residents' Perception of Destination Quality: Key Factors for Sustainable Rural Development. *Sustainability*, *11*(9), 2594. doi:10.3390u11092594

Nagaraju, L. G., & Chandrashekara, B. (2014). Rural Tourism and Rural Development in India. *International Journal of Interdisciplinary and Multidisciplinary Studies*, *1*(6), 42–48.

Nagy, E. (2001). Winners and Losers in the Transformation of City Centre Retailing in East Central Europe. *European Urban and Regional Studies*, *8*(4), 340–348. doi:10.1177/096977640100800406

Naldi, L., Nilsson, P., Westlund, H., & Wixe, S. (2015). What is smart rural development? *Journal of Rural Studies*, *40*, 90–101. doi:10.1016/j.jrurstud.2015.06.006

National Assembly of the Republic of Serbia (NARS). (2012). *Law on Social Housing*. Official Gazette of RS No. 72/09.

National Tourism Organization of Serbia. (2017). *Catalogue of village touristic households*. National tourism organization of Serbia.

Naveh, Z. (1990). Landscape ecology as a bridge between bio-ecology and human ecology. In H. Svobodova (Ed.), *Cultural Aspects of Landscape* (pp. 45–58). Wageningen, The Netherlands: Pudoc.

Naveh, Z. (1995). Interactions of landscapes and cultures. *Landscape and Urban Planning*, *32*(1), 43–54. doi:10.1016/0169-2046(94)00183-4

Naveh, Z. (2000a). What is holistic landscape ecology? A conceptual introduction. *Landscape and Urban Planning*, *50*(1-3), 7–26. doi:10.1016/S0169-2046(00)00077-3

Naveh, Z. (2000b). The Total Human Ecosystem: Integrating Ecology and Economics. *Bioscience*, *50*(4), 357–361. doi:10.1641/0006-3568(2000)050[0357:TTHEIE]2.3.CO;2

Nedović-Budić, Z., Tsenkova, S., & Marcuse, P. (2006). The urban mosaic of post-socialist Europe. In S. Tsenkova & Z. Nedović-Budić (Eds.), *The urban mosaic of post-socialist Europe. Space, institutions and policy* (pp. 3–20). Heidelberg, Germany: Physica-Verlag HD. doi:10.1007/3-7908-1727-9_1

Nejgel, E. (1974). *Struktura nauke*. Beograd: Nolit.

Nguyen, T. T., Camacho, D., & Jung, J. E. (2017). Identifying and ranking cultural heritage resources on geotagged social media for smart cultural tourism services. *Personal and Ubiquitous Computing*, *21*(2), 267–279. doi:10.100700779-016-0992-y

NHTSA. (2008). *National Motor Vehicle Crash Causation Survey, Report to congress No. DOT HS 811 059*. Washington, DC: National Highway Traffic Safety Administration.

Ni, P., & Qinghu, H. (n.d.). *Comparative Research on the Global Urban Competitiveness*. Retrieved February 14, 2019, from http://www.stats.gov.cn/english/specialtopics/iaos/papers/concurrent_session_30b.pdf

Nikezić, A., & Marković, D. (2015). Place-Based Education in the Architectural Design Studio: Agrarian Landscape as a Resource for Sustainable Urban Lifestyle. *Sustainability*, *7*(7), 9711–9733. doi:10.3390u7079711

Nikolić, M. (2011). Visitors' centers at archaeological sites in Serbia as an input for sustainable development of the country. *Journal of Applied Engineering Science*, *9*(1), 253–258.

Nilsson, K., Sick Nielsen, T., Aalbers, C., Bell, S., Boitier, B., Chery, J. P., ... Loibl, W. (2014). *Strategies for Sustainable Urban Development and Urban-Rural Linkages. European Journal of Spatial Development.*

Nizovtsev, M. I., Belyi, V. T., & Sterlygov, A. N. (2014). The facade system with ventilated channels for thermal insulation of newly constructed and renovated buildings. *Energy and Building*, *75*, 60–69. doi:10.1016/j.enbuild.2014.02.003

Noev, N. (2013, December). *EAFRD support for sustainable rural tourism 2014 – 2020*. Paper presented at "Euromeeting 2013," Brussels, Belgium.

Norberg-Schulz, C. (1980). *Genius loci: toward a phenomenology of architecture*. New York: Rizzoli.

Nunes da Silva, F., & Pereira, M. (1986). Ilusões e desilusões das periferias na área Metropolitana de Lisboa. *Sociedade e Território*, (5), 14–24.

Nunes, J. S. (2011). *Florestas de Cimento Armado. Os Grandes Conjuntos Residenciais e a Constituição da Metrópole - Lisboa, 1955-1981*. Lisboa: Fundação Calouste Gulbenkian.

Nylander, M., & Hall, D. (2005). Rural Tourism Policy: European Perspectives. In D. Hall, I. Kirkpatrick, & M. Mitchell (Eds.), *Rural Tourism and Sustainable Business* (pp. 17–40). Channel View Publications.

OECD - Organization for Economic Cooperation and Development. (2011). *OECD Regional Typology*. Directorate for Public Governance and Territorial Development.

OECD. (1994). *Tourism Strategies and Rural Development*. Paris: OECD.

Olesen, B. W. (2012). Revision of EN 15251: indoor environmental criteria. *REHVA Journal, 49*(4), 6–12. Retrieved from https://www.rehva.eu/fileadmin/hvac-dictio/04-2012/revision-of-en-15251_rj1204.pdf

One Planet Living. (2017). *Personal Action Plan Calculator*. Retrieved March 11, 2017. http://www.oneplanetliving.net/take-action/personal-action-plan-calculator

Organization for Economic Co-Operation and Development Republic Agency for Spatial Planning. (2010). Spatial plan of the Republic of Serbia for period 2010-2020. *Official Gazette of the Republic of Serbia, No. 88/2010.*

Organization for Economic Co-Operation and Development. (1994). Tourism Strategies and Rural Development. *OCDE/GD (94)49.*

Oswald, F., & Baccini, P. (2003). *Netzstad: Designing the urban*. Basel: Birkäuser.

Oswalt, P. (Ed.). (2005). Shrinking cities Volume 1 – International Research. Ostfildern-Ruit: Hatje Cantz Verlag.

Otte, D., Pund, B., Jänsch, M. (2009). Unfallursachen-Analyse ACAS für Erhebungen am Unfallort. *Zeitschrift für Verkehrssicherheit*, 122–128.

Øvstedal, L., & Ryeng, E. O. (2002). *Understanding Pedestrian Comfort in European Cities: How to improve walking conditions?* Paper presented at the European Transport Conference 2002, Homerton College, Cambridge, UK.

Özel, G., Açikkalp, E., Görgün, B., Yamik, H., & Caner, N. (2015). Optimum insulation thickness determination using the environmental and life cycle cost analyses based entransy approach. *Sustainable Energy Technologies and Assessments*, *11*, 87–91. doi:10.1016/j.seta.2015.06.004

Ozturk, A. B., Nusair, K., Okumus, F., & Singh, D. (2017). Understanding mobile hotel booking loyalty: An integration of privacy calculus theory and trust-risk framework. *Information Systems Frontiers*, *19*(4), 753–767. doi:10.100710796-017-9736-4

Pallagst, K., Wiechmann, T., & Martinez-Fernandez, C. (2014). *Shrinking Cities International Perspectives and Policy Implications*. New York: Routledge.

Palmer, M., Bernhardt, E., Chornesky, E., Collins, S., Dobson, A., Duke, C., ... Turner, M. (2004). Ecology for a crowded planet. *Scienece*, *304*(5675), 1251–1252. doi:10.1126cience.1095780 PMID:15166349

Pampuri, L., Cereghetti, N., Galbani Bianchi, P., & Caputo, P. (2017). Evaluation of the space heating need in residential buildings atterritorial scale: The case of Canton Ticino (CH). *Energy and Building*, *148*, 218–227. doi:10.1016/j.enbuild.2017.04.061

Pan, D., Chan, M., Deng, S., & Lin, Z. (2012). The effects of external wall insulation thickness on annual cooling and heating energy uses under different climates. *Applied Energy*, *97*, 313–318. doi:10.1016/j.apenergy.2011.12.009

Panerai, P. (2008). *Paris Métropole. Formes et échelles du Grand-Paris*. Paris: Éditions de la Villette.

Panyik, E., Carlos Costa, C., & Rátz, T. (2011, December). Implementing integrated rural tourism: An event-based approach. *Tourism Management*, *32*(6), 1352–1363. doi:10.1016/j.tourman.2011.01.009

Parisaki, M. (2010, May) Climate change adaptation and mitigation measures for Nordic rural areas. EU Rural Review - The Magazine from the European Network for Rural Development, 24-30.

Parker, R. E. (1978). Picture processing during recognition. *Journal of Experimental Psychology. Human Perception and Performance*, *4*(2), 284–293. doi:10.1037/0096-1523.4.2.284 PMID:660101

Park, S. (2015). Spatial assessment of landscape ecological connectivity in different urban gradient. *Environmental Monitoring and Assessment*, *187*(7), 425. doi:10.100710661-015-4645-9 PMID:26065890

Pašalić, S., & Pašalić, D. (2016). Migration losses of Bosnia and Herzegovina and demographic aging. *International E-Journal of Advances in Social Sciences*, *2*(6), 650–658. doi:10.18769/ijasos.280343

Pašalić, S., Vuković, D., Jovičić, M., & Stević, D. (2017). Dynamics and directions of demographic changes in Bosnia and Herzegovina. *International E-Journal of Advances in Social Sciences*, *3*(7), 40–48. doi:10.18769/ijasos.309482

Paul, V., & Tonts, M. (2005). Containing Urban Sprawl: Trends in Land Use and Spatial Planning in the Metropolitan Region of Barcelona. *Journal of Environmental Planning and Management*, *48*(1), 7–35. doi:10.1080/0964056042000308139

Pavia, R. (1996). *Le paure dell'urbanistica. Disagio e incertezzaa nel progetto della città contemporanea*. Gènova: Costa & Nolan.

Pavlović, T., Milosavljević, D., Aleksandar, R., & Pavlović, M. (2011). Comparison and assessment of electricity generation capacity for different types of photovoltaic solar plants of 1 MW in Sokobanja, Serbia. *Thermal Science, 15*(3), 605–618. doi:10.2298/TSCI110322065P

Pelling, M. (2011). *Adaptation to climate change: From resilience to transformation.* Abingdon, UK: Routledge.

Petovar, K. (2003). *Urbana sociologija: Наши градови између државе и грађанина* [Urban Sociology: Our Cities between State and Citizen]. Belgrade: Geografski fakultet – Arhitektonski fakultet – IAUS.

Petrovic, B., Raskovic, I. (2011). *Tradicija – tranzicija: upotreba nasleđa u arhitekturi.* Beograd, Srbija: Arhitektonski fakultet, Orion Art, Institut za arhitekturu i urbanizam Srbije.

Petrović, M. (2004). *Sociologija stanovanja: Stambena politika: Izazovi i mogućnosti* [Housing Policy: Challenges and Possibilities]. Belgrade: Institut za sociološka istraživanja.

Petrović, M. (2014). *Transformacija gradova: Ka depolitizaciji urbanog pitanja* [Transformation of Cities: To Depolitization of Urban Issue]. Belgrade: Institut za sociološka istraživanja.

Petrovic, M. (2005). *Cities after Socialism as a Research Issue.* London: LSE E-prints. Retrieved from http://eprints.lse.ac.uk/23378/

Pflieger, G., & Rozenblat, C. (2010). Urban Networks and Network Theory: The City as the Connector of Multiple Networks. *Urban Studies Journal, 47*(13), 2723–2735. doi:10.1177/0042098010377368

Pickett, S. T. A., Cadenasso, M. L., & McGrath, B. (2013). Resilience in Ecology and Urban Design – Linking Theory and Practice for Sustainable Cities. London: Springer.

Pickvance, C. (2008). Chapter 9: State State Socialism, Post-Socialism and their Urban Patterns: Theorizing the Central and Eastern European Experience. In J. Eade & C. Mele (Eds.), Understanding the City (pp. 183-203). Oxford, UK: Blackwell. DOI: 582.ch9 doi:10.1002/9780470693

Pinoncely, V., & Hartkoorn, I. (2014). *Planning Horizons no. 3 - Promoting Health Cities. The Royal Town Planning Institute*. RTPI.

Popelka, St. (2016). *Eye-tracking with dynamic stimuli.* Olomouc: Dep. of Geoinformatics, University of Olomouc.

Portas, N., Domingues, Á., & Cabral, J. (2003). *Políticas Urbanas - Tendências, estratégias e oportunidades.* Lisboa: Fundação Calouste Gulbenkian.

Porter, M. E., & Heppelmann, J. E. (2014). How smart, connected products are transforming competition. *Harvard Business Review*, 1–23.

Pratt, A. C. (2008). Creative cities: The cultural industries and the creative class. *Geografiska Annaler. Series B, Human Geography, 90*(2), 107–117. doi:10.1111/j.1468-0467.2008.00281.x

Pride, R., (2015): *Дестинација као бренд*, Београд, Srbija: Klio. Пуцар, М., Јоксимовић, Б., (2010). *Одрживи развој бањских и туристичких насеља у Србији.* Зборник радова НИ пројекта. Београд, Србија: Иаус - Посебна издања бр.64.

Prigogine, I., & Stengers, I. (1982). *Novi savez.* Zagreb: Globus.

Proffit, R. W., Fields, W. H., Ackerman, L. J., Bailey, J. L'T., & Tulloch, J. E. C. (2000). Contemporary Orthodontics (3rd ed.). Mosby Inc.

Pugh, C. (1995). Urbanization in developing countries: An overview of the economic and policy issues in the 1990s. *Cities (London, England)*, *12*(6), 381–398. doi:10.1016/0264-2751(95)00083-X PMID:12319994

Radosavljevic, U., Đorđević, A., Lalovic, K., Živkovic, J., & Đukanović, Z. (2019). Nodes and Networks: The Generative Role of Cultural Heritage for Urban Revival in Kikinda. *Sustainability*, *11*(9), 2509. doi:10.3390u11092509

Radovanović, D. (2015). Индустријско наслеђе – стамбене колоније [Industrial Heritage – Worker Colonies]. In Индустријско и техничко наслеђе Војводине [Industrial and Technical Heritage of Vojvodina]. Zrenjanin: Zavod za zaštitu spomenika kulture.

Ralević, M., & Simonović, D. (2017). Innovative approaches in the process of re-integration of city and village. In Dž. Bijedić, A. Krstić-Furundžić, & M. Zečević (Ed.), *Book of Conference Proceedings 4th International Academic Conference Places&Technologies 2017: Keeping up with technologies in the context of urban and rural synergy* (pp. 215-226). Sarajevo: Faculty of Arcitecture University of Sarajevo.

Ramon, T., Grinshpoon, A., Zusman, S. P., & Weizman, A. (2003). Oral Health and Treatment Needs of Institutionalized Chronic Psychiatric Patients in Israel. *European Psychiatry*, *18*(3), 101–105. doi:10.1016/S0924-9338(03)00023-3 PMID:12763294

Rashidi, T. H., Abbasi, A., Maghrebi, M., Hasan, S., & Waller, T. S. (2017). Exploring the capacity of social media data for modelling travel behaviour: Opportunities and challenges. *Transportation Research Part C, Emerging Technologies*, *75*, 197–211. doi:10.1016/j.trc.2016.12.008

Raspopovic, M., & Vukmirovic, M. (2018). Sentiment Analysis of Twitter Data for Exploration of Public Space Sentiments. *6th International Academic Conference on Places and Technologies*, 212-216.

Raxworthy, J., & Blood, J. (2004). Introduction. In *The MESH book: Landscape/Infrastructure* (pp. 10–17). Melbourne: RMIT University Press.

Reddy, K. S. (1993). Cardiovascular disease in India. *World Health Statistics Quarterly*, *46*, 101–107. PMID:8303903

Reddy, K. S., & Yusuf, S. (1998). Emerging Epidemic of Cardiovascular Disease in Developing Countries. *Circulation*, *97*(6), 596–601. doi:10.1161/01.CIR.97.6.596 PMID:9494031

Reinhart, C., Jakubiec, A., & Ibarra, D. (2013). Definition of a reference office for standardized evaluations of dynamic façade and lighting technologies. *Proceedings of the 13th Conference of International Building Performance Simulation Association*, 26-28.

Reitsma, R., & Vergossen, F. (1987). A Causal Typology of Migration. The Role of Commuting. *Regional Studies*, *22*(4), 331–340. doi:10.1080/00343408812331345010 PMID:12315224

Relph, E. C. (1976). *Place and Placelessness*. London: Pion Limited.

Republic Agency for Spatial Planning. (2010). Regional spatial plan for Southern Pomoravlje Municipalities. *Official Gazette of Republic Serbia No. 83/2010.*

Republic Agency for Spatial Planning. (2011). Regional spatial plan for Timočka Krajina – Bor and Zaječar districts. *Official Gazette of Republic Serbia No. 51/2011.*

Republic Agency for Spatial Planning. (2013). Regional spatial plan for Nišava, Toplica and Pirot districts. *Official Gazette of Republic Serbia No. 1/2013.*

Republic Agency for Spatial Planning. (2015). Regional spatial plan for Podunavlje and Braničevo districts. *Official Gazette of Republic Serbia No. 8/2015.*

Republika Srpska Institute of Statistics. (2002-2016). *Bulletin Demographic Statistics 5-19*. Banja Luka: Author. Retrieved August 13, 2017, from http://www.rzs.rs.ba/front/category/2/132/?&add=None

Republika Srpska Institute of Statistics. (2013b). *Estimation of the number of inhabitants by the municipalities of Republika Srpska 1996-2012*. Banja Luka: Author.

Republika Srpska Institute of Statistics. (2017a). *Results of the Census of population, households and dwellings in the Republic of Srpska 2013*. Banja Luka: Author. Retrieved from http://www2.rzs.rs.ba/static/uploads/bilteni/popis/gradovi_opstine_naseljena_mjesta/Rezultati_Popisa_2013_Gradovi_Opstine_Naseljena_Mjesta_WEB.pdf

Republika Srpska Institute of Statistics. (2017b). *Results of the Census 2013, Marital status and fertility*. Banja Luka: Author. Retrieved August 15, 2017, from http://www.rzs.rs.ba/front/article/2365/?left_mi=None&add=None

Republika Srpska Institute of Statistics. (2017c). *Population Statistics Annual Release-Marriages and divorces in Republika Srpska, No. 109/17*. Banja Luka: Author. Retrieved August 15, 2017, from http://www.rzs.rs.ba/front/article/2463/

Rieniets, T. (2006). *Shrinking cities—growing domain for urban planning*. Retrieved August 28, 2013, from http://aarch.dk/fileadmin/grupper/institut_ii/PDF/paper_presentation_EURA2005.pdf

Rifkin, J. (2014). *The zero marginal cost society: The internet of things, the collaborative commons, and the eclipse of capitalism*. New York, NY: St. Martin's Press LLC.

Ristić Trajković, J. (2016). *Architecture and nature: The application of environment-behavior theories in architectural design* (Ph.D. Thesis). University of Belgrade – Faculty of Architecture.

Ristić, S. (2010, June 21). Архитектонски вишак на скандинавски начин [Architectural Surplus on Scandinavian Way]. *Politika Online*. Assessed 28.07.2015 from http://www.politika.rs/rubrike/ Kulturni-dodatak/Arhitektonski-vishak-na-skandinavski-nachin.sr.html/

Rivera, M., Croes, R., & Zhong, Y. (2016). Developing mobile services: A look at first-time and repeat visitors in a small island destination. *International Journal of Contemporary Hospitality Management*, *28*(12), 2721–2747. doi:10.1108/IJCHM-02-2015-0052

Rosania, A. E., Low, K. G., McCormick, C. M., & Rosania, D. A. (2009). Stress, Depression, Cortisol, and Periodontal Disease. *Journal of Periodontology*, *80*(2), 260–266. doi:10.1902/jop.2009.080334 PMID:19186966

Rößler, S. (2008). Green space development in shrinking cities: opportunities and constraints. *Urbani Izziv, 19*(2), 147-152. Retrieved September 15, 2018, from https://www.jstor.org/stable/24906040?seq=1&cid=pdf-reference#references_tab_contents

Salingaros, N., & West, B. J. (1999). A universal rule for the distribution of sizes. *Environment and Planning. B, Planning & Design*, *26*(6), 909–923. doi:10.1068/b260909

Salopek, D. (1974). *Arhitektura bez arhitekta*. Zagreb, Croatia: Mala arhitektonska biblibliotека SAH-e.

Sanders, R. L. (2015). Perceived traffic risk for cyclists: The impact of near miss and collision experiences. *Accident; Analysis and Prevention*, *75*, 26–34. doi:10.1016/j.aap.2014.11.004 PMID:25460088

Santos, J. R. (2012). *Espaços de mediação infraestrutural: Interpretação e projecto na produção do urbano no território metropolitano de Lisboa*. Universidade Técnica de Lisboa. Retrieved from https://www.repository.utl.pt/handle/10400.5/5429

Santos, J. R. (2017a). AdaptPolis - Descodificar a fragmentação urbana: infraestruturas, paisagem e projeto territorial para a Área Metropolitana de Lisboa – Relatório Intermédio. Lisboa.

Santos, J. R. (2013). Planning a Metropolis from Infrastructure: Lisbon from 1940 to 1966. *Journal of Planning History, 12*(4), 312–333. doi:10.1177/1538513213487503

Santos, J. R. (2017b). Discrete Landscapes in Metropolitan Lisbon: Open Space as a Planning Resource in Times of Latency. *Planning Practice and Research, 32*(1), 4–28. doi:10.1080/02697459.2015.1028253

Saroff, J. & Levitan, A. (1969). *Survey Manual for Comprehensive Urban Planning*. University of Alaska: Institute of Social, Economic and Government Research.

Sarudi, C., Szakaly, Z., Mathe, A., & Szente, V. (2003). The Role of Organic Agriculture in Rural Development. *Agriculturae Conspectus Scientificus, 68*(3), 197–202.

Sassen, S. (2003). *Protugeografije globalizacije* [Counter-geographies of Globalization] (T. Medak, Trans.). Zagreb: Multimedijalni institut.

Sassen, S. (2010). Global inter-city networks and commodity chains: Any intersections? *Global Networks, 10*(1), 150–163. doi:10.1111/j.1471-0374.2010.00279.x

Sassen, S. (2016). Digitization And Work: Potentials and Challenges in Low-Wage Labor Markets. In J. Stollmann, K. Wolf, A. Brück, S. Frank, A. Million, P. Misselwitz, & C. Schröder (Eds.), *Beware of smart people! Redefining the smart city paradigm towards inclusive urbanism* (pp. 33–44). Berlin: Universitätsverlag der TU Berlin. doi:10.14279/depositonce-5074

Sasu, K.A., Epuran, G. (2016). An overview of the new trends in rural tourism. *Bulletin of the Transilvania University of Brasov, Series V: Economic Sciences, 9*(58), 119-126.

Satterthwaite, D., Huq, S., Reid, H., Pelling, M., & Romero Lankao, P. (2009). Adapting cities to climate change in urban areas: the possibilities and constraints in low and middle income nations. In Adapting Cities to Climate Change (pp. 3-34). London: Earthscan.

Savić, J. (2013). *Moderno i vernakularno u procesu urbane transformacije Banjaluke* [The modern and the vernacular in urban transformation of Banjaluka] (Unpublished doctoral dissertation). Beograd: Arhitektonski fakultet Univerzitet u Beogradu.

Scheer, B. C. (2004). The radial street as a timeline. A study of the transformation of elastic tissues. In B. C. S. Kiril Stanilov (Ed.), *Suburban Form: An international perspective* (pp. 102–122). London: Routledge.

Schepers, J. P., Den, B. B. P. L. M., De, W. D., Twisk, D. A. M., Schwab, A. L., & Smeets, J. B. J. (2013). Studying the role of vision in cycling: Critique on restricting research to fixation behaviour. *Accident; Analysis and Prevention, 59*, 466–468. doi:10.1016/j.aap.2013.06.027 PMID:23911618

Schmitz, A. (2003). *The New Shape of Suburbia: Trends in Residential Development*. Washington, DC: ULI-Urban Land Institute.

Schmitz, C., Díaz, P., Ruiz-Labourdette, D., Herrero-Jáuregui, C., Molina, M., Montes, C., ... Schmitz, M. (2018). Modelling of socio-ecological connectivity. The rural-urban network in the surroundings of Madrid (Central Spain). *Urban Ecosystems, 33*(6), 1119–1212.

Schmook, G. (1968). *Wiskundig Afgebakene Ommelanden en Hinterlanden van de Belgische Steden op de Basis van Geselektioneerde Diensten uit de Tertiare Sektor, Geogr*. Tijdschrift.

Schwanke, D. (2003). *Mixed-Use Development Handbook*. Washington, DC: ULI-Urban Land Institute.

Scott, A. (1988). *Flexible production systems and regional development: The rise of new industrial space in North America and Western Europe. Centre for Urban and Community Studies University of Toronto* (Research Paper N° 168, pp. 1-31). Toronto, Canada: University of Toronto Press.

Scott, A., & Storper, M. (1990). *Regional Development Reconsidered. Lewis Center for Regional Policy Studies* (Research Paper N° 1, pp. 1-50). Los Angeles, CA: University of California Los Angeles Press.

Secchi, B. (1986). Progetto di suolo. *Casabella*, (520), 19–23.

Secchi, B. (2007). Rethinking and Redesigning the Urban Landscape. *Places*, (19), 6–11.

Secchi, B., & Viganò, P. (2009). *Antwerp, Territory of a New Modernity*. Rotterdam: Sun.

Secchi, B. (2000). *Prima lezione di urbanistica. Prime lezioni*. Roma: Laterza.

Secchi, B. (2005). *La città nel ventesimo secolo*. Roma: Laterza.

Secchi, B., & Viganò, P. (2011). *La ville poreuse, un projet pour le Grand Paris et la métropole de l'après-Kyoto*. Paris: MètisPresses.

Selwitz, R. H., Ismail, A. I., & Pitts, N. B. (2007). Dental Caries. *Lancet*, *369*(9555), 51–59. doi:10.1016/S0140-6736(07)60031-2 PMID:17208642

Serebrennikov, E. N. (2018). Expansion features of Irkutsk suburban areas' development. *IOP Conference Series: Earth and Environmental Science. Sci.,* 190. 10.1088/1755-1315/190/1/012063

Shane, D. G. (2005). *Recombinant Urbanism: Conceptual Modeling in Architecture, Urban Design and City Theory*. West Sussex, UK: Wiley Academy. doi:10.1016/j.numecd.2017.10.006

Shanker, D. (2008, May). ICT and Tourism: challenges and opportunities. In *Conference on Tourism in India–Challenges Ahead* (*Vol. 15*, p. 17). Academic Press.

Shariah, A., Tashtoush, B., & Rousan, A. (1997). Cooling and heating loads in residential buildings in jordan. *Energy and Building*, *26*(2), 137–143. doi:10.1016/S0378-7788(96)01027-4

Sharing, R. (2017). *Autostop and Carpooling for commuters, free Car sharing and Hitchhiking service*. Retrieved March 11, 2017. http://www.roadsharing.com/

Shimer, J. A. (1959). *This sculptured earth*. New York: Columbia University Press.

Sieverts, T. (2003). *Cities Without Cities: Between Place and World, Space and Time, Town and Country*. London: Spon Press/Routledge. Retrieved from http://www.amazon.com/dp/0415272602

Sila, Z., & Miščević, R. (1962). Optimalne veličine stambenih zajednica. *Arhitektura urbanizam, 14*, 46.

Silva, C. N. (2008). Urban Planning in Portugal in the Twentieth Century. *The International Journal of Regional and Local Studies*, *4*(2), 23–39. doi:10.1179/jrl.2008.4.2.23

Simić, I., & Mihajlov, V. (2016). Towards the 'green city' concept: spatio - morphological development resources of small towns in Vojvodina. *Arhitektura+Urbanizam, 2016*(43), 35-45.

Šimić, Lj. (1985). Namena površina i životna sredina. *Arhitektura urbanizam, 96*, 34-35.

Simić, I., & Bajcetic, R. (2016). Small towns as mediators towards spatial resilience: The case study of network of settlements in Vojvodina region. *FACTA UNIVERSITATIS Series. Architecture and Civil Engineering*, *14*(3), 331–342.

Simon, H. (1962). The Architecture of Complexity. *Proceedings of the American Philosophical Society*, *106*(6), 467–482.

Simonović, D. (2010). *Pejzažni gradovi: poređenje razvoja urbanog identiteta gradova Banjaluke i Graca* [Landscape cities: A comparison between the development of urban identities of Banja Luka and Graz]. Banjaluka: Arhitektonsko-građevinski fakultet Univerziteta u Banjoj Luci.

Simonović, D. (2014). *Istraživanje mogućnosti obnove i unapređenja identiteta pejzažnog grada Banjaluke putem urbanističke regulacije* [Research into the possibility of rehabilitation and upgrade of the urban identity of Banja Luka as a landscape city through urban regulation] (Unpublished doctoral dissertation). Banjaluka: Arhitektonsko-građevinsko-geodetski fakultet Univerziteta u Banjoj Luci.

Simonović, D. (2015). Interweaving of Banjaluka's urban and rural landscapes. In A. Fikfak, E. Vaništa Lazarević, N. Fikfak, M. Vukmirović & P. Gabrijelčič (Eds.), *Book of Conference Proceedings of Places&Technologies 2015 - Keeping up with technologies to make healthy places book of conference proceedings* (pp. 194-200). Nova Gorica: Faculty of Architecture University of Ljubljana.

Simonović, D., Novaković, N., & Vujičić, T. (2011). Toward a strategy of regeneration of urban landscape: brownfiled as a strategic resource. In *Proceedings of I International Conference ECOLOGY OF URBAN AREAS* (pp. 439-449). Zrenjanin: Faculty of Technical Sciences Mihajlo Pupin.

Simonović, D. (1995). *Enciklopedija Niša – priroda, prostor, stanovništvo.* Niš: Gradina.

Singh, S. (2015). *Critical reasons for crashes investigated in the National Motor Vehicle Crash Causation Survey. Traffic Safety Facts Crash•Stats. Report No. DOT HS 812 115.* Washington, DC: National Highway Traffic Safety Administration.

Slowfood. (2018). *Towards an EU Common Sustainable Food Policy.* Retrieved February 11, 2019. https://www.slowfood.com/sloweurope/en/topics/food-sustainability/

Smith, S. (2012). *Urbanization and cardiovascular disease: Raising heart-healthy children in today's cities.* Geneva: The World Heart Federation. Retrieved February 20, 2019 from https://www.world-heart-federation.org/wp-content/uploads/2017/05/FinalWHFUrbanizationLoResWeb.pdf

Soja, E. W. (1989). *Postmodern geographies: The reassertion of space in critical social theory.* London, UK: Verso.

Soja, E. W. (2000). *Postmetropolis: Critical studies of cities and regions.* Oxford, UK: Blackwell Publishers.

Solgi, E., Hamedani, Z., Fernando, R., Skates, H., & Orji, N. E. (2018). A literature review of night ventilation strategies in buildings. *Energy and Building*, *173*, 337–352. doi:10.1016/j.enbuild.2018.05.052

Song, H., & Liu, H. (2017). Predicting tourist demand using big data. In *Analytics in smart tourism design* (pp. 13–29). Cham: Springer. doi:10.1007/978-3-319-44263-1_2

Stamenić, Lj. (2009). *Korišćenje solarne fotonaponske energije u Srbiji. Unecoop.* Retrieved from http://solarna.unecopn.org/studija.pdf

Stamenković, Đ. S. (2001). Geografska enciklopedija naselja Srbije, knj. I, II, III i IV [Geographical Encyclopedia of the Settlement of Serbia, Vol. I, II, III and IV]. University of Belgrade, Faculty of Geography, Belgrade:''Agena'' & ''Stručna knjiga'', Belgrade.

Stanilov, K., & Scheer, B. C. (2004). *Suburban form: an international perspective.* London: Routledge. Retrieved from http://books.google.com/books?id=fso45Od_jUYC

Stanilov, K. (2007). The restructuring of non-residential uses in the post-socialist metropolis. In K. Stanilov (Ed.), *The Post-socialist city: Urban form and space transformations in Central and Eastern Europe after socialism* (pp. 73–100). Dordrecht, The Netherlands: Springer. doi:10.1007/978-1-4020-6053-3_5

Stanilov, K. (2007a). Taking stock of post-socialist urban development: A recapitulation. In K. Stanilov (Ed.), *The Post-Socialist City: Urban Form and Space Transformations in Central and Eastern Europe after Socialism* (pp. 3–17). Dordrecht: Springer. doi:10.1007/978-1-4020-6053-3_1

Stanilov, K. (2007b). Housing trends in Central and Eastern European cities during and after the period of transition. In K. Stanilov (Ed.), *The Post-Socialist City: Urban Form and Space Transformations in Central and Eastern Europe after Socialism* (pp. 173–190). Dordrecht: Springer. doi:10.1007/978-1-4020-6053-3_9

Stanilov, K. (Ed.). (2007). *The Post-Socialist City Urban Form and Space Transformations in Central and Eastern Europe after Socialism*. Dordrecht: Springer. doi:10.1007/978-1-4020-6053-3

Stanilov, K., & Sykora, L. (2014). Postsocialist Suburbanization Patterns and Dynamics: A Comparative Perspective. In K. Stanilov & L. Sykora (Eds.), *Confronting Suburbanization: Urban Decentralization in Postsocialist Central and Eastern Europe* (pp. 256–295). Chichester, UK: Wiley-Blackwell. doi:10.1002/9781118295861.ch9

Stanilov, L., & Sykora, L. (2014). *Confronting Suburbanization. Urban Decentralization in Postsocialist Central and Eastern Europe*. Chichester, UK: Wiley Blackwell.

State Statistical Office. (2011). Statistical Yearbook 2011. Skopje, Republic of North Macedonia: State Statistical Office.

Statistical Office of the Republic of Serbia - SORS. (1982). *Municipalities in the Socialist Republic of Serbia 1981*. Belgrade: Statistical Office of the Socialist Republic of Serbia, Socialist Federal Republic of Yugoslavia.

Statistical Office of the Republic of Serbia - SORS. (2002). *Municipalities and Regions in the Republic of Serbia 2002*. Belgrade: Statistical Office of the Republic of Serbia, Republic of Serbia.

Statistical Office of the Republic of Serbia - SORS. (2011). *Municipalities and Regions in the Republic of Serbia 2011*. Belgrade: Statistical Office of the Republic of Serbia, Republic of Serbia.

Statistical Office of the Republic of Serbia - SORS. (2013). *2011 Census of Population, Households and Dwellings in the Republic of Serbia –2011 Census of population, households and dwellings in the Republic of Serbia; 23*. Belgrade: Statistical Office of the Republic of Serbia, Republic of Serbia.

Statistical Office of the Republic of Serbia - SORS. (2014). *2011 Census of Population, Households and Dwellings in the Republic of Serbia – Comparative Overview of the number of population in 1948, 1953, 1961, 1971, 1981, 1991, 2002 and 2011*. Belgrade: Statistical Office of the Republic of Serbia, Republic of Serbia.

Statistical Office of the Republic of Serbia - SORS. (2018). *Municipalities and Regions in the Republic of Serbia 2018*. Belgrade: Statistical Office of the Republic of Serbia, Republic of Serbia.

Statistical Office of the Republic of Serbia (SORS). (2013). *Dwellings by the Type of Building: Data by municipalities/ cities*. Belgrade: SORS. Retrieved from http://pod2.stat.gov.rs/ObjavljenePublikacije/Popis2011/ Knjiga27.pdf

Statistical Office of the Republic of Serbia. (2010). *Administrative territorial division and NSTJ levels 1, 2, 3*. Retrieved August 19, 2019, from http://www.stat.gov.rs/sr-Latn/oblasti/registar-prostornih-jedinica-i-gis/administrativno-teritorijalna-podela-i-nstj-nivoi-1-2-3

Statistical Office of the Republic of Serbia. (2011). *Systematic list of settlements in the Republic of Serbia*. Retrieved August 19, 2019, from http://publikacije.stat.gov.rs/G2011/pdf/G201121009.pdf

Statistical office of the Republic of Serbia. (2013). *Census of Agriculture 2012: Agriculture in Serbia* (vol. 2). Statistical office of the Republic of Serbia.

Statistical office of the Republic of Serbia. (2014). *Comparative overview of the number of the population in 1948, 1953, 1961, 1971, 1981, 1991, 2002 and 2011*. Statistical office of the Republic of Serbia.

Statistical office of the Republic of Serbia. (2016). *Municipalities and regions in the Republic of Serbia in 2015*. Statistical office of the Republic of Serbia.

Statistical office of the Republic of Serbia. (2017). *Municipalities and regions in the Republic of Serbia in 2016*. Statistical office of the Republic of Serbia.

Statistical office of the Republic of Serbia. (2018). *Municipalities and regions in the Republic of Serbia in 2017*. Statistical office of the Republic of Serbia.

Statistical office of the Republic of Serbia. (2018). *Profile of the Region of the Southern and Eastern Serbia. Statistical office of the Republic of Serbia*. Retrieved January 25, 2019.: http://devinfo.stat.gov.rs/SerbiaProfileLauncher/files/profiles/sr/1/DI_Profil_Region%20Juzne%20i%20Istocne%20Srbije_EURSRB002002.pdf

Stefanović, D. (1961). Seminar Komiteta za stanovanje Evropske ekonomske komisije u Zagrebu. *Arhitektura urbanizam, 11-12*, 46.

Stevens, Q. (Ed.). (2015). *Creative milieux: how urban design nurtures creative clusters*. New York, NY: Routledge.

Stewart, I. D., & Oke, T. R. (2012). Local climate zones for urban temperature studies. *Bulletin of the American Meteorological Society*, *93*(12), 1879–1900. doi:10.1175/BAMS-D-11-00019.1

Stewart, K. (2015). Assessing the Carbon Impact of ICT Measures: A Case Study Investigation Using Latis Model. *International Journal of Transportation Science and Technology*, *4*(3), 277–294. doi:10.1260/2046-0430.4.3.277

Stoeva, S. (2017). *Adaptive reuse of the former factory "Kuprom"* (Unpublished master thesis). Ss. Cyril and Methodius University, Skopje, Republic of North Macedonia.

Stojanović, B. (1967). Beograd na levoj obali Dunava. *Arhitektura urbanizam, 41-42*, 96-98.

Stojanović, B. (1971). Beograd na levoj obali Dunava. *Godišnjak Muzeja grada Beograda, 18*, 333-383.

Stojanović, B. (1991). Beograd na Dunavu. *Godišnjak Muzeja grada Beograda, 38*, 201-215.

Stojanović, B. (1973). Prostorni plan Beograda na levoj obali Dunava. *Urbanizam Beograda, 24*, 8–18.

Stovel, H. (2003). *Cultural landscape: The new approach to the preservation of cultural heritage*. Beograd, Srbija: Glasnik Društva konzervatora.

Streimikiene, D., Bilan, Y. (2015). Review of Rural Tourism Development Theories. *Transformations in Business & Economics, 14*(2), 21-34.

Stryjakiewicz, T., Schmeidler, K., Vajdovich Visy, E., Jaroszewska, E., Marcinczak, S., … Đukić, A. (2013). Shrinking Cities in Post-Communist Europe. In K. Pallagst (Ed.), Cities Regrowing Smaller - Fostering Knowledge on Regeneration Strategies in Shrinking Cities across Europe. CIRES Synopsis Report.

Stupar, A. (2015). Cold War vs. architectural exchange: Belgrade beyond the confines? *Urban History*, *42*(4), 622–645. doi:10.1017/S0963926815000528

Stupar, A., & Mihajlov, V. (2016). Climate change adaptation in Serbia: The role of information networks. *METU Journal of the Faculty of Architecture*, *33*(1), 37–59. doi:10.4305/METU.JFA.2016.1.3

Sustainable Aid in Agricultural Production (NITA). (2017). *Primiero Portal De Inovacoes Tecnologicas Para a Agricultura Familiar*. Retrieved December 18, 2018. http://nita.org.br/

Sykora, L., & Ourednicek, M. (2007). Sprawling post-communist metropolis: Commercial and residential suburbanisation in Prague and Brno, the Czech Republic. In M. Dijst, E. Razin, & C. Vazquez (Eds.), *Employment Deconcentration in European Metropolitan Areas: Market Forces versus Planning Regulations* (pp. 209–234). Dordrecht: Springer. doi:10.1007/978-1-4020-5762-5_8

Sykora, L., & Stanilov, K. (2014). The Challenge of Postsocialist Suburbanization. In K. Stanilov & L. Sykora (Eds.), *Confronting Suburbanization: Urban Decentralization in Postsocialist Central and Eastern Europe* (pp. 1–32). Chichester, UK: Wiley-Blackwell.

Tafuri, M. (1979). Radical Architecture and the City. In M. Tafuri (Ed.), Architecture and Utopia: Design and Capitalist Development. Cambridge, MA: MIT Press.

Tammaru, T., Leetmaa, K., Silm, S., & Ahas, R. (2009). Temporal and Spatial Dynamics of the New Residential Areas around Tallinn. *European Planning Studies*, *17*(3), 423–439. doi:10.1080/09654310802618077

Tanović Lukić, M., Pasalić, S., & Golijanin, J. (2014). Demographic Development of Bosnia and Herzegovina from the Ottoman Period Till 1991 and the Modern Demographic Problems. *Procedia: Social and Behavioral Sciences*, *120*, 238–247. doi:10.1016/j.sbspro.2014.02.101

Tasan-Kok, T., & Vranken, , J. (2011). *Handbook for Multilevel Urban Governance in Europe. Analysing Participatory Instruments for an Integrated Urban Development*. The Hague: European Urban Knowledge Network.

Tatom, J. (2004). Making the metropolitan landscape: Lyons 1812-1994. In *Suburban Form: An international perspective* (pp. 77–101). London: Routledge.

Tatom, J., & Kahn, A. (2009). Programs for a metropolitan urbanism. In *Making the Metropolitan Landscape: Standing Firm on Middle Ground* (pp. 195–202). New York: Routledge; doi:10.4324/9780203872048

Taylor, K., & Lennon, J. L. (2011). Cultural landscapes: A bridge between culture and nature? *International Journal of Heritage Studies*, *17*(6), 537–554. doi:10.1080/13527258.2011.618246

Temelová, J. (2007). Flagship Developments and the Physical Upgrading of Postsocialist Inner City: The Golden Angel Project in Prague. *Geografiska Annaler: Series B, Human Geography*, *89*(2), 169-181. DOI: .00246.x doi:10.1111/j.1468-0467.2007

The Climate Group on behalf of the Global eSustainability Initiative GeSI. (2008). *SMART 2020: Enabling the low carbon economy in the information age*. Retrieved March 11, 2017. http://www.theclimategroup.org/_assets/files/Smart2020Report.pdf

The European Network for Rural Development (ENRD)ENRD. (n.d.). *Thematic Work 2017*. Accessed March 11, 2018 https://enrd.ec.europa.eu/enrd-thematic-work_en

The World Folio. (2015). *Digital Bangladesh: an ICT revolution*. Retrieved December 18, 2018. http://www.theworldfolio.com/news/digital-bangladesh-an-ict-revolution/3603/

Theodosiou, T. G., Tsikaloudaki, A. G., Kontoleon, K. J., & Bikas, D. K. (2015). Thermal bridging analysis on cladding systems for building facades. *Energy and Building*, *109*, 377–384. doi:10.1016/j.enbuild.2015.10.037

Theodosiou, T., Tsikaloudaki, K., Tsoka, S., & Chastas, P. (2019). Thermal bridging problems on advanced cladding systems and smart building facades. *Journal of Cleaner Production*, *214*, 62–69. doi:10.1016/j.jclepro.2018.12.286

Timska voznja. (2017). *Timska vožnja*. Retrieved March 11, 2017. http://www.timskavoznja.com

Todorović, A. (1964). Ruralna i urbana sociologija. *Sociologija i prostor,* (3), 37-43.

Todorović, M., & Bjeljac, Ž. (2007). Osnove razvoja ruralnog turizma u Srbiji. Bulletin of the Serbian Geographical Society, 87(1), 135-148.

Todorović, M., & Bjeljac, Ž. (2009). Rural tourism in Serbia as a concept of development in undeveloped regions. Bulletin of the Serbian Geographical Society, Acta geographica Slovenica-geografski zbornik, 49(2), 453-465.

Tokača, M. (2012). *Bosanska knjiga mrtvih. Ljudski gubici u Bosni i Hercegovini 1991-1995* [Bosnian Book of the Dead. Human Losses in Bosnia and Herzegovina 1991-1995]. Sarajevo: Istraživačko dokumentacioni centar Sarajevo.

Tomić, V. (1980). Humanizacija života u gradu. *Urbanizam Beograda, 59-60*, 25–26.

Tošić, D. (2000). Prostorno-funkcijske veze i odnosi u urbanim regijama [Spatial-functional connections and relations in urban regions]. *Arhitektura i urbanizam, 7*, 50-58.

Tošić, D., Krunić, N. & Petrić, J. (2009). Dnevni urbani sistemi u funkciji prostorne organizacije Srbije [Daily urban systems in function of spatial organization of Serbia]. *Arhiterktura i urbanizam, 27*, 35-45.

Tošić, D. (1999). Grad u regiji [City in the region]. *Glasnik, Glasnik. Geographical Society of Republic of Srpska, 4*, 47–85.

Tošić, D. (2012). *Principi regionalizacije* [Principles of regionalization]. Belgrade: University of Belgrade - Faculty of Geography.

Toth, R. E. (1988). Theory and language in landscape analysis, planning, and evaluation. *Landscape Ecology, 1*(4), 193–201. doi:10.1007/BF00157692

Trash Track – MIT. (2017). *Trash Track*. Retrieved March 11, 2017. http://senseable.mit.edu/trashtrack/

Travasso, N., & Domingues, Á. (2015). *Território: Casa Comum*. Porto: FAUP.

TripAdvisor. (2017). Retrieved from https://tripadvisor.mediaroom.com/US-about-us

Trkulja, S., Colic, R., & Maksin, M. (2018). *Sustainable and Integrated Urban Development Strategy of the Republic of Serbia until 2030*. Novi Sad: Artprint Media.

Troll, C. (1971). Landscape ecology (geoecology) and biogeocenology–a terminology study. *Geoforum, 2*(4), 43–46. doi:10.1016/0016-7185(71)90029-7

Tsenkova, S. (2005). *Trends and Progress in Housing Reforms in South Eastern Europe*. Paris: Council of Europe Development Bank.

Tsenkova, S. (2006). Beyond transitions: Understanding urban change in post-socialist cities. In S. Tsenkova & Z. Nedovic-Budic (Eds.), *The urban mosaic of post-socialist Europe. Space, institutions and policy* (pp. 21–50). Heidelberg, Germany: Physica-Verlag Heidelberg. doi:10.1007/3-7908-1727-9_2

Tsenkova, S. (2007). *Urban futures: Strategic planning in post-socialist Europe. In The Post-Socialist City: Urban Form and Space Transformations in Central and Eastern Europe after Socialism* (pp. 447–470). Dordrecht: Springer. doi:10.1007/978-1-4020-6053-3_22

Tsikaloudaki, K., Laskos, K., & Bikas, D. (2012). On the establishment of climatic zones in Europe with regard to the energy performance of buildings. *Energies, 5*(1), 32–44. doi:10.3390/en5010032

Tsirigoti, D., & Tsikaloudaki, K. (2018). The effect of climate conditions on the relation between energy efficiency and urban form. *Energies, 11*(3), 582. doi:10.3390/en11030582

Tsoka, S., Tsikaloudaki, K., Theodosiou, T., & Dugue, A. (2018). Rethinking user based innovation: Assessing public and professional perceptions of energy efficient building facades in Greece, Italy and Spain. *Energy Research & Social Science, 38*, 165–177. doi:10.1016/j.erss.2018.02.009

Tsouros, D. A. (Ed.). (n.d.). *World Health Organization Healthy Cities Project: A Project becomes a movement, review of progress 1987 to 1990.* Sogess. Retrieved January 23, 2019, from http://www.euro.who.int/__data/assets/pdf_file/0016/101446/WA_380.pdf

Turner, M. G. (1989). Landscape ecology: The effect of pattern on process. *Annual Review of Ecology Evolution and Systematics, 20*(1), 171–197. doi:10.1146/annurev.es.20.110189.001131

Turok, I., & Mykhnenko, V. (2007). The Trajectories of European Cities, 1960-2005. *Cities (London, England), 24*(3), 165–182. doi:10.1016/j.cities.2007.01.007

Ucar, A., & Balo, F. (2010). Determination of the energy savings and the optimum insulation thickness in the four different insulated exterior walls. *Renewable Energy, 35*(1), 88–94. doi:10.1016/j.renene.2009.07.009

Ulrich, R. S. (1984). View through a window may influence recovery from surgery. *Science, 224*(4647), 420–421. doi:10.1126cience.6143402 PMID:6143402

UN HABITAT. (2016). *New Urban Agenda.* Retrieved May 15, 2019, from https://www2.habitat3.org/

UN. (2013). *World Economic and Social Survey 2013.* New York: UN.

UN. (2015). *The 2030 Agenda for Sustainable Development.* Retrieved May 15, 2019, from https://sustainabledevelopment.un.org/?menu=1300

UNESCO World Heritage Centre. (2005). *Basic Texts of the 1972 World Heritage Convention.* United Nations Educational, Scientific and Cultural Organization.

United Nations Department of Economic and Social Affairs. (2018). *68% of the world population projected to live in urban areas by 2050.* Retrieved from https://www.un.org/development/desa/en/news/population/2018-revision-of-world-urbanization-prospects.html

United Nations Development Program (UNDP). (2004). *ICT and Human Development: Towards Building a Composite Index for Asia: Realizing the Millennium Development Goals.* Retrieved December 18, 2018. http://www.undp.org/content/dam/rbap/docs/Research%20&%20Publications/human_development/RBAP-HDR-2004-Building-Composite-Index-for-Asia.pdf

United Nations Framework Convention on Climate Change (UNFCCC). (n.d.). *Cancun Adaptation Framework 2010.* Retrieved February 18th, 2019. https://unfccc.int/resource/docs/2010/cop16/eng/07a01.pdf

United Nations General Assembly. (2015). *Transforming our world: the 2030 Agenda for Sustainable Development*, 21 October 2015, A/RES/70/1. Retrieved June 14, 2019, from https://www.refworld.org/docid/57b6e3e44.html

United Nations. (1972). *Report of the United Nations Conference on the Human Environment.* United Nations Publication.

United Nations. (2015). *Transforming our world: the 2030 Agenda for Sustainable Development.* A/RES/70/1.

Ünver, R., Öztürk, L., Adigüzel, S., & Çelik, O. (2003). Effect of the facade alternatives on the daylight illuminance in offices. *Energy and Building, 35*(8), 737–746. doi:10.1016/S0378-7788(02)00227-X

Urban Eco Map. (2017). *Working together to Improve Urban Environments.* Retrieved March 11, 2017. http://urban-ecomap.org/

Urban Planning Institute of Niš. (1995). Master plan of Niš 1995-2010. *Official Gazette of the City of Niš No. 13/95.*

Urban Planning Institute of Niš. (2011). Master plan of Niš 2010-2025. *Official Gazette of the City of Niš No. 43/11.*

Urban Planning Institute of Niš. (2012). Plan of General Regulation of the Municipality of Medijana. *Official Gazette of the City of Niš No. 72/12.*

Urban Planning Institute of Niš. (2013). Plan of General Regulation of the Municipality of Crveni Krst–II phase. *Official Gazette of the City of Niš No. 96/13.*

Urban Planning Institute of Niš. (2015). Plan of General Regulation of the Municipality of Pantelej–II phase. *Official Gazette of the City of Niš No. 44/15.*

Van Der Ploeg, D., Renting, J., & Renting, H. (2000). Impact and Potential: A Comparative Review of European Rural Development Practices. *Sociologia Ruralis, 40*(4), 529–543. doi:10.1111/1467-9523.00165

Van der Poel, A. J. (1976). From landscape care to research in landscape planning. *Landscape Planning, 3*(4), 363–370. doi:10.1016/0304-3924(76)90080-0

Van Leeuwen, E. (2015). Urban-Rural Synergies: An Explorative Study at the NUTS3 Level. *Applied Spatial Analysis and Policy, 8*(3), 273–289. doi:10.100712061-015-9167-x

Vanhoutteghem, L., Skarning, G. C. J., Hviid, C. A., & Svendsen, S. (2015). Impact of façade window design on energy, daylighting and thermal comfort in nearly zero-energy houses. *Energy and Building, 102*, 149–156. doi:10.1016/j.enbuild.2015.05.018

Vansteenkiste, P., Cardon, G., & Lenoir, M. (2013). Dealing with head-mounted eye-tracking data. *Proceedings of the 2013 Conference on Eye Tracking South Africa – ETSA, 13*, 55–57. 10.1145/2509315.2509325

Vargas, C. M., Dye, B. A., & Hayes, K. L. (2002). Oral Health Status of Rural Adults in the United States. *The Journal of the American Dental Association, 133*(12), 1672–1681. doi:10.14219/jada.archive.2002.0120 PMID:12512669

Vasiljević, N. (2012). *Landscape planning as an instrument in spatial development of Serbia* (Doctoral dissertation). Belgrade: University of Belgrade - Faculty of Forestry.

Vecslir, L. (2007). Paisajes de la nueva centralidad. *Urban*, (12), 34–55. Retrieved from http://dialnet.unirioja.es/servlet/extart?codigo=2857098%5Cnhttp://dialnet.unirioja.es/descarga/articulo/2857098.pdf

Vego, J. (2000). Buildings of the Ethnographic Heritage in Blidinje Nature Park. *Prostor, 8*(2), 269–272.

Veljković, A., Jovanović, R., & Tošić, B. (1995). Gradovi Srbije centri razvoja u mreži naselja [Serbian cities development centers in the settlement network] (Special Edition of Book 44). Belgrade: Geographical Institute "Jovan Cvijić", Serbian Academy of Sciences and Arts.

Videnovic, A. (2017). *Obnova centara u ruralnim podrucjima - Novi zapisi.* Beograd, Srbija: Arhitektonski fakultet.

Videnović, A. (2017). Obnova centara u ruralnim područjima [Renewal of centers in rural areas]. University of Belgrade - Faculty of Architecture, Belgrade: PRESSIA.

Videnović, A., Aranđelović, M. (2016). Visitors' centres – New Coordinates of Serbian rural areas improvement. *Architecture and Civil Engineering, 14*(2), 191 – 200. doi:10.2298/FUACE1602191V

Viganò, P. (1999). *La città elementare. Biblioteca di Architettura, 7.* Milano: Skira.

Vranic, P., Vasilevska, Lj., & Haas, T. (2015). Hybrid spatialities: Multi-storey extensions of socialist blocks of flats under post-socialist transition in Serbia, the case of Nis. *Urban Studies (Edinburgh, Scotland)*, *53*(6), 1261–1277. doi:10.1177/0042098015571887

Vresk, M. (1990). *Grad u regionalnom i urbanom planiranju* [City in regional and urban planning]. Zagreb: Školska knjiga.

Vresk, M. (1990a). *Osnove urbane geografije* [An introduction to urban geography]. Zagreb: Školska knjiga.

Vresk, M. (2002). *Grad i urbanizacija* [City and urbanization]. Zagreb: Školska knjiga.

Vujičić, T. (2018a). *Methodological framework of integral adaptive management for shrinking cities - case study Prijedor* (Unpublished doctoral dissertation). Faculty of Architecture University of Belgrade, Serbia.

Vujičić, T. (2018b). Shifting Forward: Resilience Thinking in Out-of-Order Urban System. In A. Fikfak, S. Kosanović, M. Konjar, & E. Anguillari (Eds.), *Sustainability and resilience socio-spatial perspective* (pp. 43–72). Delft: Delft University of Technology, Netherland.

Vujičić, T. M., & Đukić, A. (2015). Methodological framework for shrinking cities case study research: Northwest region of Bosnia and Herzegovina. *Geodetski Vestnik*, *59*(3), 520–536. doi:10.15292/geodetski-vestnik.2015.03.520-536

Vujnović, R. (1975). Prognoze i tendencije u stambenoj izgradnji u narednom desetogodišnjem periodu. *Arhitektura urbanizam, 74-77*, 9-10.

Vujović, S., & Petrović, M. (2007). Belgrade's post-socialist urban evolution: Reflections by the actors in the development process. In K. Stanilov (Ed.), *The Post-Socialist City Urban Form and Space Transformations in Central and Eastern Europe after Socialism* (pp. 361–384). Dordrecht: Springer. doi:10.1007/978-1-4020-6053-3_18

Vuković, P., Kljajić, N., & Arsić, S. (2012). Multifunctional agriculture as an assumption and a condition for rural development in Serbia: Special turn to rural tourism. *International Journal of Sustainable Economies Management*, *1*(2), 24–32. doi:10.4018/ijsem.2012040103

Vuksanović-Macura, Z., & Macura, V. (2018). The right to housing: Squatter settlement in interwar Belgrade – defense and demolition of Jatagan-mala. *Journal of Urban History*, *44*(4), 755–774. doi:10.1177/0096144216632747

WeAdapt. (2018). *SDC Climate Change & Environment (CC&E) Network*. Retrieved December 18, 2018. https://www.weadapt.org/knowledge-base/sdc-climate-change-environment-network

Webster, D., & Muller, L. (2000). *Urban Competitiveness Assessment in Developing Country Urban Regions: The Road Forward.* Paper prepared for Urban Group, Washington, DC. Retrieved January 23, 2019, from https://www.adb.org/sites/default/files/competitive-cities-brochure.pdf

West Dorset District Council. (2006). *Poundbury Development Brief.* Retrieved December 02, 2013, from http://www.dorsetforyou.com/media.jsp?mediaid=116539&filetype=pdf

Wheeler, J. O. (1971). Commuting and the rural nonfarm population. *The Professional Geographer*, *23*(2), 118–122. doi:10.1111/j.0033-0124.1971.00118.x

Wheeler, J. O., & Stutz, F. P. (1971). Spatial dimensions of urban social travel. *Annals of the Association of American Geographers*, *61*(2), 371–386. doi:10.1111/j.1467-8306.1971.tb00789.x

WHO. (2011). *Burden of Disease from Environmental Noise: Quantification of Healthy Life Years Lost in Europe*. Copenhagen, Denmark: WHO Regional Office for Europe. Retrieved March 14, 2017, from http://bit.ly/2btWBXy

WHO. (2012). *Effects of urbanization on incidence of noncommunicable diseases.* WHO, Regional Office for the Eastern Mediterranean. Retrieved January 23, 2019, from http://applications.emro.who.int/dsaf/EMPUB_2012_864.pdf

Wiechmann, T. (2013, September 12). *Shrinking Cities in Europe - Evidence from COST Action "Cities Regrowing Smaller" (CIRES).* Retrieved December 9, 2014, from http://www.shrinkingcities.eu/fileadmin/Conference/Presentations/01_Wiechmann.pdf

Wiechmann, T., & Wolff, M. (2013). Urban Shrinkage in a Spatial Perspective – Operationalization of Shrinking Cities in Europe 1990 - 2010. In *AESOP / ACSP 5th Joint Congress 2013 Planning for Resilient Cities and Regions.* Dublin: Association of European Schools of Planning (AESOP).

Wiens, J. A., & Milne, B. T. (1989). Scaling of "landscape" in landscape ecology, or, landscape ecology from a beetle's perspective. *Landscape Planning, 3*(2), 87–96.

Witold, L. (2010, May) Climate change and rural areas in South East Europe. EU Rural Review - The Magazine from the European Network for Rural Development, 36-41.

Wood, J. (2007). Synergy city; planning for a high density, super-symbiotic society. *Landscape and Urban Planning, 83*(1), 77–83. doi:10.1016/j.landurbplan.2007.05.006

World Tourism Organization (UNWTO). (2004). *Rural Tourism in Europe – Experiences, Development and Perspectives. World Tourism Organization.* UNWTO.

Wu, J. (2008). Making the case for landscape ecology: An effective approach to urban sustainability. *Landscape Journal, 27*(1), 41–50. doi:10.3368/lj.27.1.41

Wu, J. (2012). Landscape Ecology. In R. Leemans (Ed.), *Ecological Systems: Selected Entries from the Encyclopedia of Sustainability Science and Technology* (pp. 179–200). Springer New York. doi:10.1007/978-1-4419-0851-3_575

Wu, X., & Zhi, Q. (2016). Impact of shared economy on urban sustainability: From the perspective of social, economic, and environmental sustainability. *Energy Procedia, 104*, 191–196. doi:10.1016/j.egypro.2016.12.033

Yao, C., Wu, Z., & Wu, J. (1993). The changing pattern of cardiovascular diseases in China. *World Health Statistics Quarterly, 46*, 113–118. PMID:8303905

Yonazi, E., Kelly, T., Halewood, N., & Blackman, C. (2012). *The transformational use of information and communication technologies in Africa.* Washington, DC: The World Bank and African Development Bank. doi:10.1596/26791

Zari, M. P. (2017). What makes a city 'biophilic'? Observations and experiences from the Wellington Nature Map project. In M. A. Schnabel (Ed.), *Back to the Future: The Next 50 Years, Proceedings of the 51st International Conference of the Architectural Science Association (ANZAScA)*, (pp. 105–114). Architectural Science Association (ANZAScA).

Zelenakova, M., Purcz, P., Hlavata, H., & Blištan, P. (2015). Climate Change in Urban Versus Rural Areas. *Procedia Engineering, 119*, 1171–1180. doi:10.1016/j.proeng.2015.08.968

Zheng, L., Ismail, K., & Meng, X. (2014). Traffic conflict techniques for road safety analysis: Open questions and some insights. *Canadian Journal of Civil Engineering, 41*(7), 633–641. doi:10.1139/cjce-2013-0558

Živković, Z. (2015). *Tradicijska kamena kuća dalmatinskog zaleđa, priručnik za obnovu i turističku valorizaciju.* Zagreb: Ministarstvo turizma Republike Hrvatske i Ministarstvo vanjskih i europskih poslova.

Zonneveld, I. S. (1989). The land unit – a fundamental concept in landscape ecology, and its applications. *Landscape Ecology, 3*(2), 67–86. doi:10.1007/BF00131171

Zonneveld, I. S. (1990). Scope and concepts of landscape ecology as an emerging science. In I. S. Zonneveld & R. T. T. Forman (Eds.), *Changing landscapes: An ecological perspective* (pp. 3–2). New York: Springer-Verlag. doi:10.1007/978-1-4612-3304-6_1

Zonneveld, W., & Stead, D. (2007). European territorial cooperation and the concept of urban–rural relationships. *Planning Practice and Research*, *22*(3), 439–453. doi:10.1080/02697450701666787

Zvizdić, D. (2012). Lessons Learned from the Bosnia and Herzegovina Urban and Architectural Heritage, As a road map for sustainable urban development. In Proceedings of Islamic Urban Heritage Research Preservation and Management, Summer School. Istambul, Turky: IRCICA.

Зборник радова стручно – научног скупа. (2004). *Планирање и уређење села и руралних подручја*. Бања Врујци: Удружење урбаниста Србије.

МаксинМ.ПуцарМ.МилијићС.КораћМ. (2011). Одрживи развој туризма у Европској унији и Србији. *Београд / Иаус, Посебна издања, 67*.

Мицковски, Г. (2015). *Трансформација индустријских простора у контексту пост-социјалистичке транзиције града Скопља (Необјављени докторски рад). Универзитет у Београду - Архитектонски факултет*. Београд, Србија.

About the Contributors

Ivana Angelova earned a master degree in architecture from the Faculty of Architecture, SS. Cyril and Methodius University in Skopje in 2017. She has a work experience in a few national and one international architecture office as well as in two municipalities in Skopje. She gained a postgraduate diploma in Urban Management and Development Theories from the IHS institute, Erasmus University in Rotterdam in 2018. Currently pursuing her doctoral research at Meiji University in Tokyo.

Branislav Antonic, Dr., is a teaching assistant at the University of Belgrade – Faculty of Architecture, Serbia. His scientific, professional and academic interests in general are urbanism and spatial planning, with a special focus on the planning of medium-size and small communities and urban dimension of housing. He has published several articles in high-ringing scientific journals and monographs on these topics. Moreover, he is an active participant at international conferences and scientific meetings, where he made more than 50 scientific contributions. He also participates in one national scientific project regarding the improvements and systematisation of housing in Serbia and in one international INTERREG project about the urban branding of cultural heritage along the Danube. As a coordinator or member in working team, he has also been involved in more than 35 works relating to urban and spatial planning and urban design, of whose several have been awarded.

Miloš Arandjelović (Belgrade, 1985), Faculty of Architecture in Belgrade (2010), doctorate (2016). Employed in teaching at the Department of Architecture Faculty of Architecture, University of Belgrade, scientific associate (2017). Engaged in teaching, scientific research, professional and artistic work. Research interests: modern regional planning, urban area, regional architecture, activation of natural resources and values, designing in rural areas, village architecture.

Dženana Bijedić, Dr., is currently employed at the University of Sarajevo, Faculty for Architecture, Architectural Structures and Building Technology Department. The focus of her interest, in last fifteen years, are environmental issues, especially environmental degradation related to spatial planning, architectural designing, and building technologies. She graduated in architecture from the University of Ljubljana, Slovenia and continued her carrier development at the University of Sarajevo. Her research included management of architectural projects (theses for Master of Science degree), and consequently management of environmental limits related to defining the architectural space (Ph.D. in Architecture and Town Planning degree). During her research for doctoral thesis she was awarded US Government, Hubert H. Humphrey scholarship and was admitted to Massachusetts Institute of Technology (MIT) where she completed Special Program for Urban and Regional Studies (SPURS). As a professor at the Faculty of Architecture in Sarajevo, she strives to enhance students' understanding of the importance of a holistic, integrated approach when defining architectural space.

Ivana Bogdanović Protić is Associate Professor at the Faculty of Civil Engineering and Architecture of Nis. She received her doctorate in the field of urban renewal in 2016. The subject of her scientific and pedagogical interests are Urban Design, Renewal and Regeneration, and Urban planning. She participated in one international and three projects of Ministry of Science of Serbia. She is the author of over 30 works of international importance, published in journals or in the conference proceedings she participated in. She has also published several works of national importance published in domestic journals and proceedings. She is the author of the monograph "Urban regeneration of high-rise housing", in 2009, published by the ''Zaduzbina Andrejevic''. She participated in several professional competitions and exhibitions, for which he receives several awards and recognitions. In addition to the scientific work, she also participated in the preparation of more than 20 professional projects.

Rada Čahtarević, Dipl. Eng. Arch., Ph.D., graduated at the Faculty of Architecture, University of Sarajevo and received PhD degree in 2003. at the Faculty of Architecture, University of Sarajevo. Currently employed at the Faculty of Architecture, University of Sarajevo as Associate professor and as a head of the department of Spatial and Graphic Representation. Presented numerous papers on scientfic conferences, and published scientific articles in publications and books, based on the research focused on the application of geometry in architecture through the new complex spatial concepts.

Christine Chaloupka is a Traffic und Clinical Psychologist, University lecturer, Coordinator of national and international research projects in traffic safety and comfort, Member -Traffic Psychology International (TPI), Founding member of INFAR (Institut for Driver Rehabilitation and Diagnostics). Founding member of ICTCT (International co-operation on theories and concepts in traffic safety).

Ratka Colic is a Town Planner with more than 25 years working in urban development practice. She graduated and later obtained MSc and PhD from Faculty of Architecture, University of Belgrade. From 2014, she works as an Assistant Professor at the Faculty of Architecture, where she teaches Urban development policies and Strategic integrated urban development planning. She has been engaged in scientific research in the field of planning theory, urban governance, integrated and sustainable urban development planning, participation and capacity development. As an architect/ town planner, she was employed in the Ministry of Construction and as a national expert in international projects being implemented in Serbia (UN-HABITAT, GIZ/AMBERO and European PROGRES). In 2016, she was engaged as international expert on projects implemented by Council of Europe.

Svetlana Cvetanović is associate professor at Faculty of Information Technology, Belgrade, Serbia. Dr, Cvetanovic has more than 30 years of experience in IT industry and education of information technology which includes: experience in software requirements engineering, software customization and implementation, Oracle RDBMS, end User Training preparation and execution, strong business consulting in banking, financial (GL, A/P, A/R, Payroll), warehousing, manufacturing, education, and library sciences. Dr Cvetanovic participated in scientific projects: Database creation and retrievals in the system for scientific and technological information, Internet and modern business and WebCat - Interactive Web Catalogue of Product Models, Adaptive eLearning funded by Serbian Ministry of Education and EUREKA, Advanced electronic services for tax and financial consultations, funded by Serbian Ministry of Science and Technological Development. Dr Cvetanovic's current areas of research are information systems (methodology for information system development - object-oriented and agile, software architecture for information system integration, enterprise information system).She actively takes part in project and research related to technics and methods for promoting effective teaching and learning by using ontology-based systems for learning objects retrieval and integration of institutional e-learning system with third-party systems. Her activities at BMU evolve management and monitoring of BMU's information system development and implementation. She also promotes ICT with BMU students through different workshops, seminars, competitions for youth in ICT areas, etc. Dr. Cvetanovic is also interested in research initiatives that analyzed presence of women in ICT sector and data analytics methods applied in different level of database systems (transaction, data warehouse, data mart).

Jelena Đekić is an architect, graduated in architecture from the Faculty of Civil Engineering and Architecture University of Niš, Serbia. She is a PhD student at the Faculty of Civil Engineering and Architecture University of Niš. Her main field of interest is the quality of life in post-socialist cities. She is the author and co-author of more than 40 papers published in scientific journals or presented at international and national scientific conferences. She works as teaching assistant at the Chair for Urban and Spatial Planning. Jelena Đekić is a member of Serbian Chamber of Engineers and she participated in the development of a large number of spatial and town planning documents.

Milena Dinić-Branković, architect, is an Assistant Professor at the University of Nis, Faculty of Civil Engineering and Architecture. She obtained her MSc in 2008 and her PhD in 2015, both in the field of urban planning. Her research mainly focuses on urban design and city centers. She is the author of the book "Mixed-use development in city center reconstruction", and an author/co-author of many scientific and professional papers. Milena Dinić-Branković is a Fellow of The Junior Faculty Development Program (JFDP) in the year 2010, which is a program of the Bureau of Educational and Cultural Affairs of the United States Department of State. She received several awards at competitions and exhibitions. Since 2017, Milena Dinić-Branković is an Editor in the field of architecture for the journal Facta Universitatis, Series: Architecture and Civil Engineering.

Aleksandra Djukić, Dr., is an Associate Professor at the University of Belgrade – Faculty of Architecture. Her field of professional activities and research is directed on urban design and planning, urban morphology, urban renewal and heritage. She has published 4 monographs, 180 articles and chapters in international and national scientific journals, books, proceedings. She has participated in numerous national and international research projects and workshops. She was a keynote speaker at three International conferences and she gave lectures at Universities in Granada, Graz, and Bari. In practice, she has done more than 50 urban plans and designs, 10 realized buildings and 5 studies in urban renewal and planning (for UNDP, Ministry, Local government). She has received many awards for competitions, in urban practice, for research project and two rewards for the best paper at International congresses. She is a vice president of Serbian Town Planner Association and a representative in ECTP-CEU.

Dijana Furundžić is a dentist; She graduated in 2007 at the Faculty of Dental Medicine, University of Belgrade, Serbia, and one-year practices she had completed at the health center "Savski venac". She is employed in the private dental practice "Furundzic ordinacija". Besides her work in dental practice, she achieved diplomas for the following courses: 1) in the field of anti-aging medicine Tesyal's Injection Course – Dermal fillers, Biorevitalisation-Mesolift, in 2009, 2) Training course in intralipotherapy, in 2014, 3) In office course for orthodontic damon system, 2014. She is Co-author of patents RS 1282 U (Instrument for tooth color determination that simulates upper front teeth) and RS 1310 U (Cartridge drawer with a system for cold sterilization UVC rays), which were awarded at the international exhibition "Inventions Belgrade 2013". There are a number of papers presented and published in international conference proceedings and the "Chapter Three – New Technology for the Promotion of Health in the Cognitive City", in an international monograph "Keeping up with Technologies to Create the Cognitive City", published by Cambridge Scholars Publishing.

Nikola Furundžić is a dentist; He graduated in 2008 at the Faculty of Dental Medicine, University of Belgrade, Serbia, and one-year practices he had completed at the same faculty. He has his own private dental practice "Furundzic ordinacija". Besides his work in dental practice, he achieved diplomas for the following courses: 1) in the field of anti-aging medicine Tesyal's Injection Course – Dermal fillers, Biorevitalisation-Mesolift, in 2009, 2) Training course in intralipotherapy, in 2014, 3) Theoretical and practical courses of ultrasonography diagnostic of cerebral cervical blood vessel in the Special hospital for cerebrovascular disease "Sveti Sava", Belgrade, 2014. He is Co-author of patents RS 1282 U (Instrument for tooth colore determination that simulates upper front teeth) and RS 1310 U (Cartridge drawer with a system for cold sterilization UVC rays), which were awarded at the international exhibition "Inventions Belgrade 2013". He participated in international projects COST TU1205-"Building integration of Solar Thermal Systems (BISTS)" and COST TU1403-"Adaptive Facades Network". There are a number of papers presented and published in international conference proceedings and the "Chapter Three – New Technology for the Promotion of Health in the Cognitive City", in an international monograph "Keeping up with Technologies to Create the Cognitive City", published by Cambridge Scholars Publishing.

Darija Gajić is an Assistant Professor at the Faculty of Architecture, Civil Engineering and Geodesy, University of Banja Luka, Bosnia and Herzegovina. Her research and publication of scientific papers in the field of architectural construction, building materials and building physics. Her current research topics are focused on the refurbishment of building by conventional, passive and active concepts in the aim to energy optimization of the building envelope. She participated, as a national expert, in drafting the Rule-books of energy efficiency of buildings and certification of buildings of the Republic of Srpska, Bosnia and Herzegovina, which will be in force since January 2016. From March 2015 to November 2016 she participated in the International project TABULA on the National Typology of Residential Buildings of Bosnia and Herzegovina. From September 2018 to June 2019 she was trained all engineers involved in the construction of buildings (designers, revidents, municipal officials, contractors, supervisors and inspectors) for proper use Energy Efficiency Rule-books in Republic of Srpska, Bosnia and Herzegovina.

Milica Igić is teaching Assistant at the Faculty of Civil Engineering and Architecture, PhD candidate at Chair for Urban and Spatial planning with research title"Defining program and project models for revitalization of rural areas with a special emphasis on climate change - the example of southeastern Serbia".

She is author and co-author of few papers dealing with rural areas and heritage. Her scientific, pedagogic and professional focus has been in the fields of urban and spatial planning. Her current research is oriented towards climate change, rural areas and renewable energy sources. Researcher within national project realized at the Faculty and financed by Ministry of Education, Science and Technological Development of Serbia. She is author and co-author of several papers published in scientific journals and presented at international and national scientific conferences. She also took part in few international exhibitions.

Milena Ivanović is a PhD student at the Faculty of Architecture of the Univeristy of Belgrade, currently working at the Republic Geodetic Authority within the National Spatial Data Infrastructure as a specialist in spatial and statistical analysis. She continued her professional development in the field of urban planning, through scientific works (19), development of urban and spatial plans (23), projects (9) and studies (2). She won the first prize in the category of "Student works" at the 23rd International urban planner exhibition, the third prize as a member of the working team in the category "Research and studies in the field of urban and spatial planning" at the 25th International urban planner exhibition, and the second prize at the 27th International urban planner exhibition in the category of „Digital techniques, design and production in urbanism". Her research interests are focused on urban planning, urban history, application of GIS tools in urban research, standardization of planning documents on national level, and management and urban planning organization. An alumnus of the Board of European Students of Technology and Youth achievements of Serbia.

Aleksandra Krstić-Furundžić, Dr., Ing.- Arch., is a Full Professor at the University of Belgrade – Faculty of Architecture, Serbia. Several times she was the head of the Department of Architectural Technology. Professional experience as an educator, architectural design practitioner, researcher, and editor. Her areas of expertise are: Architectural constructions, Innovative façade and roof technologies, Energy-efficient buildings, Passive and active solar systems, Building refurbishment technologies. She holds classes at Bachelor, Master and Doctoral studies at the Faculty of Architecture, University of Belgrade. She was a visiting professor at the Faculty of Architecture, Civil Engineering and Geodesy, University of Banja Luka, Bosnia and Herzegovina. She has participated in numerous national and international research projects and workshops. She is co-founder of the International Academic Conference on Places and Technologies, which has been recognized at the regional level and beyond for several years. She is the author of several books and a significant number of chapters in international and national monographs and scientific papers published in international journals.

Eva Vaništa Lazarević, PhD in Architecture & Urbanism (University of Belgrade, Faculty of Architecture, Serbia), MSc in Protection of Architectural Heritage (University of Zagreb, Faculty of Architecture, Croatia). Eva Vanista Lazarevic is a Full Professor of Urbanism and Urban Planning at Faculty of Architecture, University of Belgrade and has over 25 years of experience in teaching Urbanism, Urban design & Urban regeneration and planning. She was, between others: Head of Department for Urbanism and Urban planning at Faculty of Architecture, Member of City Council Commission for Urbanism of Belgrade and Councilor for Architecture and Urbanism at Ministry for Urbanism, Planning and Environment of Serbia.

Milica Ljubenović is teaching Assistant at the Faculty of Civil Engineering and Architecture, University of Niš, Chair for Town and Spatial planning. She is PhD candidate with research titled "Planning framework and development recommendations for shrinking small towns – the case of the Region of Southern and Eastern Serbia". She is author of several papers published in scientific journals and presented at national and international scientific conferences in the fields of urban and spatial planning, scenario analysis, bioclimatic design and energy efficiency. She also took part in few international exhibitions. Her professional focus is urban and regional development, spatial imbalances, shrinking cities. She is researcher on the project funded by the Ministry of Education, Science and Technological Development.

Marjan Marjanović is a Visiting Fulbright Scholar at the University of Illinois (US). He holds a triple master's degree in European spatial planning, urban planning, and regional development from universities in the Netherlands, Sweden, and Serbia. Previously, he was employed by t33 Srl, a consultancy from Ancona (Italy), where he was engaged with projects under the scope of EU Cohesion policy and with a focus on European territorial development studies. Marjan finds his professional interest in the domain of Europeanization of spatial planning, strategic spatial planning, planning theory, and planning for shrinking cities. He has been actively working on the popularization of planning profession in Bosnia and on the development of domestic spatial planning discourse.

Marija Maruna is an Associate Professor at the Faculty of Architecture, University of Belgrade, at the Department of Urban Planning. She graduated and later obtained MSc and PhD at the same faculty. She has over 20 years of experience in teaching and she is the head of the courses: Planning theory, Urban analysis, Integrated planning, and Strategic urban design and Research methodology in urban planning. Her scientific research is in the field of urban development with a focus on integrated urban development issues, public policies of urban development and planning in the postsocialist transition. She is one of the founders and the first head of the Master programme Integrated Urbanism at the Faculty of Architecture.

Goran Mickovski, Ph.D., is assistant professor at Facility of Architecture, University "Ss Cyril and Methodius" in Skopje. Holds a PhD degree in Architecture and urbanism obtained from Faculty of Architecture at University of Belgrade. Main fields of his interest are industrial heritage, transformation of production space and sustainable urban development. He is continually active in architectural design with several public objects build in recent years. He participated as author of the Macedonian exhibition at the Venice Architectural Biennale in 2014.

Mario Miličević is a GIS specialist at GDi Solutions LLD Belgrade. His main interest is application of GIS in spatial and urban planning. He is an MSc in Geo-Information Science and Earth Observation and a graduate spatial planner. Prior to GDi Solutions, Mario worked as a GIS and spatial planning intern at IUGERS Institute for Urban planning, Civil engineering and Ecology of the Republic of Srpska, where he had opportunity to participate in several spatial and urban plans.

Aleksandra Milovanović is an architect and student of doctoral studies in the field of architecture and urbanism. She is involved as Research Trainee in the national scientific project financed by the Ministry of Education and Science of the Republic of Serbia. She is also engaged as a teaching associate in architectural design and urban planning courses at the University of Belgrade – Faculty of Architecture.

Danijela Milovanović Rodić is an Assistant Professor at the Faculty of Architecture University of Belgrade. She got MSc degree on 'Improving Citizen Participation in Urban Planning with the Support of ICT' and PhD degree on 'Redefining the Model of Citizen Participation in Urban Planning in Serbia according to the Communicative-Collaborative Paradigm' at the same faculty. The focus of scientific research and professional activities is in the field of urban governance, planning and design, especially collaborative planning methodology, public participation and education of socially responsible and engaged urban planners and architects and teaches the courses Integral analysis of the territory, Planning methodology, Architects and civic initiatives for sustainable development and urban design studio at the undergraduate, master and specialist studies.

Mihailo Mitković is teaching Assistant at the Faculty of Civil Engineering and Architecture, PhD candidate at Chair for Urban and Spatial planning with research title,, Creation of a planning model for the development of the regional part of southern Serbia as an instrument for increasing the degree of urbanization, with a focus on the potential of renewable energy sources". He received several scholarships during the studies: Ministry of Education, Science and Technological Development and the Fund for Young Talents of the Republic of Serbia. Professional and scientific improvement is achieved through participation in numerous workshops, courses and campuses in the field of architecture, urban planning and design. He participated in student competitions and exhibitions where he was awarded. He is the author and co-author of many scientific-research works in the field of architecture and urban planning. His mainly research is focused towards: renewable energy sources, energy efficiency, solar power, rural areas.

Petar Mitković is a Full Professor and Dean of the Faculty of Civil Engineering and Architecture, University of Nis. He is the chief of Chair for Urban and Spatial Planning. He graduated from the Faculty of Architecture, University of Belgrade (1977) and acquired his PhD degree in the field of architecture and town planning (1989). He participated in several national scientific projects. He is the author of three books and a number of scientific papers published in scientific monographs and international scientific journals and proceedings of scientific conferences. He participated in architectonic exhibitions and competitions for which he received awards. He is a member of Engineering chamber, National Scientific Committee for traffic, urban planning and construction, and also member of Scientific Committees for several International Conferences. Author or team member of more than 100 realized projects and plans in few countries.

Ana Nikezić is an associate professor and Vice Dean for Education and Research at the University of Belgrade – Faculty of Architecture. She holds a PhD in Architecture (2006) with more than 20 years of teaching experience in the area of Architectural and Urban design and over 14 years of experience in research projects. Since 2010 she is a member and mentor for the elaborations of doctoral thesis. She has had more than 30 papers published in monographies, magazines, journals and conference proceedings of international importance. She was also a member of the organizing and science committee for a number of national and international conferences. She also participated in a great amount of international and national workshops. She is particularly interested in connecting of the theoretical and the practical dimension of designing and architecture in general. Particular academic attention has been brought to the subject of relations between architecture and nature, architecture and urban culture, as well as to the subject of socially responsible architectural education based on an interdisciplinary approach.

Adna Proho (born Šarac), born on January 16th 1992 in Sarajevo, holding academic degree master of architecture - a graduate engineer of architecture, currently works as a teaching assistant at the Faculty of Architecture in Sarajevo. She is a winner of the University Charter as the best student of her generation. In addition to working at the Faculty, she is engaged in the development of architectural and urban projects, through participation in the proposals, as a collaborator of "Gepek" studio, and through individual engagement. She is a leader of the project 'Aesthetics & Culture of Living' in Center of culture and youth Sarajevo, since 2016. Her scientific papers were presented at the conferences: "Places & Technologies 2017" and "Green Design 2017".

Miroslava Raspopovic Milic is associate professor at Faculty of Information Technology at Belgrade Metropolitan University, who is actively involved in researching new technological and pedagogical methods for different ICT platforms. Dr. Raspopovic's research focuses on networking, performance analysis of wireless spectrum sharing and wireless channel modelling, adaptive eLearning, web semantics, ontology, quality and performance of eLearning. She earned her PhD at University of Massachusetts Lowell. Dr. Raspopovic was involved in projects: Research Experience for Undergraduate Students funded by NSF, Motorola Innovation: Vibes and Waves in Action: Path to Innovation in the Classroom, Dynamic Spectrum Sharing in Wireless Networks funded by MIT Lincoln Labs and NSF, Web-based Analysis and Visualization Environment funded by Open Indicators Consortium, School-to-Work Transition for Higher education students with disabilities in Serbia, Bosnia & Herzegovina and Montenegro funded by Erasmus +, Adaptive eLearning funded by Serbian Ministry of Education, Erasmus+ School-to-Work Transition for Higher education students with disabilities in Serbia, Bosnia & Herzegovina and Montenegro, Data analysis of social networks.

Ralf Risser is a Member of Traffic Psychology International (TPI). Austrian representative in the Standing Committee of the European Federation of Psychologists EFPA since 2000. Founding member, from 1988 to 2010 Secretary, and from 2011 until 2017 President of the International Cooperation on Theories and Concepts in Traffic safety. Member of the Chairing committee of the NORBIT group (Nordic Organisation for Behaviour in Traffic). From 2007 to 2014 Visting professor at the Institute for Technology and Society, University of Lund, Sweden, since 2015 Visiting professor for Traffic psychology at the Institute of Psychology, University of Olomouc, Czech Republic.

Dušan Ristić, MSc in Spatial planning and PhD student at the Faculty of Geography of the Univeristy of Belgrade, is a teaching assistant at the Department of Geography, Faculty of Natural Sciences and Mathematics, University of Priština (temporarily settled in Kosovska Mitrovica). He has published more than 30 papers at scientific conferences and journals. Participates in the national scientific-research project Development Problems and Tendencies of Geospatial Systems of the Republic of Serbia. Previously worked in private company as expert in urban planning, spatial planning and environmental planning. He participated in the preparation of 1 spatial and more than 20 urban plans, projects and studies. His research interests are focused in spatial planning, urban planning, environmental protection, GIS (geographic information systems) and tourism planning.

Jelena Ristić Trajković is an architect, educator and author. She holds Ph.D. degree in Architecture within the field of Architectural Design and Contemporary Architecture (2016). She is an Assistant Professor at the University of Belgrade – Faculty of Architecture, where she is involved in architectural design and theory courses, as well as in several scientific research projects. She is the author of more than 30 scientific papers in reputed journals and publications.

Erdin Salihović is full Professor teaching in the field of Architectural design at the Faculty of Architecture, University of Sarajevo, where he has been employed since 1999. From 2015 to 2019, Salihović performed as the Vice-dean for student affairs, and as of academic year 2019/2020, he has been elected the Dean of the Faculty of Architecture, University of Sarajevo. He is the author of the books "Furniture design - searching for identity" (2012), "The History of the interior and furniture design at the turn of manual and industrial approach - from Arts and Crafts to Art Deco" (2016), "The testimony of the time – Hamdija Salihović, the architect" (2018), as well as a co-author of the book "Typology of residential buildings in Bosnia and Herzegovina " (2016). Salihović published numerous scientific articles with a wide range of research interest, with particular focus on the relationship between theory and practice in architectural education.

João Rafael Santos is an Architect and Assistant Professor at the Urbanism Department of Lisbon School of Architecture, University of Lisbon. He holds a Degree in Architecture (Faculty of Architecture, TU Lisbon, 2002), MSc in Urban and Environmental Regeneration (FAUTL, 2005), PhD in Urbanism (FAUTL, 2012), and Post-doctoral research (University of Tokyo, 2013/2014). Research interests include urban design and planning, especially in the field of metropolitan studies with a focus on the relationship of infrastructure with public space and socio-spatial development. Coordinator of URBinLAB research group (urbinlab.fa.ulisboa.pt) and AdaptPolis research project at CIAUD (adaptpolis.fa.ulisboa.pt) and a member of several funded research projects (NoVOID, SoftPlan, iWrecks). Has been awarded the Metrópoles Ciência 2016 Prize, by his research on Lisbon Metropolitan Area, the TU Lisbon's/Caixa Geral de Depósitos Junior Researchers Prize in 2011, the Honorable Mention for TU Lisbon's/Caixa Geral de Depósitos Junior Researchers Prize in 2010, and the Europan Europe First Prize in 2006 (as co-author).

Jasmina Siljanoska, Ph.D., is full professor of Urban Planning and Design at the Faculty of Architecture, University of the Ss. Cyril and Methodius in Skopje. Her interest is focused on theory/ history of urban development and on contemporary transformation planning processes. She has been member of numerous scientific and organizational boards of national and international projects and conferences dealing with issues of planning and education. At the moment she participates in the international projects: UN Habitat - Enhanced Right to the City for All: Skopje Public Space Profile, Horizon 2020: Regeneration and Optimisation of Cultural Heritage in Creative and Knowledge Cities (ROCK). COST-Action CA18137 - European Middle Class Mass Housing.

Ivan Simić (1983), Ph.D., received his B. Arch. and M. Arch. degree from Belgrade University. From 2009 to 2012 he attended Ph.D. course of urban planning and design on the Faculty of Architecture, University of Belgrade. During winter and summer semester 2015/2016. he received the scholarship from the Institute for advanced social-technological studies (IAS-STS) in Graz, where he worked as a research fellow. His major fields of research include resilience of cities, sustainable and resilient urban design, and topics related to cities as complex social-ecological systems. He currently holds the position of Assistant professor at The Faculty of Architecture, University of Belgrade.

Dijana Simonović teaches Town and Spatial Planning courses at the University of Banja Luka, Faculty of Architecture, Civil Engineering and Geodesy, as assistant professor, currently acting as the head of the Department of Urbanism and Spatial Planning. She gained her doctorate at the University of Banja Luka in 2014 with dissertation title: "Research into the Possibility of Rehabilitation and Upgrade of the Urban Identity of Banja Luka as a Landscape City through Urban Regulation". She is an author of monograph "Landscape Cities: A Comparison of the Development of the Urban Identities of Banja Luka and Graz" (2010). Dijana has participated in several national and international research projects and workshops and she published many papers and chapters in national and international journals and books. She was a visiting researcher at the Technical University of Graz in 2003 under Tempus Archi-Cur mobility scheme. Further, she carried out her postdoctoral research at the University of Granada, School of Architecture, Spain in 2016-2017, under Erasmus JoinEU-SEE international academic mobility exchange. Her interests include urban-morphological research, urban design, and urban renewal. She deals with the concept of landscape (cultural landscape/urban landscape/rural landscape) and its contemporary transformations.

Aleksandra Stupar is associate professor at the University of Belgrade - Faculty of Architecture, currently acting as the head of the Department of Urbanism. She has participated in numerous national and international research projects and workshops and published articles and chapters in national and international journals and books focusing on the problems of contemporary cities and their identity. In 2012 she was the guest editor of the thematic issue of the Serbian Architectural Journal ('City vs. Innovation'). During 2013-2015 she was responsible for the content and structure of the track 'City and Society' representing a part of lecture series 'Architecture-Utopia-Realism' conducted at the University of Belgrade - Faculty of Architecture. Member of ISOCARP (International Society of City and Regional Planners) and its Scientific Committee (2013-16). She is the winner of Gerd Albers Award 2008, given by ISOCARP for the best published article on urban- ism, and 2017 Ranko Radovic Award. Dr Stupar is author of two awarded books - 'Grad globalizacije - izazovi, transformacije, simboli' ('The City of Globalization - Challenges, Transformations, Symbols', 2009) and 'Grad - forme i procesi' ('The City - Forms and Processes', 2016). She was a visiting researcher at the In- stitute for Advanced Studies on Science, Technology and Society in Graz (2007, 2011), the Univer- sity of Tampere - School of Humanities and Social Sciences (2010-2013), the Aalto University - School of Built Environment (2017, 2018), as well as the guest lecturer at the Sapienza Università di Roma - Facoltà di architettura Valle Giulia (2006,2008-9).

Theodore Theodosiou is a Dr. Civil Engineer, associate professor at the Civil Engineering Department of the Aristotle University of Thessaloniki. He is also a visiting professor at the International Hellenic University (IHU). He is a member of the Laboratory of Building Construction & Building Physics teaching courses "Building Construction" I & II and "Energy Conscious Design of Buildings" at the Civil Engineering Department and "Energy Audit" in IHU. His research interests are focused mainly on building's energy efficiency, building physics, dynamic energy simulation of buildings, Low and Near

Zero Energy Buildings. He has participated in 16 European and National research programs, more than 60 national and international conferences, chapters in scientific books and he has 25 published papers in peer-reviewed scientific journals with more than 560 references. He has also participated in 3 National Technical Guidelines on the energy efficiency of buildings, as a main member of the national committee for the preparation of the new Building Energy Efficiency Regulation, and the national Near Zero Energy Buildings framework.

Katerina Tsikaloudaki is a Dr. civil engineer and has been working at the Civil Engineering Department of the Aristotle University of Thessaloniki, first as a lecturer (2006), then as an assistant professor (2013) and now as an associate professor (2017). She teaches courses related to building construction, building physics, building energy performance and sustainability at the undergraduate and postgraduate level. Her research interests are on the performance of the building envelope and elements, building physics and sustainability in the built environment. She has participated as a researcher and principal investigator in more than 25 projects funded by national and international sources. The results of her research have been published in peer review journals and conference proceedings. She has an active participation in the development of the Greek Regulation of Building Energy Performance, as well as in the preparation of the relevant technical guidelines and the manuals for the energy auditors. She is deputy member of the national committee for the recast of the energy performance legislation, as well as member of the relevant committees of the Technical Chamber of Greece and the panel working for the definition of nZEB framework in Greece. She has also served as a member of the scientific committees of several national and international conferences, a reviewer of international scientific journals and an expert evaluator for research proposals at national (GSRT) and European level (European Commission).

Dimitra Tsirigoti studied Architecture and obtained her diploma in 2004 from the Faculty of Architecture of Aristotle University of Thessaloniki (AUTH), Greece. She has a Master in Environmental Protection and Sustainable Development, from the Department of Civil Engineering (AUTH). She also holds a Ph.D. from the Faculty of Civil Engineering, (Laboratory of Building Construction & Building Physics,, A.U.Th.).Her scientific interest is in sustainable architecture, urban morphology, energy efficiency of building shells and the sustainability of urban neighborhoods. She has cooperated since 2004 with architectural and technical offices in Thessaloniki and has participated in large scale architectural projects and construction detail studies. She has also participated in 7 research projects. She has designed and supervised the construction of several residential projects in the region of central Macedonia. Since 2005, she has participated in many national architectural, urban and landscape design competitions. She has exhibited and published her architectural work, and she has presented her scientific work in several international conferences.

Stella Tsoka holds a diploma of Civil Engineer, by the Faculty of Civil Engineering of the Aristotle University of Thessaloniki, Greece. After her graduation on July 2010, she followed a joint Master 2 of Research on 'Urban Environment Sciences and Techniques" provided by ENSA Nantes, École Centrale de Nantes and Ecole des Mines de Nantes and she received her master diploma on October 2011. Since February 2019, she holds a PhD degree from the Faculty of Civil Engineering of the Aristotle University of Thessaloniki, Laboratory of Building Construction and Building Physics. Her research focuses on the urban microclimate analysis and the buildings' energy performance.

Aleksandar Videnović (Belgrade, 1963), Faculty of Architecture in Belgrade (1982), where he graduated (1989), master's degree (1995) and doctorate (2016). Employed in teaching at the Department of Architecture Faculty of Architecture, University of Belgrade, Associate (1990), teacher (2002-). Engaged in teaching, scientific research, professional and artistic work. Research interests: modern regional planning, urban area, regional architecture, activation of natural resources and values, designing in rural areas, village architecture.

Tijana Vujičić, is an Assistant Professor at University of Banja Luka Faculty of Architecture, Civil Engineering, and Geodesy. She graduated in 2005 at the same faculty and obtained PhD diploma in 2018 at the Faculty of Architecture University of Belgrade. She has 14 years of experience in teaching in the field of urban and architectural design. She is the head of the courses: Concepts and principles of sustainability and resilience of the built environment, Research methodology in urban planning, and Regeneration of suburbia. She is the co-author of two books and many scientific papers addressing the phenomenon of brownfields and the creation of innovative approaches for problem-solving. Her research interests include spatial, social and economic aspects of problems of shrinking cities and the transformation of urban planning toward the adaptive management concept adjusted to vulnerable cities facing uncertainty.

Milena Vukmirović, Dr., is an Assistant Professor at University of Belgrade - Faculty of Forestry, Department of Landscape architecture and horticulture, Chair of Planning and design and Research Associate at Urban Laboratory in Belgrade. From February 2015 she is an advisor to the Director of Urban Planning of the City of Belgrade in the domain of public space design and policies. Her research is focused on the investigation of the relationship between quality of everyday urban environment and competitiveness of the cities. She is a founder of the CSO 5km/h, co-founder of professional association Urban Laboratory, Belgrade and co-founder and executive coordinator of the International Academic Conference on Places and Technologies. At the present she is engaged in the national research project Modernization of Western Balkans, funded by the Ministry of education, science and technological development of the Republic of Serbia. She was a member of Interreg IV DANUrB (Danube Urban Brand) project, member of management committee of the COST Action 358 Pedestrian Quality Needs and substitute member of COST Action TU1203: Crime Prevention through Urban Design and Planning (2013-2016).

Nermina Zagora graduated from the Faculty of Architecture, University of Sarajevo (2005), obtained Master's degree in Product Design from the Fine Arts Academy, University of Sarajevo (2008) and doctoral degree from the Faculty of Architecture, University of Sarajevo (2012). Nermina participated in academic mobility programs in Barcelona, Spain (2004), Oslo, Norway (2005), London, UK (2007), Ghent, Belgium (2009) and Kobe, Japan (2009). As assistant professor at the Faculty of Architecture, University of Sarajevo, Nermina has been teaching courses in architectural and interior design. Nermina is a co-author of two books and twenty articles and her research included architectural education, energy efficiency, ideological dimension of architecture and correlations between interior and urban design. As a founding partner and principal architect of Sarajevo-based architectural studio Firma, Nermina has co-authored more than 90 architectural, interior and urban design projects since 2009, including several award-winning and internationally published projects. Nermina speaks English, Spanish and French.

Denis Zvizdić, graduated in 1989, obtained master of science degree in 1991 as well as doctor of sciences degree in 2007, all at the Faculty of Architecture of the University of Sarajevo. Currently serves as the chairmen of BiH Counsel of Minister, but also teaches at the UNSA, Faculty of Architecture, Department of urbanism and spatial planning. Focus of his research is related to environmental sustainability from the spatial planning perspective, i.e. a scientific and research work that scientifically, historically, thematically and methodologically investigates, analyses and defines complex matter and multifaceted aspects of our practice related to build and natural environment.

Index

F

G

H

I

J

K

L

M

N

O

P

R

S

T

U

V

W

www.ingramcontent.com/pod-product-compliance
Lightning Source LLC
LaVergne TN
LVHW080533300926
849162LV00007B/1597
* 9 7 8 1 5 2 2 5 9 9 3 2 6 *